David C. White,
10515 Research
Knoxville, TN 3
PH: 615-974-80

April 18, 1996

after Gary's meeting

ADVANCES IN LIPID METHODOLOGY - THREE

edited by

William W. Christie

The Scottish Crop Research Institute, Invergowrie, Dundee (DD2 5DA), Scotland

THE OILY PRESS

DUNDEE

ISBN 0 9514171 6 9

British Library Cataloguing-in-Publication Data. A catalogue record for this book is available from the British Library.

This is - **Volume 7** in the Oily Press Lipid Library

(**Volume 1** - "Gas Chromatography and Lipids" by William W. Christie; **Volume 2** - Advances in Lipid Methodology - One" edited by William W. Christie; **Volume 3** - "A Lipid Glossary" by Frank D. Gunstone and Bengt G. Herslöf; **Volume 4** - "Advances in Lipid Methodology - Two" edited by William W. Christie; **Volume 5**, "Lipids: Molecular Organization, Physical Functions and Technical Applications" by Kåre Larsson; **Volume 6** - "Waxes: Chemistry, Molecular Biology And Functions" edited by Richard J. Hamilton).

Printed in Great Britain by Bell and Bain Ltd, Glasgow

PREFACE

This is the third volume of an occasional series of review volumes dealing with aspects of lipid methodology to be published by the Oily Press. As with the first two volumes, topics have been selected that have been developing rapidly in recent years and have some importance to lipid analysts. For example, Arnis Kuksis presents a timely review of non-enzymatic methods for the determination of positional isomers of glycerolipids; chiral chromatography is especially important to this but various spectrometric methods must also be considered.

I trust it will not be too difficult to convince readers of the importance of high-performance liquid chromatography for analysis of phospholipids, and I have attempted to review the topic from the concept of selectivity in the choice of mobile and stationary phases. However, most of our readers will be less aware of the value of ^{31}P nuclear magnetic resonance spectroscopy as a non-destructive means for accurate analysis of phospholipids. Glonek and Merchant should convince you of the great utility and importance of this technique. I was greatly impressed by the high accuracy of the technique and with the certainty of identification of so many phospholipids. In addition, ^{31}P NMR spectroscopy is proving of great value for preparation of phase diagrams of lipids; Göran Lindblom is an internationally recognised expert in this methodology and presents an authoritative account of the topic.

One of the most difficult tasks facing lipid analysts is to isolate and quantify long-chain acyl-CoA esters. It is evident from this account by Jens Knudsen and colleagues that much remains to be done before the problem is truly solved, although valuable data can be obtained with care. This chapter will be essential reading for anyone interested in the problem.

The final chapter is an especially comprehensive account of the analysis of plant glycolipids. These are vital components of all plant membranes, but they are frequently ignored in many general review articles on the analysis of lipids. Ernst Heinz leaves us no excuse to do so in future. Here, there are detailed descriptions of structures, chromatographic separations, chemical degradation and spectrometric analyses of these fascinating compounds. This is certainly the last word on the topic.

As an appendix, I have prepared literature searches on lipid methodology for the years 1993 and 1994, continuing a feature established in the first two volumes.

The objective of the Oily Press is to provide compact readable texts on all aspects of lipid chemistry and biochemistry, and many more books are in the pipeline for The Oily Press Lipid Library. If you have suggestions or comments, please let us know. By a careful choice of authors and topics, I trust that this volume will again prove to have met all our aims.

My own contributions to the book are published as part of a programme funded by the Scottish Office Agriculture, Environment and Fisheries Dept.

William W. Christie

CONTENTS

Chapter 1

ANALYSIS OF POSITIONAL ISOMERS OF GLYCEROLIPIDS BY NON-ENZYMATIC METHODS

Arnis Kuksis

Banting and Best Department of Medical Research, University of Toronto, Toronto, Canada, M5G 1L6

A. INTRODUCTION

Natural acylglycerols exist in complex mixtures of molecular species, which differ in composition, molecular association and positional distribution of fatty acids. Although the biological significance of the acylglycerol structure is not well understood, there is evidence that it is generated and maintained by the concerted action of acyltransferases and lipases of high positional and fatty acid specificity [41]. Furthermore, there is evidence that structured acylglycerols possess different metabolic and physiological properties [12,40,59,83 and references therein]. As a result, there is growing interest in the determination of the regio- and stereospecific distribution of the fatty acids in the acylglycerol molecules, and in the quantification of the enantiomer content of each molecular species.

Historically, lipolytic degradation was first used to determine the fatty acids in the primary and secondary positions of the acylglycerol molecules [10] and to distinguish between enantiomers [9]. More recently, non-enzymatic methods have been developed for this purpose culminating in the separation of diastereomeric and enantiomeric acylglycerol derivatives on chromatographic columns. The following chapter discusses the positional analysis of natural acylglycerols using the non-enzymatic methods along with the more general physico-chemical approaches to characterizing the regio- and stereo-configuration of acylglycerol molecules. The topic has been reviewed previously in part by Takagi [95] and Christie [14].

B. PROCHIRAL NATURE OF GLYCEROLIPIDS

Acylglycerols, glycerophospholipids and glycoglycerophospholipids have glycerol as a backbone and are widely distributed as components of living cells. Glycerol possesses a plane of symmetry at C_2 and by itself is achiral or optically inactive. It becomes chiral by introduction of different substituents at C_1 and C_3. Thus, *sn*-1-monoacylglycerol is chiral, as is its enantiomer, *sn*-3-monoacylglycerol. Similarly, *sn*-1,2-diacylglycerol is chiral, as is its enantiomer, *sn*-2,3-diacylglycerol. Natural glycerophospholipids are optically active because the prochiral positions at C_1 and C_3 are occupied by different substituents. Triacylglycerols are racemic, if both *sn*-1- and *sn*-3-positions contain identical fatty acids, or enantiomeric, if the primary positions are occupied by different fatty acids. The general structure and prochiral nature of acylglycerols has been discussed further by Takagi [95] and Christie [14].

Specifically, natural triacylglycerols are mixtures of molecular species, which possess one, two or three different fatty acids of varying chain length and number and configuration of double bonds [60]. Unknown mixtures of triacylglycerols must be checked for the presence of alkyl or alkenyl groups [5]. These mixtures are generally too complex for a complete stereospecific analysis [9]. The triacylglycerols must be prefractionated on the basis of carbon number (molecular weight), degree of unsaturation, or a combination of the two chromatographic techniques before positional analysis, in order to simplify assignment of each fatty acid to the *sn*-1-, *sn*-2- and *sn*-3-positions of specific acylglycerol molecules [41].

The chiral nature of natural triacylglycerols arises mainly *via* the *sn*-1,2-diacylglycerols with a characteristic placement of saturated acids in the *sn*-1- and unsaturated acids in the *sn*-2-positions during the biosynthesis of phosphatidic acid as an intermediate in triacylglycerol biogenesis [80]. The triacylglycerol biosynthesis in the intestine proceeds largely *via* the 2-monoacylglycerol pathway, which is characterized by retention of the fatty acid composition of the secondary position of dietary triacylglycerols and also exhibits some non-randomness in the reacylation of the primary positions [58]. In addition, the *sn*-2-position contains largely unsaturated fatty acids, while the polyunsaturated acids are mainly confined to the *sn*-3-position of natural triacylglycerols, although there are important exceptions [7,41]. During the subsequent transport in plasma and clearance by tissues the triacylglycerols undergo extensive transformation, which results in complex exchanges of fatty acids, but considerable specificity is retained. Thus, lipoprotein lipase [1,67] and hepatic lipase [1] are known to attack preferentially the *sn*-1-position and lingual lipase [1] the *sn*-3-position of the triacylglycerol molecule. These enzymes also exhibit specificity during the lipolysis of the *sn*-1,2- and *sn*-2,3-diacylglycerol and 2-monoacylglycerol intermediates (*e.g.* [67]).

C. SPECTROMETRIC METHODS OF ANALYSIS

Physical methods for regio- and stereochemical characterization of triacylglycerols were first explored by Schlenk [88]. The methods tested included the measurement of optical rotation, determination of piezoelectric effect, melting point depression and X-ray diffraction. Since that time optical rotatory dispersion, proton and ^{13}C nuclear magnetic resonance (NMR) spectroscopy and mass spectrometry have been extensively explored for this purpose. The success of these methods has varied with the asymmetry of the molecules examined and the extent of the prefractionation of the sample. The spectrometric methods are considered here only to the extent to which they have contributed to practical positional analysis of glycerolipids.

1. Optical Rotatory Dispersion

Schlenk [88] found that triacylglycerols in which the three acids differed greatly in chain lengths show measurable optical rotation. The small differences in the optical rotation of most asymmetrical natural triacylglycerols can be detected best by the highly sensitive techniques of optical rotatory dispersion and circular dichroism (CD), which allow measurements in the vacuum ultraviolet region. By means of this technique Gronowitz *et al.* [24] observed that saturated triacylglycerols with the greater chain length at *sn*-1- than at *sn*-3-position possess negative rotation, while triacylglycerols with the greater chain length at *sn*-3-position possess positive rotation. For pure triacylglycerols, therefore, optical rotatory dispersion can indicate the presence or absence of asymmetry in the molecule, as well as the stereochemical configuration, if appropriate standards are available. Natural triacylglycerols with fatty acids differing greatly in chain length also show readily measurable optical rotation, and optically active triacylglycerols have been identified in the seed oil of *Euonymus verrucosus* [39], where an acetyl group, and in bovine milk fat [33], where an acetyl and a butyryl group are located specifically in the *sn*-3-position.

Uzawa *et al.* [104] have performed CD studies and NMR spectroscopy on a series of chiral *sn*-1,2-dibenzoylglycerols, and have proposed a general method to correlate the sign of the exciton CD curves with the absolute stereochemistry of the chiral dibenzoylglycerols: *sn*-1,2-dibenzoylglycerols give positive exciton CD, which is independent of the type of substituents at C-3. Therefore, it should be possible to determine the optical purity and the absolute configuration of a diacylglycerol, if it can be converted to the chiral dibenzoate.

Uzawa *et al.* [105] have applied this CD method for determining the optical purity and absolute configuration of *sn*-1,2-(or *sn*-2,3-)-dibenzoylglycerol *via* the corresponding *tert*-butyldimethylsilyl (*t*-BDMS) ether. Initially, significant racemization occurred during the conversion of the chiral diacylglycerol into the intermediate silyl ether. A detailed study of the procedure revealed that migration of the silyl group occurred during the NaOMe/MeOH deacylation. Subsequently, Uzawa *et al.* [106] developed an improved method to determine the optical purity of *sn*-1,2 (or *sn*-2,3)-diacylglycerols *via* the silyl ether intermediate. The chiral diacylglycerols were first silylated and the acyl groups were removed by Grignard degradation to yield the *sn*-3- or *sn*-1-*t*-BDMS ether and subsequent benzoylation led to the corresponding dibenzoylsilyl ether without racemization. The optical purity was determined from the strong exiton Cotton effect, which was positive for *sn*-3-*t*-BDMS-1,2-dibenzoylglycerol and negative for the *sn*-1-*t*-BDMS-2,3-dibenzoylglycerol at 238 nm at a concentration of about 1 mM. The *t*-BDMS ether of *sn*-1,2-dipalmitoylglycerol, which had caused difficulty with alkaline deacylation, was readily deacylated by ethyl or methylmagnesium bromide in diethyl ether at room temperature in a few minutes, while the benzoylation was smoothly achieved by treatment *in situ* with excess benzoyl chloride to give the *sn*-1,2-dibenzoylglycerol-3-*t*-BDMS ether. To avoid isomerization during the initial silylation of the diacylglycerol, the diacylglycerols were added to the silylating reagents already in pyridine. The method was successfully applied to determine the stereo-selectivities of lipases from three sources, *Pseudomonas*, porcine pancreatin and *Candida* using tripalmitoylglycerol as substrate. Figure 1.1 shows the CD and UV spectra of dibenzoyl-*sn*-glycerol *t*-BDMS ethers in methanol as derived from lipolysis of an achiral triacylglycerol [105]. The advantage of Uzawa's method is the ability to determine the optical purity and the absolute configuration of diacylglycerols of different acyl groups by CD without authentic samples of known configuration.

2. Nuclear Magnetic Resonance Spectroscopy

High-resolution proton magnetic resonance (PMR) spectroscopy has been used only to a limited extent because of the small number of signals, which limits the

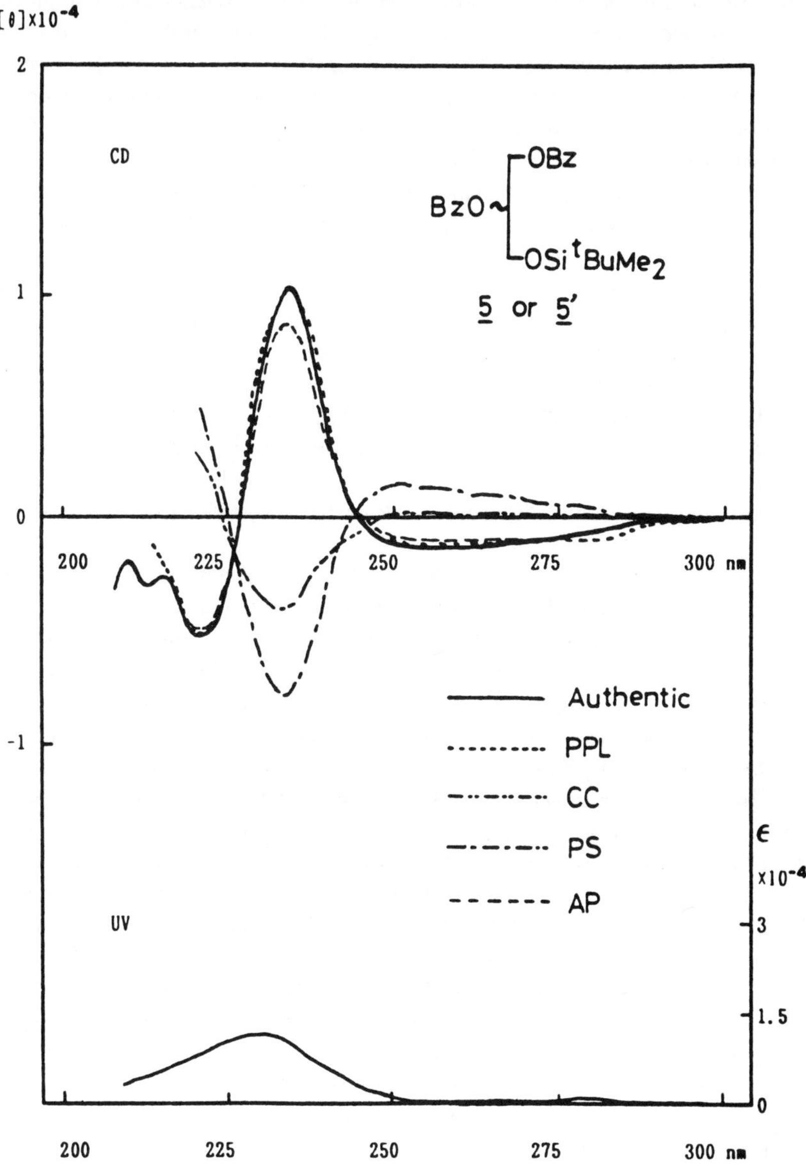

Fig. 1.1. CD and UV spectra (in methanol) of the *t*-BDMS ethers of *sn*-1,2- and *sn*-2,3-dibenzoylglycerols as obtained following lipolysis of tribenzoylglycerol [105]. PPL, porcine pancreatic lipase; CC, *Candida cylindracea* lipase; PS, *Pseudomonas* lipase; and AP, *Amano* P lipase. (Reproduced by kind permission of the authors and of *Biochemical and Biophysical Research Communications*).

amount of information. Bus *et al.* [11] employed PMR spectroscopy in combination with a chiral shift reagent to resolve the signals from ester groups near the center of chirality of triacylglycerols. The chiral shift reagents form complexes *via* the free electron pairs of the ester groups of triacylglycerols. The two enantiomers of a triacylglycerol molecule form diastereomeric associations, which give basically different PMR spectra. Using synthetic model enantiomers to assign the signals, the absolute configuration of the main triacylglycerol of the seed oil *Euonymus alatus* was found to be 3-acetyl-1,2-distearoyl-*sn*-glycerol [39] and that of a monobutyryl triacylglycerol fraction from hydrogenated bovine butter fat was confirmed to be mainly 1,2-diacyl-3-butyryl-*sn*-glycerol [8,79]. Other studies by means of PMR spectroscopy in combination with achiral and chiral chemical shift reagents have been employed to verify the primary versus secondary positioning of fatty acids in synthetic triacylglycerols that contain saturated chains in combination with either unsaturated [78] or branched [107] chain fatty acids. Lok *et al.* [62] have shown, by PMR spectroscopy with a shift reagent, that the predominant enantiomer of the diacylglycerols present in fresh milk has the *sn*-1,2-configuration. Recently, Rogalska *et al.* [86] have demonstrated characteristic differences between the PMR spectra of *sn*-1,2-diacylglycerol (*R,R*-carbamate) and *sn*-2,3-diacylglycerol (*S,R*-carbamate) derivatives, which are depicted in Figure 1.2.

In contrast, high resolution ^{13}C NMR spectroscopy is emerging as a powerful technique for lipid analysis. Gunstone [25] has recently reviewed the application of this methodology to lipids and has noted the potential for positional analyses. Pfeffer *et al.* [78] were first to demonstrate the primary positioning of butyric acid in unaltered bovine milk triacylglycerols by ^{13}C NMR spectroscopy. The primary positioning of butyric and caproic acids in butterfat was later confirmed by ^{13}C NMR spectroscopy by Gunstone [26]. Ng [73,74] has shown that the positional distribution of acyl groups in palm oil can be defined from the carbonyl ^{13}C NMR spectra region, and the positions of the unsaturated fatty acids can be discerned also [73,75]. Wollenberg [108,109] has discussed in detail the high resolution ^{13}C NMR location of the 1(3)-acyl and 2-acyl groups in vegetable oils. The positional distribution data for corn, peanut, canola and sunflower oils indicated that polyunsaturates are replaced in the 1,3-glycerol position exclusively by saturates, while the oleoyl distribution remains random. In a subsequent study, Wollenberg [109] has shown that a distortionless enhancement by polarization transfer techniques significantly improves sensitivity to the extent that whole rapeseeds can be examined within an hour of acquisition time. Furthermore, some 1(3)- or 2-acyl groups could be identified leading to a partial estimation of the positional distribution of the fatty acids.

The above techniques have removed much of the uncertainty concerning the isomeric positioning of fatty acids in natural triacylglycerols as derived by enzymatic methods. The chemical methods for determination of the enantiomer content and absolute configuration of natural triacylglycerols will hopefully show a similar good agreement with the results of the physico-chemical approaches.

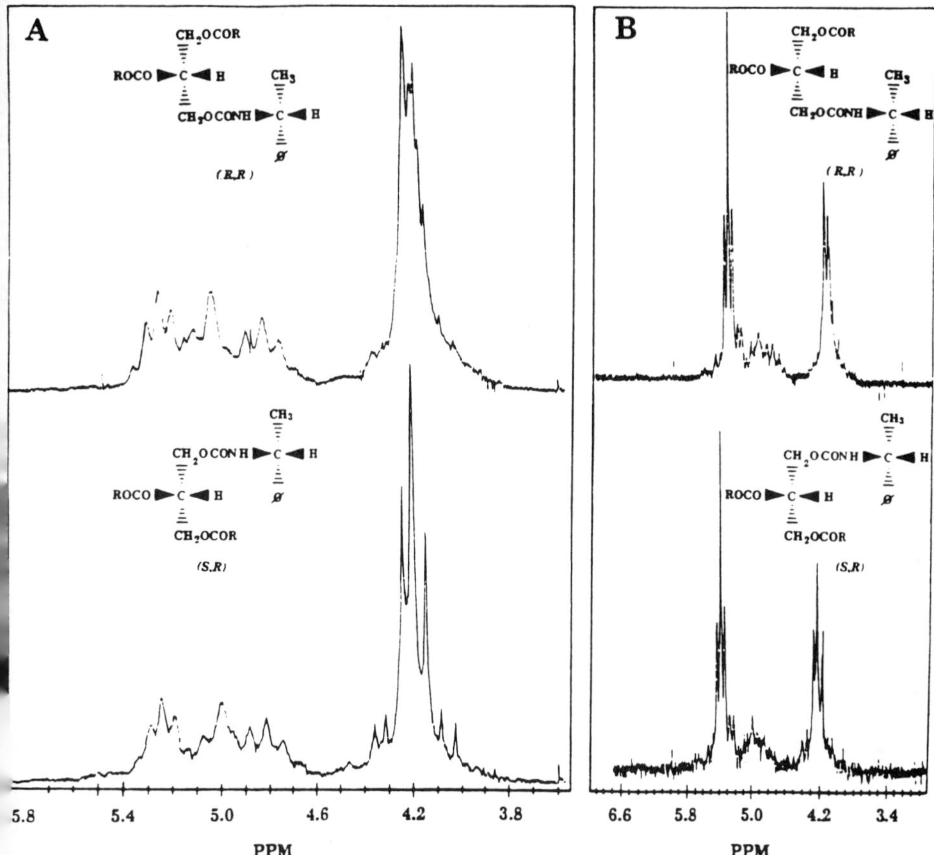

Fig 1.2. Proton NMR spectra showing characteristic differences between R,R and S,R carbamate diastereomers in 4.2-ppm region. A, carbamate derivatives of 1,2-dioctanoyl-*sn*-glycerol (*R,R*) and 2,3-dioctanoyl-*sn*-glycerol (*S,R*). B, carbamate derivatives of 1,2-dioleoyl-*sn*-glycerol (*R,R*) and 2,3-dioleoyl-*sn*-glycerol (*S,R*) [86]. (Reproduced by kind permission of the authors and of *Journal of Biological Chemistry*).

3. Mass Spectrometry

Mass spectrometry has provided one of the earliest methods of distinguishing between regio-isomers of acylglycerols and the subject has been reviewed [35]. Figure 1.3 shows the mass spectra of the *t*-BDMS ethers of two saturated *rac*-1,2- and *rac*-2,3-diacylglycerols [70]. There is a difference in the spectra of the 1-palmitoyl-2-stearoyl- (Figure 1.3A) and the 1-stearoyl-2-palmitoyl-*rac*-glycerol (Figure 1.3B), which is manifested in the abundance ratio of the ions due to losses of the acyloxy radicals from position 1 and position 2. For the above two reverse

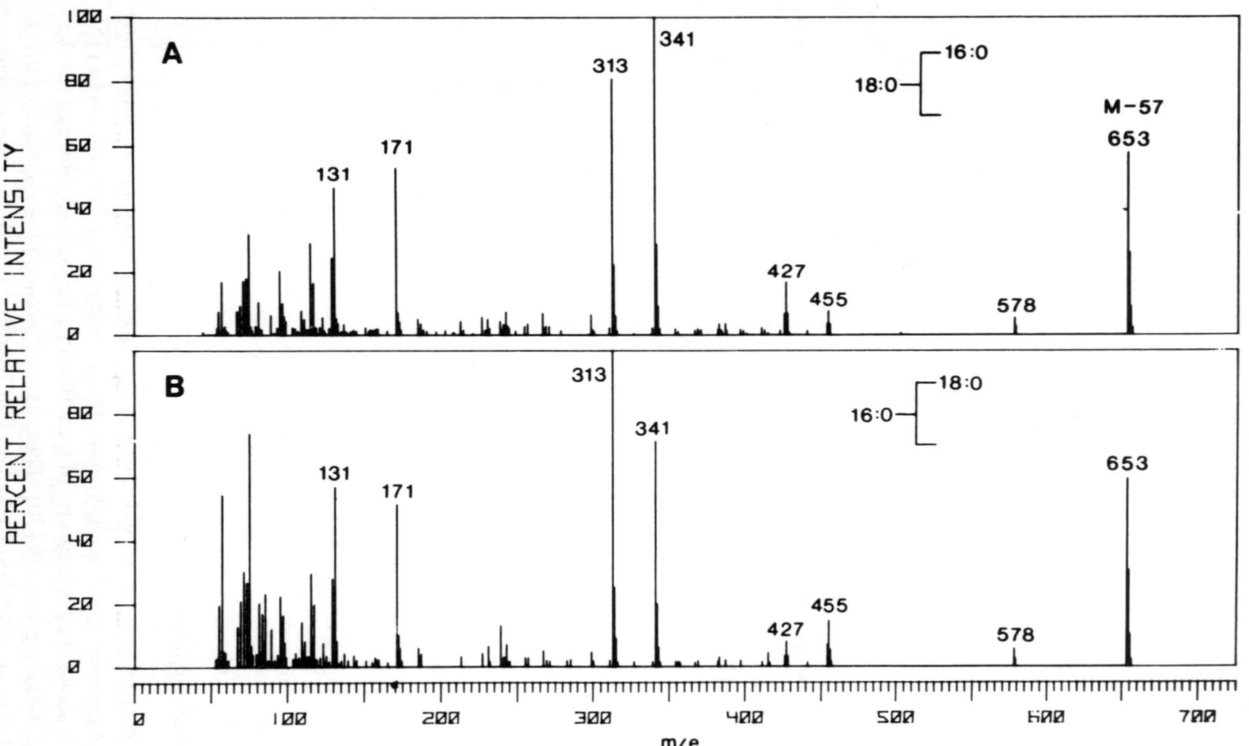

Fig. 1.3. GC/MS spectra of the *t*-BDMS ethers of *rac*-1,2-diacylglycerols: A, 1-palmitoyl-2-stearoyl-*rac*-glycerol; B, 1-stearoyl-2-palmitoyl-*rac*-glycerol [70]. (Reproduced by kind permission of the authors and of *Analytical Chemistry*).

isomers, the ratios of *m/z* = 427 (loss of stearoyl) to that of *m/z* = 455 (loss of palmitoyl) are 2.0 and 0.6, respectively. There is also a smaller change in the intensity ratio for the $[acyl + 74]^+$ ions at *m/z* = 313 and 341. Comparable differences between the mass spectra of these reverse isomers have been previously observed for the direct probe spectra of the TMS ethers, but their GC/MS spectra are indistinguishable.

The above method of determination of the reverse isomers of *sn*-1,2-diacylglycerols is satisfactory also for analyses of oligoenoic diacylglycerols. The polyunsaturates, which show progressively decreasing amounts of the $[M\text{-acyloxy}]^+$ fragment ions with increasing unsaturation, can be examined after $AgNO_3$-TLC separation and hydrogenation, provided the fatty chains in the *sn*-1 and *sn*-2-positions differ in carbon number [70]. Thus, it has been possible to estimate the reversed isomer content for all 16/18, 16/20, 18/20, 16/22 and 18/22 fatty carbon species [19]. The most significant source of error in the GC/MS method involves the tendency of the fatty acids bound to the *sn*-1-position of the glycerol moiety to fragment more readily than those bound at the *sn*-2-position [70]. This method of differentiation between positional isomers can be applied also to intact glycerophospholipids of known enantiomeric nature [45,46].

An investigation of the mass spectra of the nicotinoyl derivatives of mono- and diacylglycerols has revealed that they permit differentiation between regioisomers [114,115]. The nicotinoyl derivatives were examined by MS in case of the pure substances and as mixtures by GC/MS. The electron spectra of X-1- and 2-monoacylglycerols and of X-1,2- and X-1,3-diacylglycerols exhibit the same fragmentation patterns. Therefore, the distinction between positional isomers is only possible by characteristic differences in the relative abundances of fragment ions. Monoacylglycerols produce ions at *m/z* = 166 and *m/z* = 179, which are strikingly more abundant in the spectra of 2-monoacylglycerols. In the case of X-1,3-diacylglycerols the fragment at *m/z* = 164 is more abundant than the ion at *m/z* = 166 and the ion at *m/z* = 180 is more abundant than that at *m/z* = 179. In contrast, X-1,2-diacylglycerols prefer to generate the fragment ions *m/z* = 166 and *m/z* = 179 [114]. These authors have later shown [115] that the nicotinoyl derivatives are also well suited for the accurate determination of double bond positions in the regioisomers of mono- and diacylglycerols by characteristic spacings between abundant diagnostic acids in their fragmentation pattern.

The fatty acids associated with the primary and secondary positions of the triacylglycerol molecule can also be distinguished in this way, although no differentiation is possible between the *sn*-1- and *sn*-3-positions [20]. Kallio and Currie [36] have demonstrated that tandem mass spectrometry (MS/MS) with negative-ion chemical ionization (NICI) allows improved distinction between the *sn*-1(3)- and *sn*-2-positions of the triacylglycerols. From an examination of a large number of reference compounds, it could be seen in the daughter ion spectra that the formation of $[M\text{-}H\text{-}RCO_2H\text{-}100]^-$ from the *sn*-2-position is strongly reduced but not entirely eliminated. They have suggested that the limited formation of $[M\text{-}H\text{-}RCO_2H\text{-}100]^-$ from the *sn*-2-position is a result of acyl migration, presumably caused by the method of sample introduction into the mass spectrometer (*i.e.* volatilization from a heated probe) [36]. Kallio and Currie [35] have shown that it is possible to determine the proportion of triacylglycerols (*i.e.* the degree of preference that a fatty acid has for the *sn*-2-position) in simple binary mixtures of the type A-A-B (*rac*-AAB) and ABA (*rac*-ABA), where A and B represent different acyl groups in a triacylglycerol. Figure 1.4 shows an empirical plot of the ratio of the ions $[M\text{-}H\text{-}B\text{-}100]^-/[M\text{-}H\text{-}A\text{-}100]^-$ in a mixture of triacylglycerols of the type AAB and ABA [35]. The theoretical plot of $[M\text{-}H\text{-}B\text{-}100]^-/[M\text{-}H\text{-}A\text{-}100]^-$, if no cleavage of the fragment (RCO_2H-100) from position *sn*-2 occurs, is also shown in Figure 1.4. The ratios of $[M\text{-}H\text{-}B\text{-}100]^-/[M\text{-}H\text{-}A\text{-}100]^-$ from the pure triacylglycerol *rac*-AAB and *sn*-ABA/ are 0.784±0.015 and 0.145±0.016, regardless of the fatty acid comprising the triacylglycerols in the mixture. The preference of fatty acids for the *sn*-2-position in selected molecular weight fractions of triacylglycerols in fats and oils is generally more difficult or impossible to determine. Errors in the calculations of the regio-specific positions of the acyl groups increased with increasing complexity of the triacylglycerol mixture, due to deviation in each ion measured [35].

Kallio and Currie [35] have shown that the MS/MS data indicate a clear preference for the placement of the 18:2 acid in the secondary position in the major molecular weight species of the evening primrose oil, black currant seed oil, and turnip seed oil. Borage oil had an 18:3 acid also located abundantly at the *sn*-2-position. This was in good agreement with the results of Lawson and Hughes [56]. Similarly, a good agreement with previous results was found for the preferential association with the *sn*-2-position of certain other fatty acids, including palmitic acid, in human milk fat [17,38] as shown using rapid analysis by ammonia NICI MS/MS. Attempts to calculate the regiospecific distribution of fatty acids in some triacylglycerols from the Baltic herring (*Clupea harengus membras*) were not immediately successful because of a lack of reference compounds and thus appropriate correction factors for the $[RCO_2]^-$ ions. There was a clear need for a preliminary chromatographic fractionation of the triacylglycerols. Either argentation HPLC or supercritical fluid chromatography (SFC), or both have shown promise for complementary peak resolution, collection of samples and improved MS/MS. SFC with electron impact (EI) MS/MS was used by Kallio *et al.* [37] to determine the fatty acid distribution in *sn*-2- and *sn*-1(3)-positions of butterfat triacylglycerols, while HPLC prefractionation with silver ions has been applied by Laakso and Kallio [53,54] to the investigation of the monoenoic butterfat triacylglycerols containing oleic and vaccenic acids. The two fractions with the same degree of unsaturation showed quite different molecular weight profiles and differences in the distribution of the *cis*- and *trans*-monoenoic acids between the primary and secondary positions [53,54]. The MS/MS method is powerful and fast, when proportions of fatty acid combinations from relatively simple mixtures are to be compared. Eventually, a further refinement could be introduced by combining reversed-phase HPLC of intact triacylglycerols with on-line MS/MS.

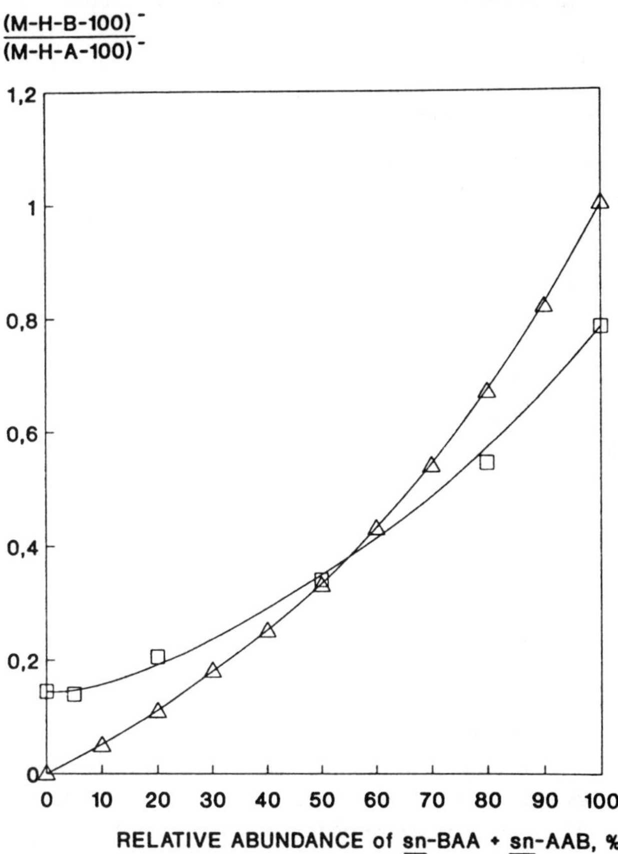

Fig. 1.4. Dependence of the intensity of the $[M\text{-}H\text{-}RCO_2H\text{-}100]^-$ ion on the regiospecific position of the corresponding fatty acyl group in the triacylglycerol [35]; □, empirical correlation; △, theoretical curve. (Reproduced by kind permission of the authors and of *CRC Handbook of Chromatography*).

D. CHROMATOGRAPHIC RESOLUTION OF REGIOISOMERS

Chromatographic resolution of regioisomers of glycerolipids is based on differences in polarity between the primary and secondary alcohol or ester groups and to a lesser extent on the shielding of the secondary relative to the primary positions. Each regioisomer may be made up of one or more molecular species depending on additional resolution in the system. The extent of resolution of regioisomers can be increased by differential complexing of the primary and secondary hydroxyl groups, while derivatization usually reduces the separation factor. A special case arises from the inversion of the fatty acids on the adjacent primary and secondary carbon atoms of regioisomers, when they are referred to as reverse isomers.

1. Adsorption chromatography

Adsorption chromatographic resolution of regioisomers of acylglycerols has been reviewed by Myher [68]. The separation is best observed with the X-1,3- and X-1,2-diradylglycerols, which possess a secondary and a primary hydroxyl group, respectively. The X-1,3-isomers are less polar and migrate ahead of the X-1,2-isomers. This difference in polarity is lost upon acetylation, as the X-1,3-diacylglycerol acetates overlap with the X-1,2-diacylglycerol acetates. The separations are less extensive or absent for the X-1- and 2-monoradylglycerols, which possess a combination of a primary and a secondary, and two primary hydroxyl groups, respectively. The extent of separation of the regioisomeric monoradylglycerols can be increased by borate complexation TLC. As a result, the 2-isomers migrate ahead of the X-1-isomers using a variety of solvent systems [103]. Derivatization of the monoradylglycerols results in an overlap of the isomers during adsorption chromatography [68].

Likewise, normal-phase HPLC retains the X-1,2-diacylglycerol isomers longer than the X-1,3-isomers. Normal-phase HPLC also allows the resolution of X-1- and 2-monoradylglycerols, with the X-1-isomer being eluted ahead of the 2-isomer [43]. Derivatization results in an overlap of the regioisomers of monoacylglycerols also during normal-phase HPLC.

Conventional adsorption chromatography on either thin-layer plates or high-performance columns fails to resolve regioisomers of triacylglycerols. However, triacylglycerols containing oxygenated fatty chains can be resolved by normal-phase HPLC. Thus, regiospecific resolution of saturated and oligounsaturated triacylglycerols is possible following epoxidation [27]. The triacylglycerol containing the epoxidized monounsaturated fatty acid in the 2-position is eluted ahead of the isomeric triacylglycerol containing the epoxidized fatty acid in the X-1-position. It has been noted that epoxidized triacylglycerol containing linoleic acid yields a double peak which approximates a constant ratio of 3:2 in area. This is apparently due to the production of the 9,10-, 12,13-diepoxy stearate derivatives of linoleic acid in the ratio of 3:2. These peaks represent resolution of two sets of diastereomers. Hammond [27] has obtained similar effects with the trienes. The preparation of epoxides requires reaction with an excess of *m*-chloroperoxybenzoic acid in order to ensure complete reaction. Comparable separations are obtained with isomeric triacylglycerols containing hydroperoxy fatty acids [23,72]. Thus, the regioisomers of monohydroperoxy trilinoleoylglycerols have been resolved by normal-phase HPLC with a mixture of 2-propanol/hexane (0.5:99.5, v/v) using two conventional 5 micron silica Zorbax columns in series. The 2-isomer was eluted ahead of the X-1-isomer [72]. These separations were obtained for both 9-hydroperoxy and 13-hydroperoxy isomers and their *cis,trans* and *trans,trans*-derivatives. The regioisomers of *cis,trans*-13-hydroperoxy derivatives were resolved more extensively than the *cis,trans*-9-hydroperoxy derivatives [72]. Similar separations were obtained for the monohydroperoxy derivatives of trilinolenoylglycerol on normal-phase HPLC. The 2-isomer of the *cis,trans,cis*-12(13)-hydroperoxy derivative was eluted ahead of the *cis,trans,cis*-12(13)-hydroperoxy trilinolenate and the 2-isomer of *trans,cis,cis*-9-hydroperoxy derivative was eluted ahead of the X-1-isomer of the *trans,cis,cis*-9-hydroperoxy derivative of autoxidized trilinolenoylglycerol [23]. Likewise, normal-phase HPLC separations were obtained between regioisomers of monohydroperoxy epidioxide mixtures from autoxidized trilinolenoylglycerol [23].

2. Silver ion chromatography

Silver ion TLC effectively resolves acylglycerols according to number of double bonds as well as regioisomers. Nikolova-Damyanova [76] has published an extensive review of these separations. The pairs SSO-SOS and SSL-SLS, where S stands for saturated and O and L for oleic and linoleic acid, respectively, are resolved with relative ease at ambient temperature as already noted by early workers. The symmetrical isomers SOS and SLS migrate ahead because the unsaturated acid in the 2-position complexes less strongly with silver ions than the corresponding acid in the X-1-position. However, the pair OOS-OSO have also been resolved, with the OOS isomer, which contains a combination of primary and secondary ester groups, migrating ahead of the OSO isomer, which contains two unsaturated primary ester groups. Presumably, similar separations could be obtained with the regioisomers of sn-1,2-diacylglycerols derived from natural glycerophospholipids following appropriate derivatization, if necessary. This would allow separation of such reverse isomers as SO and OS or SL and LS. Likewise, the sn-2,3-diacylglycerols, derived from Grignard degradation of natural triacylglycerols and chiral-phase separation of the enantiomeric diacylglycerols, could then be resolved into their reverse isomers.

More durable columns are produced by loading the silver ions on sulfonic acid anion-exchange columns [13]. These columns, however, do not permit resolution of regioisomers of triacylglycerols [27].

3. Reversed-phase chromatography

Regiospecific resolution of X-1,3- and X-1,2-diacylglycerols is also obtained on reversed-phase HPLC, with the X-1,3-isomers emerging ahead of the X-1,2-isomers [43]. This resolution is lost upon derivatization. There is marginal resolution of the isomeric 1- and 2-monoacylglycerols, with complete overlapping resulting from derivatization [92]. Reversed-phase HPLC usually does not resolve regioisomers of triacylglycerols. However, regioisomers of triacylglycerols containing epoxy [18], hydroxy and hydroperoxy [94] fatty acid groups are readily resolved. The isomers containing the oxo-fatty acids in the 2-position are eluted ahead of those containing the oxo-fatty acids in the X-1-position. Thus, epoxidation of triacylglycerols with subsequent reversed-phase HPLC allowed rapid separation of monounsaturated acylglycerols according to isomeric position (X-1,2- and X-1,3-) and molecular weight. The X-1,3-dipalmitoyloleoylglycerol was eluted ahead of the X-1,2-dipalmitoyloleoylglycerol using ethylenedichloride-acetonitrile in varying proportions from 20:80 to 40:60 in 1 hour [18]. Likewise, regioisomers of triacylglycerols containing the hydroxy (hydroperoxy) fatty acid in the 2-position were eluted ahead of those with the hydroxy (hydroperoxy) fatty acid in the X-1-position [94]. Furthermore, triacylglycerols containing core aldehydes were resolved into regioisomers by reversed-phase HPLC using a linear gradient of 20-80% isopropanol in methanol [94]. The isomers with the core aldehyde in the 2-position were eluted later than those with the aldehyde group in the primary position.

4. Gas chromatography

Because of isomerization at the elevated temperatures, GLC is not suitable for the analysis of free mono- and diacylglycerols. However, certain regioisomers of acylglycerols can be resolved by GLC after derivatization. Myher [68] has discussed the GLC resolution of acylglycerol regioisomers on both polar and non-polar columns. Non-polar GLC columns of minimal length resolve primary and secondary monoacylglycerols, when run as the TMS ethers, but not as the acetates. The 2-isomer emerges ahead of the X-1-isomer. Likewise, the 2-isomer is eluted ahead of the X-1-isomer when run as the TMS ether on a polar capillary column [111]. Both non-polar and polar capillary GLC is capable of resolving X-1,2- and X-1,3-isomers of diacylglycerols using columns of increased length and appropriate derivatives. Thus, Lohninger and Nikiforov [61] have reported the resolution of the TMS ethers of X-1,2- and X-1,3-diacylglycerols on a 12 m column in the temperature range 240-300°C using hydrogen as carrier gas. The X-1,3-isomers migrate ahead of the X-1,2-isomers of identical carbon number. Similar resolution has been obtained on much shorter columns containing polar liquid phase [68]. Recently GLC has been employed to introduce the nicotinoyl-diacylglycerols into the mass spectrometer for identification of regioisomers, which apparently were not chromatographically resolved [114,115]. As for other chromatographic systems, GLC has not been observed to resolve the reverse isomers of saturated X-1,2-diacylglycerols, except in cases of unequal chain length derivatives [J.J. Myher and A. Kuksis, 1990,. Unpublished results]. Thus, the *rac*-1-acetyl(butyryl)-2-palmitoylglycerol emerge ahead of the respective *rac*-1-palmitoyl-2-acetyl(butyryl)glycerol from both non-polar and polar capillary columns.

E. CHROMATOGRAPHIC RESOLUTION OF STEREOISOMERS

Enantiomeric triacylglycerols cannot physically be resolved by chromatographic or other physico-chemical methods even when the substituent groups are markedly different [22,33]. Likewise, the positional distribution of fatty acids in a triacylglycerol molecule cannot be reliably determined by differential chemical release of the acids from the primary and secondary positions, although some difference in the reactivity has been demonstrated with certain reagents [7,9]. A complete separation of the positional isomers, including enantiomers, however, is possible in the form of partial acylglycerols, following preparation of diastereomeric [14] or enantiomeric [95] derivatives by means of normal- or chiral-phase HPLC, respectively. An analysis of the fatty acid composition of the enantiomeric mono- and diacylglycerols then allows a direct or indirect determination of the fatty acids associated with each stereospecific position of the glycerol molecule. Foglia and Maeda [22] have shown that such racemic glycerol derivatives as 1,2-isopropylidene-*rac*-glycerol, its 3-benzoyl derivative, and 1-hexadecyl-2-benzoyl-*rac*-glycerol can be resolved into enantiomers by a chiral stationary phase containing cyclodextrin tribenzoate. This method has not been investigated for positional analysis of fatty acids in acylglycerols, however.

1. Random Generation of Acylglycerols

A critical aspect of this type of stereospecific analysis is the random generation of the mono- and diacylglycerols and absence of isomerization of the degradation products during subsequent work-up [9]. Brockerhoff [9 and references therein] first reported the random or near random cleavage of the primary and secondary ester bonds in triacylglycerols by means of ethyl or methylmagnesium bromide in diethyl ether (Grignard degradation), and many workers have employed this method subsequently. Under the usual conditions the yield of diacylglycerols is less than 25%. The fatty acids are released in the form of tertiary alcohols with the formation of *sn*-1,2-, *sn*-2,3- and *sn*-1,3-diacylglycerols as well as significant amounts of *sn*-1-, *sn*-2- and *sn*-3-monoacylglycerols. There is evidence that the *sn*-1,3-diacylglycerols are compromized by significant isomerization of *sn*-1,2- and *sn*-2,3-diacylglycerols, which may account for a contamination of the *sn*-1- and *sn*-3-positions with up to 10% of the fatty acid in the *sn*-2-position [15].

The Grignard degradation is conveniently performed by the following modification [69] of the original large-scale procedure of Brockerhoff [9]. About 10-15 mg of the triacylglycerol mixture is dissolved in 0.4 mL of anhydrous diethyl ether, and 0.1 mL of 1 M ethylmagnesium bromide in diethyl ether is added. After 25 seconds, 0.4 mL of glacial acetic acid-diethyl ether (1:9, v/v) is added, and the mixture is stirred for 30 seconds. Then 4 mL of diethyl ether and 0.5 mL of water are added and the mixture is stirred for 2 minutes. After removing the aqueous layer, the diethyl ether phase is washed once with 0.5 mL of 2% aqueous $NaHCO_3$ and twice with 0.5 mL aliquots of water. The extract is dried with anhydrous Na_2SO_4 and then evaporated under nitrogen. The mixture of the reaction products is resolved into the *sn*-1,2(2,3)- and *sn*-1,3-diacylglycerols and separately recovered by TLC using 5% borate impregnated silica gel and chloroform-acetone (97:3, v/v) as the developing solvent [69]. The diacylglycerol bands are visualized under UV light after spraying with aqueous 0.01% Rhodamine 6G or with 0.005% methanolic 2,7-dichlorofluorescein. The diacylglycerols are recovered by scraping the silica gel into Pasteur pipettes containing small cotton plugs and eluting the gel with 100 mL of diethyl ether. The yield of the diacylglycerols is usually 2-5 mg.

Santinelli *et al.* [87] have described a further miniaturized procedure for producing random diacylglycerols from silver ion subfractions of triacylglycerols. In this instance 1-2 mg of triacylglycerol were dissolved in dry diethyl ether (1 mL), freshly prepared 0.5 M ethyl magnesium bromide in dry diethyl ether (250 μL) was added, and the mixture was shaken for 1 min before glacial acetic acid (6 μL) in pentane (5 mL) and water (2 mL) were added to stop the reaction. The organic layer was washed twice with water (2 mL) and dried over anhydrous sodium sulfate. After evaporating the solvent in a stream of nitrogen at room temperature, the mixture of hydrolysis products was derivatized directly. Laakso and Christie [50] used preparative HPLC for a rapid isolation of the diacylglycerols using hexane-tetrahydrofuran-2-propanol (100:3:1.5 by volume) as the mobile phase.

In these routines, the *sn*-2-monoacylglycerols may also be recovered, as they contain little isomerization products. The *sn*-1- and *sn*-3-monoacylglycerols are usually discarded because of actual or potential contamination of isomerization products from *sn*-2-monoacylglycerols. However, Takagi and Ando [97] and Ando *et al.* [3] have successfully employed essentially the original Brockerhoff [9] routine for the generation of *sn*-1-, *sn*-2- and *sn*-3-monoacylglycerols without significant isomerization. After removal of the solvent at ambient temperature, the monoacylglycerols were recovered by preparative TLC on boric acid (10%) impregnated silica gel plates developed with chloroform-methanol (98:2, v/v) under nitrogen. The yields of 1- and 2-monoacylglycerols were generally 5-7 mg and 1-2 mg, respectively, starting with 100 mg total triacylglycerol. The method allows direct access to all three positions of the glycerol molecule, which is an advantage over other more established methods of positional fatty acid analyses, where some of the positions are being estimated indirectly by calculation. Taylor *et al.* [102], who also performed stereospecific analyses of seed oil triacylglycerols by the method of Takagi and Ando [97] did all the ether extractions during the monoacylglycerol preparation and recovery with ether saturated with 10% (wt/vol) boric acid solution to minimize potential isomerization. Later Ando and Takagi [2] have described a micromethod (down to 1 mg level) for stereospecific analysis of triacylglycerols *via* chiral-phase HPLC of the monoacylglycerol diurethanes.

Recently, Becker *et al.* [4] have shown that the more reactive Grignard reagent, allylmagnesium bromide, yields more representative *sn*-2-monoacylglycerols allowing the composition of the *sn*-2-position to be estimated directly. For the allyl magnesium bromide degradation, 6-6.5 mg triacylglycerol was dissolved in diethyl ether (5 mL) in a Teflon-capped reaction tube with stirring. The allylmagnesium bromide (200 μL) was added with a pipette that had been flushed with nitrogen. The ether solution became opaque, indicating a spontaneous reaction.

After one minute, additional diethyl ether (5 mL) was added. Then the organic phase was washed, first with boric acid buffer (4 mL) prepared by adding 37% HCl (1 mL) to a 0.4 M boric acid solution (36 vol) giving a final concentration of 0.27 M. This neutralized the $Mg(OH)_2$ formed in the reaction mixture upon addition of water. Additional washings (2 X 4 mL) with 0.4 M boric acid solution removed the excess HCl and remaining magnesium salts. The monoacylglycerols were resolved using boric acid impregnated plates and chloroform-acetone (96:4, v/v) as the solvent. While the *sn*-1(3)-monoacylglycerols from the Grignard deacylation were slightly contaminated with isomerized *sn*-2-monoacylglycerols, the *sn*-2-monoacylglycerols were essentially free of contamination with the isomerization products from *sn*-1(3)-monoacylglycerols. This was explained by the less favourable equilibrium between the *sn*-1(3) and *sn*-2-isomers [55]. The method proved suitable for the regiospecific differentiation of fatty acids between the *sn*-1(3) and *sn*-2-positions in the common seed oils, in fish oils, and in seed oils containing medium chain fatty acids.

In order to take full advantage of the pre-purification of intact triacylglycerols, these degradations must be scaled down to submicrogram levels, using appropriate carrier triacylglycerols, if necessary. An effective preliminary segregation of triacylglycerols can be readily obtained by argentation TLC [6,84] or argentation HPLC [51,87]. It provides triacylglycerol subfractions of uniform degree of unsaturation ranging from zero to as many as nine double bonds per molecule. The triacylglycerol fractions are readily recovered and, depending on their nature, may be sufficiently large and simple enough for a meaningful structural analysis. Detailed descriptions of the procedures and solvent systems for $AgNO_3$-TLC of triacylglycerols have been published by several groups of workers [6,84] and various applications have been reviewed [76]. A somewhat more difficult method of preliminary segregation of natural triacylglycerols is provided by preparative GLC on non-polar liquid phases [44]. Recently, supercritical fluid chromatography (SFC) has been shown to provide triacylglycerol subfractions similar to those obtained by preparative GLC [37,49]. This technique yields triacylglycerol subfractions of uniform molecular weight and is best suited for work with the more saturated shorter-chain triacylglycerols. Capillary SFC could be used for a further segregation of the triacylglycerol fractions recovered from argentation or reversed-phase HPLC or both, in which case the triacylglycerol fractions of uniform molecular weight and degree of unsaturation (or even uniform geometry of the double bonds) would be obtained. Sempore and Bezard [89,93] have demonstrated the experimental feasibility of combining argentation TLC and reversed-phase HPLC for a pre-fractionation of triacylglycerol mixtures prior to stereospecific analysis.

It is possible to effect also a preliminary resolution of the diacylglycerol or indeed the monoacylglycerol mixture on the basis of molecular weight or number of double bonds prior to separation of the urethane derivatives. Pre-fractionation of the monoacylglycerol di-3,5-DNPU derivatives prior to the chiral-phase HPLC has already been shown to be advantageous and necessary in some circumstances [28,93,96].

2. Preparation of Derivatives

It is important that the derivatives for the chromatographic resolution of the diastereomers or enantiomers be prepared without isomerization of the acylglycerols. Laakso and Christie [50] prepared the 1-(1-naphthyl)ethyl urethane derivatives by dissolving 1-2 mg of the diacylglycerols in dry toluene (300 μL) and reacting them with the (*R*) or (*S*)-forms of 1-(1-naphthyl)ethyl isocyanate (10 μL, Aldrich Chemical Co.) in the presence of 4-pyrrolidinopyridine (approx. 10 μmole) overnight at 50°C. The products were extracted with hexane-diethyl ether (1:1, v/v) and washed with 2 M HCl and water. The organic layer was taken to dryness under a stream of nitrogen and the sample was purified on a short column of Florisil eluted with diethyl ether. In subsequent preparations Christie *et al.* [16] and Santinelli *et al.* [87] the reaction mixture, after overnight standing, was evaporated to dryness and the residue dissolved as far as possible in methanol-water (95:5, v/v) by warming. A Bond-Elut ODS solid-phase extraction column (500 mg; Jones Chromatography, Hengoed, Wales) was solvated by passing 10 mL of this solvent through it. The reaction mixture was filtered through a small cotton-wool plug onto the column and washed through with a further 15 mL of solvent. The required products were then eluted with acetone (10 mL). Further purification, if necessary, was achieved by an elution through a short Florisil column with hexane-acetone (96:4, v/v, 10 mL). Rogalska *et al.* [86], who also prepared the diastereomers of diacylglycerols for normal phase HPLC resolution, performed the derivatization with *R*-(+)-phenylethylisocyanate (2 mmol/22 μmol dioleoylglycerol) under stirring in a sealed vial at room temperature for 48 hours (80°C, 5 h in the case of dioctanoylglycerol). The excess isocyanate was removed from the reaction mixture by eluting it with heptane on a small silica gel column. The carbamates were then recovered with ethyl acetate, evaporated, and injected into the HPLC column in 0.5 or 0.7% ethyl alcohol in heptane.

The 3,5-dinitrophenylurethane (3,5-DNPU) derivatives of diacylglycerols and dialkylglycerols for chiral column separation were prepared as described by Takagi and Itabashi [98]. The samples of less than 1 mg were reacted with 3,5-dinitrophenyl isocyanate (Sumitomo Chemical Co., Osaka, Japan) of about 2 mg in dry toluene (4 mL) in the presence of dry pyridine (40 μL) at ambient temperature for 1 hour. The crude 3,5-DNPU derivatives of diacyl- and dialkylglycerols were purified by TLC on silica gel GF plates using hexane-ethylene dichloride-ethanol (40:10:3 and 40:10:1 by volume, respectively).

The *sn*-1(3)-monoacylglycerols were converted to their corresponding di-3,5-DNPU derivatives as described by Takagi and Ando [97]. The monoacylglycerol fraction (1-3 mg) was dissolved in dry toluene (500 μL) and was reacted overnight at ambient temperature with 3,5-dinitrophenyl isocyanate (10-20 mg) in the presence of dry pyridine (50 μL). The reaction products were isolated by spotting a chloroform extract of the reaction mixture on silica gel plates and developed with n-hexane-1,2-dichloroethane-ethanol (40:15:6 by volume) under nitrogen. Taylor *et al.* [102], who employed a similar technique for stereospecific analysis of triacylglycerols, purified the di-DNPUs by a second TLC step on silica gel plates developed with chloroform-methanol (98:2, v/v). They recovered the derivatives by extraction with diethyl ether saturated with water. The di-3,5-DNPU monoacylglycerol derivatives were dissolved in 1 mL n-hexane-dichloroethane-ethanol (40:12:3 by volume) and filtered through a 0.2 µm Spin-X Nylon membrane (Costar, Cambridge, MA) before HPLC. The 3,5-DNPU derivatives of diacylglycerols can be effectively prefractionated by reversed-phase HPLC [90,113].

3. Resolution of Diastereomers

Michelsen *et al.* [65] first reported a partial resolution of enantiomeric *sn*-1,2- and *sn*-2,3-diacylglycerols as the diastereomeric (S)-(+)-1-(1-naphthyl)ethyl urethanes by HPLC using a normal-phase HPLC column. Subsequently, Rogalska *et al.* [86] and Laakso and Christie [50] showed that diacyl-*sn*-glycerols, as their (*S*)- or (*R*)-1-(1-naphthyl)ethyl urethanes, could be completely resolved by HPLC on a silica gel column by modifying the original conditions, and the latter have investigated the potential of the method for the stereospecific analysis of triacylglycerols. Laakso and Christie [50] resolved the diastereomeric diacylglycerol derivatives using a series of two columns of silica gel (Hypersil 3 µm, 25 cm x 4.6 mm id) with 0.5% 2-propanol in hexane as the mobile phase at a flow rate of 0.8 mL/min. The diacylglycerols derivatized with (*S*)-(+)-1-(1-naphthyl)ethyl isocyanate eluted in the order of X-1,3-, followed by *sn*-1,2- and finally *sn*-2,3-isomers. When the (*R*)-form of 1-(1-naphthyl)ethyl isocyanate was used for derivatization, the elution order changed so that *sn*-2,3-isomers eluted before *sn*-1,2-isomers. The X-1,3-isomers were eluted in approximately half the time needed for elution of the *sn*-1,2- and *sn*-2,3-isomers. Molecular species of the single-acid diacyl-*sn*-glycerol derivatives studied were generally well resolved, although the order of elution was unusual, *e.g.* 18:1 < 18:0 < 18:2 < 16:0. It was neither that expected for normal-, nor that for reversed-phase partition chromatography. The molecular species were identified on the basis of comparison of retention times to synthetic standards, or by peak overlap with standards coinjected with the unknowns. Nevertheless, the resolution achieved for the *sn*-1,2- and *sn*-2,3-enantiomers of the model compounds was complete, which was sufficient for a stereospecific analysis of natural triacylglycerols provided the molecular species resolution did not result in overlap between the enantiomers (diastereomers in this case). Thus, Christie *et al.* [16] applied this method for the stereospecific analysis of natural fats and oils containing primarily C_{16} and C_{18} fatty acids, *i.e.* safflower, sunflower, olive and palm oils, tallow, egg and rat adipose tissue, which yielded good agreement with previous analyses [7]. Only data for myristic acid appeared dubious, because of a carry-over of molecular species enriched in this component from the *sn*-1,2-diacylglycerol group of peaks into the *sn*-2,3-group. As before [15], the results for position *sn*-2- obtained by analysis of the contaminated 1,3-diacylglycerols were appreciably in error. The stereospecific analysis procedure described here did not appear to be suitable for natural triacylglycerol samples containing a wider range of fatty acids, such as those present in milk fat, coconut oil, palm kernel oil and fish oil. Although the *sn*-1,2- and *sn*-2,3-diacylglycerol derivatives were separated into a number of distinct peaks, there was overlap of different diastereomers. Clearly, there was a need for a prefractionation of the triacylglycerol mixture prior to the preparation and separation of the diastereomers. In a subsequent publication, Santinelli *et al.* [87] used this method of structure determination of triacylglycerols for the major silver ion fractions obtained from olive oil triacylglycerols. The study resulted in an unexpected but reasonable demonstration that a prefractionation of a triacylglycerol mixture prior to the resolution of the diastereomers may be necessary to reach valid conclusions about the positional distribution of the fatty acids within specific acylglycerol molecules. Thus, although stereospecific analysis of the intact triacylglycerols indicated that the fatty acids in positions *sn*-1- and *sn*-3- were similar, the stereospecific analyses of individual molecular species demonstrated marked asymmetry. The data showed that the 1,3-random-2-random or the 1-random-2-random-3-random distribution theories [60,69,71] were not applicable to olive oil.

A comparable resolution of the diastereomers of both synthetic and natural diacylglycerols was reported by Rogalska *et al.* [86] using 0.5 or 0.7% ethyl alcohol in heptane as eluent on a Beckman Ultrasphere 5 μm column (10 x 2.5 cm) and 3.5 mL/min flow rate. The use of (*R*) derivatives led to elution of the *sn*-2,3-isomers ahead of the *sn*-1,2-isomers. The elution order was determined by reference to standard *sn*-1,2-diacylglycerol derivatives. The diastereomeric carbamates separated by HPLC were characterized by proton NMR and mass spectrometry. The method was applied to establish the stereobias of lipases by means of substrates that were chemically alike but contained sterically non-equivalent ester groups within one single triacylglycerol molecule. Figure 1.5 shows the chromatograms obtained for the diastereomeric carbamate mixtures derived from *sn*-1,2- and *sn*-2,3-diacylglycerols generated by different lipases from the corresponding achiral triacylglycerol substrates [86].

4. Resolution of Enantiomers

Itabashi and Takagi [29] were the first to report a complete resolution of the enantiomeric diacylglycerols as the 3,5-DNPU derivatives by HPLC on a chiral liquid phase consisting of *N*-(*S*)-2-(4-chlorophenyl)isovaleroyl-(*R*)-phenylglycine (Sumipax OA-2100). Subsequently, other chiral liquid phases have been shown to be better suited for resolution of enantiomeric diacylglycerols from natural sources and for the resolution of enantiomeric monoacylglycerols. Takagi [95] has published an extensive review of the development and early applications of this methodology.

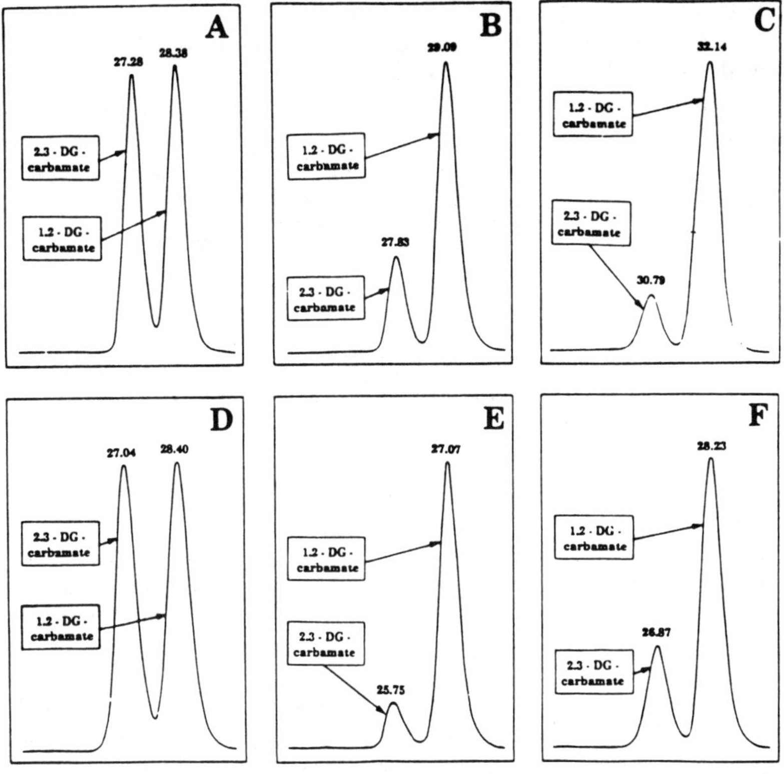

Fig. 1.5. Chromatograms of diastereomeric carbamate mixtures from lipolysates [86]. A (5 min) and D (16 min) lipolysis with porcine pancreatic lipase; B (11 min) and E (16 min) lipolysis with human gastric lipase; C (11 min) and F (20 min) lipolysis with rabbit gastric lipase. A,B,C, carbamates of 2,3-dioctanoyl-*sn*-glycerol (first peak, *S,R*) and 1,2-dioctanoyl-*sn*-glycerol (second peak, *R,R*). D,E.F, carbamates of 2,3-dioleoyl-*sn*-glycerol and 1,2-dioleoyl-*sn*-glycerol; elution order as above. (Reproduced by kind permission of the authors and *Journal of Biological Chemistry*).

In the early work Itabashi and Takagi [29] used the Sumipax OA-2100 packing in a 75 cm-long column and several hours were needed to effect a complete resolution of synthetic diacylglycerols. Later, Takagi and Itabashi [98] reported an improved separation of enantiomeric diacylglycerols as the 3,5-DNPU derivatives using another type of chiral phase, *N*-(*R*)-1-(1-naphthyl)ethylaminocarbonyl-(*S*)-valine (Sumipax OA-4100). This phase permitted complete enantiomer resolution within 10 min on a 25 cm-long column, with the *sn*-1,2-enantiomers being eluted ahead of the *sn*-2,3-enantiomers. Figure 1.6 shows the postulated three point interaction between the liquid phase and the enantiomeric diacylglycerol derivatives necessary for resolution [47]. Apparently, the *sn*-2,3-enantiomer interacts with this phase more strongly than the *sn*-1,2-enantiomer and is eluted later. Sempore and Bezard [89] have utilized the Sumipax OA-4100 chiral phase in stereospecific analysis of model triacylglycerols isolated from peanut and cotton-

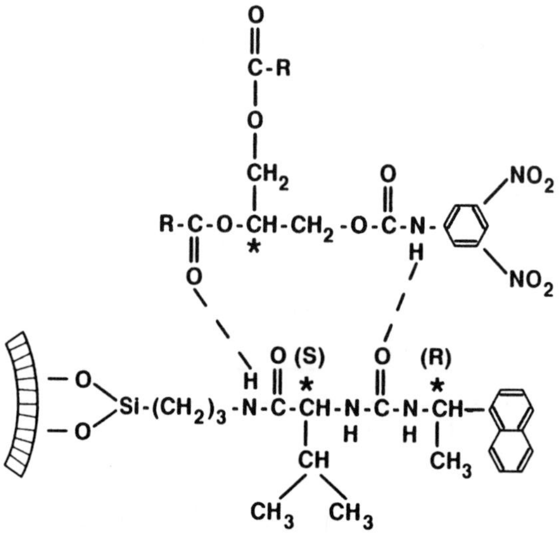

Fig. 1.6. Possible hydrogen bonding between a 3,5-DNPU derivative of an enantiomeric diacylglycerol and the *N*-(*R*)-1-((-naphthyl)ethylaminocarbonyl-(*S*)-valine chemically bonded to α-aminopropyl silanized silica (Sumipax OA-4100, Sumitomo Chemical Co., Osaka, Japan) [47]. (Reproduced by kind permission of the authors and the American Oil Chemists' Society, and redrawn from the original).

seed oils by combined application of argentation TLC and reversed-phase HPLC. The diacylglycerol DNPU derivatives resolved by the chiral column were further separated into molecular species by reversed-phase HPLC. The excellent results obtained suggested that the method could be applied to other oils and fats.

However, Itabashi *et al.* [30] have shown that the 3,5-DNPU derivatives of enantiomeric diacylglycerols derived from natural triacylglycerols were much better resolved on a (*R*)-1-(1-naphthyl) ethylamine (YMC-Pack A-KO3) column, which apparently possesses a higher enantioselectivity than previously used chiral phases [28,29,98]. A highly satisfactory separation was obtained between the enantiomeric diacylglycerols derived by Grignard degradation from corn, linseed, and menhaden oil triacylglycerols. Figure 1.7 shows the chiral-phase HPLC resolution of the diacylglycerol moieties of lard triacylglycerols, as the 3,5-DNPU derivatives on the YMC-Pack A-KO3 column [31]. The *sn*-l,2(2,3)-diacylglycerols are resolved into two groups with equal total peak areas, representing the *sn*-l,2- and *sn*-2,3-enantiomers. Three major peaks can be recognized in the *sn*-1,2-enantiomer region and two in the *sn*-2,3-enantiomer region. The three peaks in the *sn*-l,2-enantiomer correspond to equivalent carbon number (ECN) values of 34, 32 and 30, and those in *sn*-2,3-enantiomer to ECN of 32 and 30, in order of appearance. In D and E of Figure 1.7 are shown the polar capillary GLC profiles of the *sn*-1,2-and *sn*-2,3-diacylglycerols as the TMS ethers obtained by silolysis of the urethanes. The elution order of the molecular species of the diacylglycerol urethanes from the chiral column has been examined in detail by on-line mass

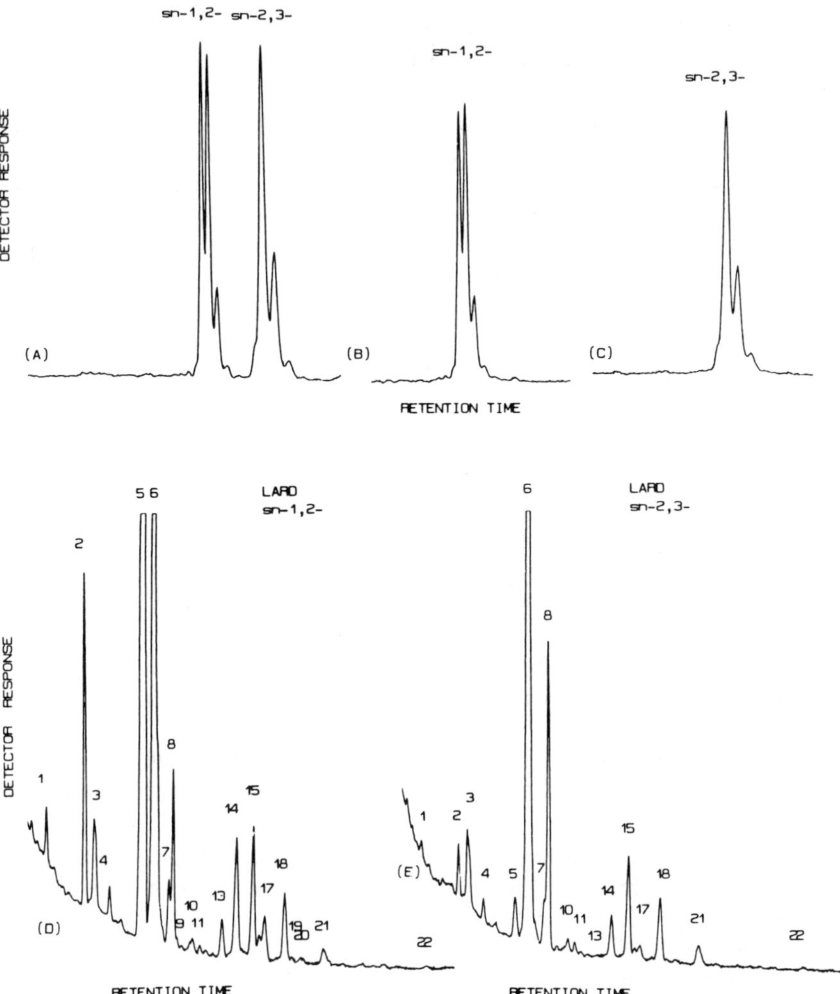

Fig. 1.7. Chiral-phase HPLC of the DNPU derivatives of original lard *rac*-1,2-diacylglycerols (A) and of the *sn*-1,2- (B) and *sn*-2,3-enantiomers (C) after collection from the chiral HPLC column, along with polar capillary GLC of TMS ethers of the *sn*-1,2- (D) and *sn*-2,3-diacylglycerols (E) regenerated from DNPU derivatives of the enantiomers [31]. Peak identification: 1, 14:0-16:0; 2, 14:0-18:0 + 16:0-16:0; 3, 14:0-18:1 + 16:0-16:1; 4, 14:0-18:2; 5, 16:0-18:0; 6, 16:0-18:1; 7, 16:1-18:1; 8, 16:0-18:2; 9. 17:0-18:1; 10, 16:1-18:2; 11, 16:0-18:3; 13, 18:0-18:0; 14, 18:0-18:1; 15, 18:1-18:1; 17, 18:0-18:2 + 16:0-20:2; 18, 18:1-18:2; 21, 18:2-18:2. Column: 25 cm x 4.6 mm (i.d.), packed with 5 μm particles of YMC-Pack A-KO3; Solvent system: heptane-1,2-dichloromethane-ethanol (40:10:1, by volume). (Reproduced by kind permission of the authors and *Journal of Lipid Research*).

spectrometry [32,45,64] and a complex order of elution recognized. The palmitoyl (16:0) and oleoyl (18:1) residues are nearly equivalent in their retention volume (time) and one double bond increases the retention time by nearly two methylene units on the chiral column when using hexane-l,2-dichloroethane-ethanol (40:10:1 by volume) as the mobile phase. Accordingly, the diacylglycerols as the 3,5-DNPUs elute from the chiral column in order of decreasing ECN values.

Although the X-l,3- and the *sn*-1,2-diacylglycerols of the C_{16} and C_{18} oligounsaturated fatty acids are completely resolved on the A-K03 column, mixtures of molecular species made up of longer chain and more unsaturated fatty acids give some peak overlap between the X-l,3-diacylglycerols with a high degree of unsaturation and *sn*-1,2-diacylglycerols of low degree of unsaturation. Therefore, a preliminary resolution of the X-l,3- and the *rac*-l,2-diacylglycerols is necessary before chiral-phase PLC [30,31,110,112,113]. There was no enantiomer resolution among the X-l,3-diacylglycerols under the present working conditions, but Takagi *et al.* [100] have demonstrated chiral-phase HPLC resolution of l-hexadecyl-3-hexadecanoyl-*rac*-glycerols as the 3,5-DNPU derivatives on the Sumipax OA-4100 chiral liquid phase. Like the resolution between the X-l,3- and the *sn*-l,2-isomers, the resolution between the *sn*-l,2- and *sn*-2,3-enantiomers depends on the qualitative and quantitative composition of the molecular species. Itabashi *et al.* [30] have shown that during the resolution of the *sn*-l,2- and *sn*-2,3-diacylglycerols derived from linseed oil the major 18:3-18:3 species of the *sn*-1,2-enantiomers overlaps completely with the minor 18:0-18:1 and partly with the minor 16:0-18:1 species of the *sn*-2,3-enantiomer. On the other hand, the *sn*-l,2- and *sn*-2,3-diacylglycerols derived from menhaden oil triacylglycerols did not appear to give a serious overlap between the last peak of the *sn*-l,2- and the first peak of the *sn*-2,3-diacylglycerols. It was later shown [110] that the mixture of the *sn*-1,2(2,3)-diacylglycerols derived from the menhaden oil could be resolved more completely if the separation was performed at lower temperature [101]. Figure 1.8 shows the separation of the *sn*-1,2- and *sn*-2,3-diacylglycerol moieties of the chylomicron triacylglycerols from menhaden oil feeding [110]. An essentially complete resolution is obtained as shown by rechromatography by chiral-phase HPLC and by calculation of the positional distribution of the fatty acids in the original chylomicron triacylglycerols. The molecular species making up the *sn*-l,2- and *sn*-2,3-diacylglycerols were identified by polar capillary GLC following silolysis of the diacylglycerol urethanes [112] and the compositions were shown to be closely similar. The A-KO3 chiral phase has also been shown to be well suited for the separation of diacylglycerols containing both short- and long-chain saturated fatty acids. Thus, Itabashi *et al.* [33] have employed this system to demonstrate the exclusive location of the acetic and butyric acid residues in the *sn*-3-position of bovine milk triacylglycerols. In this instance both hydrogenation and TLC prefractionation were used to reduce the complexity of the triacylglycerol mixture to effect maximum resolution of the enantiomers and their X-l,3-isomers of short, medium and long chain-length. The identification of peaks was accomplished by LC/MS of the DNPU derivatives [64). The chiral YMC-Pack A-KO3 column is able to separate also the reverse isomers of *rac*-1,2-diacylglycerols of uneven fatty chain length (Y. Itabashi and A. Kuksis, 1990. Unpublished results). Figure 1.9 shows the resolution of the reverse isomers of *rac*-1,2-acetylpalmitoylglycerols as

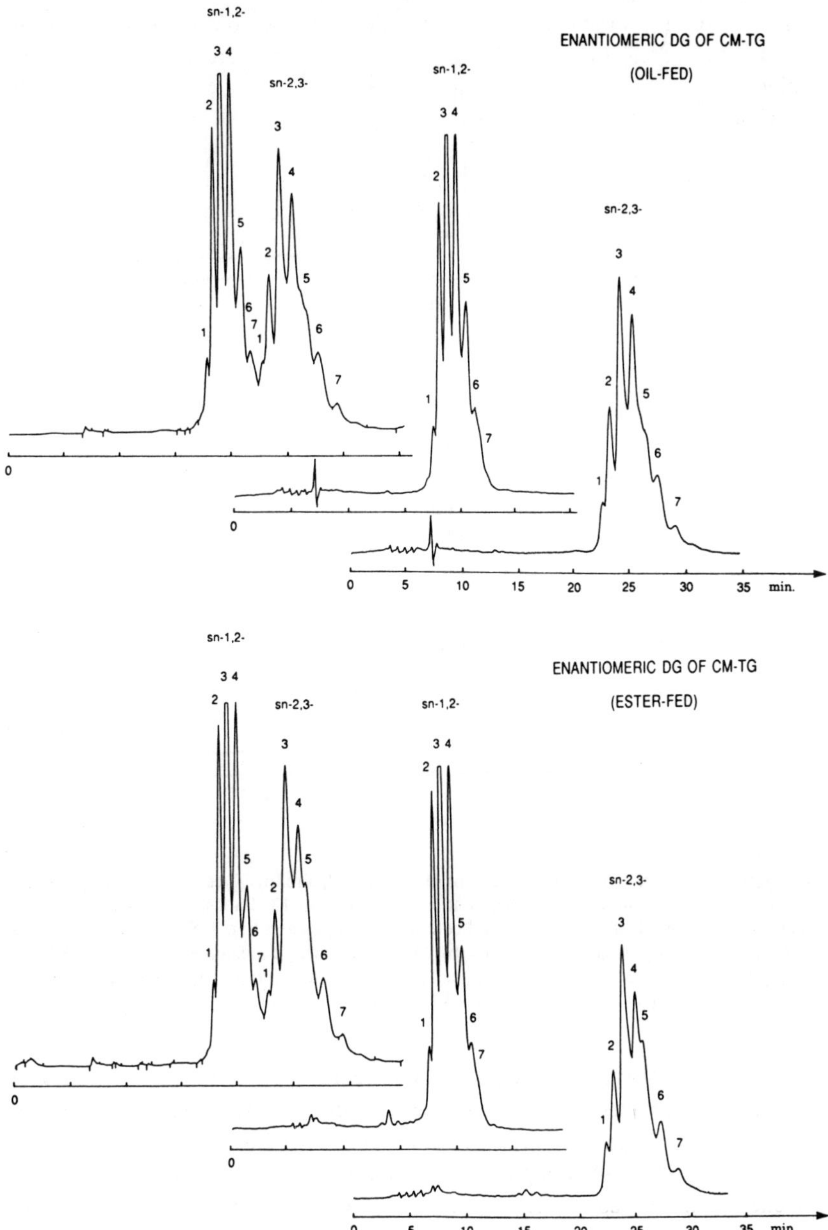

Fig. 1.8. Initial resolution and rechromatography by chiral-phase HPLC of *sn*-1,2- and *sn*-2,3-diacylglycerol moieties of chylomicron triacylglycerols as the 3,5-DNPU derivatives [110]. A, menhaden oil feeding; B, menhaden oil fatty acid ethyl ester feeding. Enantiomer identification as given in figure. Column and solvents as in Figure 1.7. (Reproduced by kind permission of the authors and *Journal of Lipid Research*).

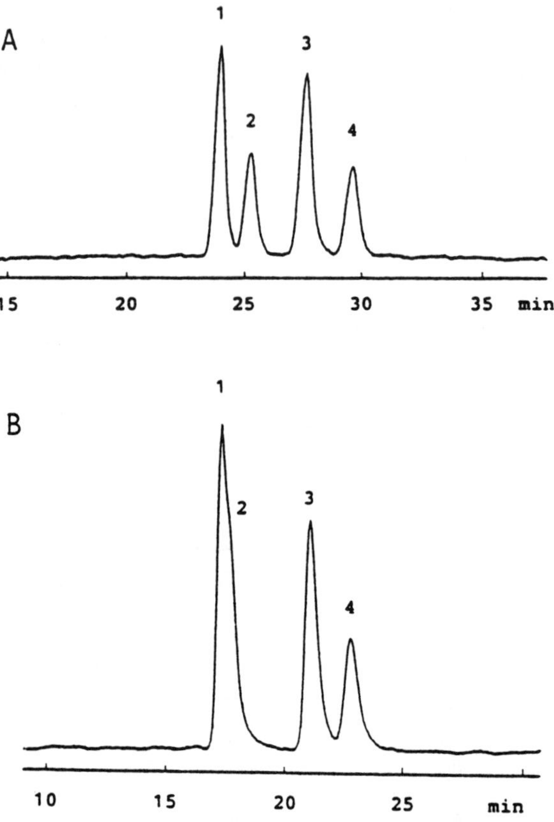

Fig. 1.9. HPLC resolution of reverse isomers of *rac*-1,2-acetylpalmitoylglycerols as 3,5-DNPU derivatives on a chiral column using two different solvents systems [Y. Itabashi and A. Kuksis. 1990. Unpublished results]. Instrument and chiral column was as given in Fig. 7. Solvent system: A, hexane/1,2-dichloroethane/acetonitrile (85:10:5, by vol.); B, hexane/1,2-dichloroethane/ethanol (80:20:2, by vol.); flow rate, 1 mL/min at 25°C. Peak identification: 1, *sn*-1-acetyl-2-palmitoylglycerol; 2, *sn*-1-palmitoyl-2-acetylglycerol; 3, *sn*-2-palmitoyl-3-acetylglycerol; 4, *sn*-2-acetyl-3-palmitoylglycerol.

the 3,5-DNPU derivatives in two solvent systems. There is a base-line resolution of the reverse isomers within each enantiomer class when the mobile phase contains acetonitrile (Figure 1.9A). In contrast, only a shoulder is seen for the sn-1,2-enantiomers, when acetonitrile is replaced by ethanol (Figure 1.9B), yet the reverse isomers of the sn-2,3-enantiomers were resolved. In both enantiomer classes the isomers with the acetyl chain in the primary position were eluted first.

The separation of enantiomeric monoacylglycerols was first reported by Itabashi and Takagi [28] using the di-3,5-DNPU derivatives and a chiral HPLC column containing the Sumipax OA-2100 phase. Complete separation of the urethane derivatives of racemic monoacylglycerols with C_{12} - C_{18} saturated fatty chains was achieved using isocratic elution with a mixture of hexane-ethylene dichloride-ethanol (40:12:3 by volume) as the mobile phase. The *sn*-l-enantiomers were eluted ahead of the corresponding *sn*-3-enantiomers. Complete separation of the *sn*-2-isomers from the corresponding enantiomers and partial separation of the enantiomer homologues differing by two acyl carbons was also observed. The *sn*-2-isomer was eluted ahead of the *sn*-l-isomer. The retention volume of an *sn*-1-enantiomer was approximately equal to that of an *sn*-3-enantiomer with six more carbons, *e.g.* *sn*-l-12:0 and *sn*-3-18:0. Such critical pairs could not be separated on the column under the conditions used. The separation of homologues differing by two carbons was poor compared to the enantiomer separation. Complete separation on this chiral column was also achieved for the corresponding monoalkylglycerol enantiomers [98]. An improved separation of the enantiomer mixtures of monoacylglycerols was later obtained [96] on the Sumipax OA-4100 liquid phase. This column also had the capability of separating homologues based on number of carbon atoms and number of double bonds, when developed with hexane-ethylene dichloride-ethanol (40:10:1 by volume) as mobile phase, All the racemic C_{12}-C_{22} monoacylglycerols were separated into enantiomers with separation factors of 1.31-1.35 and peak resolutions of 5.19-7.10. Complete enantiomer separations were obtained for the unsaturated monoacylglycerols of 18:2, 18:3, 20:4, 20:5 and 22:6 acids using a 50 cm x 4 mm (i.d.) column of Sumipax OA-4100. However, the peaks tailed rather badly and the addition of one double bond or two carbons to an enantiomeric homologue resulted in an overlapping pair. Thus, *sn*-l-16:0 and *sn*-l-18:1 were not resolved and nor were the corresponding enantiomers. A preliminary silver ion TLC fractionation on boric acid-treated plates was necessary to overcome this problem.

The Sumipax OA-4100 liquid phase is also suitable for the separation of natural mixtures or enantiomeric monoacylglycerols, as demonstrated by Takagi and Ando [97]. An excellent enantiomer resolution was obtained for the *rac*-l-monoacylglycerol homologue mixtures derived from soybean, macadamia nut and yew seed oil triacylglycerols based on carbon and double bond number as discussed above. The *rac*-l-monoacylglycerols from high erucic acid-rapeseed oil triacylglycerols required a 50 cm-column for a clear resolution of the enantiomers under similar conditions. However, when fish oils were used, complete enantiomer separation of *rac*-l-monoacylglycerols was more difficult. The first peak in the *sn*-3-group, *i.e.* 22:1 or 24:1 monoacylglycerol derivatives, and the last peak in the *sn*-1-group, *i.e.* 16:4 or 22:6 monoacylglycerol derivatives, overlapped on the OA-4100 column at ambient temperature. Lowering the column temperature to -9°C and using a 50 cm-column brought about sufficient improvement for a practical enantiomer separation. The time required for the resolution was now increased from 30 to 200 min. As a result it became possible to apply this method to the enantiomer resolution of the *rac*-l-monoacylglycerols prepared from saury, herring, capelin, sardine, menhaden and dolly warden fish oils [3]. MacKenzie *et al.* [63] and Taylor *et al.* [102] have recently applied the method of Takagi and Ando [97] in the stereospecific analyses of the seed oil triacylglycerols with apparently excellent results. Taylor *et al.* [102] demonstrated that within the

Brassicaceae, there exists *B. oleracea* germplasm for seed oils with substantial concentrations of erucic acid (30-35 mol%) at the *sn*-2-position, which confirmed previous results obtained with lipases. Despite taking the usual precautions to avoid high temperatures, to limit exposure to polar, highly acidic or basic solvents, and to perform extractions and TLC separations in the presence of boric acid, some isomerization appeared unavoidable [4,102]. The method of Ando *et al.* [3] has been modified recently by Ota *et al.* [77] for work with very-long-chain fatty esters.

Analysis of synthetic l,3-dioleoyl-2-stearoyl-*sn*-glycerol by Takagi and Ando [97] had shown that acyl shifts from *sn*-2- to the *sn*-l- and *sn*-3 positions were less than 3% of the *sn*-2-acyl group and Takagi [95] has suggested that increases in the percentage at the *sn*-1- and *sn*-3-position should be limited to less than 1%, when the acyl group in the *sn*-2-position makes up 33%. To minimize isomerization, the monoacylglycerols must be converted immediately into the urethane derivatives [2].

5. Calculation of Fatty Acid Distribution

The positional distribution of the fatty acids in the three positions of the acylglycerol molecule are determined either directly from the isolated individual monoacylglycerols [2,3,101] or indirectly by calculation from a knowledge of the fatty acid composition of the total triacylglycerol and the enantiomeric diacylglycerol composition [16,30,50,110,112,113]. The fatty acid analyses are performed by GLC on polar capillary columns following preparation of the fatty acid methyl esters by transmethylation [14,42,95]. The relative molecular weight contributions of individual fatty acid methyl esters differ greatly according to their chain length and degree of unsaturation. Therefore, in comparisons between the hydrolysis products and original triacylglycerols, fatty acid composition is expressed on a mol% basis. Furthermore, of the 30-40 fatty acids (including isomers) observed routinely, the fatty acid profiles are usually summarized according to the 14 major fatty acids in marine oils, and to less than 10 acids for seed oils. For this, purpose all isomers of a given monounsaturated components are assumed to behave identically, and are combined and expressed as a single value. This assumption may not be valid, as Kallio and Laakso [53,54) have shown that the positional distribution and molecular association of the *cis*- and *trans*-monoenoic acids differ in butterfat triacylglycerols. In any event, detailed analyses of molecular species must consider separately all detectable fatty acid components.

Brockerhoff [9] first recognized that the most accurate method of obtaining the positional distribution of fatty acids is to determine it directly by isolating the fatty acids while still associated with the glycerol molecule. Since the *sn*-3-position could not be directly analysed by the method, it contained the cumulative error of the analysis. Takagi and Ando [97] and Ando *et al.* [3] originally showed and others since have confirmed [2,93,101], that this problem can be avoided by chiral-phase HPLC resolution of the enantiomeric *sn*-l- and *sn*-3-monoacylglycerols, provided their isomerization is minimized during generation, isolation and the preparation of derivatives.

Direct analysis of the methyl esters from *sn*-l-, *sn*-2- and *sn*-3-monoacylglycerols, however, can involve considerable experimental error. Therefore, Takagi and Ando [97] recommend that the amount of each acyl group in the *sn*-l- and *sn*-3-positions is calculated from the concentration of each component in the *sn*-l(3)-positions and GLC analyses of *sn*-l- and *sn*-3-monoacylglycerols resolved by HPLC on the chiral column:

%[*sn*-l] = 2 x [*sn*-l(3)-] x ([*sn*-l]/[*sn*-l] + [*sn*-3])

%[*sn*-3] = 2 x [*sn*-l(3)-] x ([*sn*-3]/[*sn*-l] + [*sn*-3]).

The mean composition of each acyl group in the *sn*-l- and *sn*-3-positions [*sn*-l(3)] was calculated from the GLC analyses of the original triacylglycerols and those of the 1- and 2-monoacylglycerols isolated by boric acid TLC, *e.g.* for each component:

% [*sn*-l(3)] = 3 x TG x [l-MG]/(2 x [1-MG] + [2-MG])

where TG, 1-MG and 2-MG represent the proportions of each acyl group in the original triacylglycerols, 1-monoacylglycerols and 2- monoacylglycerols, respectively. The composition of each acyl group in the *sn*-2-position was calculated in a similar manner:

% [*sn*-2] = 3 x TG x [2-MG]/(2 x [1-MG] + [2-MG])

The simplest indirect calculation based on the Grignard degradation of triacylglycerols to diacylglycerols involves the estimation of the fatty acid composition of the *sn*-2-position [4,34]. For this purpose it is important to isolate only pure *sn*-1,2(2,3)-diacylglycerols. Thus,

[*sn*-2] = 4 x [*sn*-l,2(2,3)-DG] - 3 x [TG]

For each fatty acid, the percent of the total amount present in the original oil or fat, which is located at the *sn*-2-position is then:

% [*sn*-2] = 100 x [*sn*-2]/3 [TG]

The average composition of the *sn*-l(3)- positions can then be calculated as follows:

[*sn*-l, *sn*-3] =(3 x [TG] - [*sn*-2]) / 2

The calculations are performed using values obtained for the *sn*-1,2(2,3)-diacylglycerols rather than the X-l,3-diacylglycerols since the latter tend to be more highly contaminated by acyl migration than the *sn*-l,2(2,3)-components [9,15].

Becker *et al.* [4], who used the allylmagnesium bromide for the Grignard degradation and isolated four different classes of partial acylglycerols and triacylglycerols, compared different approaches of calculating the composition of the fatty acids in positions *sn*-1(3) and *sn*-2 of the original triacylglycerol. For the *sn*-2-position, the best estimate was the direct determination of the fatty acid composition of 2-monoacylglycerol. Mole percentages of stearic acid and decanoic acid in the *sn*-l(3)-positions of 1,3-distearoyl-2-oleoylglycerol and 1,3-didecanoyl-2-palmitoylglycerol were most accurately estimated from the fatty acid compositions of triacylglycerol and 2-monoacylglycerol according to the formula:

$$\% [sn\text{-}1,sn\text{-}3] = 1.5 \times [TG] - 0.5 \times [2\text{-}MG].$$

Using the allylmagnesium bromide degradation and the present method of calculation gave results, which were more accurate and showed smaller standard deviations than those obtained using other common deacylating agents, such as ethylmagnesium bromide or pancreatic lipase.

The indirect calculation of the fatty acid composition of the *sn*-l- and *sn*-3-positions is based on the comparison of the fatty acid composition of the *sn*-l,2- and *sn*-2,3-diacylglycerols resolved as the diastereomers on normal or as enantiomers on the chiral-phase HPLC columns and the original triacylglycerols. Laakso and Christie [50] determined the distribution of the fatty acids in positions *sn*-1-, *sn*-2- and *sn*-3- of the triacylglycerols from the compositions of the combined X-1,3-, *sn*-1,2- and *sn*-2,3-diacylglycerol fractions. Thus, the composition of fatty acids in the *sn*-1-position was calculated for each component by subtracting twice its concentration in the *sn*-2,3-diacylglycerols from three times its concentration in the intact triacylglycerols. The calculations for all positions can be summarized as follows:

$$[sn\text{-}1] = 3 \times [TG] - 2 \times [sn\text{-}2,3\text{-}DG]$$

$$[sn\text{-}2] = 3 \times [TG] - 2 \times [X\text{-}1,3]$$

$$[sn\text{-}3) = 3 \times [TG] - 2 \times [sn\text{-}1,2\text{-}DG]$$

Since the composition of the X-1,3-diacylglycerols may be less representative than those of the *sn*-1,2- or *sn*-2,3-isomers, the *sn*-2-position can be calculated by subtracting three times the fatty acid composition of the triacylglycerol from four times the fatty acid composition of the *sn*-1,2(2,3)-diacylglycerols, as discussed above (or by subtracting the results for positions *sn*-1 plus position *sn*-2 from three times the triacylglycerols [16]). Alternatively, the fatty acid composition of the *sn*-2-position has been derived from the 2-monoacylglycerol released by pancreatic lipase [16]. The results of stereospecific distribution of the fatty acids determined by the separation of the diastereomeric diacylglycerols gave results, which agreed closely to those obtained previously by enzymatic methods [7], where parallel data were available.

The compositions of the fatty acids in the *sn*-1-, *sn*-2- and *sn*- 3-positions of triacylglycerols analysed by chiral-phase separation of the *sn*-1,2- and *sn*-2,3-diacylglycerols are also calculated indirectly [30,89,110,112,113]. Thus, Yang and Kuksis [110] obtained best results for the stereospecific analyses of the chylomicron triacylglycerols from fish oil feeding by a combination of the above methods of calculation. The fatty acid composition of the *sn*-2-position was obtained by subtracting the fatty acid composition of chylomicron triacylglycerols from the fatty acid composition of the *sn*-1,2(2,3)-diacylglycerols, while the fatty acid compositions of the *sn*-1- and *sn*-3-positions were obtained by subtracting the composition of the *sn*-2-position from the fatty acid composition of the appropriate diacylglycerol enantiomer. The latter values were cross-checked by subtracting the fatty acids of the appropriate enantiomer from those of the original triacylglycerol. The overall correctness was assessed by comparing the reconstituted total to the total of the original triacylglycerols. In addition, the fatty acid composition of the *sn*-2-position of the triacylglycerols was determined directly by pancreatic lipase hydrolysis [69]. The chiral-phase HPLC method and the above calculations gave stereospecific distributions of fatty acids from synthetic and natural triacylglycerols, which compared closely to those derived by enzymatic methods [30,69].

Kuksis [41] has discussed how fatty acid compositions of the various classes of the acylglycerols derived by Grignard degradation and enzymatic degradation of glycerophospholipids, prepared as intermediates in stereospecific enzymatic analyses, can be effectively cross-checked for validity. These cross-checks are also important in non-enzymic analyses in order to avoid pitfalls from incomplete resolution of enantiomers, losses of polyunsaturates due to peroxidation during lengthy processing and losses due to insolubility of derivatives, as well as those due to non-representative transmethylation of the fatty acids for GLC.

E. BIOLOGICAL SIGNIFICANCE

The importance of positional analysis arises from the recognition that natural triacylglycerols are mixtures of enantiomeric and racemic molecular species of non-random molecular association of fatty acids [7,41]. It has been assumed that knowledge of the positional distribution and molecular association of the fatty acids will help to understand the biological function and the chemical and physico-chemical properties of natural fats and oils. Although methods of chromatographic resolution and stereospecific assays of triacylglycerols have been known and applied to positional analyses for some 30 years, the hopes of understanding the physiological significance of the triacylglycerol structure have been slow in realization. This has been due to the cumbersome nature of the methodology and requirement for skilled analysts. The recent development of more robust methods of stereospecific analysis of natural triacylglycerols promises rapid advances in a number of areas of application.

Thus far, the diastereomer resolution of diacylglycerols has helped to characterize the lipolysis products of various lipases using synthetic achiral triacylglycerols and to make novel observations about the specificity of these enzymes [82,85,86]. Santinelli *et al.* [87] have used this method to demonstrate marked asymmetry between *sn*-1- and *sn*-3-positions of individual triacylglycerol species, although the stereospecific analysis of the total triacylglycerol mixture had indicated symmetry. The data suggest that the 1-random 2-random 3-random distribution theory may not adequately predict the structure of vegetable oils.

The chiral-phase HPLC separation of enantiomeric diacylglycerols also has allowed significant observations to be made. Yang and Kuksis [110,112] have used the method for stereospecific analysis of menhaden oil and of the chylomicron triacylglycerols derived from it to propose a convergence of the phosphatidic acid and the monoacylglycerol pathways of triacylglycerol biosynthesis during fat absorption. Yang *et al.* [113] have employed the chiral-phase HPLC method to demonstrate that hepatic triacylglycerols are not directly transferred to VLDL but first undergo lipolysis to acylglycerols before being re-synthesised into VLDL triacylglycerols. In a separate study [33] the chiral-phase HPLC method was employed to demonstrate the exclusive location of the acetyl and butyryl groups in the *sn*-3-position of bovine milk fat, which previously had been claimed on the basis of incomplete enzymatic evidence [7,8,79]. Lehner and Kuksis [57] have used the method to establish the enantiomeric nature of the biosynthetic course of action of a new enzyme, the *sn*-l,2(2,3)-diacylglycerol transacylase, while Lehner *et al.* [58] utilized it to demonstrate the preferential formation of *sn*-l,2-diacylglycerols during acylation of *sn*-2-monoacylglycerols by purified triacylglycerol synthetase.

Ando *et al.* [3] have employed the chiral-phase HPLC separation of enantiomeric *sn*-l- and *sn*-3-monoacylglycerols to characterize the positional distribution of fatty acids in fish oils, which has led to the introduction of the concept of fatty acid interaction during biosynthesis. This idea challenges the earlier notion that each position of the triacylglycerol molecule is esterified independently of the other two from separate pools of fatty acids [60,69]. Taylor *et al.* [102] have utilized the stereospecific analysis of the seed oil triacylglycerols from high-erucic acid *Brassicaceae*, which does not lend readily to the lipolytic methods of analysis. Their findings demonstrated that, within *Brassicaceae*, there exists *B. oleraceae* germplasm yielding seed oils with substantial amounts of erucic acid in the *sn*-2-position.

The analytical importance of the development of the diastereomeric and enantiomeric diacylglycerol resolution is also of great importance because of the advantages the two methods offer for the analysis of molecular species of enantiomeric diacylglycerols by means of HPLC with UV absorption and LC/MS with NICI [48,113]. The small amounts of material required for these analyses permit the application of the stereospecific method to progressively simpler and smaller groups of molecular species of triacylglycerols to yield direct evidence of the positional distribution and molecular association of the fatty acids in individual triacylglycerol molecules and thus of the specificity of the acyltransferases [57,58,110,112,113] and lipases [34,81-83,85] modulating their composition during metabolism. The ability to remove the 3,5-DNPU groups by silolysis provides a simple route to the preparation and isolation of diacylglycerols of high enantiomeric purity [31,110,112,113].

F. CONCLUSIONS AND FUTURE PROSPECTS

This review of the methodology and applications of the non- enzymatic methods of positional analysis of triacylglycerols has demonstrated that the methods are robust, simple and subject to validation. They are free of the type of bias seen in the older enzymatic techniques of stereospecific analyses of natural triacylglycerols. Both diacylglycerol and monoacylglycerol approaches to the stereospecific analysis appear highly effective, with the diacylglycerol approach possibly being less subject to problems arising from isomerization. However, for detailed positional analysis, the new methodology must be applied to individual triacylglycerol species or small groups of species. Furthermore, the general theories of semi-random fatty acid distribution in natural triacylglycerols now may have to be abandoned in favor of specific structures governed by the molecular association as well as the positional distribution and the nature of the fatty acid, all of which are being recognized by the acyltransferases and lipases involved in the metabolism of acylglycerols.

The stereospecific analysis of individual triacylglycerol species will require further miniaturization of the methodology of random degradation of triacylglycerols, possibly by means of carriers, and the combination of chiral-phase HPLC with on-line MS/MS of the diacylglycerol urethanes. There is also a need to re-examine the conventional Grignard degradation for bias with a greater selection of standard triacylglycerols, and for increasing the yield of the diacylglycerols or monoacylglycerols, as required, and minimizing isomerization, especially of the monoacylglycerols. Perhaps, the ethylmagnesium bromide needs to be replaced by the more reactive allylmagnesium bromide as the Grignard reagent. Finally, the prefractionation of the triacylglycerols advocated for the stereospecific analyses by non-enzymatic methods might lead also to more meaningful positional analysis of the intact triacylglycerols by purely physical methods.

ACKNOWLEDGMENTS

The studies by the author and his collaborators referred to in this review were done with funds provided by the Heart and Stroke Foundation of Ontario, Toronto, Ontario and the Medical Research Council of Canada, Ottawa, Ontario, Canada.

REFERENCES

1. Akesson,B., Gronowitz,S., Herslof,B., Michelsen,P. and Olivecrona,T., *Lipids*, **18**, 313-318 (1983).
2. Ando,Y. and Takagi,T., *J. Am. Oil Chem. Soc.*, **70**, 1047-1049 (1993).
3. Ando,Y., Nishimura,K., Aoyanagi,N. and Takagi,T., *J. Am. Oil Chem. Soc.*, **69**, 417-424 (1982).
4. Becker,C.C., Rosenquist,A. and Holmer,G., *Lipids*, **28**, 147-149 (1993).
5. Blank,M.L. and Snyder,F.L., in *Lipid Chromatographic Analysis*, pp. 291-316 (1994) (edited by T. Shibamoto, Marcel Dekker, N.Y.).
6. Bottino,N.R., *J. Lipid Res.*, **12**, 24-30 (1971).
7. Breckenridge,W.C., in *Handbook of Lipid Research*, pp. 197-232 (1978) (edited by A. Kuksis, Plenum Press, New York).
8. Breckenridge,W.C. and Kuksis,A., *J. Lipid Res.*, **9**, 388-393 (1968).
9. Brockerhoff,H., *Lipids*, **6**, 942-956 (1971).
10. Brockerhoff,H. and Jensen,R.G., *Lipolytic Enzymes*, pp. 10-24 (1974) (Academic Press, Orlando, FL).
11. Bus,J., Lok,C.M. and Groenewegen,A., *Chem. Phys. Lipids*, **16**, 123-132 (1976).
12. Christensen,M.S., Mullertz,A. and Hoy,C.E., *Lipids* (1995), in press.
13. Christie,W.W., *J. Chromatogr.*, **454**, 273-284 (1988).
14. Christie,W.W., in *Advances in Lipid Methodology - One*, pp. 121-148 (1992) (edited by W.W. Christie, Oily Press, Ayr, Scotland).
15. Christie,W.W. and Moore,J.H., *Biochim. Biophys. Acta*, **176**, 445-452 (1969).
16. Christie,W.W., Nikolova-Damyanova,B., Laakso,P. and Herslof,B., *J. Am. Oil Chem. Soc.*, **68**, 695-701 (1991).
17. Currie,G.J. and Kallio,H., *Lipids*, **28**, 217-222 (1993).
18. Deffense,E., *Rev. Fr. Corps Gras*, **40**, 40-47 (1993).
19. Dickens,B.F., Ramesha,C.S. and Thompson,G.A., *Anal. Biochem.*, **127**, 37-48 (1982).
20. Dubis,E., Poplawski,J., Wrobel,J.T., Kusmierz,J., Malinski,E. and Szafranek,J., *Lipids*, **21**, 434-439 (1986).
21. Foglia,T., Vail,P.D. and Iwama,T., *Lipids*, **22**, 362-365 (1987).
22. Foglia,T.A. and Maeda,K., *Lipids*, **26**, 769-773 (1991).
23. Frankel,E.N., Neff,W.E. and Miyashita,K., *Lipids*, **25**, 40-47 (1990).
24. Gronowitz,S., Herslof,B., Ohlson,R. and Toregard,B., *Chem. Phys. Lipids*, **14**, 174-188 (1975).
25. Gunstone,F.D., in *Advances in Lipid Methodology - Two*, pp. 1-68 (1993) (edited by W.W. Christie, Oily Press, Ayr, Scotland).
26. Gunstone,F.D., *J. Am. Oil Chem. Soc.*,**70**, 361-366 (1993).
27. Hammond,E.G., *Chromatography for the Analysis of Lipids*, pp. 113-168 (1993).
28. Itabashi,Y. and Takagi,T., *Lipids*, **21**, 413-416 (1986).
29. Itabashi,Y. and Takagi,T., *J. Chromatogr.*, **402**, 257-264 (1987).
30. Itabashi,Y., Kuksis,A., Marai,L. and Takagi,T., *J. Lipid Res.*, **31**, 1711-1717 (1990).
31. Itabashi,Y., Kuksis,A. and Myher,J.J., *J. Lipid Res.*, **31**, 2119-2126 (1990).
32. Itabashi,Y., Marai,L. and Kuksis,A., *Lipids*, **26**, 951-956 (1991).
33. Itabashi,Y., Myher,J.J. and Kuksis,A., *J. Am. Oil Chem. Soc.*, **70**, 1177-1181 (1993).
34. Iverson,S.J., Sampugna,J. and Oftedal,O.T., *Lipids*, **27**, 870-878 (1992).
35. Kallio,H. and Currie,G., in *CRC Handbook of Chromatography - Analysis of Lipids*, pp. 435-457 (1993) (edited by K.D. Mukherjee and N. Weber, CRC Press, Boca Raton, FL.).
36. Kallio,H. and Currie,G., *Lipids*, **28**, 207-215 (1993).
37. Kallio, H., Laakso,P., Huopalahti,R., Linko,R.R. and Oksman,P., *Anal. Chem.*, **61**, 698-700 (1989).
38. Kallio,H. and Rua,P., *J. Am. Oil Chem. Soc.*, **71**, 985-992 (1994).
39. Kleiman,R., Miller,R.W., Earle,R.R. and Wolff,I.A., *Lipids*, **1**, 286-287 (1966).
40. Kritchevsky,D., *Nutr. Revs.*, **46**, 177-181 (1988).
41. Kuksis,A., in *Lipid Research Methodology*, pp. 77-131 (1984) (edited by J.A.Story, Alan R. Liss, New York).
42. Kuksis,A., in *Chromatography*, 5th Edition, pp. B171-B227 (1992) (edited by E. Heftmann, Elsevier, Amsterdam).
43. Kuksis,A., in *Lipid Chromatographic Analysis*, pp. 177-222 (1994) (edited by T. Shibamoto, Marcel Dekker, Inc., New York).
44. Kuksis,A. and Ludwig,J., *Lipids*, **1**, 202-208 (1966).

45. Kuksis,A. and Myher,J.J., in *Mass Spectrometry*, pp. 255-351 (1989) (edited by A.M. Lawson, Walter deGruyter, Berlin).
46. Kuksis,A. and Myher,J.J., *J. Chromatogr*. (1995), in press.
47. Kuksis,A., Marai,L., Myher,J.J., Itabshi,Y. and Pind,S., in *Analysis of Fats, Oils and Lipoproteins*, pp. 214-232 (1991) (edited by E.G. Perkins, American Oil Chemists' Society, Champaign, IL).
48. Kuksis,A., Myher,J.J., Yang,L-Y. and Steiner,G., in *Biological Mass Spectrometry, Present and Future*, pp. 480-494 (1994) (edited by T. Matsuo, R.M. Caprioli, M.L. Gross and Y. Seyama, John Wiley, New York).
49. Laakso,P., in *Advances in Lipid Methodology - One*, pp. 81-119 (1992) (edited by W.W. Christie, Oily Press, Ayr, Scotland).
50. Laakso,P. and Christie,W.W., *Lipids*, **25**, 349-353 (1990).
51. Laakso,P. and Christie,W.W., *J. Am. Oil Chem. Soc*., **68**, 213-222 (1991).
52. Laakso,P., Christie,W.W. and Pettersen,J., *Lipids,* **25,** 284-291 (1990).
53. Laakso,P. and Kallio,H., *J. Am. Oil Chem. Soc.,* **70**, 1161-1171 (1993).
54. Laakso,P. and Kallio,H., *J. Am. Oil Chem. Soc*., **70,** 1173-1176 (1993).
55. Larsson,K., in *The Lipid Handbook*, pp. 364-367 (1986) (edited by F.D. Gunstone, J.L. Harwood and F.B. Padley, Chapman and Hall, London).
56. Lawson,L.D. and Hughes,G.B., *Lipids*, **25**, 313-317 (1988).
57. Lehner,R. and Kuksis,A., *J. Biol. Chem.*, **268,** 8781-8786 (1993).
58. Lehner,R., Kuksis,A. and Itabashi,Y., *Lipids,* **28,** 29-34 (1993).
59. Leray,C., Raclot,T. and Groscolas,R., *Lipids*, **28**, 279-284 (1993).
60. Litchfield,C., *Analysis of Triglycerides*, Academic Press, New York, 1972.
61. Lohninger,A. and Nikiforov,A., *J. Chromatogr.,* **192**, 185-192 (1980).
62. Lok,C.L., *Receuil. J. Roy. Neth. Chem. Soc.*, **98**, 92-95 (1979).
63. MacKenzie,S.L., Giblin,E.M. and Mazza,G., *J. Am. Oil Chem. Soc.*, **70,** 629-631 (1993).
64. Marai,L., Kuksis,A., Myher,J.J. and Itabashi,Y., *Biol. Mass Spectrom*., 21, 541-547 (1992).
65. Michelsen,P., Aronsson,E., Odham,G. and Akesson,B., *J. Chromatogr*., **350**, 417-426 (1985).
66. Miyashita,K., Frankel,E.N., Neff,W.E. and Awl,R.A., *Lipids,* **5,** 48-53 (1990).
67. Morley,N.H., Kuksis,A., Buchnea,D. and Myher,J.J., *J. Biol. Chem*., **250,** 3414-3418 (1975).
68. Myher, J.J., In *Handbook of Lipid Research,* Vol. 1, *Fatty Acids and Gklycerides*, pp. 123-196 (1978) (edited by A. Kuksis, Plenum Press, Inc, New York).
69. Myher,J.J. and Kuksis,A., *Can. J. Biochem*., **57**, 117-124 (1979).
70. Myher,J.J., Kuksis,A., Marai,L. and Yeung,S.K.F., *Anal. Chem*., **50**, 557-561 (1978).
71. Myher,J.J., Kuksis,A., Yang,L-Y. and Marai,L., *Biochem. Cell Biol*., **65**, 811-821 (1987).
72. Neff,W.E., Frankel,E.N. and Miyashita,K., *Lipids,* **25,** 33-39 (1990).
73. Ng,S., *Lipids*, **19**, 56-59 (1984).
74. Ng,S., *Lipids*, **20,** 778-782 (1985).
75. Ng,S., *Lipids*, **23**, 140-143 (1988).
76. Nikolova-Damyanova,B., in *Advances in Lipid Methodology - One,* pp. 181-237 (1992) (edited by W.W.Christie, Oily Press, Ayr, Scotland).
77. Ota,T., Kawabata,Y. and Ando,Y., *J. Am. Oil Chem. Soc.*, **71,** 475-478 (1994).
78. Pfeffer,P.E., Sampugna,J., Schwartz,D.P. and Schoolery,J.N., *Lipids*, **12**, 869-871 (1977).
79. Pitas,R.E., Sampugna,J. and Jensen,R.G., *J. Dairy Sci*., **50**, 1332-1337 (1967).
80. Possmayer,F., Scherphof,G.L., Dubbelmann,R.M.A., Van Golde,L.M.G. and Van Deenen,L.L.M., *Biochim. Biophys. Acta*, **176**, 95-100 (1969).
81. Raclot,T. and Groscolas,R., *J. Lipid Res*., **34**, 1515-1526 (1993).
82. Ransac,S., Rogalska,E., Gargouri,Y., Dveer,A.M.T.J., Paltauf,F., de Haas,G.H. and Verger,R., *J. Biol. Chem.*, **265,** 20263-20270 (1990).
83. Redden,P.R., Lin,X., Fahey,J. and Horrobin,D.F., *J. Chromatogr. A*, **704,** 99-111 (1995).
84. Rhoem,J.N. and Privett,O.S., *Lipids*, **5**, 353-358 (1970).
85. Rogalska,E., Cudrey,C., Ferrato,F. and Verger,R., *Chirality*, **5**, 24-30 (1993).
86. Rogalska,E., Ransac,S. and Verger,R., *J. Biol. Chem.*, **265**, 20271-20276 (1990).
87. Santinelli,F., Damiani,P. and Christie,W.W., *J. Am. Oil Chem. Soc*., **69**, 552-556 (1992).
88. Schlenk,W., *J. Am. Oil Chem. Soc*., **42**, 945-957 (1965).
89. Sempore,G. and Bezard,J., *J. Am. Oil Chem. Soc*., **68**, 702-709 (1991).
90. Sempore,G. and Bezard,J., *J. Chromatogr.,* **547,** 89-103 (1991).
91. Sempore,B.G. and Bezard,J.A., *J. Chromatogr.,* **557**, 227-240 (1991).

92. Sempore,B.G. and Bezard,J.A., *J. Chromatogr.*, **596,** 185-196 (1992)
93. Sempore,B.G. and Bezard,J.A., *J. Liquid Chromatogr.* **17,** 1679-1694 (1994).
94. Sjovall,O. and Kuksis,A., *INFORM*, **6,** 508 (1995). Abs. F.
95. Takagi,T., *Progr. Lipid Res.*, **29**, 277-298 (1990).
96. Takagi,T. and Ando,Y., *Lipids*, **25**, 398-400 (1990).
97. Takagi,T. and Ando,Y., *Lipids*, **26**, 542-547 (1991).
98. Takagi,T. and Itabashi,Y., *J. Chromatogr.*, **366**, 451-455 (1986).
99. Takagi,T. and Itabashi,Y., *Lipids*, **22**, 596-600 (1987).
100. Takagi,T.,Okamoto,J., Ando,Y. and Itabashi,Y., *Lipids*, **25**, 108-110 (1990).
101. Takagi,T. and Suzuki,T., *J. Chromatogr.*, **625,** 163-168 (1992).
102. Taylor,D.C., MacKenzie,S.L., McCurdy,A.R., Mcvetty,P.B.E., Giblin,E.M., Pass,E.W., Stone,S.J., Scarth,R., Rimmer,S.R. and Pickard,M.D., *J. Am. Oil Chem. Soc.*, **71**, 163-167 (1994).
103. Thomas,A.E.III, Scharoun,J.E. and Ralston,H., *J. Am. Oil Chem. Soc.*, **42**, 789-792 (1965).
104. Uzawa,H., Noguchi,T., Nishida,Y., Ohrui,H. and Meguro,H., *J. Org. Chem.*, **55**, 116-122 (1990).
105. Uzawa,H., Nishida,Y., Ohrui,H. and Meguro,H., *Biochem. Biophys. Res. Commun.*, **168**, 506-511 (1990).
106. Uzawa,H., Noguchi,T., Nishida,Y., Ohrui,H. and Meguro,H., *Biochim. Biophys. Acta*, **1168**, 253-260 (1993).
107. Wedmid,Y. and Litchfield,C., *Lipids*, **11**, 189-193 (1976).
108. Wollenberg,K.F., *J. Am. Oil Chem. Soc.*, **67**, 487-494 (1990).
109. Wollenberg,K.F., *J. Am. Oil Chem. Soc.*, **68**, 391-400 (1991).
110. Yang,L-Y. and Kuksis,A., *J. Lipid Res.*, **32**, 1173-1186 (1991).
111. Yang,L-Y., Kuksis,A. and Myher,J.J. *J. Lipid Res.*, **31**, 137-147 (1990).
112. Yang,L-Y., Kuksis,A. and Myher,J.J., *J. Lipid Res.*, **36**, 1046-1057 (1995).
113. Yang,L-Y., Kuksis,A., Myher,J.J. and Steiner,G., *J. Lipid Res.*, **36,** 125-136 (1995).
114. Zollner,F., Lorbeer,E. and Remberg,G., *Org. Mass Spectrom.*, **29,** 253-259 (1994).
115. Zollner,F. and Lorbeer,E., *J. Mass Spectrom.*, **30**, 432-437 (1995).

Chapter 2

^{31}P NUCLEAR MAGNETIC RESONANCE PROFILING OF PHOSPHOLIPIDS

Thomas Glonek[1] and Thomas E. Merchant[2]

[1]MR Laboratory, Midwestern University, Chicago, Illinois 60615; [2]Department of Radiation Oncology, Memorial Sloan-Kettering Cancer Center, New York, New York 10021, USA.

A. INTRODUCTION

Crude total-lipid extracts from tissues can be analysed by ^{31}P nuclear magnetic

Correspondence: Thomas Glonek, MR Laboratory, Midwestern University, 5200 S. Ellis Ave., Chicago, IL 60615, U.S.A.; Telephone: 1-312-947-4697; Fax: 1-708-386-1980; E-mail: MarHus@aol.com resonance (NMR) to provide well-resolved quantitative phospholipid profiles that may be used to investigate diseases, monitor pollution, classify organisms, or glean insights into the fundamental properties of biological membranes. The basic preparation methods are simple, requiring only common laboratory chemicals and equipment. The only unusual requirement for the analysis is access to a high-field (4.7 Tesla or greater) NMR spectrometer equipped to detect ^{31}P

Two aspects of phospholipid preparation and spectroscopy that are particular to ^{31}P NMR analysis involve the inorganic coordination chemistry of the ionic phosphates. The first is the scrubbing of crude extracts with potassium-EDTA in the back-washing step [50]. The second is the use of a hydrated chloroform-methanol solvent containing a caesium salt of EDTA for the preparation of the analytical sample [48]. Phosphates, particularly inorganic orthophosphate and its corresponding monoesters and diesters [76], coordinate to metal cations wherever these may be found, regardless of whether the molecules containing these groups are anionic, zwitterionic and neutral, or net cationic. Metal cations, which in tissue extracts always include some fraction of divalent, trivalent, and paramagnetic transition-element species, must be displaced from the phosphate functional group and replaced by a single common alkali-metal cation. The best cation has been determined [48] to be Cs, the next best K. When the only metal cations present in the NMR sample solution are either of these, ^{31}P NMR spectroscopy yields narrow high-resolution NMR phospholipid signals that may be quantified precisely to the third significant figure [51,53]. Regarding instrumentation, most modern multinuclear NMR spectrometers can detect phosphorus. More problematic is sequestering sufficient instrument time.

From the analytical point of view, the techniques described in this chapter are designed to be as simple as possible. The idea is to minimize the number of separate chemical manipulations and, thereby, minimize errors. Further, the phospholipid profiles obtained contain all members of the broad class of phospholipid biochemicals, the familiar, the new, the bizarre, and the previously unrecognized. Such an analytical strategy not only saves time, more importantly, it provides a total picture of the phospholipid component from which new and useful natural products may be obtained and from which insights into the operant tissue biochemistry may be provided. Our intention is to examine tissue phospholipids as an integrated chemical set from which functional biochemical systems may be elucidated. These efforts should result in an improvement in the understanding of biomembrane function and control as it is mediated and modulated by the phospholipid population.

B. TOTAL LIPID EXTRACTION

Although methods of analysis *in vivo* are advancing rapidly, detailed analysis of phospholipids in cellular materials still requires extraction of a phospholipid-containing fraction followed by chemical analysis using chromatographic and/or spectrographic techniques. For ^{31}P NMR analysis of most tissue phospholipids, a modified procedure [50] derived from that of Folch *et al.* [16] can be used in which potassium-(ethylenedinitrilo)-tetraacetic acid (K-EDTA) replaces KCl in the back-washing step: Samples are homogenized with chloroform/methanol (2:1, v/v), using a 20:1 (ml solvent/g sample) ratio of solvent to sample. The homogenate is allowed to settle (*ca.* 2 hr, depending upon the sample), after which it is filtered into a separatory funnel. K-EDTA (0.2 M, pH 6.0, 0.2 ml per ml chloroform/methanol) is added, the funnel shaken vigorously, and the two liquid phases allowed to separate overnight. The lower chloroform phase is collected and evaporated on a rotary evaporator at 37°C. The concentrated and dried lipid sample is then taken up in chloroform and an appropriate quantity used for ^{31}P NMR analysis. The extraction of body fluids, such as human urine [11], amniotic fluid, cyst fluid, plasma, synovial fluid, or cerebrospinal fluid, may require further modification.

Because urine and other body fluids are water solutions containing large quantities of dissolved solutes and cells, the phospholipid extraction procedure must be modified from that used for tissue phospholipids [11]. Example: fresh urine (fasted overnight) is successively passed through a Whatman #2 filter paper, a Gelman 0.45 mm vinyl/acrylic filter and a Gelman 0.2 mm polysulfone membrane filter to remove cells and cellular fragments. To the filtrate, K-EDTA is added to a concentration of 0.2 M, and the pH adjusted to 6. Chloroform/methanol (2:1) is then added to this EDTA-treated urine filtrate in a ratio of 5:1 ($CHCl_3$-MeOH/H_2O), shaken vigorously, and the mixture subsequently worked up for NMR analysis.

^{31}P NMR spectroscopy is particularly sensitive to the presence of polyvalent cations coordinated to the phosphodiester functional groups of phospholipids. For example, Mg^{++} [54], Ca^{++} [54] and Al^{+++} [65] may quench the resonance signal as will Mn^{++} [32] and all other coordinated paramagnetic cations. Minimally, coordination of such cations will result in a broadening of phospholipid resonances that degrades the ^{31}P NMR spectrum [48,50]. All such cations, therefore, must be purged from the sample.

C. A ^{31}P NMR PHOSPHOLIPID REAGENT

1. A hydrated chloroform-methanol reagent for ^{31}P NMR analysis

A chloroform-methanol-aqueous-Cs-EDTA phospholipid NMR reagent has been formulated for the analysis of phospholipids by ^{31}P NMR spectroscopy [19,48,51]. The analytical medium consists of two reagents: (A) Reagent-grade chloroform, containing 5 % benzene-d_6 and an appropriate concentration of trimethylphosphate. (B) Reagent-grade methanol containing 0.2 M aqueous Cs-EDTA, pH 6 (4:1, v/v). The final methanol reagent is prepared by dissolving 1 ml of the aqueous EDTA solution in 4 ml methanol.

Benzene-d_6 is used for internal deuterium field-frequency stabilization; the concentration of the trimethylphosphate chemical-shift-and-quantification reference [19,50] may be varied within wide limits, depending on the phospholipid concentration range to be determined. The Cs-EDTA is generated by titrating an aqueous suspension of the EDTA free acid with CsOH to pH 6, at which point an EDTA solution is obtained. The EDTA should be titrated from the free acid with care so that, at the end point, only the required number of cation equivalents are present for each equivalent of EDTA, and the solution contains no excess extraneous salts, such as chlorides. If, in a given preparation, the final pH is greater than desired, the solution should be back-titrated with EDTA free acid. It is important not to compromise the cation-scrubbing action of the EDTA with excess cations. Because the solution is strongly buffered, final pH adjustment may be carried out efficiently using solid CsOH, which is recommended.

The chemical shift of the trimethylphosphate, at 1.97 ppm, is conveniently located just to the low-field side of the phospholipid resonance band. Under the NMR conditions usually employed, its proton broad-band decoupled resonance consists of a single resonance signal, having a line width of < 0.1 Hz under optimal conditions.

For ^{31}P NMR analysis, a lipid sample preparation (0.01-100 mg) is dissolved in 2 ml reagent A and transferred to a 10 mm NMR sample tube; 1 ml reagent B is added, and the combination mixed thoroughly. A small aqueous phase separates and is allowed to rise to the top of the sample (1 min). Ordinarily, there is no need to remove this small aqueous phase, which contains no detectable phospholipids, and its presence above the sample ensures that the proper equilibria are maintained and that the organic solvents will not evaporate significantly during the NMR analysis.

The formulation described does contain both methanol and water in the lower (chloroform) phase. The aqueous layer is usually EDTA-saturated, and, depending on the nature and quantity of the sample lipid, a third phase, precipitated EDTA, may appear within the aqueous phase. It is the aqueous-phase EDTA that contains the alkaline-earth and transition-metal cations purged from the sample, in addition to the bulk of the Cs [14,54]. Small amounts of EDTA, however, always remains within the chloroform phase [54].

Figure 2.1 shows a ^{31}P NMR phospholipid profile obtained from a fresh water sponge using the described reagent. Nineteen phospholipids were detected, and eighteen of these were identified. The signal resolution shown in the figure can be obtained with any properly prepared tissue phospholipid extract. This figure may be compared to earlier attempts to use ^{31}P NMR for phospholipid analysis. In 1970 [18], only two broad signals could be resolved in the spectral region shown in Figure 2.1.

2. *Effect of phospholipid concentration on phosphorus chemical shifts*

The chemical shifts of some of the phospholipids vary to a small degree, depending on the concentration of total lipid in the reagent. These shift differences are particularly noticeable when analysis is attempted of extremely dilute

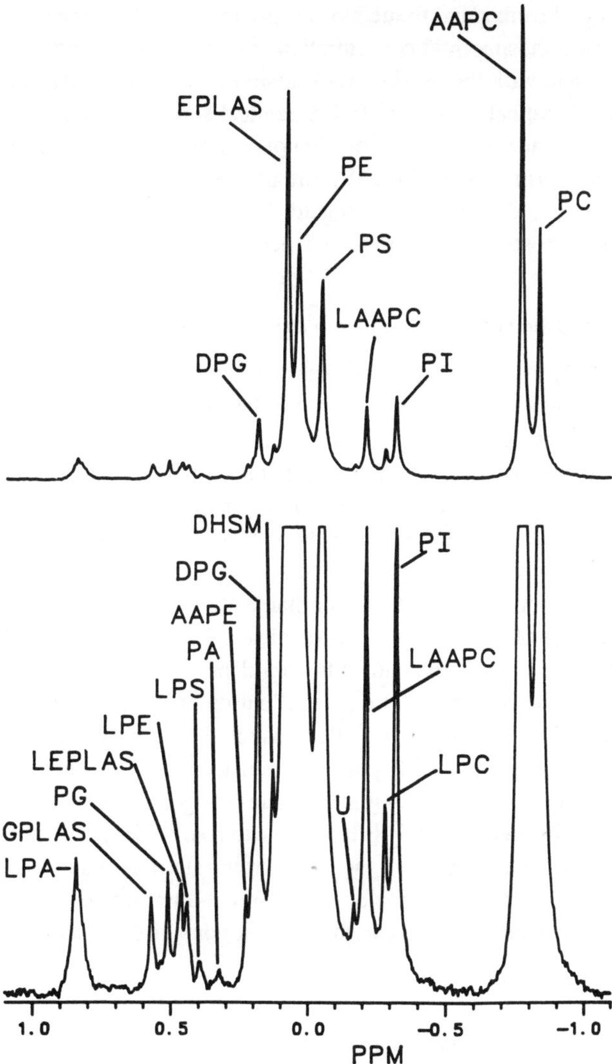

Fig. 2.1 ^{31}P NMR spectral phospholipid profile of a fresh water finger-form sponge, *Eunapius fragilis*. The top spectrum is displayed with all of the signals on scale; the bottom spectrum is the top spectrum expanded vertically to display the minor resonances: LPA, lysophosphatidic acid; GPLAS, glycerol plasmalogen; PG, phosphatidylglycerol; LEPLAS, lyso-ethanolamine plasmalogen; LPE, lysophosphatidylethanolamine; LPS, lysophosphatidylserine; PA, phosphatidic acid; AAPE, unknown resonance tentatively assigned to the alkylacylphosphatidylethanolamine; DPG, diphosphatidylglycerol; DHSM, dihydrosphingomyelin; EPLAS, ethanolamine plasmalogen; PE, phosphatidylethanolamine; PS, phosphatidylserine; U, uncharacterized phospholipid; LAAPC, lysoalkylacylphosphatidylcholine; LPC, lysophosphatidylcholine; PI, phosphatidylinositol; AAPC, alkylacylphosphatidylcholine; PC, phosphatidylcholine. In addition, this sponge contains a phosphono-phospholipid. The phosphono-phospholipid resonance band is displaced 20 ppm downfield (in the positive direction) from the corresponding phosphate-phospholipid band and is not shown in the figure. In this sponge, only one phosphono-phospholipid was detected, the diacylglyceryl-(2-aminoethyl)phosphonate (DAG-AEP).

samples derived from small tissue biopsy specimens. The chemical shift is linear with the log of phospholipid concentration, within the usable phospholipid concentration range, with the maximum excursion of approximately 0.6 ppm being exhibited by phosphatidylserine (PS). Chemical shifts decrease with increasing phospholipid concentration for all phospholipids. With the exception of the ethanolamine plasmalogen (EPLAS), the phospholipids that show a chemical-shift-concentration dependence are negatively charged under the conditions of the determination; those that show no such dependence are zwitterionic and neutral.

3. Effect of acyl side chains on phosphorus chemical shifts

Regardless of the nature of the phospholipid polar head group, there is essentially no contribution to the chemical shift of the length of the esterified fatty-acid side chains, unless these chains are under 6 carbons. For example, the shifts of synthetic phosphatidylglycerol (PG) and phosphatidylcholine (PC) preparations were determined where the fatty-acid side chains were varied from dilignoceric through dioctanoic. The ^{31}P NMR shifts of these are identical. The shift of the platelet activating factor, 1-O-alkyl-2-acetyl-*sn*-glyceryl-3-phosphorylcholine, however, is -0.65 ppm, 0.19 ppm higher than that of PC at -0.84 ppm. This essential independence of the phospholipid chemical shifts in the chloroform-methanol reagent is, however, in sharp contrast to their dependence on carbon chain length in aqueous detergents [66], where a small but distinct logarithmic shift-dependence on acyl chain length has been demonstrated. Similarly, unsaturation in the acyl chains has no effect on ^{31}P shifts in the reagent. Once again, this is in very sharp contrast to the shift behavior in aqueous detergents, particularly in aqueous 5 % cholate [66], where, not only is it possible to resolve the resonances of di-(saturated)-acyl phospholipids from their di-(unsaturated)-acyl analogues, but it is possible to resolve the resonance of the saturated-unsaturated mixed acyl phospholipid from the resonances of both the analogous di-saturated and di-unsaturated species.

4. Effect of pH on phosphorus chemical shifts

In the chloroform-methanol reagent, only one phospholipid, PA, shows a chemical-shift pH-dependence between pH values of 5 and 13, the only segment of the pH range thus far examined rigorously. (Presumably, the two phosphomonoesters of the phosphatidylinositol bisphosphate, PIP_2, would exhibit a weak acid pH-dependence of the monoester chemical shifts also). The ionization curve exhibits the typical features of such data [26,76]. There are two well-defined endpoints, corresponding to the monoprotonated acid, at pH values < 7, and the conjugate dianionic base, at pH values > 10, and a single transition segment between these two extremes. An unusual feature is the pK_a, which at 8.44 is rather basic for a simple orthophosphate monoester [24,26,27]. Phosphatidic acid pK_as reported by

Table 2.1
Phospholipid ^{31}P NMR chemical shifts in the hydrated chloroform-methanol-Cs-EDTA NMR reagent

Phospholipid	Common Abbreviation	Chemical Shift[a] (ppm)
Diacylglyceryl-(2-aminoethyl)phosphonates	DAG-AEP	21.19
U		21.01
Ceramide-(2-tetramethylammonium-ethyl)phosphonates	CER-AEP	20.38
U		2.01
U		1.97
Trimethyl phosphate (quantitative and qualitative reference)		1.97
U		1.71
Lysophosphatidylglycerol	LPG	1.10
Phosphatidylmethanol	PM	1.01
Lysophosphatidic acid[b]	LPA	0.91
U		0.84
Monolysodiphosphatidylglycerol	(mono)LDPG	0.14; 0.74
Dilysodiphosphatidylglycerol	(di)LDPG	0.69
U		0.64
Glycerol plasmalogen	GPLAS	0.55
Phosphatidylglycerol	PG	0.52
Lysoethanolamine plasmalogen	LEPLAS	0.47
Lysophosphatidylethanolamine	LPE	0.43
Phosphatidic acid	PA	0.29
Mono-oleyl phosphate		0.23
Alkylacylphosphatidylethanolamine[c]	AAPE	0.22
Diphosphatidylglycerol (cardiolipin)	DPG (CL)	0.18
Ethanolamine sphingomyelin[c]	NH2SM	0.16
Dihydrosphingomyelin	DHSM	0.13
Ethanolamine plasmalogen	EPLAS	0.07
Phosphatidylethanolamine	PE	0.03
Lysophosphatidylinositol	LPI	0.00
Phosphatidylserine	PS	-0.06
Di-oleyl phosphate		-0.08
Sphingomyelin	SM	-0.09
U		-0.17
Dimethylphosphatidylethanolamine	DiMePE	-0.18
Lysoalkylacylphosphatidylcholine	LAAPC	-0.20
Lysophosphatidylcholine	LPC	-0.28
Phosphatidylinositol	PI	-0.37
Phosphatidylinositol monophosphate	PIP	1.88 (4); -0.38
Sphingosylphosphorylcholine	SPC	-0.40
Phosphatidylinositol *bis*phosphate	PIP_2	1.98 (4); 1.09 (5); -0.54
Platelet activating factor	PAF	-0.65
Alkylacylphosphatidylcholine	AAPC	-0.77
Phosphatidylcholine	PC	-0.84

[a]Phosphatidylcholine (PC) is the internal chemical-shift reference; benzene-d_6 provided the reference for field-frequency stabilization. Chemical-shift values are reported relative to 85 % orthophosphoric acid.
[b]At pH 6.
[c]Tentative assignment.

others for the second ionization are: 8.78, 8.0, 9.0, 8.5, 8.9, 8.25 and 8.1 [45], and 8.0 [66].

By contrast, in aqueous detergent systems [7,44,66,72], phosphatidylethanolamine (PE) and PS also exhibit a chemical-shift pH dependence on the alkaline side in the region of pH 9.5, corresponding to the ionization of the ethanolamine and serine residue amino groups.

Sample pH can be used to distinguish between phospholipid monoesters and diesters. For example, to distinguish between lysophosphatidic acid (LPA) and lysophosphatidylglycerol (LPG), raise the pH of the solvent to 8. The resonance of LPA will migrate downfield; that of LPG will remain fixed.

5. Effect of water on phosphorus chemical shifts

The degree to which the chloroform phase is hydrated affects the relative chemical shifts of the phospholipids slightly. Hydration is influenced by the nature of the lipid sample, the quantity of sample used, any other reagents that might be added, *e.g.* aqueous salt solutions, organic neutral lipid solutions, and, of course, the precise ratio of the chloroform reagent A to the aqueous methanol reagent B. This behavior of the phospholipid chemical shifts with sample hydration in the chloroform-methanol reagent has been examined by Edzes *et al.* (their Figure 3) [14]. The relative sensitivity of the phospholipid chemical shifts to sample hydration can be used as a tool for the identification of individual phospholipid species, since each phospholipid has its own characteristic response to the degree of sample hydration.

6. Concluding remarks regarding reagent chemistry

For amphiphilic molecules, such as the phospholipids, it is difficult to find a solvent in which the solute exists as discrete solvated entities and is not aggregated in some manner. The chloroform-methanol-aqueous-Cs-EDTA NMR reagent was formulated with the expectation that, in this solvent system, the phospholipid molecules would be individually solvated and isolated by solvent from each other, because clear solutions that did not scatter light were obtained and because NMR signals derived from phospholipids dissolved in this solvent were quite narrow, with some being less than 1 Hz at half-height [19,48,51], indicating reasonably efficient averaging of local field gradients.

The data indicate, however, that aggregate formation of some type remains a confounding feature of phospholipid solvation. To be sure, the aggregates are not very large and may be as small as simple dimers, since the chloroform phase of the solvent appears transparent to the unaided eye, indicating a particle size having a maximum radius-of-gyration of < 100 nm [15,74], or, more particularly by laser particle sizing, < 5 nm. It is safe to state that multilamellar structures are precluded. Moreover, since the solvent is chloroform, the aggregates must be inverted, *i.e.* the hydrocarbon tails of the phospholipids are solvated by the continuous organic phase, and the polar phospholipid head groups are solvated by either each other or by a few molecules of water and/or methanol that, in aggregate, constitute a dispersed phase. It appears further that the gross aspects of these aggregates depend little on the nature of the individual polar head groups or on the nature of the counter cation or any salt present.

Further, a concentration-dependence of the ^{31}P NMR phospholipid chemical shift has become apparent from the application of the method to the routine chemical analysis of tissue-specimen samples. PI, PS, DPG, PA, EPLAS and PG show this characteristic. With PC, sphingomyelin (SM), and PE, chemical shift is independent of concentration changes. The concentration dependence of the chemical shift indicates the presence of an intermolecular equilibrium, most probably among aggregates. At some point, as the solute concentration diminishes, the chemical shifts of those phospholipids that change with concentration ought to either become constant or approach some asymptotic value. This end point appears to lie beyond the sensitivity range of current NMR instrumentation, however.

D. PHOSPHOLIPID IDENTIFICATION

1. Chemical shifts

PC gives rise to a well resolved, narrow resonance signal that, in complex mixture samples, is usually the narrowest signal observed in the spectrum and almost always the highest field resonance detected. This signal characteristic, added to the stability of the signal's chemical shift, makes the PC resonance the perfect internal reference for these preparations and avoids the addition of new molecules to the sample. Further, PC is present at significant concentrations in almost all cellular preparations.

Table 2.1 gives the nominal ^{31}P NMR chemical shifts of phospholipids thus far detected in tissue extracts using the extraction and detection procedures described. When known, phospholipids are identified in the table along with a commonly used abbreviation. For unidentified phospholipids (U), only the chemical shift is given. The table is arranged in the order of decreasing chemical shifts (left to right as ^{31}P spectra are ordinarily obtained). CAUTION: In the laboratory, chemical shifts will vary slightly, depending on the factors discussed above; therefore, the data of Table 2.1 are intended only as a guide to the relative relationships among the phospholipid chemical shifts.

Regarding the unidentified phospholipids presented in Table 2.1, the ^{31}P NMR method is a rather discriminating analytical method that is about ten times better at detecting new phospholipids than HPLC when sample quantity is not a problem. It has a higher signal-from-signal resolving power and a large dynamic range. It detects the isolated phosphorus heteroatoms in each phospholipid and detects only phospholipids, enabling analysis to be performed on crude preparations, while reducing the number of error-generating chemical manipulations.

Moreover, because the ^{31}P NMR method readily detects minor phospholipid components and can resolve the signals from phospholipids that are chemically similar (Figure 2.1), the method is quite useful in uncovering new phospholipids. Thus, the method should prove useful in discovering phospholipids having pharmacological value or in uncovering specific metabolic errors associated with disease states. As a survey tool for new phospholipid natural products, the ^{31}P NMR method has formidable power and an excellent track record.

Identifying unknown phospholipids is not trivial. With the possible exception of dihydrosphingomyelin (DHSM), newly detected phospholipids are uncommon phospholipids for which commercial standards are not available and for which even the use of academic standards provided by friends and colleagues usually proves fruitless. Like all newly discovered natural products, rigorous characterization depends upon isolation, purification and systematic chemical analysis [6].

Known phospholipids can be identified with considerable certainty using ^{31}P NMR. Spectra with reasonable signal-to-noise should be acquired and a line-width at half-height of the resonance in question determined. An appropriate quantity of known material, usually equivalent to a signal half the intensity of the resonance in question, is added to the sample and coalescence of the sample and added resonances determined. Some perturbing factor discussed above, such as changing counter cations, changing concentration, or changing the degree of sample hydration, should be applied to be certain that the observed signal coalescence is real by the line-width test. Such procedures are usually sufficient to determine the identity of the more common phospholipids for which appropriate standards are available.

2. *Acetone precipitation of crude extracts*

Crude lipid extracts containing large fractions of neutral lipids, *e.g.* human female breast extracts, will give rise to an annoying variability in the phospholipid chemical shifts that lends a degree of uncertainty to signal identification, although such variability usually does not compromise quantification. For the purpose of signal identification, the evaporated extracted lipid sample can be treated with from 5 to 10 volumes of acetone. The phospholipids are only sparingly soluble in acetone, and by this process the acetone-insoluble phospholipids may be purged of the bulk of the neutral lipid component. ^{31}P NMR analysis of the acetone-insoluble phospholipid fraction will yield precise chemical-shift values that are reliable for signal identification, although, in general, these acetone-treated samples are not usable for quantitative tissue profiling.

Acetone scrubbing is also of value in those instances where the presence of sample neutral lipid fractions causes resonance signals to coalesce. For example, coalescence of the PS and SM resonances is frequently observed in profiles of mammalian nerve tissue samples. Occasionally, coalescence is so complete that the signals appear as a single resonance in the shift position of SM. Reduction of the neutral lipid fraction through acetone scrubbing permits resolution of the PS and SM signals, and if carefully carried out, can be used to quantify the ratio between the rather acetone-insoluble PS and SM.

3. Saponification and polar head-group determination

Saponification of extracted tissue phospholipids yields a set of isolated glycerol-3-phosphoryl phospholipid polar head groups from which semi-quantitative ^{31}P NMR spectral profiles can be obtained [53]. Many of the signals from these phospholipid polar head groups are the same signals as those detected by ^{31}P NMR in perchloric acid extracts of tissues. The resonance signals from these phospholipid head group fragments can be used as an aid in the characterization of isolated phospholipids, of tissue phospholipid ^{31}P NMR profiles, and of the phosphorus resonances typically seen in tissue ^{31}P NMR *in vivo*, *ex vivo*, and in perchloric acid extract profiles.

E. QUANTIFICATION

The use of trimethylphosphate as an internal concentration reference permits the determination of phospholipid concentrations in terms of mg/ml. Further, since, even in generic phospholipid preparations of unknown fatty acid composition, the number of P atoms per molecule is known, generic phospholipid molarities also may be determined with accuracy, and this is a distinct advantage of the ^{31}P NMR method over high-performance liquid chromatography.

By formulating solvents with appropriate trimethylphosphate concentrations, a large range of phospholipid concentrations may be determined. Quantification is limited, on the low end, by the sensitivity of the spectrometer employed and the available signal-averaging time and, on the high end, by total lipid concentrations not exceeding about 100 mg/3 ml. Given sufficient signal-to-noise in the accumulated spectrum, a range of concentrations encompassing a thousand fold between the least and most concentrated phospholipid components can be determined to three significant figures.

When using NMR as a quantitative modality, signals may undergo disproportionate magnetic resonance saturation, depending on whether signal-averaging process are used and, if used, on the length of time elapsing between repetition pulses. In general, disproportionate saturation is of little consequence [48,67] if, for example, repetition rates are kept longer than 1.8 sec in an 11.75 Tesla field using a spin-flip angle of 45° [48]. Further, signal intensity errors are consistent from sample to sample as long as the repetition rate and the solvent medium are not changed, permitting the use of calibration factors. With NMR, the advantage of using a high repetition rate in conjunction with calibration factors is that signal-averaging time can be reduced substantially without sacrificing analytical precision significantly.

Factors that limit the accuracy of ^{31}P NMR measurements [1,2,5,75], ^{31}P NMR spectral profile measurements [35], and NMR measurements of extracted phospholipids [14,19,48,51,53,67] have been discussed. In general, properly calibrated NMR spectra yield accurate quantitative data [50,51,53], with errors usually attributable to the formulation of inappropriate standards that deviate in matrix composition from those of the analytical samples in the nature and quantity of solutes, solvents, and for phosphates, counter cations and pH.

The precision achievable with biological specimens has also been evaluated [47,50,51,53]. The variation in pig lens phospholipid profiles [51] is typical of biological samples. For the pig lens, the major phospholipid components, defined as those signals accounting for > 10 % of the total detected P, in this case PC, PE, EPLAS, PS and SM, exhibit standard deviations in the range of ± 5 % of each individual resonance's signal area value. In the case of the perfused rat heart [42,53], the standard deviations are somewhat better (± 3.6 %). For statistical purposes, a minimum experimental number value (n) of about 12 is generally required because of biological variability [51,53]. To determine significant differences among experimental groups, for example, among malignancies and their normal host tissues, the familiar one-way analysis of variance for multi-group analysis is generally performed, followed, in those instances where significant differences are observed ($F < 0.05$), by Scheffé simple and/or complex contrasts employing a statistical range of $P > 0.05$ or 0.01 [64].

F. TISSUE PHOSPHOLIPID PROFILES

The ease with which phospholipids can be determined using ^{31}P NMR makes the procedure ideal for determining tissue phospholipid profiles. To a first approximation, such profiles reflect the generic phospholipid composition of cellular membranes. Further, each tissue's profile is unique, rendering it useful for characterizing tissues and tissue metabolic states. These broad features are apparent in Tables 2.2 to 2.5, which present ^{31}P NMR phospholipid profiles from a variety of biological materials. The data are listed as the mole-percent phosphorus detected as a given generic species in the total phosphorus detected, which is the simplest and most straight-forward manner of presenting the spectroscopic data. This method of computing concentration values does not require exhaustive phospholipid extraction, only representative extraction, and it corrects internally for quantification errors developed during sample work-up.

Some general observations regarding phospholipid profiles are the following:

1. PC is the most common and usually the most abundant phospholipid.
2. SM and EPLAS are the next most abundant phospholipids in animals, PE and/or PG in plants.
3. PI and lysophosphatidylcholine (LPC) are almost always detectable.
4. PS is prominent in animals and detectable in plants at about the 3 % level.
5. PG is usually detectable in plants, sometimes at elevated levels, and it is often detectable in animals.
6. DHSM and AAPC are detectable in a wide range of tissue types.

Table 2.2
Phospholipid metabolic indexes

Index set 1, reduction at the glycerol 1-carbon

INDEX	Index expression	*INDEX*	Index expression
PLASA	EPLAS + AAPC	*PLASGC*	GPLAS + DHSM + EPLAS + AAPC
PLASB	PE + PC		
UNSAT	*PLASA/PLASB*	*UNSATGSM*	*PLASGC/PLASGD*
PLASGA	GPLAS + EPLAS + AAPC	*PLASC*	DHSM + EPLAS + AAPC
PLASGB	PG + PE + PC	*PLASD*	PE + SM + PC
UNSATG	*PLASGA/PLASGB*	*UNSATSM*	*PLASC/PLASD*
PLASGD	PG + PE + SM + PC		

Index set 2, polar headgroup functionality

INDEX	Index expression	*INDEX*	Index expression
PLANTHIN	GPLAS + PG	*OL-B*	DAG-AEP + CER-AEP + PA + DHSM + EPLAS + PE + PS + SM + AAPC + PC
CEPHALIN	EPLAS + PE		
LECITHIN	AAPC + PC		
CEPHLEC	*CEPHALIN/LECITHIN*	*POLYOL*	*OL-A/OL-B*
PLANLEC	*PLANTHIN/LECITHIN*	*CHOL-EOL*	*CHOL-ETH/OL-A*
PLANCEPH	*PLANTHIN/CEPHALIN*	*SPHINGOA*	CER-AEP + DHSM + SM
PLCELECI	*(PLANTHIN + CEPHALIN)/LECITHIN*	*SPHINGOB*	DAG-AEP + GPLAS + PG + PA + DPG + EPLAS + PE + PS + PI + AAPC + PC
PLLECEPH	*(PLANTHIN + LECITHIN)/CEPHALIN*	*SPHINGO*	*SPHINGOA/SPHINGOB*
CHOLA	CER-AEP + DHSM + SM + AAPC + PC	*DIANIONA*	PA + DPG + PS
		DIANIONB	DAG-AEP + CER-AEP + GPLAS + PG + DHSM + EPLAS + PE + PI + AAPC + PC
CHOLB	DAG-AEP + GPLAS + PG + PA + DPG + EPLAS + PE + PS + PI	*DIANION*	*DIANIONA/DIANIONB*
CHOLINE	*CHOLA/CHOLB*	*DIAN-CHO*	*DIANIONA/CHOLA*
CHOL-ETH	*CHOLA*/(DAG-AEP + EPLAS + PE)	*DIAN-SPH*	*DIANIONA/SPHINGOA*
		SPH-CHO	*SPHINGOA/CHOLA*
ETH-SM	(DAG-AEP + EPLAS + PE)/(CER-AEP + DHSM + SM)	*ANION*	GPLAS + PG + PA + DPG + PS + PI
		NEUTRAL	DAG-AEP + CER-AEP + DHSM + EPLAS + PE + SM + AAPC + PC
CHOL-SM	(AAPC + PC)/(CER-AEP + DHSM + SM)		
OL-A	GPLAS + PG + DPG + PI	*AN-NEUT*	*ANION/NEUTRAL*

Index set 3, membrane asymmetry

INDEX	Index expression	*INDEX*	Index expression
PGOUT	CER-AEP + PG + DHSM + SM + AAPC + PC	*INSIDE*	DAG-AEP + EPLAS + PE + PS
		PGLEAF1	*PGOUT/INSIDE*
PGIN	DAG-AEP + GPLAS + PG + EPLAS + PE + PS	*PGLEAF2*	*OUTSIDE/PGIN*
		LEAFLET	*OUTSIDE/INSIDE*
OUTSIDE	CER-AEP + DHSM + SM + AAPC + PC		

Index set 4, ratios

INDEX	Index expression	*INDEX*	Index expression
GPLASPC	GPLAS/PC	*DAGAEPPS*	DAG-AEP/PS
LPCPC	LPC/PC	*etc.* for all combinations of two phospholipids	

Index set 5, hydrolysis

INDEX	Index expression	*INDEX*	Index expression
PA-RATIO	PA/(PG + PE + PS + PI + PC)	*LYSIS*	*LYSO/NLYSO*
LYSO	LDAG-AEP + LPA + LPG +	*LETHER*	LAAPC + SPC
	LDPG + LEPLAS + LPE + LPS +	*LPLAS*	LGPLAS + LEPLAS
	LPI + LAAPC + LPC	*LESTER*	LDAG-AEP + LPA + LPG +
NLYSO	DAG-AEP + GPLAS + PG + PA		LDPG + LPE + LPS + LPI + LPC
	+ DPG + EPLAS + PE + PS + PI	*LETHEST*	*LETHER/LESTER*
	+ AAPC + PC	LPLASEST	LPLAS/LESTER

There exist notable exceptions, however, to the above general rules. For example, PC represents only 3 % of the human crystalline lens profile, while the dominant phospholipid in fresh water sponge profiles is AAPC. In some algae AAPC can be detected.

The ^{31}P NMR method is particularly useful for discovering rare and uncharacterized phospholipids, because no fractionation procedures are involved other than extraction and metal-ion scrubbing and because all lipids containing a phosphorus atom are detected.

Probably the most important findings to date are the quantification of the plasmalogens, including the choline and glycerol plasmalogens, the detection of dihydrosphingomyelin (DHSM in Table 2.1) [6], which appears to be a rather common phospholipid unusually enriched in the human crystalline lens, and the detection of the alkylacylphosphatidylcholine (AAPC).

The lyso-phospholipids (here we are speaking of deacylation at the 2-position of glycerol or its equivalent), including the deacylated sphingomyelin, are also easily detected and quantified. In animal tissues, LPC is usually detectable in the range of 1-3 % of the total phospholipid profile. The others, in healthy tissues, are not detectable ordinarily. In diseased tissues or tissues roughly handled during surgical procedures, lyso-phospholipids other than LPC may be detected. LPA, LDPG, LEPLAS, LPE and LPS are frequently detected. LPG, LPI and LAAPC (in invertebrates rich in AAPC) have been detected. Even the deacylated SM (sphingosylphosphorylcholine or SPC) commonly appears in some animal tissues, and this analytical method is extremely useful for quantifying sphingomyelinase activity in lipoproteins [78] and cancers [36]. (In those studies involving cancerous surgical tissue specimens, care was taken to obtain tissues from areas of least necrosis. The lyso phospholipids are prominent in necrotic tissue or tissues rendered non-viable by therapeutic intervention [52]).

Table 2.3
Phospholipid profiles of animal tissues.

Animal Tissue	Phospholipid (mole %)															Ref.
	LPA	PG	LPE	PA	DPG	DHSM	EPLAS	PE	PS	SM	U	LPC	PI	AAPC	PC	
HUMAN TISSUES																
Blood																
erythrocytes		3.0			2.5	3.8	11.1	4.4	13.4	30.3		2.2	3.2	2.1	27.0	
plasma						1.2	1.1	0.8	0.4	18.3		6.6	3.5	5.1	63.0	
platelets					1.2	3.0	15.0	9.3	11.2	15.0		1.7	3.8	5.2	34.6	
polymorpho-nuclear cells					1.0	3.8	20.3	8.1	10.6	11.3		1.3	4.9	13.6	25.1	
Breast																
female		1.2		0.9	3.7	3.6	11.8	7.2	9.2	12.6		2.0	4.3	3.7	39.8	62
male					2.0	1.3	14.4	6.7	9.4	14.8		1.8	3.8	2.7	43.1	
milk				0.9		2.4	10.1	23.6	11.0	25.1			6.4		20.5	
Colon			0.9	0.8	3.0	3.3	15.6	8.3	9.5	12.5		0.5	5.6	3.0	37.0	57
Esophagus					3.5		15.9	8.2	10.2	15.1			4.8	1.8	40.5	56
Joint capsule																
Synovial fluid (OA)						1.3	1.6	0.7	2.7	21.7		5.8	4.2	7.5	54.5	
Synovial fluid (RA)						1.4	1.9	0.7	2.8	21.8		6.1	3.1	8.8	53.4	

Table 2.3 (continued)

Animal Tissue	Phospholipid (mole %)															Ref.
	LPA	PG	LPE	PA	DPG	DHSM	EPLAS	PE	PS	SM	U	LPC	PI	AAPC	PC	
HUMAN TISSUES																
Synovium (RA)					2.2	2.2	13.7	8.7	10.4	13.1			5.8	4.1	39.8	
Kidney		0.3		0.3	6.4	2.2	10.6	18.4	8.1	10.3		0.8	5.1	0.9	36.6	
Lens[a]		9.9	1.0	2.6		43.7	14.5	6.1	1.9	9.7		1.5	0.8	1.1	2.8	33,60
Lipoprotein-A						0.5	1.4	0.5		23.8		3.0	1.7	2.1	67.0	13
Lipoprotein, low-density						0.8	2.0	0.8		24.7		3.1	1.7	2.5	64.4	13
Lung	1.3	3.5		0.2	1.2	1.7	11.3	8.0	9.8	14.0		0.8	3.4	2.6	42.2	55
Lymph node		0.3		0.3	3.1	2.0	12.0	12.9	8.2	8.6			6.4	3.7	42.5	
Prostate		0.6		0.2	2.3	1.9	15.7	10.3	8.0	11.8		1.3	5.6	3.9	38.4	
Sciatic nerve						4.8	27.4	7.0	18.2	25.5		0.7	1.6	1.7	13.1	12
Stomach					5.7		11.5	14.3	7.8	10.4			5.7	1.7	42.9	56
Thyroid				0.2	1.6	0.9	9.6	8.9	6.2	16.5	0.8	2.3	5.7	1.9	45.4	
CULTURED HUMAN CORNEAL CELLS																
	LPA	PG	LPE	PA	DPG	DHSM	EPLAS	PE	PS	SM	U	LPC	PI	AAPC	PC	
Epithelium	0.4	0.3		0.6		17.3	13.3	0.4	7.6	6.4			7.5	1.1	45.1	61
Fibroblast	0.5		0.1	0.1	1.6	10.3	16.2		6.6	6.5		0.8	9.7	1.8	45.8	61
Endothelium	0.4	0.1		0.6	3.6	9.9	11.7		8.3	9.0		1.4	7.5	1.3	46.1	61

HUMAN CANCERS

	LPA	PG	LPE	PA	DPG	DHSM	EPLAS	PE	PS	SM	U	LPC	PI	AAPC	PC	
Brain																
glioblastoma multiforme	0.1	1.1		2.6	1.9	1.3	6.9	7.4	6.3	11.9		1.1	4.9	2.8	50.5	63
oligodendroglioma		1.9		1.8	6.5		6.2	7.0	10.8	6.4		3.3	4.3	0.7	46.4	63
meningioma, type 1[b]		3.6		7.2	6.5	3.4		4.8	8.6	19.4		6.8	2.2	0.7	42.0	73
meningioma, type 2[b]		3.9		4.3	2.5	3.0	2.5	6.1	6.5	12.7		3.5	4.6	0.5	48.7	73
meningioma, type 3[b]		3.4		2.0	3.0	2.6	9.3	9.5	5.4	19.1		0.8	3.9	0.9	39.3	73
neurilemmoma				7.2				11.6		24.7			2.5	4.2	49.7	63
neurofibroma		2.2	1.7	2.7	5.7		10.9	10.1	14.1	9.0			1.1	4.1	38.9	63
Breast																
female, benign		0.8	1.5	1.5	3.0	2.4	11.3	7.9	8.3	12.8		1.2	4.2	3.6	41.5	62
female, malignant		0.9	0.6	1.4	3.1	3.0	10.9	9.9	8.1	10.3		1.2	5.8	4.6	40.2	62
Colon cancer[b]	0.6	0.8	0.1	0.9	3.5	3.9	13.7	8.9	8.5	7.6	1.0	0.9	5.6	5.4	38.0	59
Embryonal metastasis		1.2		4.8	2.2	2.9	6.3		1.6	19.3		2.7	3.1	4.0	49.4	
Esophagus cancer					3.00		12.0	10.1	8.1	11.8			6.7	2.3	46.0	56
Esophagus metastasis		4.9		5.0	2.3	3.0	3.0	4.8		30.3	1.8	2.8	2.3	1.0	38.9	
Kidney cancer	0.5			0.4	2.1	1.9	13.7	10.1	7.6	11.9		1.1	5.7	1.9	43.1	
Lung cancer	0.7	0.8		0.2	2.0	1.7	10.3	12.4	7.3	9.3	0.3	0.7	6.4	2.9	45.2	55
Lung metastasis		0.9		2.2	3.8		7.2	16.0	5.3	8.5		0.6	8.5	2.5	43.4	
Murine mammary carcinoma		3.3	1.7	4.0	1.0	3.4	12.7	10.0	7.2	15.3		3.2	3.3	4.7	30.3	52
Thyroid metastasis		3.0		3.6	7.8		8.3	8.9	16.7	8.7		0.9	3.6	1.4	36.2	

Table 2.3 (continued)

Animal Tissue	Phospholipid (mole %)															Ref.
	LPA	PG	LPE	PA	DPG	DHSM	EPLAS	PE	PS	SM	U	LPC	PI	AAPC	PC	
OTHER ANIMAL TISSUES																
Beef																
brain							30.8	8.8	18.5	17.1			2.4	0.5	21.9	
lens[c]		0.2		0.4	0.8	7.7	15.9	12.7	11.9	20.5	0.3	0.5	1.1	2.0	24.3	33
mitochondria, heart[b]		10.3	3.7	2.9	12.7	8.3	9.0	2.9	2.4	0.5		7.0	2.5	5.8	18.6	
mitochondria, liver[b]		1.3		1.2	2.2		7.2	2.3	2.8	6.9		4.5	5.7	2.1	60.7	
Chicken																
egg lecithin	2.5		0.9				2.4	11.8	1.2	2.2		4.2	1.1		73.7	
pectoralis		1.4		0.7	3.9	0.2	9.0	7.3	5.1	5.9		1.2	8.5	8.4	48.4	
Calf																
brain							32.6	7.1	17.5	13.6			1.9	1.4	25.9	
lens[c]				0.2	0.5	5.6	16.6	13.9	11.2	16.6		0.6	1.2	1.7	31.3	33
Chinook salmon lens				0.4	0.4	6.7	5.0	18.9	9.4	1.7			3.3	7.3	46.9	33
Coho salmon																
cornea				0.7	1.3	0.5	5.7	16.0	6.2	5.5			4.2	5.1	54.8	4
lens				0.5	1.0	8.2	5.5	17.9	7.4	1.0		0.1	3.2	7.1	48.1	4
Crayfish[b]																
tail muscle				0.6	1.0	1.6	10.3	14.3	6.4	4.6			5.1	7.2	47.0	

hepatopancrease					2.9	1.7	10.2	9.6	3.5	1.5	0.8		5.8	12.2	50.3	
Dog lens[c]				0.5	0.8	5.0	17.7	11.1	14.5	22.2		0.5	0.8	1.4	24.1	33
Fathead minnow lens				1.5	0.6	0.9	13.7	20.0	8.2	3.1		0.7	3.4	2.7	45.2	4
Golden roach fish lens				1.5		1.6	16.2	17.7	8.3	3.0			3.8	6.6	41.3	33
Guinea pig lens		0.4		0.2	0.8	2.0	18.8	10.8	15.8	8.9		0.6	2.9	2.1	36.7	33
Lake trout lens		1.3		0.6	1.0	7.1	5.7	18.0	8.3	0.9			3.5	5.5	48.1	4
Lamb lens[c]				0.5	0.8	7.0	17.0	12.7	10.9	14.3		0.4	0.9	2.2	33.0	33
Mouse																
brain, cerebrum	1.4		0.4		4.2	1.7	12.1	13.9	8.4	8.9		1.2	5.4	2.4	40.0	73
lens					1.7	3.2	22.1	8.9	18.5	20.5		0.7	0.9	2.3	21.2	33
Pig																
cornea		0.6	1.4		2.1	1.5	7.6	10.5	9.4	11.1		1.1	4.9	1.6	48.2	71
corneal stroma			0.8		1.2		8.4	8.4	7.2	13.7		1.4	4.7	1.6	52.6	71
lens		1.9	3.2		2.0	6.7	12.2	12.0	11.1	24.2		2.2	1.6	1.4	21.5	33,49,51
sclera			2.7		1.9	3.7	16.5	6.3	11.3	12.6		0.7	3.7	3.1	37.5	71
Rabbit																
aorta				0.9	1.8	1.3	9.3	7.1	8.0	27.1		2.7	5.2	1.3	35.3	
choroid	0.3	0.2	0.2	0.8	1.4	2.1	16.5	9.2	12.1	18.2	0.5	1.2	4.5	2.9	29.9	41
ciliary body	0.4	0.2		0.6	2.5	1.6	15.1	12.4	10.3	12.0	0.2	0.9	5.6	1.6	36.6	41
conjunctiva		0.4		0.4	3.5	0.9	13.0	12.8	10.1	10.7		1.4	5.3	1.9	39.6	30
cornea			1.2		2.1	1.5	11.2	11.5	8.9	10.2		0.9	5.3	2.2	45.0	30

Table 2.3 (continued)

Animal Tissue	Phospholipid (mole %)															Ref.
	LPA	PG	LPE	PA	DPG	DHSM	EPLAS	PE	PS	SM	U	LPC	PI	AAPC	PC	
iris	0.3	0.1		0.2	2.6	2.0	14.9	13.6	10.4	11.8		0.9	5.2	1.8	36.2	41
lens[c]			0.6		0.3	5.0	21.2	12.2	11.5	13.6		0.5	0.5	2.6	31.7	33,49
lens, capsule + epi			1.6		4.5		11.2	16.3	9.0	8.9		1.5	3.9	2.9	40.2	23
lens, cortex					0.4	6.9	20.1	12.2	13.5	12.2		0.6	1.3	2.2	30.6	23
lens, nucleus					0.6	5.8	24.2	5.8	13.1	18.6				1.9	30.0	23
optic nerve				1.6		1.5	33.9	8.7	15.1	9.8			1.6	1.4	26.4	20,21
optic nerve head				2.9	0.7	1.0	31.0	9.6	14.8	9.6			2.0	0.8	27.6	21
recti muscle		0.2		0.2	6.4	1.5	11.2	16.0	6.2	5.4		0.6	5.5	1.4	45.4	31
retina, neural				0.5	1.7	1.1	11.1	21.2	12.6	2.9			3.3	1.1	44.5	21
retina, neural, peripheral				0.6	2.7	1.3	11.0	18.4	10.5	3.8		0.3	4.1	0.6	46.7	22
retina, neural, central				1.1	2.4	3.1	16.8	14.7	11.9	5.6		0.2	2.9	1.6	39.7	22
sclera				0.7	2.6	1.3	13.2	8.6	9.5	16.6	2.1	1.0	4.7	1.6	38.1	31
Tenon's capsule		0.1		0.2	2.6		10.1	15.3	9.7	11.2		1.3	7.0	3.8	38.7	31
vitreous	0.6		0.3	1.5	2.9	1.7	10.5	14.0	9.6	9.5		0.4	5.4	1.8	41.8	31
Rainbow trout																
cornea				0.5	0.9	1.4	5.0	12.6	5.0	4.6			3.7	6.8	59.5	4
lens				0.6	0.5	10.2	4.7	17.6	8.3	0.9		0.8	2.0	7.6	46.8	4

Rat																
brain							17.7	16.0	13.4	6.1			3.9	1.3	41.6	
heart		1.4	0.2		16.2		9.8	22.6	4.4	3.4			4.6		37.4	42
kidney																
lens					0.4		25.8	10.7	13.7	7.9	0.7	0.2	1.6	1.1	37.9	33
liver																
plantaris muscle		0.5	0.3	1.0	7.5		6.5	9.7	4.0	5.8		1.2	7.5	0.3	55.7	43
sciatic nerve			1.1			4.1	31.2	5.3	20.5	11.2		0.8	1.6	2.3	21.9	68
soleus muscle		0.8	0.7	0.5	2.0		3.0	3.2	6.1	8.8		3.2	4.7	0.5	66.5	43
							PRIMITIVE ANIMALS									
	LPA	PG	LPE	PA	DPG	DHSM	EPLAS	PE	PS	SM	U	LPC	PI	AAPC	PC	
Anemones[d]																
Bunodosoma granulifera		0.7			2.8		22.7	4.4	11.7		2.5	0.9	2.8	13.7	13.3	46
Condylactis gigantea		0.5		0.6	1.9		26.0	1.8	5.9			3.5	4.2	20.4	9.5	46
Palithoa caribbea[a]		2.2			2.9		24.5	3.2	5.6		4.9	3.7		23.2	3.6	46
Sp1		0.6		0.5	2.3		22.9	3.7	10.5			2.7	3.3	16.8	12.7	46
Sp2		3.8			1.8		26.7	5.6	9.6		2.7		3.0	16.0	8.6	46
Stoichactis helianthus		1.7			1.8		24.6	3.4	12.7		3.4		2.7	18.1	8.4	46
Zoanthus sociatus					2.1		19.6	10.1	6.0		1.3	0.9	2.9	31.3	8.1	46
Clam[d]		1.0	2.6		3.3	2.3	24.4	7.0		4.9	1.9		7.2	7.2	21.2	48
Earthworm[a]	0.6			2.3	0.7	6.9	21.8	6.9	0.7	2.1	0.4		4.1	23.4	11.8	48
Zebra mussels[d]	0.7				3.0		24.4	16.1	2.7				5.0	6.6	24.4	3

Abbreviations: LPA, lysophosphatidic acid; PG, phosphatidylglycerol; LPE, lysophosphatidylethanolamine; PA, phosphatidic acid; DPG, diphosphatidylglycerol; DHSM, dihydrosphingomyelin; EPLAS, ethanolamine plasmalogen; PE, phosphatidylethanolamine; PS, phosphatidylserine; SM, sphingomyelin; U, uncharacterized phospholipid; LPC, lysophosphatidylcholine; PI, phosphatidylinositol; AAPC, alkylacylphosphatidylcholine; PC, phosphatidylcholine.

Taxonomic names of less familiar species: chinook/king salmon, *Oncorhynchus tshawytscha*; coho salmon, *Onchorhynchus kisutch*; crayfish, *Orchynectes rusticus*; earthworm, *Lumbricus terrestris*, fathead minnow, *Pimephales promelas*; golden roach, *Notemigonus crysoleucas*; lake trout, *Salvelinus namaycush;* rainbow/steelhead trout, *Oncorhynchus mykiss*, zebra mussel, *Dreissena polymorpha*.

[a]Other uncharacterized phospholipids detected (tissue, ^{31}P NMR chemical shift, mole %): earthworm, -0.46 ppm, 0.4, -0.53 ppm, 1.6, -0.59 ppm, 0.50, 1.71 ppm, 15.8; human lens, 1.31 ppm, 2.6; *Palithoa caribbea*, -0.88 ppm, 2.9.

[b]Other phospholipids detected (tissue, phospholipid, mole %): colon tumor, lyso ethanolamine plasmalogen, 0.6; crayfish tail muscle, phosphatidylmethanol, 1.9; crayfish hepatopancrease, sphingosylphosphorycholine (SPC), 1.5; meningioma, type 1, lysophosphatidylinositol (LPI), 3.2, SPC, 2.1; meningioma, type 2, LPI, 2.3, SPC, 1.2; meningioma, type 3, SPC, 0.8; mitochondria (beef heart), SPC, 4.6; mitochondria (beef liver), SPC, 3.1.

[c]Tissue profile also contains an unidentified phospholipid at 1.20 ppm (mole %): beef lens, 1.7; calf lens, 0.6: dog lens, 1.4; human lens, 1.8; lamb lens, 0.3; rabbit lens, 0.3.

[d]Tissue profile also contains phosphonic acids: (1) diacylglycerol-aminoethylphosphonates (DAG-AEP): *Bunodosoma granulifera*, 20.9; *Condylactis gigantea*, 23.5; *Palithoa caribbea*, 0.0; *Sp*1, 22.9; *Sp*2, 21.7; *Stoichactis helianthus*, 20.3; *Zoanthus sociatus*, 4.8; clam, 17.0; zebra mussel, 14.9. (2) ceramide-[2-(tetramethylammonium-ethyl)]-phosphonates (CER-AEP): *Bunodosoma granulifera*, 3.6; *Condylactis gigantea*, 2.2; *Palithoa caribbea*, 23.3; *Sp*1, 1.1; *Sp*2, 0.5; *Stoichactis helianthus*, 2.9; *Zoanthus sociatus*, 12.9; zebra mussel, 1.7 (3) Unknown phosphonic acid at 21.01 ppm: *Stoichactis helianthus*, 4.1.

The limit of detection in most instances was approximately 0.03 % of the total detected phospholipid profile.

G. PHOSPHOLIPID METABOLIC INDEXES

Phospholipid metabolic indexes facilitate the interpretation of phospholipid profile data. Their use is analogous to the use of similar indexes formulated for the interpretation of the phosphatic metabolites of intermediary metabolism [17,25,26,28,29,34,38,39,40,52,57, 58,70]. Phospholipid indexes compare and contrast phospholipids or groups of phospholipids, providing more system-specific metabolic information than is possible from the examination of individual phospholipids isolated from the context in which they functioned. Indexes are computed from phospholipid profile data, which are usually in the form of mole-fractions or mole-percents, and, subsequently, are treated statistically for inter-group comparisons in a manner parallel to that used in the statistical analyses of phospholipid profiles [20,21,23,43,52,56,57,59-62,71].

A great variety of phospholipid indexes can be formulated from 15 or more phospholipids that are ordinarily obtained in a tissue phospholipid profile. The number of indexes can be reduced by selecting only those formulations that make biochemical sense, *i.e.* that can be interpreted in terms of biochemical pathways, biochemical processes, or biochemical systems.

Because metabolic indexes are formulated from combinations of phospholipids, ratios of phospholipids and ratios of combinations of phospholipids that are related to metabolic systems, the indexes are more sensitive to small changes in tissue composition than are the gross concentrations of the individual phospholipids from which they are constructed. Thus, an index may differ in a statistically significant manner between tissue types or processes even though the components of the index may exhibit no statistically significant variations among experimental groups. Indexes monitor specific biochemical systems.

Further, because of the complexity of membrane phospholipid composition, the types of insights into the operant biochemistry that metabolic indexes provide are not readily apparent from an examination of the total phospholipid profile, without the use of metabolic indexes. Thus, even without knowledge of the function a given index may have in a given tissue, the relative numerical value of the index can be used as a parameter that monitors metabolic activity.

The fact that a given index cannot be interpreted in terms of known biochemical processes does not detract from its value as a metabolic indicator, since the index may represent a function or process unique to the tissue under study or an unrecognized biochemical pathway. Many new pathways have yet to be uncovered. The ubiquitous occurrence of DHSM and AAPC and the uncharacterized phospholipids, such as AAPE, are all relatively new findings for which no firm biological functions have been established.

Certain biochemical properties ought to be common to all living membranes, for example, the property of net negative charge. Such properties, which arise from phospholipid functional group chemistry, can be incorporated into formulations or sets of formulations that consistently yield the same numerical index from tissue to tissue. (An example in the system of high-energy phosphate metabolism is Daniel Atkinson's Energy Charge, a formulation for describing, mathematically, the metabolic steady state for ATP, the intracellular reservoir of all chemical, mechanical, and electrical energy. For healthy cells, this index has a value in the range of 0.80 to 0.95). Formulations that vary widely among tissues can probably be discarded as inappropriate descriptors of fundamental membrane properties. Some formulations, because of their borderline behavior, will be intriguing. These formulations are viewed as extremely important, because they may provide clues for linking together disparate segments of the total system. Further, it should be mentioned that the simplest composition is not necessarily a sufficient composition, because it may lack essential functional elements common to all other living membranes. Undoubtedly, there will be a continuum of compositions, ranging from the simple, and functionally limited, to the complex, and functionally diversified. There will also be compositions that are more or less competent in accomplishing a variety of tasks, as well as compositions that may be very efficient at any given task but that are not very adaptable to changing environments. All of these factors impart to any given membrane a range of properties that, among other considerations, render a degree of membrane adaptability, or a degree of membrane specialization, that is useful to a cell in the particular environment in which it operates.

"It is ideas that are important, not facts [77]." When formulating metabolic indexes, creative interpretation, let us say, artistic license, is absolutely required and must be exercised if progress towards understanding is to be realized. No single set of equations is sacrosanct. The index definitions that follow, therefore, represent only one set of interpretations, only one model set. Investigators with other backgrounds and other prejudices will surely formulate other indexes and other models. Eventually, a consensus will begin to form as to what is truly relevant and useful. It is in this spirit that this exercise was undertaken.

The metabolic indexes defined in Table 2.2 are grouped in sets, with each set representing a system based essentially upon current knowledge but also containing some innovation. The lyso (hydrolysed) phospholipids are included only within the sets describing simple phospholipid ratios and phospholipid hydrolysis. Index names are presented in italics; phospholipid abbreviations are presented with upright (roman) characters. Some indexes are formulated using both phospholipids and other indexes.

The first set addresses the nature of the polar head group at the glycerol 1-carbon position, where chemical reductions transform the ester function to the plasmalogen en-ol ether or to the corresponding alkyl ether. Indexes *PLASA* through *PLASGA*/*PLASGB* are a measure of the relationship of the more-reduced en-ol-ether- and alkyl-ether-containing phospholipids to their more oxidized ester-containing analogues. *PLASA* and *PLASGA* represent the sums of detected (reduced) plasmalogen. *PLASB* and *PLASGB* represent the sums of their corresponding (oxidized) ester-containing analogues. The ratios of these are a measure of the balance resident within the total ester/en-ol-ether+ether metabolic system. The ratio magnitudes indicate the relative positions of the metabolic equilibrium points. The restricted glycerol, ethanolamine, and choline ratios, *GPLAS/PG*, *AAPE/PE*, *EPLAS/PE*, and *AAPE/PC*, which are defined in set 4, ratios, identify the contribution each generic phospholipid system makes to the overall metabolic balance point. Collectively, these indexes define the overall glycerol 1-carbon metabolic system.

The discovery of the alkyl-acyl derivatives, particularly the wide-spread occurrence of the alkylacylphosphatidylcholine, is rather recent [14,48]. It is not known what role these ether-containing phospholipids play in animal cells. Most probably they have roles that are different from those of the corresponding plasmalogens and cannot be freely interchanged with the plasmalogens. Further, these ether phospholipids are the least polar members in their respective families and will have the lowest capacity to interact with water at the phospholipid-water interface. These phospholipids will have a more compact cross-sectional profile at the polar head group, making them good candidates for creating curved phospholipid structures and for inducing lamellar to hexagonal-II phase transitions. The commonest alkylacyl phospholipid, however, is the choline-containing AAPC, which is not the type of head group associated with facile lamellar to hexagonal phase transitions.

PG is included within this set, because PG (and its plasmalogen) plays some role in protein translocation across the inner membrane of *E. coli* [10], even though PG is a bilayer-forming phospholipid having a cylindrical shape [8]. The dihydrosphingomyelin, DHSM, is another reduction to the level of the alkanes at a sphingolipid molecular position equivalent to that of the phosphatide glycerol 1-carbon ester. Most probably, the roles of these various reduced phospholipids in membrane function are analogous but not equivalent.

The second set of indexes examines the chemical functionality of the polar head groups, *i.e.* whether choline, ethanolamine, or glycerol, alcohol-containing, anionic, dianionic, neutral ionic, and including the sphingolipids as a subset. The classical polar head-group pair is choline-ethanolamine [8]. The combination choline-ethanolamine-alcohol has rarely been taken seriously, yet the alcohols are always present in tissue phospholipid profiles, and PG, like PE, promotes phase transitions to rounded membrane structures. The index *LECITHIN* represents the sum of detected phosphatidylcholines exclusive of the lyso derivatives. *CHOLA* is the sum of all choline-containing phospholipids; *CHOLB* is the sum of all other phospholipids detected, exclusive of the lyso compounds, and *CHOLINE* is the ratio of these. The index *CHOL-ETH* is the ratio of all cholines to all ethanolamines. *CHOL-EOL* incorporates the alcohol-containing phospholipids into the denominator of *CHOL-ETH*. *OL-A* represents alcohol-containing functional groups, such as PG, DPG, and PI.

Electric charge, particularly the presence of multiple charges, is another concept to which too little attention has been given, save that membrane charge exists. Charges, particularly those involving the phosphate functional group, represent coordination sites for metal cations. It is the charges, both positive and negative, that impart the properties of an ion-exchanger to the biological membrane.

Complexation by the cholines and the ethanolamines must be fundamentally different. Choline is a quaternary amine that is repelled by metal cations because of its formal positive charge, its bulk, and its lack of an appropriate coordination residue. The primary ammonium function on ethanolamine, by contrast, readily coordinates metal cations because it contains the nitrogen lone-pair coordinating electrons, it is small, and it does not possess a formal positive charge.

AN-NEUT is the ratio of all anionic phospholipids to all neutral-ionic phospholipids, with the exception of the lyso derivatives. *AN-NEUT* is an index of membrane surface charge, since the greater the anionic phospholipid content, the greater will be the negative charge density and, with this property, the greater will be the capacity to bind metallic cations electrostatically. (Naturally occurring net-positively-charged phospholipids can be detected by this ^{31}P NMR method; however, none have been as yet).

Three phospholipids have two negative charges per molecule, PA, which is a phosphomonoester possessing an unusually high weak-acid pK_a, PS, which contains the α-carboxylic anion, and DPG, which contains two phosphates linked by a glycerol moiety possessing a free hydroxyl group at the glycerol 2-carbon position. All are very good metal ion complexers. Moreover, they differ among which metal cations each prefers to complex [54]. From the point of view of coordination chemistry, these three phospholipids represent strongly negatively charged sites of greatly different physical geometries. (Note that these three phospholipids also are cone-shaped phospholipids that promote the formation of hexagonal lipid phases [8]). The sphingolipids possess an amide nitrogen that represents a potential metal ion coordination site also, and they contain a free hydroxyl group that is also a potential metal ion coordination site. *DIANIONA* is the index of doubly negatively charged phospholipids. It appears to be constant among experimental groups.

The question of membrane asymmetry [69] is addressed in the third index set. It is fairly well demonstrated now that the two leaflets of the cellular membrane are dissimilar and represent an asymmetric distribution of phospholipids. It is clear that for some of the phospholipids, *e.g.* PC, there is a distinct preference for one of the leaflets, the outer leaflet in this case. It also appears fairly certain that any phospholipid may be found, to some mole fraction, in either leaflet. For most of the phospholipids, a distinct preference for either leaflet has not been established. Further, although membrane phospholipid asymmetry is believed to be a fundamental property of all membranes, in only a few membranes has the veracity of such a concept been adequately verified.

The indexes *PGOUT*, *OUTSIDE*, *INSIDE*, *LEAFLET*, *etc.*, address the relationships among the phospholipids principally responsible for membrane asymmetry [69]. The index *OUTSIDE* is the sum of PC and SM, the index *INSIDE*, the sum of PE and PS, the index *LEAFLET*, the ratio of these two. Other indexes, such as *PGOUT* and *PGIN*, are more comprehensive bilayer-leaflet indicators and include phospholipids peculiar to lower life forms, such as DAG-AEP and AAPE.

A fourth comprehensive index set involves all possible ratios of two phospholipids. For example, GPLAS/PC, EPLAS/SM, DAG-AEP/PS, LPC/PC, *etc.* The point of this set is to look for equivalence among phospholipid pairs and for possible phospholipid substitutions.

The fifth and final index set addresses phospholipid hydrolysis of the polar head groups and hydrolysis at the glycerol 2-carbon to the lysophospholipids. *PA-RATIO* is an index of the phospholipase-D equilibrium associated with the removal of the polar-head-group ester to generate PA or, if there is extensive deacylation, LPA. (It is not, however, an index of the phospholipase-D activity responsible for the polar head group exchange that is known to occur in plants [9,79]. Such an index, for example, might be PI/PC. The phospholipid that most prominently displays this activity is phosphatidylcholine).

With the possible exception of LPC, which is found to levels in the range of 0.3 to 3 mole % in almost all tissues thus far examined, the lyso phospholipids represent membrane degradation and are rare in healthy tissue profiles. The lyso-derivatives (LPG, LPA, LDPG, LGPLAS, LEPLAS, LPE, LPS, LPI, LAAPC and LPC) are created through the action of a phospholipase-A_2 activity acting on the parent phospholipids (PG, PA, DPG, GPLAS, EPLAS, PE, PS, PI, AAPC and PC). Such a lysosomal activity is among the first steps in a digestive process.

Lyso phospholipids that are frequently detected in low concentrations are LPA, LPE, the various possible LDPGs, LEPLAS, and, rarely, LPS. Sphingosylphosphorylcholine (SPC, lysoSM) is known and has been detected in some tissues in very small amounts. The lysophosphatidic acid spectral signal is usually rather broad and exhibits fine-structure in the ^{31}P NMR spectrum, indicating the presence of a group of LPA derivatives. These derivatives are most probably created by the actions of both a phospholipase A_1 activity, which deacylates phospholipids at the glycerol C_1-position, and a phospholipase A_2 activity.

The indexes *LETHER*, *LPLAS*, and *LESTER* define the respective equilibrium points of lysosomal activity directed against the alkyl ether phospholipids, the enol-ether plasmalogens, and the ester phospholipids (the phosphatides). The indexes *LPLASEST* and *LETHEST* are ratios, respectively, of the plasmalogen lipase activity to that of the ester lipase activity and the alkyl ether lipase activity to that of the ester lipase activity.

1. Statistical treatment of tissue phospholipid profiles

Using the data of Tables 2.3, 2.4 and 2.5, a multivariate discrimination analysis was performed with the individual phospholipids LPA, PG, LPE, PA, DPG, DHSM, EPLAS, PE, PS, SM, U, LPC, PI, AAPC, and PC, to determine which lipids were independently significant in predicting tissue type. The discrimination analysis enters the individual lipids in a forward stepwise manner, where a significance level of 0.05 is used as a criterion for inclusion in the model. The data of Tables 2.3, 2.4 and 2.5 were divided into 5 groups: vegetables (92 cases), healthy vertebrate tissues (77 cases), healthy invertebrate tissues (10 cases), human cancer tissues (18 cases), and algae (4 cases). In the discrimination analysis, the statistical model represents a linear combination of the independent variables (phospholipid concentrations), whose purpose is to assign a specific case (one phospholipid profile) to one of the different groups. The model acts as a summary of the individual phospholipids into a single index through the calculation of coefficients which weight the independent variables in a multiple linear regression equation.

All the phospholipids met the criteria for inclusion in the model, which means that the profiles of the five groups defined above are quantitatively distinct and different, to the extent that the ^{31}P NMR phospholipid profiles were able to discriminate among the groups 98.9 % for vegetables, 98.7 % for vertebrate tissues, 100 % for invertebrates, 33 % for human cancers and 100 % for algae. The predictability of the model was least successful for the human cancers, which may be attributable to their inherent variability. Human cancers were incorrectly classified 66.7 % of the time as vertebrate tissues. The remaining cancer cases were correctly classified.

In practical terms for the biochemist or the clinical biologist, a phospholipid profile can be classified for group membership based solely on the concentrations of the 15 discriminating phospholipids, *i.e.* without recourse to any other information. It is important to note that almost every bit of quantitative information derived from these phospholipid profiles is useful in discriminating among the five statistical groups tested.

Similarly, if a univariate one-way analysis of variance is performed to discriminate among the 5 tissue groups at the level of the individual phospholipids, as opposed to the entire profile, the mean values of the individual phospholipids are distinctly different by group. Only LPE and DHSM were found not to differ significantly among the groups.

The lipid data on a case by case basis is characteristic for a particular group. Of interest to the life scientist, however, are relationships (biochemical pathways, systems) among the phospholipid profiles that are common to all of the tissue groups. Such common properties would represent biochemical systems that are fundamental to the function of the living membrane.

One possible strategy for identifying such fundamental pathways/systems is to examine groupings or ratios of individual phospholipids that represent both known and/or hypothetical pathways involving the phospholipids. The indexes previously discussed represent one method for examining these pathways or systems. Indexes computed for the five groups in which no statistically significant differences were observed among the groups and that also exhibited a non-significant F-statistic when compared by a univariate one-way analysis of variance represent pathways that may be common among the groups.

Again the data of Tables 2.3, 2.4 and 2.5 were used to compute 49 indexes on a case by case basis for the five groups. The mean values were compared by a one-way analysis of variance. A Scheffé comparison procedure was performed as a *post hoc* test, accepting a $P < 0.05$ as significant. A Scheffé comparison procedure was not performed when differences between the groups were not identified. These non-significant indexes identify invariant pathways among the five groups.

Four computations were carried out, involving four different combinations of the five designated tissue groups:

5 groups - plants, animals, primitive animals, tumors and algae.

4 groups - plants, animals, primitive animals, and tumors.

3 groups - plants, animals, and primitive animals.

3 groups - plants, animals + tumors, and primitive animals.

Indexes found to be invariant among the five groups in the various combinations computed were the following: *CEPHLEC*, *DIANIONA*, *DIAN-CHO*, *INSIDE*, *PGLEAF1*, and *LEAFLET* (Table 2.6). In addition, two simple ratio indexes, LPE/PE and LPG/PG, also were found to be invariant. These, however, exhibit not-detectable or very low values for the lysophospholipid component of each ratio, weakening their predictive value.

Subsequent to the above determination of invariant indexes, index means for three test cases were computed (Table 2.6):

1. A pooled group of all healthy tissues, including plants, animals, primitive animals, and algae (183 cases).
2. A group consisting only of tumors (18 cases).
3. A pooled group consisting of all living tissues, including plants, animals, primitive animals, algae and tumors (201 cases).

From the above, the following dimension-less membrane function, f_{mem}, may be formulated:

$$f_{mem} = PGLEAF1/LEAFLET + 2(CEPHLEC + DIAN\text{-}CHO) - INSIDE/DIANIONA.$$

This function represents the invariant phospholipid combinations derived from the phospholipid profiles. It is a mathematical model of that portion of the living membrane defined by the phospholipid component. The model incorporates the following phospholipids: DAG-AEP, CER-AEP, PG, PA, DPG, DHSM, EPLAS, PE, PS, SM, AAPC, PC. Note that every major phospholipid is represented in the model except one, PI.

It is not known whether this equation represents a complete formulation or some component of a larger total solution. In a perfect and totally descriptive model, and assuming a sufficiently large sample size, the residual value would equal zero. The residuals presented in Table 2.6 of 0.11, although small, indicate that further refinement of the model is possible.

Considering the cancer-group data set of Table 2.6, the residual value is statistically equivalent to that obtained from the healthy-tissue group or from a pooled set of all tissues, even though there is marked variation in the value of the component indexes of the cancer group when compared to the corresponding values of the other two groups. This equality indicates the presence of compensatory changes in the composition of the cancer group phospholipid profiles.

Table 2.4
Phospholipid profiles of plant tissues.

Plant (common name)	Phospholipid (mole %)													Taxonomic Name
	LPG	PG	LPE	PA	DPG	EPLAS	PE	PS	SM	U	LPC	PI	PC	
Alfalfa sprouts	0.3	7.5		1.1	2.1	0.6	29.9	3.8		0.8	0.3	10.1	43.5	*Medicago sativa*
Artichoke heart	0.1	4.5	0.1	1.7	2.1	0.5	25.5	3.9		0.4	0.4	12.5	48.3	*Cynara scolymus*
Asparagus	0.2	7.6		1.2	3.2	0.5	29.9	3.4				9.4	44.5	*Asparagus officinalis altilis*
Avocado	0.1	2.2		3.8	2.9	25.1		1.7			0.7	8.5	55.0	*Persea americana*
Beet[a]	0.6[b]	5.3		4.5	3.6	1.4	29.8	4.8		0.6	0.5	6.1	42.5	*Beta vulgaris*
Bok choy		3.4		4.5	2.6		35.3	5.0		0.7	1.3	8.5	38.7	*Brassica chinesis*
Broccoli	0.2	6.8	0.2	2.0	2.3	0.4	24.1	4.1		1.6	0.2	12.8	45.3	*Brassica oleracea*
Broccoliflower	0.2	3.4	0.3	3.6	2.1	0.4	23.8	2.6		3.0	0.4	13.8	46.4	*Brassica oleracea*
Brussels sprouts	0.5	6.7	0.4	1.5	1.8	0.4	27.4	3.4		0.9	0.6	11.5	44.9	*Brassica oleracea gemmifera*
Cabbage		1.6	0.1	2.8	1.9		28.1	5.7		5.1	1.0	12.0	41.7	*Brassica oleracea capitata*
Cactus (edible)		7.3		6.3	2.5		25.7	3.1				11.5	43.6	*Opuntia engelmannii*
Carrot		6.3		3.6	2.4		25.8	3.7		0.6		13.6	44.0	*Daucus carota sativa*
Cauliflower	0.1	2.1		1.5	1.4		28.8			2.5		15.7	48.0	*Brassica oleracea botrytis*
Celeriac	0.2	2.9	0.3	5.5	2.8		31.5	3.3		1.7	0.9	12.0	38.9	*Apium graveolens rapaceum*
Celery	0.4	3.3	0.3	6.1	2.8	0.6	28.0	3.7			1.3	8.9	44.6	*Apium graviolens dulce*
Celery cabbage, Chinese	0.3	4.0		4.1	2.7		33.2	5.4		1.5	0.6	8.8	39.4	*Brassica pekinensis*

Chayote		3.2	0.3	2.1	2.3	0.2	32.9	4.9	0.4			12.1	41.6	*Sechium edule*
Chengensa		5.0		4.8	2.5		33.2	5.8			0.9	7.0	40.8	
Chicory	0.1	5.5		1.2	3.0	0.9	29.2	5.0		0.5		11.3	43.3	*Chichorium endivia*
Chive	0.1	24.5		2.6	2.4	0.5	23.9	3.8				5.9	36.3	*Allium schoenoprasum*
Cilantro (coriander)		30.6		1.6	2.8		8.5	3.5				10.4	42.6	*Coriandrum sativum*
Collards	0.3	21.0		1.5	2.8		21.5	3.8		0.4		9.4	39.3	*Brassica oleracea acephala*
Cucumber		4.1		4.8	2.9	0.6	30.6	7.0		1.6		8.9	39.5	*Cucumis sativus*
Dandelion	0.1	20.3		1.7	2.4		20.6	5.2				9.0	40.7	*Taraxacum officinale*
Dill		21.0		2.8	3.3		24.8	2.9				7.9	37.3	*Anethum graveolens*
Eggplant[a,c]	0.9	2.9	1.6	5.9	3.0		31.5	4.0	2.2		1.7	7.5	37.6	*Solanum melongena esculentum*
Endive (Belgian)	0.4	18.0		1.8	2.8		23.6	2.9		0.5	0.1	9.8	40.1	*Chichorium endivia*
Escarole		6.8		1.7	2.3	0.2	31.0	3.7		0.5	0.5	9.1	44.2	*Chichorium endivia*
Fennel	0.4	6.5		2.7	2.2		27.8	4.8			0.8	7.6	38.2	*Foeniculum vulgare dulce*
Garlic	0.3	5.7		1.2	3.3	0.4	29.0	3.0				10.9	46.2	*Allium sativum*
Ginger[b]	0.2	2.3		4.2	2.2		5.1	1.7				18.7	47.7	*Zingiber officinale*
Gobo		1.5		4.3	2.5		30.5	5.5				14.9	40.8	
Horseradish[a]		4.4		6.3	1.1		26.1	2.2		0.7		16.3	42.3	*Armoracia rusticana*
Jerusalem artichoke[e]		1.6		1.6	1.7		25.1	3.4			0.4	15.2	50.3	*Helianthus tuberosus*
Jicama[a]	0.4	6.3	1.6	7.5	1.4	0.7	30.9	5.3		0.5	0.4	13.1	31.3	*Exogonium bracteatum*
Kabu		3.4		5.1	3.0		33.4	4.8				9.4	40.9	
Kohlrabi		2.7		4.3	2.5		32.6	3.0		1.2		11.2	42.5	*Brassica caulorapa*

Table 2.4 (continued)

Plant (common name)	Phospholipid (mole %)													Taxonomic Name
	LPG	PG	LPE	PA	DPG	EPLAS	PE	PS	SM	U	LPC	PI	PC	
Leek	0.4	9.3	0.7	2.6	2.9		27.6	2.2			0.6	7.2	46.5	*Allium porrum*
Lettuce, green leaf		14.1		1.8	3.0		27.9	3.4				8.7	41.1	*Lactuca sativa crispa*
Lettuce, iceberg, the white	0.4	7.9		4.3	2.4		26.9	4.1		0.6	0.4	7.5	45.5	*Lactuca sativa capitata*
Lettuce, iceberg, the green	0.2	12.2		2.6	2.3		24.1	4.0		1.6		10.4	42.6	*Lactuca sativa capitata*
Mint	0.9	14.4		1.7	2.3	1.0	18.2	5.4		0.8		13.6	41.7	*Mentha piperita*
Mushroom, common[a]				3.8	1.6		46.9	14.5				2.4	30.8	*Agaricus campestris*
Mushroom, oyster		0.5		2.2	3.3		35.5	17.5				4.1	36.9	*Pleurotus ostreatus*
Mushroom, wood ear[f]	1.8	11.2		11.6	9.5		57.0					1.4	4.0	
Mustard greens		18.9		1.5	2.3		25.4	3.4				8.6	39.7	*Brassica juncea*
Nagaimo		2.3		4.5	1.7		28.3	3.6		0.5		16.8	42.3	
Nappa		2.6		4.4	1.8		35.8	4.1				9.9	40.8	*Brassica napus*
Nira		15.2		1.9	2.9		24.7	4.2			0.8	8.4	41.9	
Ohba		22.0		1.5	2.2	0.6	18.1	5.7				11.2	38.7	
Okra (gumbo)	0.2	3.3		6.4	3.0	0.7	26.2	4.0		0.6	0.4	11.6	43.6	*Hibiscus esculentus*
Onion, red	0.2	7.9		4.0	1.9	0.5	31.0	3.8			0.2	6.3	44.2	*Allium cepa*
Onion, white	3.1[b]	4.3		2.8	2.2	17.5	20.8	4.4		1.3		8.8	34.8	*Allium cepa*
Onion, yellow	3.1	5.4		3.1	2.0	10.4	26.2	5.3				7.9	36.6	*Allium cepa*

Parsley	0.2	22.6		0.7	2.6		19.4	3.6		1.2	0.5	9.0	40.2	*Petroselinum hortense*
Parsnip		4.1		2.9	1.6		27.7	3.4				14.7	45.6	*Pastinaca sativa*
Pea pod	0.3	7.1	0.2	2.5	2.4	0.3	25.9	4.6		0.6	0.4	8.2	47.5	*Pisum sativum macrocarpon*
Pepper, bell		5.1	0.2	2.1	2.9	0.8	22.6	2.5	1.8		0.6	13.8	47.6	*Capsicum annuum grossum*
Pepper, jalapeño	0.2	5.6		2.4	2.7		27.4	4.8			1.3	11.7	43.9	*Capsicum frutescens longum*
Pepper, long, cayenne	0.5	10.2		4.2	3.4	1.1	24.1	3.2			1.1	13.0	39.4	*Capsicum frutescens longum*
Pepper, shishito		3.1		2.3	3.7	1.6	23.4	3.8			1.3	13.0	47.8	*Capsicum frutescens longum*
Potato, red		1.8		1.4	2.2	0.5	32.7	4.4	0.9			14.2	41.9	*Solanum tuberosum*
Potato, sweet		6.4		2.1	2.3		30.2	3.2	1.8			18.6	35.4	*Ipomoea batatas*
Potato, white		2.5	0.9	1.9	2.6	0.2	29.1	3.6				15.8	43.4	*Solanum tuberosum*
Radiccio (chicory)		5.8		1.8	2.4	0.9	23.6	5.1		0.8		13.6	46.9	*Chichorium intybus*
Radish, black[a]		1.6	0.3	5.6	1.6		38.7	4.0		1.1		11.0	35.6	
Radish, red[a]	0.2	5.2		1.7	2.4	0.2	34.8	4.6		1.1	0.3	9.1	39.9	*Raphanus sativus*
Rapini	0.2	15.2		1.4	2.8	0.5	24.0	3.7		1.6	0.5	10.0	40.1	*Brassica oleracea*
Renkon		2.1		3.0	2.0		35.0	5.3				13.5	39.1	
Rhubarb	0.9	2.7		7.6	2.9	0.6	25.8	9.0		4.0		12.1	34.4	*Rheum rhaponticum*
Rutabaga	0.2	2.2		4.9	2.0		29.8	4.9		2.0		12.5	41.5	*Brassica napobrassica*
Savoy, salad	0.3	7.2		2.0	2.7	0.1	27.2	3.2		0.7	0.3	11.2	45.1	*Brassica oleracea*
Satoimo		1.9		2.2	1.0		25.7	3.5				17.9	47.8	
Scallion	0.4	17.9		1.1	3.7		29.4	3.8		0.3	0.5	6.8	36.1	*Allium cepa (ascalonia)*
Sea kale	0.6	20.3		1.6	2.6		19.6	3.3				9.8	42.2	*Crambe maratima*
Shallots	1.4[b]	6.2		1.9	3.5	3.0	24.8	3.7				9.7	45.8	*Allium ascalonicum*

Table 2.4 (continued)

Plant (common name)	Phospholipid (mole %)													Taxonomic Name
	LPG	PG	LPE	PA	DPG	EPLAS	PE	PS	SM	U	LPC	PI	PC	
Shun giku	0.5	18.3		2.4	3.2		20.9	4.1				8.9	41.7	
Sorrel (dock)	0.4	21.9		2.1	2.5	0.9	20.7	4.6				9.2	37.7	*Rumex acetosa*
Spinach	0.8	18.5		1.0	2.8		24.2	2.4		0.7	0.3	9.9	39.4	*Spinacia oleracea*
Squash, acorn		1.8		6.9	1.6		36.2	5.3				10.2	38.0	*Cucurbita pepo*
Squash, butternut		3.5		9.1	2.0		34.6	5.5	0.6	0.7		10.1	33.9	*Cucurbita pepo moschata*
Squash, zucchini	0.1	4.6	0.5	6.7	2.2		36.1	6.4			0.3	6.1	37.0	*Cucurbita pepo*
Swiss chard[g]		21.2		1.4	2.7	0.7	22.3	3.1		1.0	0.5	7.5	38.6	*Beta vulgaris cicla*
Tomatillo	0.2	3.9		4.7	2.7	0.9	26.4	4.2			1.3	12.9	42.8	*Physalis ixocarpa*
Tomato, common, pulp		3.8	2.2	19.8	1.2	1.5	16.1	2.2	2.7		2.2	12.2	36.1	*Lycopersiconesculentumcommun*
Tomato, common, seeds		1.7		15.2	1.3	1.2	10.9	2.2		0.2		22.0	45.3	*Lycopersiconesculentumcommune*
Tomato, cherry, pulp	0.7	4.9		5.9	1.8	0.8	23.7	7.1		1.1	0.7	11.3	42.0	*Lycopersicon esculentum cerasiform*
Tomato, plum, pulp	0.8	5.1		11.5	2.2	1.6	20.3	7.0		1.4	1.3	10.5	38.3	*Lycopersicon esculentum cerasiform*
Turnip	0.2	3.9		7.1	3.1		32.2	3.8		1.8	0.6	9.9	37.4	*Brassica rapa lorifolia*
Turnip greens	0.4	20.0	0.5	2.5	2.8		23.5	4.2				8.8	37.3	*Brassica rapa lorifolia*
Watercress	1.8	18.0		1.9	2.3	0.4	22.9	2.6		0.4		10.4	39.3	*Roripa nasturtium-aquaticum*
Yam	0.1	3.8	0.4	4.0	2.3	0.8	25.7	4.2		0.8	0.1	19.0	38.8	*Dioscorea alata*

Abbreviations: LPG, lysophosphatidylglycerol; PG, phosphatidylglycerol; LPE, lysophosphatidylethanolamine; PA, phosphatidic acid; DPG, diphosphatidylglycerol; EPLAS, ethanolamine plasmalogen; PE, phosphatidylethanolamine; PS, phosphatidylserine; SM, sphingomyelin; U, uncharacterized phospholipid; LPC, lysophosphatidylcholine; PI, phosphatidylinositol; PC, phosphatidylcholine.

[a]Tissue profile also contains LPA at 0.91 ppm: beet, 0.3 %; eggplant, 0.3 %; horseradish, 0.6 %; jicama, 0.4 %; mushroom, field, 0.5 %; radish, black, 0.5 %; radish, red, 0.5 %.

[b]Two signals are present at chemical shifts of 0.72 and 0.68 ppm respectively: beet, 0.3 %, 0.3 %; onion, white, 2.3 %, 0.8 %; shallots, 0.8 %, 0.6 %.

[c]Eggplant also contains an unknown phospholipid at 0.46 ppm (0.9 %.

[d]Ginger also contains a group of 4 unknown phospholipids having the following respective chemical shifts (ppm) and mole percentages: -0.40, 3.3; -0.42, 2.5; -0.45, 8.9; -0.48, 3.2.

[e]Jerusalem artichoke also contains an unknown phospholipid at 0.31 ppm (0.7 %).

[f]Mushroom, wood ear, contains an unknown phospholipid at 0.22 ppm (3.5 %).

[g]Swiss chard also contains an unknown phospholipid at -0.19 ppm (1.0 %).

Table 2.5
Phospholipid profiles of algae [47].

Alga	Phospholipid (mol %)												
	LPA	GPLAS	PG	PA	DPG	EPLAS	PE	SM	U	LPC	PI	SPC	AAPC	PC
Gracilaria verrucosa	3.6	3.2	11.4	4.8	6.7		3.6		5.8	4.3	2.6	1.8		52.2
Bryothamnion triquetrum	2.8		25.3	6.1	3.9	2.7	4.3	3.2	3.0	1.2	4.2		4.5	38.8
Padina gymnospora			31.3	7.4	9.5	10.3	11.1	3.0		4.9	13.3		2.6	6.6
Caulerpa sertularioides			24.8	3.1	8.9			8.9		3.2	13.3			37.8

Abbreviations: LPA, lysophosphatidic acid; GPLAS, glycerol plasmalogen; PG, phosphatidylglycerol; PA, phosphatidic acid; DPG, diphosphatidylglycerol; EPLAS, ethanolamine plasmalogen; PE, phosphatidylethanolamine; SM, sphingomyelin; U, uncharacterized phospholipid; LPC, lysophosphatidylcholine; PI, phosphatidylinositol; SPC, sphingosylphosphorylcholine, AAPC, alkylacylphosphatidylcholine; PC, phosphatidylcholine.

Table 2.6
Membrane model parameters.

Test sample group (included sub-groups)	Number of cases	Index values (means + S.E.)						Residual (means ± S.E.)
		CEPHLEC	*DIANIONA*	*DIAN-CHO*	*INSIDE*	*PGLEAF1*	*LEAFLET*	
Healthy tissues (plants, animals, primitive animals, algae)	183	0.78 ± 0.08	11.11 ± 0.31	0.28 ± 0.03	32.00 ± 0.81	2.70 ± 0.40	2.49 ± 0.40	0.11 ± 0.26
Cancers	18	0.41 ± 0.04	12.80 ± 1.50	0.23 ± 0.04	25.30 ± 2.37	3.38 ± 0.76	3.29 ± 0.73	0.22 ± 0.20
All tissues	201	0.75 ± 0.08	11.24 ± 0.31	0.28 ± 0.03	31.50 ± 0.78	2.75 ± 0.37	2.55 ± 0.37	0.11 ± 0.24

For the residual, no significant differences were observed at the $P < 0.05$ level.

ACKNOWLEDGEMENT

Supported by NIH ES05987.

ABBREVIATIONS

AAPC, alkylacylphosphatidylcholine; AAPE, alkylacylphosphatidyl-ethanolamine; CER-AEP, ceramide-(2-(tetramethylammonium-ethyl))phosphonates; DAG-AEP, diacylglyceryl-(2-aminoethyl)phosphonates; DHSM, dihydrosphingomyelin; DiMePE, dimethylphosphatidylethanolamine; PIP_2, phosphatidylinositolbisphosphate (1,2-diacyl-*sn*-glycero-3-phospho(1-*D*-myoinositol 4,5-bisphosphate)); DPG, diphosphatidylglycerol (cardiolipin); EDTA, (ethylenedinitrilo)-tetraacetic acid; EPLAS, ethanolamine plasmalogen; GPLAS, glycerol plasmalogen; LAAPC, lysoalkylacylphosphatidylcholine; LDPG, lysodiphosphatidylglycerol; LEPLAS, lysoethanolamine plasmalogen; LPA, lysophosphatidic acid; LPC, lysophosphatidylcholine; LPE, lysophosphatidylethanolamine; LPG, lysophosphatidylglycerol; LPI, lysophosphatidylinositol; NH2SM, ethanolamine sphingomyelin; PA, phosphatidic acid; PAF, platelet activating factor; PC, phosphatidylcholine; PE, phosphatidylethanolamine; PG, phosphatidylglycerol; PI, phosphatidylinositol; PIP, phosphatidylinositol monophosphate (L-α-phosphatidylinositol 4-monophosphate); PM, phosphatidylmethanol; PS, phosphatidylserine; PSM, phytosphingomyelin; SM, sphingomyelin; SPC, sphingosylphosphorylcholine, U, uncharacterized phospholipid

REFERENCES

1. Bárány,M. and Glonek,T., in *Phosphorus-31 NMR: Principles and Applications*, pp. 511-545 (1984) (edited by D. Gorenstein, Academic Press Inc., New York).
2. Bárány,M. and Glonek,T., in *Methods in Enzymology*, Vol. 85B, pp. 624-676 (1982) (edited by D.L. Frederiksen and L.W. Cunningham, Academic Press Inc., New York).
3. Bardygula-Nonn,L.G. and Glonek,T., *The 4th International Zebra Mussel Conference*, Madison, 1994.
4. Bardygula-Nonn,L.G., Iwata,J.L. and Glonek,T., with permission.
5. Burt,T.C., Glonek,T. and Bárány,M., *J. Biol. Chem.*, **251**, 2584-2591 (1976).
6. Byrdwell,W.C., Borchman,D., Porter,R.A., Taylor,K.G. and Yappert,M.C., *Invest. Ophthalmol. Vis. Sci.*, **35**, 4333-4343 (1994).
7. Capuani,G., Aureli,T., Miccheli,A., Di Cocco,M.E., Ramacci,M.T. and Delfini,M., *Lipids*, **27**, 389-391 (1992).
8. Cullis,P.R., Hope,M.J., de Kruijff,B., Verkleij,A.J. and Tilcock,C.P.S., in *Phospholipids and Cellular Regulation*, Vol. 1, pp. 1-59 (1985) (edited by J.F. Kuo, CRC Press, Boca Raton).
9. Dawson,R.M.C., *Biochem. J.*, **102**, 205-210 (1967).
10. de Vrije,T., de Swart,R.L., Dowhan,W., Tommassen,J. and de Kruijff,B., *Nature*, **334**, 173-175 (1988).
11. Driscoll,D.M. and Glonek,T., *J. Am. Osteopath. Ass.*, **92**, 1184-EOA (1992).
12. Driscoll,D.M., Ennis,W., Meneses,P., *Int. J. Biochem.*, **26**, 759-767 (1994).
13. Edelstein,C., Scanu,A.M. and Glonek,T., with permission.
14. Edzes,H.T., Teerlink,T., van der Knapp,M.S. and Valk,J., *Magn. Reson. Med.*, **26**, 46-59 (1992).
15. Entenman,C., *J. Am. Oil Chem. Soc.*, **38**, 534-538 (1961).
16. Folch,J., Lees,M. and Sloane Stanley,G.H., *J. Biol. Chem.*, **226**, 497-509 (1957).
17. Glonek,T., Greiner,J.V. and Lass,J.H., in *NMR: Principles and Applications to Biomedical Research*, pp. 157-203 (1990) (edited by J.W. Pettegrew, Springer-Verlag, New York).
18. Glonek,T., Henderson,T.O., Hilderbrand,R.L. and Myers,T.C., *Science*, **169**, 192-194 (1970).
19. Glonek,T., in *P-31 NMR Spectral Properties in Compound Characterization and Structural Analysis*, Chapt. 22, pp. 283-294 (1994) (edited by L.D. Quin and J.G. Verkade, VCH Publishers, New York).
20. Greiner,C.A.M., Greiner,J.V., Hebert,E., Berthiaume,R.R. and Glonek,T., *Ophthalmic Res.*, **26**, 264-274 (1994).
21. Greiner,C.A.M., Greiner,J.V., Leahy,C.D., Auerback,D.B., Marcus,M.D., Davies,L.H., Rodriguez,W. and Glonek,T., *Int. J. Biochem. Cell Biol.*, **27**, 21-28 (1995).

22. Greiner,J.V. and Glonek,T., with permission.
23. Greiner,J.V., Auerbach,D.B., Leahy,C.D. and Glonek,T., *Invest. Ophthalmol. Vis. Sci.*, **35**, 3739-3746 (1994).
24. Greiner,J.V., Kopp,S.J. and Glonek,T., *Invest. Ophthalmol. Vis. Sci.*, **23**, 14-22 (1982).
25. Greiner,J.V., Kopp,S.J. and Glonek,T., *Survey Ophthalmol.*, **30**, 189-202 (1985).
26. Greiner,J.V., Kopp,S.J., Sanders,D.R. and Glonek,T., *Invest. Ophthalmol. Vis. Sci.*, **21**, 700-713 (1981).
27. Greiner,J.V., Kopp,S.J., Sanders,D.R. and Glonek,T., *Invest. Ophthalmol. Vis. Sci.*, **22**, 613-624 (1982).
28. Greiner,J.V., Lass,J.H. and Glonek,T., *Arch. Ophthalmol.*, **102**, 1171-1173 (1984).
29. Greiner,J.V., Lass,J.H. and Glonek,T., in *Corneal Surgery, Theory, Technique, and Tissue*, pp. 637-644 (1986) (edited by F.S. Brightbill, C.V. Mosby Company, St. Louis).
30. Greiner,J.V., Leahy,C.D. and Glonek,T., *Invest. Ophthalmol. Vis. Sci.*, **35**, 1941-EOA (1994).
31. Greiner,J.V., Leahy,C.D. and Glonek,T., *Ophthalmic Res.*, in press.
32. Henderson,T.O., Kruski,A.W., Davis,L.G., Glonek,T. and Scanu,A.M., *Biochemistry*, **14**, 1915-1920 (1975).
33. Iwata,J.L., Bardygula-Nonn,L.G., Glonek,T. and Greiner,J.V., *Curr. Eye Res.*, in press (1995).
34. Kasimos,J.N., Merchant,T.E., Gierke,L.W. and Glonek,T., *Can. Res.*, **50**, 527-532 (1990).
35. Klunk,W.E., Xu,C.-J., Panchalingam,K., McClure,R.J. and Pettegrew,J.W., *Neurobiology of Aging*, **15**, 133-140 (1994).
36. Kolesnik,R., *Trends Cell Biol.*, **2**, 232-236.
37. Kuo,J.F. (Editor), *Phospholipids and Cellular Regulation*, Vols. 1 and 2, CRC Press Inc., Boca Raton, 1985.
38. Lass,J.H., Greiner,J.V. and Glonek, T., in *Corneal Surgery, Theory, Technique, and Tissue*, pp. 650-657 (1986) (edited by F.S. Brightbill, C.V. Mosby Company, St. Louis).
39. Lass,J.H., Greiner,J.V., Medcalf,S.K., Kralik,M.R., Meneses,P. and Glonek,T., *Ophthalmic Res.*, **20**, 368-375 (1988).
40. Lass,J.H., Greiner,J.V., Reinhart,W.J., Medcalf,S.K. and Glonek,T., *Cornea*, **10**, 346-353 (1991).
41. Leahy,C.D., Greiner,J.V., Davies,L. and Glonek,T., *Invest. Ophthalmol. Vis. Sci.*, **35**, 1466-EOA (1994).
42. Liang,M.T.C., Glonek,T., Meneses,P., Kopp,S.J., Paulson,D.J., Gierke,L.W. and Schwartz,F.N., *Int. J. Sports Med.*, **13**, 417-423 (1992).
43. Liang,M.T.C., Meneses,P., Glonek,T., Kopp,S.J., Paulson,D.J., Schwartz,F.N. and Gierke,L.W., *Int. J. Biochem.*, **25**, 337-347 (1993).
44. London,E. and Feigenson,G.W., *J. Lipid Res.*, **20**, 408-412 (1979).
45. Marsh,D., *CRC Handbook of Lipid Bilayers*, pp. 81-85 (1990) (CRC Press, Boca Raton).
46. Meneses,P. and Navarro,J.N., *Comp. Biochem. Physiol.*, **102B**, 403-407 (1992).
47. Meneses,P. and Navarro,J.N., *Phosphor. Sulfur Silic.*, **51/52**, 403-EOA (1990).
48. Meneses,P., Glonek,T., *J. Lipid Res.*, **29**, 679-689 (1988).
49. Meneses,P., Greiner,J.V. and Glonek,T., *Exp. Eye Res.*, **50**, 235-240 (1990).
50. Meneses,P., Navarro,J.N. and Glonek,T., *Int. J. Biochem.*, **25**, 903-910 (1993).
51. Meneses,P., Para,P. and Glonek,T., *J. Lipid Res.*, **30**, 458-461 (1989).
52. Merchant,T.E., Alfieri,A.A., Glonek,T. and Koutcher,J.A., *Radiation Res.*, in press (1995).
53. Merchant,T.E. and Glonek,T., *J. Lipid Res.*, **31**, 479-486 (1990).
54. Merchant,T.E. and Glonek,T., *Lipids*, **27**, 551-559 (1992).
55. Merchant,T.E. and Glonek,T., with permission.
56. Merchant,T.E., de Graaf,P.W., Minsky,B.D., Obertop,H. and Glonek,T., *NMR Biomed.*, **6**, 187-193 (1993).
57. Merchant,T.E., Diamantis,P.M., Lauwers,G., Haida,T., Kasimos,J.N., Guillem,J., Glonek,T. and Minsky,B.D., *Cancer*, in press (1995).
58. Merchant,T.E., Gierke,L.W., Meneses,P. and Glonek,T., *Can. Res.*, **48**, 5112-5118 (1988).
59. Merchant,T.E., Kasimos,J.N., de Graaf,P.W., Minsky, B.D., Gierke,L.W. and Glonek,T., *Int. J. Colorect. Dis.*, **6**, 121-126 (1991).
60. Merchant,T.E., Lass,J.H., Meneses,P., Greiner,J.V. and GlonekT., *Invest. Ophthalmol. Vis. Sci.*, **32**, 549-555 (1991).
61. Merchant,T.E., Lass,J.H., Roat,M.I., Skelnik,D.L. and Glonek,T., *Curr. Eye Res.*, **9**, 1167-1176 (1990).

62. Merchant,T.E., Meneses,P., Gierke,L.W., Den Otter, W. and Glonek,T., *Br. J. Cancer*, **63**, 693-698 (1991).
63. Merchant,T.E., van der Ven,L.T.M., Minsky,B.D., Diamantis,P.M., Delapaz,R., Galicich,J. and Glonek,T., *Brain Res.*, **649**, 1-6 (1994).
64. Norušis,M.J., *The SPSS Guide to Data Analysis for SPSS/PC+*, 2nd Ed., pp. 148-416, SPSS Inc., Chicago, 1991.
65. Panchalingam,K., Sachedina,S., Pettegrew,J.W. and Glonek,T., *Int. J. Biochem.*, **23**, 1453-1469 (1991).
66. Pearce,J.M., and Komoroski,R.A., *Magn. Reson. Med.*, **29**, 724-731 (1993).
67. Pearce,J.M., Shifman,M.A., Pappas,A.A. and Komoroski,R.A., *Magn. Reson. Med.*, **21**, 107-116 (1991).
68. Romano,F. and Driscoll,D.M., with permission.
69. Rothman,J.E. and Lenard,J., *Science*, **195**, 743-753 (1977).
70. Sachedina,S., Greiner,J.V. and Glonek,T., *Exp. Eye Res.*, **52**, 253-260 (1991).
71. Sachedina,S., Greiner,J.V. and Glonek,T., *Invest. Ophthalmol. Vis. Sci.*, **32**, 625-632 (1991).
72. Sappey Marinier,D., Letoublon,R. and Delmau,J., *J. Lipid Res.*, **29**, 1237-1243 (1988).
73. Seijo,L., Merchant,T.E., van der Ven,L.T.M., Minsky,B.D. and Glonek,T., *Lipids*, **29**, 359-364 (1994).
74. Singh,H. and Privett,O.S., *Lipids*, **5**, 692-697 (1970).
75. Van Wazer,J.R. and Glonek T., in *Analytical Chemistry of Phosphorus Compounds*, pp. 151-188 (1972) (edited by M. Hallmann, John Wiley & Sons Inc., New York).
76. Van Wazer,J.R., *Phosphorus and Its Compounds*, Vol. 1, Interscience, New York, 1961.
77. Watson,J.D., oft delivered from the podium.
78. Xiang-Xi,X. and Tabas,I., *J. Biol. Chem.*, **266**, 24849-24858 (1991).
79. Yang,S.F., Freer,S. and Benson,A.A., *J. Biol. Chem.*, **242**, 477-484 (1967).

Chapter 3

SEPARATION OF PHOSPHOLIPID CLASSES BY HIGH-PERFORMANCE LIQUID CHROMATOGRAPHY

William W. Christie

The Scottish Crop Research Institute, Invergowrie, Dundee, Scotland DD2 5DA

A. INTRODUCTION

Although thin-layer chromatography was for many years the most convenient and widely used method for separation and analysis of phospholipid classes, high-performance liquid chromatography (HPLC) has rapidly supplanted it, especially for the common range of phospholipid classes found in animal tissues. HPLC is much more expensive in terms of both equipment and running costs, but can be automated to a considerable degree and gives much cleaner fractions in micro-preparative applications. Adsorption ("normal-" or "straight-phase") chromatography with silica gel has been used as the stationary phase in most instances, although bonded phases are finding increasing applications. Whilst partial separation of molecular species of phospholipids can sometimes be effected during class separation, this is rarely of analytical value and indeed can be a nuisance. Separation of molecular species of lipids is best accomplished by reversed-phase chromatography of distinct phospholipid classes isolated by the methods described below.

The other important component of the chromatographic process is the mobile phase. Regretfully, the composition of this may be restrained severely by the type of detector available to the analyst. This has governed the development of the technique and still has an appreciable influence on the approach to problems of analysis, as most lipids lack chromophores that would facilitate spectrophotometric detection. For example, detection in the ultraviolet range between 200 and 210 nm (for isolated double bonds and 'end-absoption') was used in some of the first published HPLC separations of phospholipids and it is still widely used. Differential refractometry has been employed to a lesser extent as it can only be utilized with isocratic elution. More recently, transport-flame ionization detectors and those that operate on the evaporative light-scattering principle have proved their worth, and these "mass" and other "universal" detectors are now being used in increasing numbers of laboratories. The literature to 1987 on this subject was reviewed elsewhere by the author [28], while the properties of detectors, especially those operating on the evaporative light-scattering principle, have been reviewed more recently [29]. The reader is referred to these reviews for detailed discussion of such vital components of the chromatographic system, although the topic will be touched on frequently below, especially in relation to quantification (Section G). However, mass spectrometric (MS) detection is such a specialised and distinctive methodology that I cannot do justice to it here; it is the subject of reviews elsewhere [41,77,97]. It is also a topic that is in a state of flux as new types of HPLC-MS interface are becoming available.

It is worth noting that analysts do not always require a perfect analytical system that resolves every single lipid class in a sample. For example, if the interest is in the properties of a specific phospholipid class, it may simply be necessary to optimise an elution scheme so that the compound of interest is isolated in a relatively pure state; resolution of other components can be ignored. However, an ideal analytical system should give sharp well-resolved peaks for all the important phospholipids in tissue extracts, especially the acidic and choline-containing components, and it should be stable and reproducible for months in continuous use. Ideally, it should be adaptable to the simultaneous analysis of simple lipids and of glycolipids in addition to phospholipids (see below).

Nowadays, it is a relatively easy matter to separate the common choline- and ethanolamine-containing phospholipids by HPLC, but acidic lipids such as phosphatidic acid, phosphatidylinositol and phosphatidylserine can still present difficulties. Relatively little work has been done on HPLC analysis of the phospholipids (and other complex lipids) in plants and microorganisms. As analysts tackle these areas, they are likely to encounter many more problems.

This review is not intended as a comprehensive historical account of the separation of phospholipids by HPLC, nor will recipes of useful practical systems be published, as this was the approach in an earlier book by this author [28]. Rather, the intention here has been to present a critical discussion of the principles that govern such separations, especially the concept of "selectivity" (see Section B2), in the hope that it will inspire further research. Many published procedures have been devised in a rather *ad hoc* manner, rather than by employing chemometric methodology (reviewed by Kaufmann [72]). Although this approach cannot be described in detail again here, it is recommended to anyone interested in developing novel separations.

B. FACTORS INFLUENCING THE CHROMATOGRAPHI SEPARATION

1. Stationary Phases

In HPLC applications, the silica gel used as a stationary phase consists of porous spheres (3 to 10 μm in diameter) with a surface area inversely related to the size of the pores, 80 to 120Å being the standard in analytical applications. The adsorptive properties are due to silanol groups (or the hydroxyl moieties of these) on the surface that are free or hydrogen-bonded. In addition, water of hydration plays a part and this exists first in a strongly bound layer and then in one or more loosely bound layers on the surface. The loosely bound water-layer can have a marked effect on the repeatability of separations, as it is readily removed in an irreproducible manner by elution with dry solvents. Although this is not usually a problem in the analysis of polar lipids, it can be troublesome if simple lipids are to be separated simultaneously. Silica gel by its nature presents a rather heterogeneous surface to an analyte, and control of its properties requires some skill and experience on the part of the analyst.

A further constraint on the use of silica gel is its stability to acid and base. With mobile phases containing water at a pH of less than 2 or greater than 7.5, the surface of the silica will slowly dissolve. Above a pH of 8.5 the process can be quite rapid and chromatographic resolution will quickly deteriorate. The effects can be ameliorated by passing the solvent through a silica pre-column, but this is rarely compatible with gradient elution.

In order to provide a more uniform adsorbent surface therefore, commercial suppliers have introduced several stationary phases with organic moieties bonded chemically to silica gel. Those with diol, nitrile, amine and sulphonic acid residues as the functional group are of particular value, and some interesting applications to phospholipid class separations have been published. Amino phases are especially useful in that the selectivity is altered dramatically so that choline-containing lipids are eluted first. However, acidic lipids such as phosphatidylinositol and phosphatidylserine are very strongly retained. Columns containing bonded phenylsulphonic acid have been used less widely, but they do not appear to require counter ions in the mobile phase thus simplifying the isolation of pure components. Bonded propanediol, polymeric vinyl alcohol and cyclodextrin phases give a type of separation closer to that obtained with silica gel. On the other hand, the more uniform surface means a more rapid equilibration with the mobile phase and elution with less polar solvents than is possible with silica gel. Propylnitrile bonded phases offer yet another change in the selectivity of elution in that the acidic phospholipids especially tend to elute earlier than from silica gel. Practical examples of all of these are listed below (Section D).

With most chemically bonded-phases, as with silica gel *per se*, it is important to maintain the pH of the mobile phase in the range 4 to 7.5, otherwise the bonds between the bonded moieties and the silica gel may be hydrolysed and the organic layer stripped from the silica surface.

2. *Solvent Selectivity in Mobile Phases*

Analysts usually aim to effect a separation such that each component is eluted as a single distinct band in the chromatogram. To achieve this it is necessary to optimise the "selectivity" of the mobile phase, *i.e.* the capacity to affect relative retention or spacing of analytes and even the order of elution. Snyder and co-workers, in particular, have studied the properties of solvents used for chromatographic separations in a systematic manner and have formulated a "solvatochromically based solvent-selectivity triangle" to describe related groups [119-121]. The properties considered to be of special importance are acidity (*e.g.* ethanol), basicity (*e.g.* dioxane) and dipolarity (*e.g.* nitromethane). No two solvents will be identical in all of these properties, but they can be grouped according to overall similarities. Thus, aliphatic ethers and amines fall into one group, aliphatic alcohols in a second, tetrahydrofuran and amides in a third, acetic acid in a fourth, dichloromethane and dichloroethane in a fifth, ketones in a sixth, aromatic hydrocarbons in a seventh, and water and chloroform in an eighth (aliphatic hydrocarbons fall out with the scheme).

These classifications were not arrived at with separations of lipids in mind, and it is not possible to predict effects on the elution of lipids in chromatographic systems from first principles. Any effects will depend on the mode of chromatography (adsorption, reversed-phase, ion exchange, *etc.*) and the precise nature of the lipids. Indeed, there is a need for systematic studies of the influence of Snyder's selectivity groups on lipid separations. There is a body of knowledge from trial and error over many years on what solvent combinations are of particular value, and Kaufmann *et al.* have published valuable data on some solvents and solvent mixtures used widely in lipid research [74]. However, an understanding of Snyder's solvent groups can be very helpful in developing novel separations, and the concept of selectivity is useful in discussing different published elution schemes.

In order to overcome detection difficulties at UV wavelengths, two basic solvent systems transparent in the range of 200 to 210 nm were developed as mobile phases for HPLC separation of phospholipids in 1976-77, and these still find almost universal application today, *i.e.* hexane-propan-2-ol-water and acetonitrile-water (sometimes with added methanol) mixtures. With acetonitrile-methanol-water (61:21:4 by volume) and silica gel as stationary phase, phosphatidylethanolamine elutes before phosphatidylcholine and then sphingomyelin, and indeed all the choline-containing phospholipids are well resolved [68]. An additional virtue is that the acidic lipids are eluted with relative ease, ahead of phosphatidylethanolamine.

Similarly, mobile phases based on hexane-propan-2-ol-water have been used with silica gel in very many laboratories since their introduction [44,54]. Phosphatidylethanolamine elutes before phosphatidylcholine in this instance also, but the latter and the other choline-containing lipids, such as sphingomyelin and lysophosphatidylcholine, tend to be less well resolved. The acidic lipids, including phosphatidic acid, are separated from each other, but now they emerge between phosphatidylethanolamine and phosphatidylcholine. By adding further solvents to the basic mixture to modify the selectivity or by using gradients, it has been possible to improve the resolution of the choline-containing lipids. This type of mobile phase has proved easier to adapt to simultaneous separation of simple lipids and glycolipids than has that based on acetonitrile (see Section E). Table 3.1 summarizes these results. However, it should be noted that the pH of the mobile phase and the ionic strength of the aqueous component can change the order of elution appreciably (see next section, for example).

The only published mobile phases to differ substantially from the two basic systems above are based on chloroform-methanol-water usually with ammonia added as a key component. This appears to give a similar order of elution to hexane-propan-2-ol-water mixtures, but the acidic phospholipids emerge as especially sharp peaks. It seems clear that the presence of ammonia is a key to the sharpness of the peaks, but this may also have disadvantages (see next section). Privett and co-workers appear to have been the first to use elution systems of this type [40,105-107,122].

The selectivities of the solvents used in the mobile phase can exert marked effects on the separation of individual phospholipids, and in particular they can change the order of elution of specific components. While other standard solvent mixtures for mobile phases will no doubt be devised in time, most work at present

Table 3.1
The order of elution of phospholipids on HPLC with silica gel as stationary phase and mobile phases based on either acetonitrile or propan-2-ol.*

Acetonitrile-based	Propan-2-ol-based
phosphatidic acid	cardiolipin
cardiolipin	phosphatidylethanolamine
phosphatidylinositol	phosphatidylinositol
phosphatidylserine	phosphatidylserine
phosphatidylethanolamine	phosphatidic acid
phosphatidylcholine	phosphatidylcholine
sphingomyelin	sphingomyelin
lysophosphatidylcholine	lysophosphatidylcholine

*There may be some modification to this order (especially of cardiolipin), depending on the nature of the other solvents and of any ionic species in the mobile phase.

seems to be concentrated on the interaction of the two widely used systems described above (with minor modifications and including ionic species discussed in the following section) with the newer stationary phases with chemically bonded functional groups. The advent of light-scattering detectors has opened up many new opportunities, since the primary limitation is now the volatility of the components of the mobile phase. There are many different solvent types - ethers, alcohols, ketones, aromatic, and halogen- or other hetero-atom-containing compounds that should be tried. Appreciable potential must remain to change the selectivity of separations yet further and to enhance the opportunities for isolation of specific phospholipid components.

3. *Addition of Ionic Species to the Mobile Phase*

Phospholipids are ionic molecules and require a counter ion in solution. If lipid samples are obtained from tissues by a conventional chloroform-methanol extraction, including a wash with sodium chloride (saline) solution, sodium will be the main counter ion. Special precautions are only necessary in rare circumstances to ensure that small amounts of other ions do not remain, although different ionic forms of acidic phospholipids may have different mobilities on chromatography. The practical method for avoiding such difficulties during HPLC is to add counter ions or acids or bases to the mobile phase. Inorganic species have been used most frequently with spectrophotometric detection, often at relatively high concentrations, while organic ions at lower levels are necessary with evaporative light-scattering detectors. The nature of the ionic species (especially the pH) and its concentration may have a marked effect on the selectivity of the separation.

If ionic species are not added to the mobile phase, there is a steady accumulation of polar lipids on the surface (and within the pores?) of the stationary phase. Excellent resolution will often be achieved with a new column straight from the manufacturer, but separations will deteriorate steadily with time, usually a matter of weeks. It is the author's opinion that any published methods for separation of phospholipids that do not utilise ionic species have simply not been tested over a sufficiently long period. Fortunately, columns that have begun apparently to degrade in this way are rapidly regenerated on elution with mobile phases containing ions. Some workers were reportedly able to restore the activity of silica gel columns after every 15 to 30 runs by eluting with methanol at above ambient temperatures [38].

Provided that the longer term disadvantages are recognised, methods that avoid the use of ionic species in the mobile phase may be useful for the preparative-scale isolation of highly pure lipids (see Section F).

Sulphuric [128] and phosphoric [24] acids have often been used as counter ions in mobile phases, but in addition to the substantial drawback of dissolving HPLC equipment they bring about complete destruction of any plasmalogens present in animal or microbial lipids [12,24,60,69,114,115,128]. Lysophospholipids, and lysophosphatidylethanolamine especially, then appear in spuriously large amounts. Although this problem was reported in the original publications, it still seems to take some analysts by surprise. When sulphuric acid is used in the mobile phase, it appears to stick to the column for a considerable time after changing to neutral solvents, continuing to exert harmful effects on plasmalogens (Author, unpublished). The same is true of phosphoric acid [111]. With plant lipid extracts where plasmalogens do not occur, strong acids in mobile phases present less of a problem.

Trifluoroacetic acid has been employed to acidify a mobile phase and has the virtue of being sufficiently volatile to be used with an evaporative light-scattering detector [15]. No information on its effect with plasmalogens was recorded. Relative to phosphoric acid, it appeared to change the order of elution of acidic phospholipids on a diol column at least [116]. Acetic acid has also been utilised in mobile phases both with silica [13,98,99] and diol columns [57,75], and seems to have no deleterious effects either on the lipids or on the stationary phase.

At the opposite extreme, ammonia has been much used in mobile phases, especially with chloroform-methanol mixtures. As discussed earlier, silica gel will dissolve rapidly at a pH value of 8.5 or more. The same type of problem tends to arise with chemically bonded-phases, where the organic moiety can be quickly stripped from the column as discussed above. Use of a "silica-saturation" pre-column may prolong the life of the analytical column, and indeed it has been suggested that this may obviate the requirement for ionic species in the mobile phase [88], but I have doubts about this.

Strong acids or bases are simply not necessary for HPLC analysis of phospholipids and neutral ionic species are perfectly satisfactory. For example, a phosphate buffer (25 mM, pH 7.0), together with propan-2-ol and hexane, gave excellent results in a method described by Patton *et al.* [103,104], that has been adopted in many laboratories (see Section C1). Phosphate buffers were also used in very different elution schemes with aminopropyl stationary phases [1,17,43,46,117]. However, such buffers cannot be used with evaporative light-scattering detection. Acetate buffers have been used similarly, usually in conjunction with UV detection, with silica [7,50,98,99], aminopropyl [65,83] and cyanopropyl [113] columns. The author has used serine (0.5 mM, pH 7.5) [27] and triethylamine/acetate buffers [32], which are relatively innocuous, and appear to give well-shaped peaks with difficult analytes such as phosphatidylserine and phosphatidylinositol. As the ionic species in the mobile phase were at rather low concentrations and were organic in nature, it was still possible to use evaporative light-scattering detection.

Ikeda *et al.* [63] carried out systematic studies of the effect of various neutral salts (lithium, potassium, calcium and magnesium chlorides) in hexane-propan-2-ol-water on separation of selected phospholipids, and they obtained substantial effects on retention times, especially with the divalent cations. There must be scope to test a greater range of ionic species, both organic and inorganic, at controlled pH values for their suitability as counter ions and better conditions will surely be found. There may also be opportunities to change the selectivity of the mobile phase by such means to effect specific separations, especially of the acidic phospholipids.

There have been hints that elevated column temperatures may reduce the requirement for ionic species in the mobile phase, but there are few systematic studies. However, in a chemometric optimisation with a diol stationary phase, the best results were obtained at a column temperature of 75°C [5].

C. HPLC SEPARATION OF PHOSPHOLIPIDS ON SILICA GEL

1. Elution Schemes Based on Hexane-Propan-2-ol-Water

UV-transparent hexane-propan-2-ol-water mixtures were first utilised with columns of silica gel in gradient elution schemes. In the first published application [44,54], these solvents were present initially in the mobile phase in the ratio 6:8:0.75 (by volume), and they were changed by a linear programme to 6:8:1.4. The acidic phospholipids eluted between phosphatidylethanolamine and phosphatidylcholine. Many other workers have used this or closely related systems to separate, isolate and quantify (by a variety of methods) the main phospholipid components from tissue extracts of both animal and plant origin. Often minor modifications have been made to suite local conditions or specific brands of adsorbent, and it is not possible to describe these comprehensively here. In one systematic study [110] of how many different phospholipids (and their monoacyl forms) eluted with differing proportions of hexane, propan-2-ol and water (but no ionic species) in the mobile phase, it was concluded that one limiting factor would always be the inherent heterogeneity of the acyl group compositions in natural lipids, since this tended to cause broadening of peaks. However, in my opinion, this need not be troublesome if ionic species are present in the solvents.

Sulphuric acid was reported to be an excellent additive in that it gave very sharp peaks even with acidic phospholipids, although there are compelling reasons for seeking less troublesome mobile phases (see Section B3 above). It appeared to bring about a marked change in selectivity in that it caused the acidic lipids to

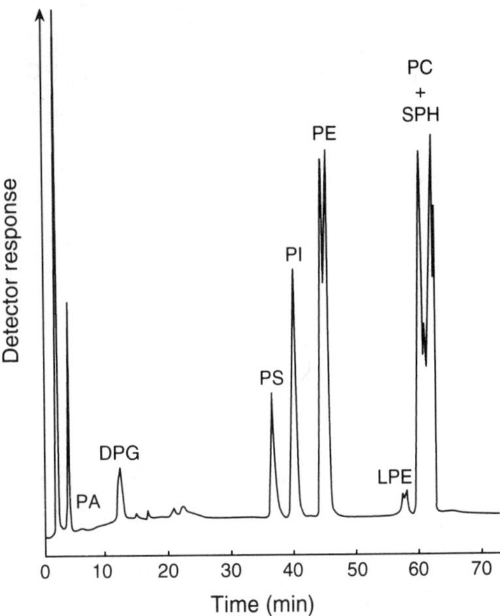

Fig. 3.1. Gradient elution of phospholipids from silica gel with hexane-propan-2-ol-water-sulphuric acid mixtures as the mobile phase, and detection at 205 nm [128]. (Reproduced by kind permission of the authors and of *Analytical Biochemistry*, and redrawn from the original publication).

Abbreviations; NL, neutral lipids; CE, cholesterol esters; TG, triacylglycerols; C, cholesterol; PG, phosphatidylglycerol; DPG, diphosphatidylglycerol; PE, phosphatidylethanolamine; LPE, lysophosphatidylethanolamine; PA, phosphatidic acid; PI, phosphatidylinositol; PS, phosphatidylserine; PC, phosphatidylcholine; SPH, sphingomyelin; LPC, lysophosphatidylcholine; CMH, ceramidemonohexoside.

elute ahead of phosphatidylethanolamine in the order phosphatidic acid > diphosphatidylglycerol > phosphatidylserine > phosphatidylinositol as shown in Figure 3.1 [128], *i.e.* the same as that obtained with acetonitrile-water-sulphuric acid mixtures (see next section).

Good results have been achieved with much less corrosive ionic species. For example, with a column of Partisil™ silica gel and a similar solvent gradient to the above, but with a 1 mM acetate buffer (pH 6.0) as an ion-suppressant in the aqueous component, good resolution of isotopically labelled phospholipids (containing ^{3}H-arachidonic acid) was achieved [50]. In this instance, aliquots of the eluent were removed for liquid-scintillation counting and also for detection. The order of elution of individual phospholipid classes was as listed in Table 3.1. Similarly, a small amount of acetic acid (0.005%) was incorporated into the aqueous component of the mobile phase to assist in order to isolate platelet-activating factor (1-*O*-alkyl-2-acetyl-*sn*-glycero-3-phosphorylcholine) especially, although good separations of the other phospholipids were obtained at the same time [13]. (This was also accomplished with a complex gradient mixture of aromatic solvents, propanol, water and acetic acid [2]). A related isocratic elution scheme has proved

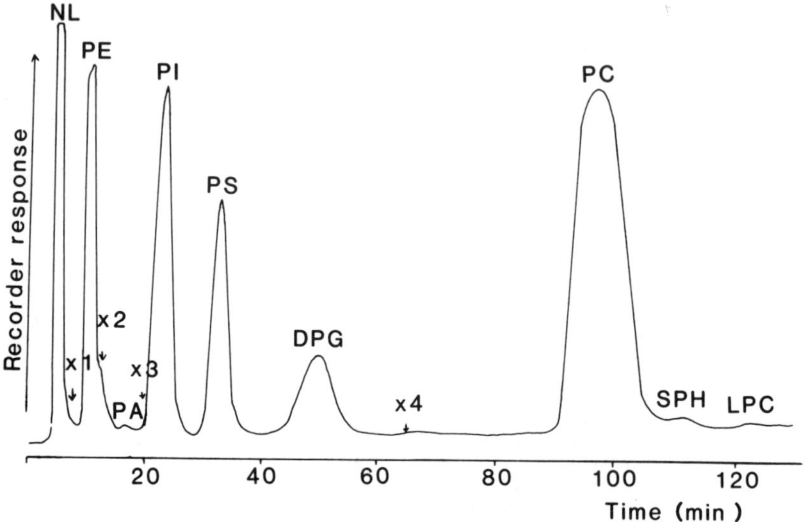

Fig. 3.2. Isocratic elution of rat liver phospholipids from a column of silica gel with hexane-propan-2-ol-25 mM phosphate buffer-ethanol-acetic acid (367:490:62:100:0.6 by volume) as mobile phase at a flow-rate of 0.5 mL/min for the first 60 minutes then of 1 mL/min, and with spectrophotometric detection at 205 nm [103]. (Reproduced by kind permission of the authors and of the *Journal of Lipid Research*, and redrawn from the original publication). The legend to Figure 3.1 contains a list of abbreviations. X1, X2, X3 and X4 are unknown phospholipids.

of value for the analysis of plant "lecithins", *i.e.* a column of Lichrosorb™ SI 60 silica gel, maintained at 30°C, and eluted with hexane-propan-2-ol-acetate buffer (pH 5.8) (8:8:1 by volume) as the mobile phase. In this instance phosphatidylethanolamine, phosphatidylcholine, phosphatidylinositol and phosphatidic acid were eluted in sequence and were partially resolved [7,98,99]. With hexane-propan-2-ol-0.2M acetic acid (8:8:1 by volume) as the mobile phase, the acidic lipids eluted with phosphatidylethanolamine, and well ahead of phosphatidylcholine.

Of the large number of procedures of this kind described, that of Patton and co-workers [103,104] is particularly convincing, and has been adopted by many others - always a recommendation. It also has the merit of employing isocratic elution, so reducing the requirements in terms of costly equipment. The nature of the separation is illustrated in Figure 3.2. Hexane-propan-2-ol-25 mM phosphate buffer-ethanol-acetic acid (367:490:62:100:0.6 by volume) was the mobile phase with silica gel as stationary phase, and detection was at 205 nm. With a lipid extract from rat liver, phosphatidylethanolamine eluted just after the neutral lipids, and was followed by each of the acidic lipids, *i.e.* phosphatidic acid, phosphatidylinositol and phosphatidylserine, then by diphosphatidylglycerol and the individual choline-containing phospholipids, with only phosphatidylcholine and sphingomyelin overlapping slightly. Although there was a fairly long elution time, most of the important phospholipid classes did emerge eventually. As each component was eluted, it was collected, washed to remove the buffer, and determined by phosphorus assay. In addition, the fatty acid composition of each lipid class was obtained with relative ease, by GLC analysis after transmethylation.

Mobile phases based on hexane-propan-2-ol-water have proved more useful than others as the basis of more comprehensive elution schemes designed to separate both simple and complex lipids in a single chromatographic run (see Section E).

Chloroform-propanol-acetic acid-water gradients gave remarkably good resolution of phospholipids, although the authors did not have a good detection system and the work appears to have been overlooked [14]. Phosphatidic acid and diphosphatidylglycerol eluted ahead of phosphatidylethanolamine, with the remaining phospholipids in the expected order.

2. Elution Schemes Based on Acetonitrile-Water

Also among the more popular elution systems are those based on acetonitrile-methanol-water or simple acetonitrile-water mixtures, transparent again in the 200 to 210 nm region of the spectrum. With such mobile phases, the acidic lipids elute ahead of phosphatidylethanolamine and phosphatidylcholine, and each of the choline-containing phospholipids is especially well resolved. A number of valuable isocratic separations have been described.

In the first published application by Jungalwala *et al.* in 1976 [68], a column of MicroPak™ SI-10 silica gel, elution with acetonitrile-methanol-water (61:21:14 by volume) at a flow-rate of 1 mL/min, and detection at 203 nm were employed. Phosphatidylethanolamine and phosphatidylserine emerged together near the solvent front, but phosphatidylcholine and sphingomyelin were particularly well resolved. Although this separation does not look impressive now (or even in relation to the earlier work by Privett's group), it represents a milestone in lipid chemistry since it enabled many more laboratories to adapt HPLC for lipid separations. Resolution of the choline-containing components is important in that the ratio of phosphatidylcholine and sphingomyelin in amniotic fluid is of special relevance as an indicator of fetal lung maturity, and a rapid quantitative procedure based on this mobile phase has been described [36].

As in the previous section, it is not possible to describe comprehensively every minor modification or application of the technique. However, a wider range of individual phospholipid classes were separated by isocratic elution with acetonitrile-methanol-water (50:45:6.5 by volume) at a flow-rate of 0.4 mL/min as in Figure 3.3 [70]. Phosphatidylinositol was eluted ahead of phosphatidylethanolamine and each of the choline-containing phospholipids. By means of a rather similar elution system, it was clearly established that both phosphatidylserine and phosphatidylinositol were eluted separately (in this order) ahead of phosphatidylethanolamine [18]. Phosphonolipids, such as 2-aminoethylphosphonate linked to ceramide or diacylglycerols, also elute ahead of phosphatidylethanolamine [118].

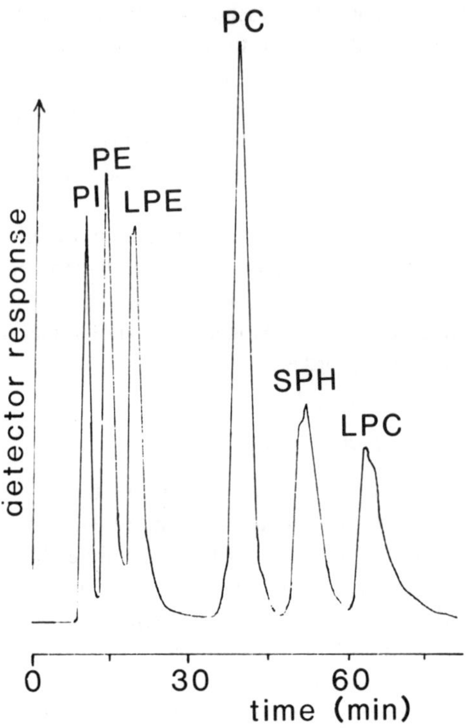

Fig. 3.3. Separation of a reference phospholipid mixture on a silica gel column, and by isocratic elution with acetonitrile-methanol-water (50:45:6.5 by volume) at a flow-rate of 0.4 mL/min. Detection was by phosphate analysis [70]. (Reproduced by kind permission of the authors and of *Analytical Chemistry*, and redrawn from the original publication). The legend to Figure 3.1 contains a list of abbreviations.

A more comprehensive separation of the acidic lipids made use of acetonitrile-methanol-sulphuric acid (100:2.1:0.05 by volume) as mobile phase, and enabled a separation of phosphatidic acid, cardiolipin, phosphatidylinositol and phosphatidylserine before phosphatidylethanolamine emerged [64]. Better results might be attained with gradient elution. Part of the selectivity here may have been conferred by the sulphuric acid component, as similar results have been obtained with hexane-propan-2-ol-sulphuric acid (see previous section) [128]. Sulphuric acid is not recommended as a mobile phase and better conditions to make use of the remarkable selectivity properties of acetonitrile-containing mixtures could surely be devised.

The use of strong acids in the mobile phase as ion suppressants has had a number of devotees, and for example, acetonitrile-methanol-85% phosphoric acid (130:5:1.5 by volume), as described by Chen and Kou [24], has been adopted in many laboratories. In this instance, a rather surprising result was a change in selectivity in that sphingomyelin eluted after rather than before lysophosphatidyl-

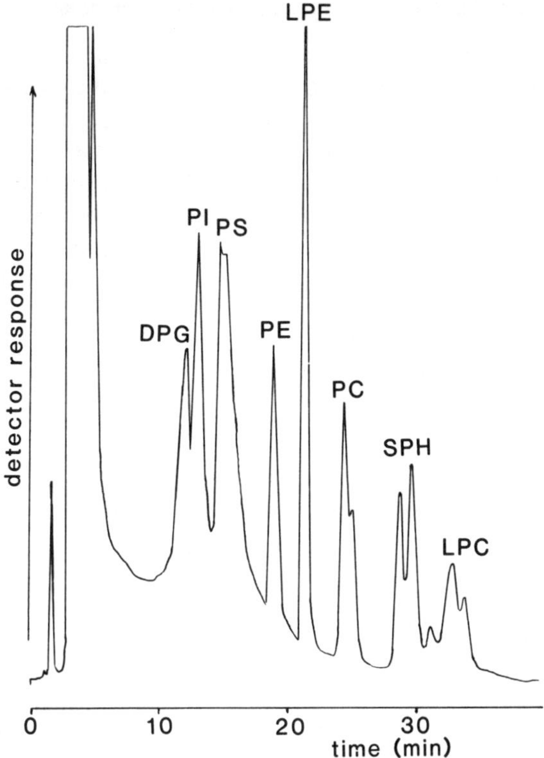

Fig. 3.4. HPLC separation of phospholipid standards on a column of silica gel, eluted with a gradient of water into acetonitrile, and with spectrophotometric detection at 203 nm [100,101]. (Reproduced by kind permission of the authors and of *Fette Seifen Anstrichm.*, and redrawn from the original publication). The legend to Figure 3.1 contains a list of abbreviations.

choline. The corrosive nature of the mobile phase is not always troublesome. For example, isocratic elution with acetonitrile-water-sulphuric acid (135:5:0.2 by volume) was used to give useful separations of both glycolipids and phospholipids of plant origin, that do not contain plasmalogens [86,87]. In addition, the excellent resolution of the choline-containing phospholipids has led to the use of a procedure of this type for the isolation of 1-*O*-alkyl-2-acetyl-*sn*-glycero-3-phosphorylcholine (platelet-activating factor) from lipid extracts; this lipid class tended to migrate just ahead of lysophosphatidylcholine and sphingomyelin [66,126].

One considerable advantage of methods in which acetonitrile-methanol-water are employed is that they lend themselves to isocratic elution so that less costly equipment can be used. Nonetheless, it is evident that gradient elution will give sharper peaks and resolve more components in relatively shorter times [100,101], as illustrated in Figure 3.4, for example, which can be compared with separation in the previous figure.

3. Elution Schemes Based on Chloroform-Methanol-Ammonia

Complicated gradients of chloroform, methanol and ammonia and columns of silica gel were used in the pioneering work of Privett and co-workers [40,105-107,122] discussed in Section E below. The selectivity of the separation was similar to that obtained with hexane-propan-2-ol-water mixtures, although phosphatidylinositol was reported to elute after phosphatidylcholine in one study [45]. That chloroform-methanol-ammonia mixtures do indeed give excellent resolution of acidic phospholipids, which can emerge as very sharp peaks is attested by more recent work [1,8,71,95]. However, the problem of column life with mobile phases of such high pH is seldom addressed.

Highly acidic phospholipids, such as phosphatidylinositol and the polyphosphoinositides, can probably only be separated at present by mobile phases containing ammonia (both with silica gel and amine phases), and in such circumstances a reduced column life can probably be tolerated [34,81,127].

D. HPLC SEPARATION OF PHOSPHOLIPIDS ON CHEMICALLY BONDED STATIONARY PHASES

1. Diol and Related Phases

It is possible to change the selectivity of separations by utilising a stationary phase other than silica gel. Many stationary phases with organic moieties bonded chemically to silica gel have been used in phospholipid separations, because these afford a more uniform surface that equilibrates relatively rapidly with the mobile phase. However, it should be remembered that there are usually some residual silanol groups present that exert an effect on resolution. That phase closest to silica in its properties is a "diol" phase, *i.e.* with propane diol groups bonded by some means to a silica backbone (the chemistry of the bonding process is usually a commercial secret).

Useful separations of phospholipids were first reported with a column of LiChrosorb™ DIOL, maintained at 35°C, and eluted with a gradient of water into acetonitrile at a flow-rate of 2 mL/min [16]. The method was later adopted by others for the analysis of phospholipids from amniotic fluid [55,56,80], or with phosphoric acid as the ionic species for cellular phospholipids [123]. Subsequently, it was shown that much cleaner separations could be obtained with such columns by using 0.005 M sodium dihydrogen phosphate buffer (pH 5) as the aqueous component; in this way, sharper peaks were obtained for the acidic phospholipid, as illustrated in Figure 3.5 [79]. In less than 20 minutes, phosphatidylglycerol, phosphatidylinositol, phosphatidylserine, phosphatidylethanolamine, phosphatidylcholine, sphingomyelin and lysophosphatidylcholine were clearly separated from each other. This is the same order of elution as with silica gel, and similarly the order expected from Table 3.1 was obtained with hexane-propan-2-ol-water mixtures as the mobile phase in one study [4]. However, it proved possible to change the selectivity of the column such that phosphatidylinositol was eluted after phos-

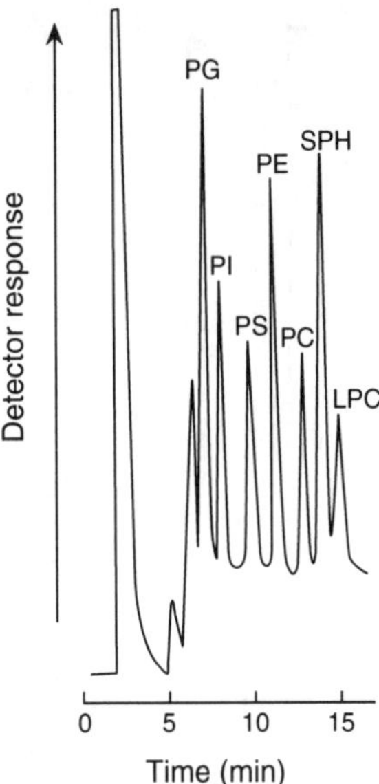

Fig. 3.5. HPLC separation of phospholipid standards on a column of LiChrosorb™ DIOL, eluted with a gradient of 0.005M sodium dihydrogen phosphate buffer (pH 5.0) into acetonitrile, and with spectrophotometric detection at 203 nm [79]. (Reproduced by kind permission of the authors and of *Journal of Chromatography*, and redrawn from the original publication). The legend to Figure 3.1 contains a list of abbreviations.

phatidylcholine by changing the relative proportions of the solvents and adding ammonium acetate as an ionic species [20]. This result was also obtained in the work of Herslöf *et al.* [57], who have published some of the more interesting separations with this type of column (see Section E2). Platelet activating factor (and its lyso-derivative) eluted after sphingomyelin with a mobile phase consisting of methyl*tert*butyl ether-methanol-water-ammonia [85].

In addition to the YMC PVA-Sil™ column of polymerized vinyl alcohol (discussed in Section E2), a related phase used for phospholipid separations consists of β-cyclodextrin bonded to silica. With a mobile phase of a hexane-propan-2-ol-ethanol-aqueous tetramethylammonium phosphate (5mM, pH 6.3), phosphatidylinositol and phosphatidic acid eluted after phosphatidylcholine [1].

2. Aminopropyl Phases

A very different selectivity can be obtained by utilising stationary phases which have a surface layer of amine groups bound covalently via a propyl spacer to a silica gel support. With such a stationary phase, the separation involves ion exchange mechanisms as well as adsorption. For example, with a column of Bondapak™-NH_2 and a mobile phase consisting of various chloroform-methanol-water mix-

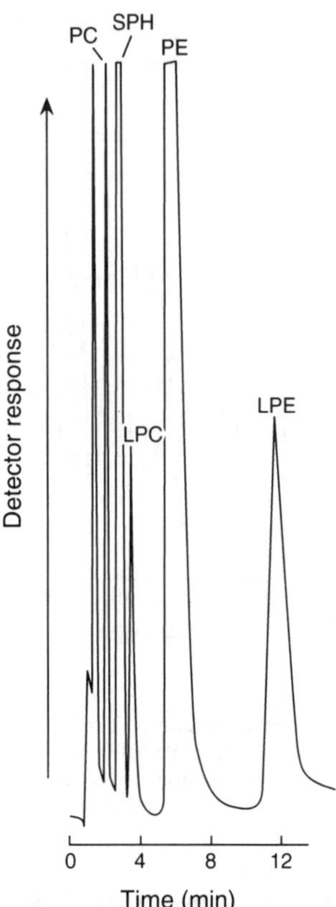

Fig. 3.6. Separation of phospholipid standards by HPLC on a column containing an amine-bonded phase, eluted isocratically with acetonitrile-methanol-0.05M acetic acid (10:5:0.4 by volume) at 2 mL/min, and spectrophotometric detection at 206 nm [20]. (Reproduced by kind permission of the authors and of *Annali di Chimica*, and redrawn from the original publication). The legend to Figure 3.1 contains a list of abbreviations.

tures, it was noted that phosphatidylcholine eluted before phosphatidylethanolamine, reversing the order found with adsorption chromatography on silica gel [78]. It was demonstrated subsequently that the choline-containing lipids, phosphatidylcholine, sphingomyelin and lysophosphatidylcholine, eluted in turn before the ethanolamine-containing lipids, phosphatidylethanolamine and lysophosphatidylethanolamine, but that the acidic lipids were retained very strongly indeed. Similar results were obtained with acetonitrile-methanol-water [20,25,43,46,65,117] and with a hexane-propan-2-ol-(methanol)-water [53] mixture, as illustrated in Figure 3.6. Phosphatidylglycerol was found to elute just ahead of phosphatidylcholine in one system [46] and just after it in another [65]; phosphatidyl-*N*-methylethanolamine eluted before phosphatidylethanolamine while phosphatidyl-*N,N*-dimethylethanolamine emerged just after it [25].

Initially, it proved difficult to find conditions suited for the elution of the acidic phospholipids, as these are probably retained strongly by ionic bonds on such columns. For example, phosphatidylmannose and related lipids could only be recovered by elution with gradients of up to 10 mM aqueous ammonium acetate as the mobile phase [35]. Somewhat milder conditions have now been devised for phospholipids of this type. Phosphatidylserine was purified on an amino column by elution with ethanol-0.9M phosphoric acid (88:12, v/v) [21,22]. In a more practical elution scheme, an acetonitrile-methanol-water-methylphosphonic acid mixture (pH 6.3) was the mobile phase [9]. Two similar amine columns (50 mm and 175 mm long were used in series, and after elution of phosphatidylethanolamine from the second column, a switching device was used to enable elution of phosphatidylinositol and phosphatidylserine which had not yet cleared the first of the columns. Acidic phospholipids have also been recovered by elution with a gradient of aqueous ammonium acetate into chloroform-methanol-water [34,83], and with acetonitrile-methanol-aqueous ammonium dihydrogen phosphate (10 mM, pH 4.76) [117]. Perhaps surprisingly with the latter system, diphosphatidylglycerol required an elution time three times that of phosphatidylserine.

An amine-bonded stationary phase gradually lost its activity in continued use, and phospholipids were eluted with solvents of lower polarity, possibly because peroxides, ketones or aldehydes reacted slowly with the amine groups, rendering them inactive, or because of a steady accumulation of material which was strongly adsorbed [25]. For a similar reason, acetone and other ketones should not be used as components of mobile phases.

Although other commercial amino columns are available, they do not appear to have been tested for phospholipids. However, home-made acetamido- and diamino- [20] and Cobalt(III)-amino-phases [43] did not appear to offer any advantage over conventional aminopropyl columns.

3. A Phenylsulphonate Phase

Gross and co-workers have made use of a silica-based column with bonded benzene sulphonate residues as the functional group [47-49]. A column of Partisil™ SCX and elution with acetonitrile-methanol-water were used to effect separation of the main ethanolamine- and choline-containing phospholipids of animal tissues. Phosphatidylinositol eluted at the solvent front. There was a particularly good separation of phosphatidylcholine, sphingomyelin and lysophosphatidylcholine and no ions appear to be required in the mobile phase. However, in micro-preparative applications it became apparent that the phosphatidylethanolamine fraction was contaminated, limiting the value of the method. However, it would be of interesting to investigate the capabilities of this type of column further with gradient elution and different mobile phases. The method has been adapted to LC-MS in one laboratory [102].

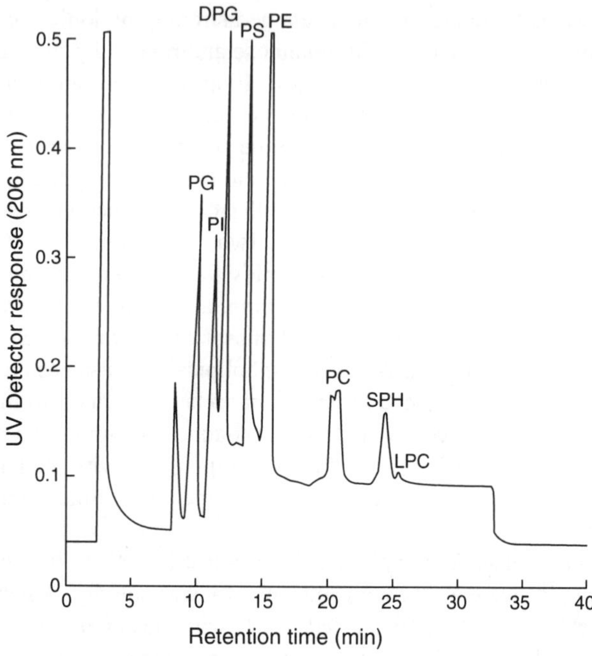

Fig. 3.7. Separation of a reference phospholipid mixture on a nitrile column, and a gradient of 5 mM aqueous sodium acetate (pH 5) in acetonitrile with UV detection at 205 nm [113]. (Reproduced by kind permission of the authors and of *Analytical Biochemistry*, and redrawn from the original publication). The legend to Figure 3.1 contains a list of abbreviations.

4. *Cyanopropyl Phases*

Cyanopropyl phases offer a useful alternative to silica as lower proportions of water are required to elute components. Samet *et al.* [113] used a column of this type (manufacturer and dimensions not specified, unfortunately) and a gradient of 5 mM aqueous sodium acetate (pH 5) in acetonitrile for phospholipid separations, as illustrated in Figure 3.7. The acidic phospholipids gave especially sharp peaks, in the order phosphatidylglycerol, phosphatidylinositol, diphosphatidylglycerol (cardiolipin) and phosphatidylserine, before phosphatidylethanolamine and then the choline-containing phospholipids eluted. In the author's laboratory, a cyanopropyl column was utilised for separation of the simple and complex lipids of plant tissues in one chromatographic run (see Section E2 below). With elution systems based on hexane-propan-2-ol-water, some change in selectivity was observed [32]. Some retention data have also been reported for bacterial lipids (hopanoids and glycolipids) [93].

Free silanol groups on the silica matrix may play a part in the separation, as the author had difficulties in obtaining reproducible separations with a fully end-capped column of this type (unpublished data).

E. SEPARATION OF SIMPLE AND COMPLEX LIPIDS IN A SINGLE CHROMATOGRAPHIC RUN

1. Lipids from Animal Tissues

It is possible to separate individual simple lipids followed by each of the main phospholipid classes in a single chromatographic run with gradient elution, as was first demonstrated in the laboratory of the late Orville Privett and co-workers in important publications commencing as long ago as 1973 [40,105-107,122] (and reviewed in greater detail elsewhere [28]). In many ways, this work was ahead of its time, since a home-made detector of the transport-flame ionization type, never commercialised or made available to others, was employed together with columns of silica gel of a type much inferior to those available now. No one appears to have attempted to repeat this work with similar solvent systems and newer commercial columns and detectors, and Privett's work set a standard that is only now being surpassed.

The author obtained a less comprehensive but more practical separation of this type by using a complex ternary gradient system and evaporative light-scattering detection [26,27]. With this type of detector (a proprietary model much inferior to those available now), there was the capacity to use complicated gradients, and the available options were greatly increased.

For example, it was possible to take the lipids of an extract of a tissue such as rat kidney and separate components ranging in polarity from cholesterol esters to lysophosphatidylcholine in a single chromatographic run on a short column of 3 μm silica gel (Figure 3.8). All the more abundant components were resolved (on the 0.2 to 0.4 mg scale) in only 20 min. The base-line was absolutely steady, although there were abrupt changes of solvent at some points. In essence, isooctane (or hexane) was used to elute first the cholesterol esters and then the triacylglycerols, before propan-2-ol (with 20% chloroform to modify the selectivity for the choline-containing phospholipids) was bled in to bring off cholesterol and other simple lipids, and then water was introduced to elute each of the phospholipids in turn. Finally, the polarity of the gradient was reversed to remove most of the water from the stationary phase (addition of a small proportion of tetrahydrofuran or methyl*tert*butyl ether to the hexane component greatly aided this) and to restore the column to its original activity over a further 10 min. The precise conditions are listed in Table 3.2. Propan-2-ol was an essential component of the system as few alternative solvents have the capacity to mediate the transition from hexane- to water-based mixtures.

It was also determined that better column life and sharper peaks for the acidic components were obtained if ionic species, *i.e.* serine buffered with ethylamine to the correct pH (7.5) [27], were added to the aqueous component. At a level of 0.5 mM and with light-scattering detection, this had virtually no effect on the base-line. In an 8 hour day, it was possible to run 15 samples, and we were also able to inject more than 1000 samples onto a column without significant loss of resolu-

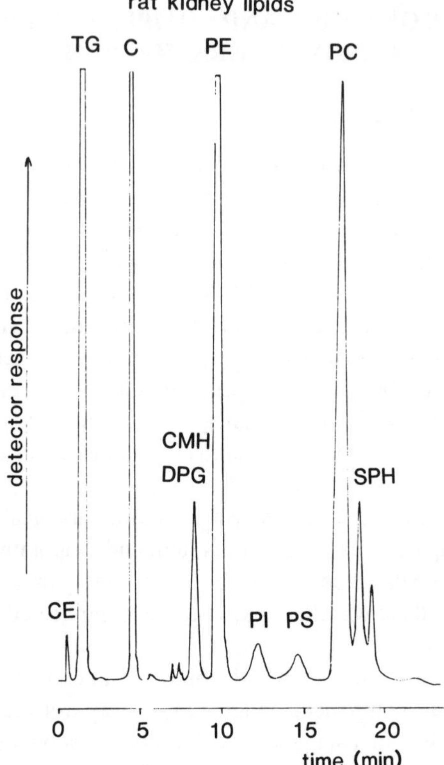

Fig. 3.8. Separation of rat kidney lipids (0.35 mg) by HPLC on a column (5 x 100 mm) of Spherisorb™ silica gel (3 μm particles) with evaporative light-scattering detection, with a ternary gradient elution scheme based on hexane, propan-2-ol and water [26]. (Reproduced by kind permission of the *Journal of Chromatography*). The legend to Figure 3.1 contains a list of abbreviations.

Table 3.2

Gradient elution conditions for separation of simple and complex lipids in a single chromatographic run on a column (5 x 100 mm) of Spherisorb™ silica gel (3 μm particles) [26,27].

Time (min)	Mobile phase		
	%A*	%B	%C
0	100		
1	100		
5	80	20	
5.1	42	52	6
20	32	52	16
20.1	30	70	
25	100		
30	100		

Solvents:A: hexane-methyl*tert*butyl ether (98:2, v/v), B: propan-2-ol-chloroform (80:20, v/v), C: propan-2-ol-0.5 mM aqueous serine (pH 7.5, 1:1, v/v) at a flow-rate of 2 mL/min.

tion. It has been suggested, however, that the addition of ionic species is not necessary in this system if silica-saturation pre-columns are used [88]. In this work, tissue samples from humans with metabolic diseases were analysed, using the more modern and sensitive Varex™ evaporative light-scattering detector.

The plasmalogen components of the sample were determined before and after acidic hydrolysis, by using the cholesterol content of the sample as a constant factor. This approach to the analysis of plasmalogens was first adapted to HPLC by Chen and Kou [25], and has subsequently been used by others (with simpler HPLC elution schemes [96,115,129]).

This basic separation procedure was adapted for fluorescence detection by adding synthetic lipids containing pyrene fatty acids into the samples [59]. Others have modified the method for use with a dual (as opposed to ternary) pumping system, while also incorporating a degree of automation both for injection and quantification [108]. In addition, the original procedure has been adapted somewhat for the analysis of phospholipids alone [67], cholesterol being employed as internal standard for quantification purposes. Yet others have modified this procedure, mainly by changing the timing of the steps and flow-rates, to improve the resolution of phospholipid and cerebroside components [84]. Also in this last work, the Varex light-scattering detector was employed with *N*-oleoylethanolamine as an internal standard. This detector was used in a "simplified" modification to the method [82] as was the older ACS detector [89]. A simple binary gradient of hexane-propan-2-ol-water has been used to obtain a total neutral lipid fraction, followed by free fatty acids and each of the phospholipids in turn [23].

2. *Lipids from Plant and Microbial Tissues*

Plant lipids present added difficulties to analysts because of their high content of glycolipids of various kinds in addition to phospholipids. However, good progress has now been made in developing methods for resolving individual lipid classes here also. We developed a somewhat altered elution scheme to that described in the previous section to separate the glycolipids in cereal lipids on silica gel [30]. By eluting with hexane-butanone-acetic acid before starting the phospholipid gradient described earlier, both groups of compounds were successfully resolved. Others adapted the ternary elution system, developed in my laboratory, for comprehensive analysis of cereal lipids, including the galactosyldiacylglycerols and *N*-acylphosphatidylethanolamine [32]. In essence, the modification involved using acetic acid in the mobile phase and 10 μm rather than 3 μm silica in the column, so permitting a more rapid and complex gradient. Moreau and colleagues developed a somewhat different modification of the basic procedure with a similar goal, involving a longer HPLC column, extended gradients and reduced flow-rates. Initially they used a transport flame-ionization detector [92], although subsequently it was shown that much more stable base-lines were possible with evaporative light-scattering detection (Figure 3.9) [90,91,93]. In this instance with

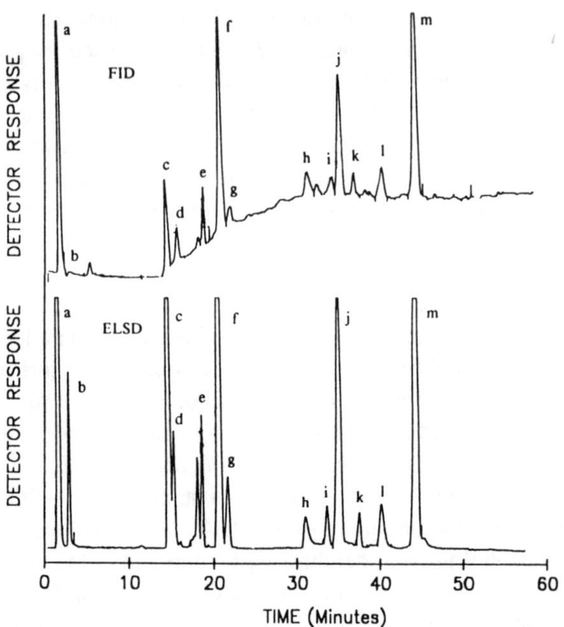

Fig. 3.9. Comparison of the analysis of lipid classes from corn coleoptiles by HPLC with transport-flame ionization detection (FID) (Tracor model) and evaporative light-scattering detection (ELSD) (Varex model) (B), each with 125 micrograms of lipid in total [90]. (Reproduced by kind permission of the author and of Portland Press). Abbreviations a, sterol esters; b, triacylglycerols; c, sterols; d, free fatty acids; e, acylated sterol glycosides; f, monogalactosyldiacylglycerols; g, sterol glycosides; h, digalactosyldiacylglycerols; i, cardiolipin; j, phosphatidylethanolamine; k, phosphatidylglycerol; l, phosphatidylinositol; m, phosphatidylcholine.

extracts from potato tissues, not only were the main simple lipids, mono- and digalactosyldiacylglycerols and phospholipids separated, but also the sterolglycosides and cerebrosides. The value of this approach has now been confirmed in another laboratory in analyses of wheat flour lipids, where there is a rather different spectrum of phospholipids especially from many other plant tissues [33]. An analogous approach was used in a further laboratory [19].

A simple binary gradient system, from hexane-propan-2-ol to hexane-propan-2-ol-water, has been used to separate neutral lipids as a single class, followed by each of the glycolipids and phospholipids in turn from an algal lipid extract [109].

Moreau and colleagues have also adapted an elution scheme of this type for the analysis of the complex lipids, including hopanoids (pentacyclic triterpenes), found in certain bacterial species [10,11,94,112]. In the most recent paper [94], a ternary gradient elution scheme, utilising hexane-propan-2-ol-water-triethylamine in the mobile phase, was employed with a silica column and the Tracor™ transport-flame ionization detector.

Hammond [51,52] has published brief details of a very different quaternary gradient elution scheme for the analysis of wheat flour lipids, based on toluene-ethyl

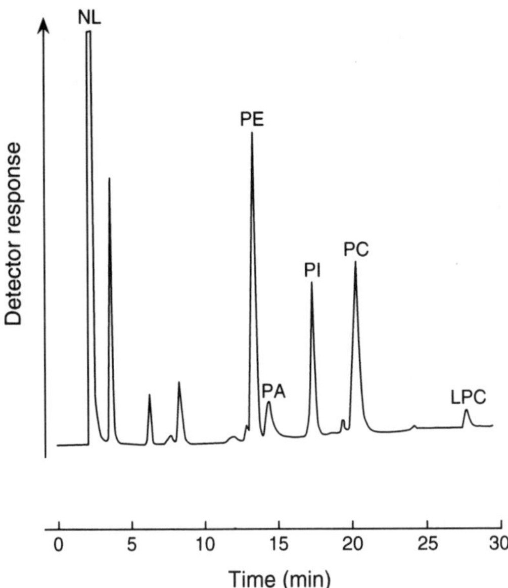

Fig. 3.10. Separation of soybean lipids by HPLC with evaporative light-scattering detection [5,73]. A multi-step binary gradient scheme was utilised with hexane-propan-2-ol-butanol-tetrahydrofuran-aqueous ammonium acetate mixtures of increasing polarity, and a column of LiChroCART™ 100 DIOL maintained at 75°C. The figure was adapted from a specimen chromatogram kindly supplied by the authors. The legend to Figure 3.1 contains a list of abbreviations.

acetate-methanol-water as mobile phase and a silica gel column; the Tracor™ transport-flame ionization detector was used. This unusual combination of solvents in a series of gradients conferred a distinctive selectivity on the separation and an order of elution quite different from that reported by others, i.e. monogalactosyldiacylglycerols, sterol glycosides, *N*-acyl phosphatidylethanolamine, digalactosyldiacylglycerols, phosphatidylserine, phosphatidylethanolamine and phosphatidylcholine.

It seems likely that silica gel *per se* will be less used for this purpose in future, as much better recoveries of glycolipids especially were obtained with chemically bonded stationary phases [32], a phenomenon for which no explanation is currently available. Total neutral lipids and individual phospholipid classes from plant "lecithins" have been well separated on a diol phase (see Section D1 above), Lichrospher™ 100 diol, by gradient elution with hexane-propan-2-ol-water-acetic acid mixtures of increasing polarity [57]. This elution system seems to have quality control applications, and it has been combined with plasmaspray tandem mass spectrometry for analysis of lipids [124]. Recently, the method has been improved by a chemometric optimization procedure [5,73], as illustrated in Figure 3.10. A multi-step binary gradient scheme was devised, utilising hexane-propan-2-ol-butanol-tetrahydrofuran-aqueous ammonium acetate mixtures of increasing polarity, and a column of LiChroCART™ 100 DIOL maintained at a temperature

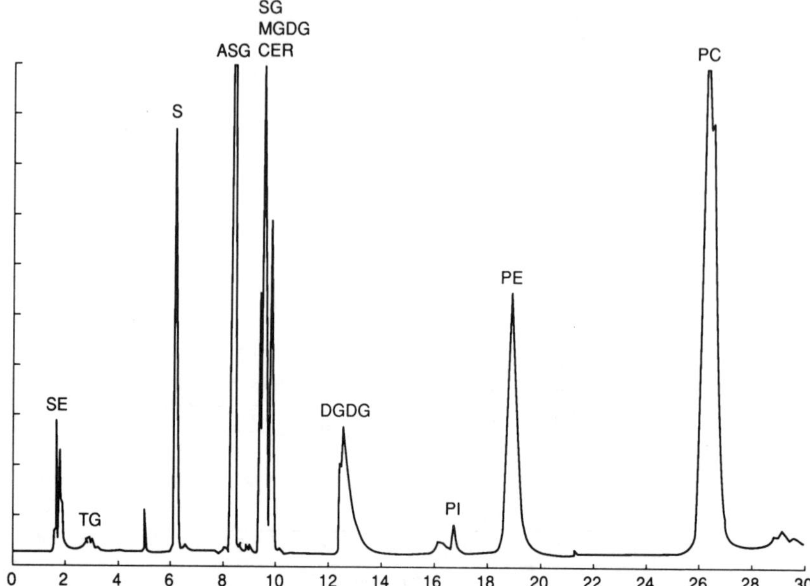

Fig. 3.11. Separation of a potato lipid extract on a column of Spherisorb™ 3CN (100 x 3.2 mm i.d.) with evaporative light-scattering detection [32]. The mobile phase was a complex ternary gradient involving mixtures of isohexane-methyltertbutyl ether-propan-2-ol-chlorofom-acetic acid-triethylamine-water. (Reproduced with the permission of the *Journal of High Resolution Chromatography*). The legend to Figure 3.1 contains a list of abbreviations. In addition: ASG, acylsteryl glycosides; SG, sterylglycosides; CER, cerebrosides; MGDG, monogalactosyldiacylglycerols; DGDG, digalactosyldiacylglycerol.

of 75°C. In this instance phosphatidic acid and phosphatidylinositol eluted between phosphatidylethanolamine and phosphatidylcholine. Unfortunately, I have been unable to find conditions suited to the separation of the important simple lipids, sterol esters and triacylglycerols, on a diol column (unpublished observation).

A stationary phase that is chemically similar to diol and silica gel in its properties has been manufactured by polymerising and cross-linking vinyl alcohol to silica gel, *i.e.* PVA Sil™ (YMC Co., Japan). In this instance, the whole surface is covered and deactivated so that the mobile phase and analytes interact with a uniform layer of hydroxyl groups only. A ternary elution gradient scheme was devised as before with *iso*hexane-methyl*tert*butyl ether (98:2, v/v), propan-2-ol-acetonitrile-chloroform-acetic acid (84:8:8:0.025 by volume) and propan-2-ol-water-triethylamine (50:50:0.2 by volume) as the three components. It proved possible to elute each of the simple lipids, glycolipids and phospholipids sequentially under milder conditions and in apparently better yields than with columns of silica gel [32].

Similarly, cyanopropyl columns, which had shown earlier promise (see Section D4), proved suitable for separations of this kind as illustrated in Figure 3.11 [32]. In this instance, a column of smaller dimensions than normal (100 x 3.2 mm, of

Spherisorb™ CN) was used to permit a sufficiently slow flow-rate (0.5 mL/min) for potential application to HPLC-MS.

F. PREPARATIVE-SCALE SEPARATIONS OF PHOSPHOLIPIDS

Normally, analysts tend to fractionate phospholipids on about the 0.1 mg scale, and this is sufficient for direct quantification or even for isolation of major components for say fatty acid analysis. It is often possible to use exactly the same equipment and conditions, including the analytical column, to separate as much as 2 mg of lipids, although the resolution will deteriorate. Most HPLC pumps are sufficiently flexible to be able to accommodate flow rates as high as 10 mL/min, and they can be used with longer and wider columns than are typical for analytical scale applications to enable separation of as much as 200 mg phospholipids in a single chromatographic run. Standard analytical grade adsorbents (spherical) can also be used, although they are not ideal for the purpose, and particle sizes of 20 μm or more are recommended to reduce the pressure drop along the column. It is also advisable to avoid the use of ionic species, which would contaminate fractions.

In what appears to have been the first published preparative scale separation, up to 10 g of egg lecithin was fractionated into three main components, phosphatidylethanolamine, lysophosphatidylethanolamine and phosphatidylcholine, on columns (eight 100 cm x 1 cm i.d. columns in series) packed with 20-40 μm silica gel particles, and step-wise elution with chloroform-methanol mixtures [42]. Very large volumes of solvent were required so the work is now of academic interest only. Others have used more conventional instrumentation, with columns 20-25 cm in length and diameters of 1 to 5 cm, and with established mobile phases for preparative fractionation of phospholipids and other complex lipids on the 10 to 200 mg scale [3,6,39,58,61,112].

Van der Meeren and co-workers [62,125] studied the problem of optimising preparative-scale HPLC of phospholipids in terms of both column efficiency and solvent usage rather systematically, and have described what appear to be the best practical separations to date. They obtained the most useful results with irregular silica gels (15-35 μm) packed in conventional analytical-size columns (250 x 4.6 mm). By using a simple solvent switching system to generate step-wise gradients of hexane-propan-2-ol-water mixtures, they were able to make use of a single pump and obtain excellent separations of a wide range of phospholipids on the 100 mg scale.

G. QUANTIFICATION OF PHOSPHOLIPIDS SEPARATED BY HPLC

The topic of quantification of lipids separated by HPLC has been discussed in detail by the author elsewhere, and readers are referred thence [28,29]. The topic is discussed briefly here, partly for those who do not have immediate access to these publications but mainly for the influence the availability of specific detectors has on the choice of mobile phase and therefore on the selectivity of separations (see Section B2).

1. UV Detection

Many times more UV detectors are sold than all other types put together, so it is hardly surprising that lipid analysts have shown considerable ingenuity in adapting these to lipid analysis. Most natural lipids exhibit a weak absorbance in the range 200 to 206 nm of the UV spectrum, sometimes termed "end absorption", that is the result of the presence of isolated double bonds predominantly, although carbonyl, carboxyl, phosphate, amino, quaternary ammonium and other functional groups contribute also. This narrow window permits a limited range of solvents to be employed (see Section B2). In the important early paper of Jungalwala *et al.* [68], it was shown that the response of a UV detector both to phosphatidylcholine and sphingomyelin was rectilinear up to about 60 micrograms, and that direct quantification was possible if the apparent extinction coefficient for each component was determined accurately.

The absorption spectra of individual lipid classes can be very different, mainly because of the varying degrees of unsaturation of each, so it is essential that the standards used for calibration should resemble the compounds to be analysed in the composition of their fatty acids as closely as possible. Relatively saturated lipids, such as sphingomyelin or lysophosphatidylcholine, might even be missed on chromatographic traces as they give such a poor response. With careful calibration, it is possible to use the UV response for direct quantification. This is probably easiest with samples such as soybean "lecithin" [7,98,99], but the technique has also been used for platelet [76], amniotic fluid and plasma [55,56,80], pulmonary surfactant [37,80], erythrocytes [115] and reproductive tissue [100,101] phospholipids, for example.

There are many more published papers in which UV detection has been used to enable collection of fractions for analysis by other methods, such as phosphorus assay or gas chromatography of the fatty acid derivatives following transmethylation with an internal standard.

2. Transport-Flame Ionization Detection

At first glance, most lipid analysts would consider transport-flame ionization detectors to be the ideal since they are universal in applicability and the response should be linear over a wide range of samples and lipid classes, without the need for substantial correction factors. The first commercial model made by Pye Unicam was not a success, and the author knows of three companies who have announced models which were never actually put into production. Only the Tracor™ Model 945 (Trimetrics, Houston, TX) continues to have a tenuous hold on the market. Its use for the analysis of analysis of lipids has recently been reviewed [52,91]. It is evident that there are disadvantages, especially that traces

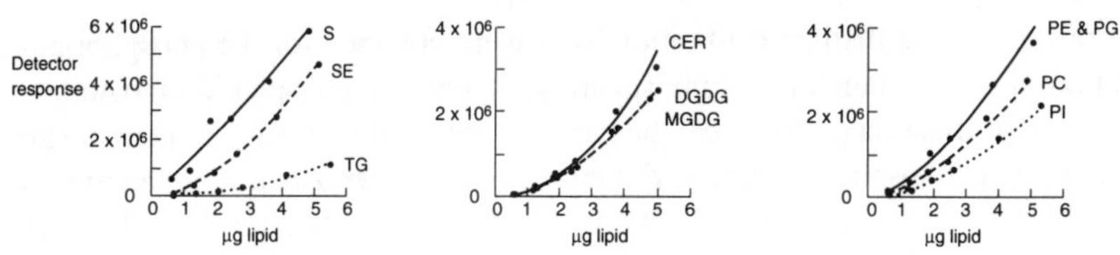

Fig. 3.12. Calibration curves for lipid standards (µg of lipid injected versus detector response in integrator counts), obtained with the Varex Mk III evaporative light-scattering detector [32]. (Reproduced with the permission of the *Journal of High Resolution Chromatography*). The legends to Figure 3.1 and 3.11 contain a list of abbreviations.

of inorganic ions, such as sodium and potassium, play havoc with the response, but a few groups in the USA and UK use it regularly and successfully.

Most solvents that can be volatilized appear to be usable with transport-flame ionization detection, so that there are good opportunities to vary the selectivity of separation. The problem is the choice of an appropriate ionic species to ensure sharp peaks for phospholipids. Hammond [52] and others [45] have used ammonia successfully for the purpose, while Moreau and colleagues [94] now employ triethylamine. Although the base-line stability does not appear to be particularly good with the latter, a linear response and good quantification are claimed.

3. Evaporative Light-Scattering Detection

Evaporative light-scattering detectors are now being used in a large number of laboratories as they tend to be robust instruments, and of course can be used with most solvents that can be volatilized at reasonable temperatures. Thus, there are excellent opportunities to change or modify the selectivity of separations. Many examples are discussed above, and I have discussed the topic in greater detail elsewhere [29].

The instrument is linear over a reasonable range, depending on the specific commercial model, but the sensitivity does tend to fall off rapidly for small components. The diameters of the lipid droplets formed in the stream of air when the solvent is evaporated depend on the amount of material in a peak. When the component is too small, these droplets no longer reflect or refract the light, and the sensitivity of the detector falls off rapidly. In addition, it is evident that the sensitivity of the detector depends on the nature of the lipid and of the mobile phase. However, my experience is that with careful calibration the evaporative light-scattering detector can give as good results as most alternative methods. Some calibration curves obtained for plant lipids are illustrated in Figure 3.12 [32]. As had been reported earlier for an older model of detector [26], the response differed for each lipid class and was not rectilinear over a significant concentration range. The highest relative response was for sterols and sterol esters, which were also the sharpest peaks in the chromatographic traces. The relatively poor response for triacylglycerols was probably due to the fact that this was a broad band, because of a partial separation of molecular species. Individual glycolipids tended to have similar responses, but distinct calibration lines were obtained for the phospholipids, *i.e.* phosphatidylethanolamine, phosphatidylcholine and phosphatidylinositol.

The quantification problem can be partially resolved by using an internal standard, such as phosphatidyldimethylethanolamine, as described in an analysis of the phospholipids of milk [31]. Others have used cholesterol [67] and *N*-oleoylethanolamine [84] as internal standards.

H. CONCLUSIONS

A great deal of useful work has now been published on separation of complex lipid classes from plant and animal tissues by means of HPLC. The wider availability of evaporative light-scattering detectors gives a much greater scope for testing novel elution systems, since now the only limitation is the volatility of the solvents. Similarly, there are many new types of chemically bonded stationary phases available from commercial suppliers, whose use should be explored systematically. The major problem remaining appears to be in reproducible analysis of acidic phospholipids, especially phosphatidylserine, phosphatidylinositol and phosphatidic acid. The answer may be to find a suitable non-corrosive and volatile ionic species to add to the mobile phase. While considerable progress has been made in analytical methods for plant lipids, the last word has still to be said on the subject. Polyphosphoinositides still represent a challenge and microbial complex lipids have scarcely been touched by HPLC.

ACKNOWLEDGEMENT

This review is published as part of a programme funded by the Scottish Office Agriculture and Fisheries Dept.

REFERENCES

1. Abidi,S.L., Mounts,T.L. and Rennick,K.A., *J. Liqu. Chromatogr.*, **17**, 3705-3725 (1994).
2. Alam,I., Smith,J.B., Silver,M.J. and Ahern,J., *J. Chromatogr.*, **234**, 218-221 (1982).
3. Amari,J.V., Brown,P.R., Grill,C.M. and Turcotte,J.G., *J. Chromatogr.*, **517**, 219-228 (1990).
4. Andrews,A.G., *J. Chromatogr.*, **336**, 139-150 (1984).
5. Arnoldsson,K.C. and Kaufmann,P., *Chromatographia*, **38**, 317-324 (1994).
6. Bahrami,S., Gasser,H. and Redl,H., *J. Lipid Res.*, **28**, 596-598 (1987).
7. Beare-Rogers,J.L., Bonekamp-Nasner,A. and Dieffenbacher,A., *Pure Appl. Chem.*, **64**, 447-454 (1992).
8. Becart,J., Chevalier,C. and Biesse,J.P., *J. High Resolut. Chromatogr.*, **13**, 126-129 (1990).
9. Bernard,W., Linck,M., Creutzberg,H., Postle,A.D., Arning,A., Martin-Carrera,I. and Sewing,K.-F., *Anal. Biochem.*, **220**, 172-180 (1994).
10. Berry,A.M., Harriott,O.T., Moreau,R.A., Osman,S.F., Benson,D.R. and Jones,A.D., *Proc. Natl. Acad. Sci. USA*, **90**, 6091-6094 (1993).
11. Berry,A.M., Moreau,R.A. and Jones,A.D., *Plant Physiol.*, **95**, 111-115 (1991).
12. Bjerrum,O.W., Nielsen,H. and Borregaard,N., *Scand. J. Clin. Lab. Invest.*, **49**, 613-622 (1989).
13. Blank,M.L. and Snyder,F., *J. Chromatogr.*, **273**, 415-420 (1983).
14. Blom,C.P., Deierkauf,F.A. and Riemersma,J.C., *J. Chromatogr.*, **171**, 331-338 (1979).
15. Breton,L., Serkiz,B., Volland,J-P. and Lepagnol,J., *J. Chromatogr.*, **497**, 243-249 (1989).
16. Briand,R.L., Harold,S. and Blass,K.G., *J. Chromatogr.*, **223**, 277-284 (1981).
17. Burdge,G.C., Kelly,F.J. and Postle,A.D., *Biochem. J.*, **290**, 67-73 (1993).

18. Caboni,M.F., Lercker,G. and Ghe,A.M., *J. Chromatogr.*, **315**, 223-231 (1984).
19. Carr,N.O., Daniels,N.W.R. and Frazier,P.J., in *Wheat End-Use Properties*, Proceedings of ICC Meeting, University of Helsinki, 1989, pp. 151-172 (1989).
20. Carunchio,V., Nicoletti,I., Frezza,L. and Sinibaldi,M., *Annali di Chimica*, **74**, 331-339 (1984).
21. Chen,S., Kirschner,G. and Traldi,P., *Anal. Biochem.*, **191**, 100-105 (1990).
22. Chen,S., Menon,G. and Traldi,P., *Org. Mass Spectrom.*, **27**, 215-218 (1992).
23. Chen,S.F. and Chan,P.H., *J. Chromatogr.*, **344**, 297-303 (1985).
24. Chen,S.S.-H. and Kou,A.Y., *J. Chromatogr.*, **227**, 25-31 (1982).
25. Chen,S.S.-H. and Kou,A.Y., *J. Chromatogr.*, **232**, 237-249 (1982).
26. Christie,W.W., *J. Lipid Res.*, **26**, 507-512 (1985).
27. Christie,W.W., *J. Chromatogr.*, **361**, 396-399 (1986).
28. Christie,W.W., *High Performance Liquid chromatography and Lipids* (1987) (Pergamon Press, Oxford).
29. Christie,W.W., *in Advances in Lipid Methodology - One*, pp. 239-271 (1992) (edited by W.W. Christie, Oily Press, Dundee).
30. Christie,W.W. and Morrison,W.R., *J. Chromatogr.*, **436**, 510-513 (1988).
31. Christie,W.W., Noble,R.C. and Davies,G., *J. Soc. Dairy Technol.*, **40**, 10-12 (1987).
32. Christie,W.W. and Urwin,R.A., *J. High Resolut. Chromatogr.*, **18**, 97-100 (1995).
33. Conforti,F.D., Harris,C.H. and Rinehart,J.T., *J. Chromatogr.*, **645**, 83-88 (1993).
34. Coté,G.G., DePass,A.L., Quarmby,L.M., Tate,B.F., Morse,M.J., Satter,R.L. and Crain,R.C., *Plant Physiol.*, **90**, 1422-1428 (1989).
35. Creek,C.E., Rimoldi,D., Clifford,A.J., Silverman-Jones,C.S. and De Luca,L.M., *J. Biol. Chem.*, **261**, 3490-3500 (1986).
36. D'Costa,M., Dassin,R., Bryan,H. and Joutsi,P., *Clin. Biochem.*, **18**, 27-31 (1985).
37. Dethloff,L.A., Gilmore,L.B. and Hook,G.R., *J. Chromatogr.*, **382**, 79-87 (1986).
38. Dugan,L.L., Demediuk,P., Pendley,C.E. and Horrocks,L.A., *J. Chromatogr.*, **378**, 317-327 (1986).
39. Ellingson,J.S. and Zimmerman,R.L., *J. Lipid Res.*, **28**, 1016-1018 (1987).
40. Erdahl,W.L., Stolyhwo,A. and Privett,O.S., *J. Am. Oil Chem. Soc.*, **50**, 513-515 (1973).
41. Evershed,R.P., in *Developments in the Analysis of Lipids*, pp. 123-160 (1994) (edited by J.H.P. Tyman and M.H. Gordon, Royal Soc. Chem., Cambridge).
42. Fager,R.S., Shapiro,S. and Litman,B.J., *J. Lipid Res.*, **18**, 704-709 (1977).
43. Federici,F., Lionetti,P., Messina,A., Nicoletti,I. and Sinibaldi,M., *Anal. Letts.*, **23**, 1265-1277 (1990).
44. Geurts van Kessel,W.S.M., Hax,W.M.A., Demel,R.A. and de Gier,J., *Biochim. Biophys. Acta*, **486**, 524-530 (1977).
45. Grieser,M.D. and Geske,J.N., *J. Am. Oil Chem. Soc.*, **66**, 1484-1487 (1989).
46. Grit,M., Crommelin,D.J.A. and Lang,J., *J. Chromatogr.*, **585**, 239-246 (1991).
47. Gross,R.W., *Biochemistry*, **22**, 5641-5646 (1983).
48. Gross,R.W., *Biochemistry*, **24**, 1662-1668 (1985).
49. Gross,R.W. and Sobel,B.E., *J. Chromatogr.*, **197**, 79-85 (1980).
50. Guichardant,M. and Lagarde,M., *J. Chromatogr.*, **275**, 400-406 (1983).
51. Hammond,E.W., *Trends Anal. Chem.*, **8**, 308-313 (1989).
52. Hammond,E.W., *Chromatography for the Analysis of Lipids* (CRC Press. Boca Raton, FL) (1993).
53. Hanson,V.L., Park,J.Y., Osborn,T.W. and Kiral,R.M., *J. Chromatogr.*, **205**, 393-400 (1981).
54. Hax,W.M.A. and Geurts van Kessel,W.S.M., *J. Chromatogr.*, **142**, 735-741 (1977).
55. Heinze,T., Kynast,G., Dudenhausen,J.W. and Saling,E., *J. Perinat. Med.*, **16**, 53-60 (1988).
56. Heinze,T., Kynast,G., Dudenhausen,J.W., Schmitz,C. and Saling,E., *Chromatographia*, **25**, 497-503 (1988).
57. Herslöf,B., Olsson,U. and Tingvall,P., in *Phospholipids*, pp. 295-298 (1990) (edited by I. Hanin and G. Pepeu, Plenum Press, New York).
58. Holte,L.H., Van Kuijk,F.J.G.M. and Dratz,E.A., *Anal. Biochem.*, **188**, 136-141 (1990).
59. Homan,R. and Pownall,H.J., *Anal. Biochem.*, **178**, 166-171 (1989).
60. Hoving,E.B., Prins,J., Rutgers,H.M. and Muskiet,F.A.J., *J. Chromatogr.*, **434**, 411-416 (1988).
61. Hurst,W.J., Martin,R.A. and Sheeley,R.M., *J. Liqu. Chromatogr.*, **9**, 2969-2976 (1986).
62. Huys,M., Van der Meeren,P., Vanderdeelen,J. and Baert,L., *Med. Fac. Landbouww. Rijksuniv. Gent*, **53**, 1667-1677 (1988).

63. Ikeda,M., Hattori,M. and Matsumotu,U., *Anal. Sci.*, **2**, 379-383 (1986).
64. Islam,A., Smogorzewski,M., Pitts,T.O. and Massry,S.G., *Minor Electrolyte Metab.*, **15**, 209-213 (1989).
65. Jääskeläinen,I. and Urtti,A., *J. Pharm. Biomed. Anal.*, **12**, 977-982 (1994).
66. Jackson,E.M., Mott,G.E., Hoppens,C., McManus,L.M., Weintraub,S.T., Ludwig,J.C. and Pinckard,R.N., *J. Lipid Res.*, **25**, 753-757 (1984).
67. Juaneda,P., Rocquelin,G. and Astorg,P.O., *Lipids*, **25**, 756-759 (1990).
68. Jungalwala,F.B., Evans,J.E. and McCluer,R.H. *Biochem. J.*, **155**, 55-60 (1976).
69. Kaduce,T.L., Norton,K.C. and Spector,A.A., *J. Lipid Res.*, **4**, 1398-1403 (1983).
70. Kaitaranta,J.K. and Bessman,S.P., *Anal. Chem.*, **53**, 1232-1235 (1981).
71. Kaneko,K., Ohta,Y. and Machita,Y., *Agric. Biol. Chem.*, **51**, 2023-2024 (1987).
72. Kaufmann,P., *in Advances in Lipid Methodology - One*, pp. 149-180 (1992) (edited by W.W. Christie, Oily Press, Dundee).
73. Kaufmann,P., *Chemometr. Intell. Lab. Systems*, **27**, 105-114 (1995).
74. Kaufmann,P., Kowalski,B.R. and Alander,J., *Chemometr. Intell. Lab. Systems*, **23**, 331-339 (1994).
75. Kaufmann,P., Olsson,U. and Herslof,B.G., *J. Am. Oil Chem. Soc.*, **67**, 537-540 (1990).
76. Kawasaki,T., Kambayashi,J., Mori,T. and Kosaki,G., *Thromb. Res.*, **36**, 335-344 (1984).
77. Kim,H.Y. and Salem,N., *Prog. Lipid Res.*, **32**, 221-245 (1993).
78. Kiuchi,K., Ohta,T. and Ebine,H., *J. Chromatogr.*, **133**, 226-230 (1977).
79. Kuhnz,W., Zimmermann,B. and Nau,H., *J. Chromatogr.*, **344**, 309-312 (1985).
80. Kynast,G. and Schmitz,C., *J. Perinat. Med.*, **17**, 203-212 (1989).
81. Lester,R.L., Wells,G.B., Oxford,G. and Dickson,R.C., *J. Biol. Chem.*, **268**, 645-856 (1993).
82. Letter,W.S., *J. Liqu. Chromatogr.*, **15**, 253-266 (1992).
83. Low,M.G., in *Methods in Inositide Research*, pp. 145-151 (1990) (edited by R.F. Irvine, Raven Press, New York).
84. Lutzke,B.S. and Braughler,J.M., *J. Lipid Res.*, **31**, 2127-2130 (1990).
85. Mallet,A.I., Cunningham,F.M. and Daniel,R., *J. Chromatogr.*, **309**, 160-164 (1984).
86. Marion,D., Douillard,R. and Gandemer,G., *Rev. Franc. Corps Gras*, **35**, 229-234 (1988).
87. Marion,D., Gandemer,G. and Douillard,R., in *Structure, Function and Metabolism of Plant Lipids*, pp. 139-143 (1984) (edited by P.-A. Siegenthaler and W. Eichenberger, Elsevier, Amsterdam).
88. Markello,T.C., Guo,J. and Gahl,W.A., *Anal. Biochem.*, **198**, 368-374 (1991).
89. Melton,S.L., *J. Am. Oil Chem. Soc.*, **69**, 784-788 (1992).
90. Moreau,R.A. in *Plant Lipid Biochemistry, Structure and Utilization*, pp. 20-22 (1990) (edited by P.J. Quinn and J.L. Harwood, Portland Press, London).
91. Moreau, R.A. in *Lipid Chromatographic Analysis*, pp. 251-252 (1994) (edited by T. Shibamoto, Marcel Dekker, N.Y.).
92. Moreau,R.A., Asmann,P.T. and Norman,H.A., *Phytochemistry*, **29**, 2461-2466 (1990).
93. Moreau,R.A. and Gerard,H.C., in *CRC Handbook of Chromatography. Analysis of Lipids*, pp. 41-55 (1993) (edited by K.D. Mukherjee and N. Weber, CRC Press, Boca Raton).
94. Moreau,R.A., Powell,M.J., Osman,S.F., Whitaker,B.D., Fett,W.F., Roth,L. and O'Brien,D.J., *Anal. Biochem.*, **224**, 293-301 (1995).
95. Mounts,T.L., Abidi,S.L. and Rennick,K.A., *J. Am. Oil Chem. Soc.*, **69**, 438-442 (1992).
96. Murphy,E.J., Stephens,R., Jurkowitz-Alexander,M. and Horrocks,L.A., *Lipids*, **28**, 565-568 (1993).
97. Murphy,R.C., *Mass Spectrometry of Lipids (Handbook of Lipid Research, Vol. 7)* (1993) (Plenum Press, N.Y.).
98. Nasner,A. and Kraus,Lj., *J. Chromatogr.*, **216**, 389-394 (1981).
99. Nasner,A. and Kraus,Lj., *Fette Seifen Anstrichm.*, **83**, 70-73 (1981).
100. Nissen,H.P. and Kreysel,H.W., *J. Chromatogr.*, **276**, 29-36 (1983).
101. Nissen,H.P., Topfer-Petersen,E., Schill,W.B. and Kreysel,H.W., *Fette Seifen Anstrichm.*, **85**, 590-595 (1983).
102. Odham,G., Valeur,A., Michelsen,P., Aronsson,E. and McDowall,M., *J. Chromatogr.*, **434**, 31-41 (1988).
103. Patton,G.M., Fasulo,J.M. and Robins,S.J., *J. Lipid Res.*, **23**, 190-196 (1982).
104. Patton,G.M., Fasulo,J.M. and Robins,S.J., *J. Nutr. Biochem.*, **1**, 493-500 (1990).
105. Phillips,F.C., Erdahl,W.L. and Privett,O.S., *Lipids*, **17**, 992-997 (1982).

106. Phillips,F.C. and Privett,O.S., *J. Am. Oil Chem. Soc.*, **58**, 590-594 (1981).
107. Privett,O.S., Dougherty,K.A., Erdahl,W.L. and Stolyhwo,A., *J. Am. Oil Chem. Soc.*, **50**, 516-520 (1973).
108. Redden,P.R. and Huang,Y.S., *J. Chromatogr.*, **567**, 21-27 (1991).
109. Rezanka,T. and Podojil,M., *J. Chromatogr.*, **463**, 397-408 (1989).
110. Rivnay,B., *J. Chromatogr.*, **294**, 303-315 (1984).
111. Robinson,N.C., *J. Lipid Res.*, **31**, 1513-1516 (1990).
112. Roth,L.H., Moreau,R.A., Powell,M.J. and O'Brien,D.J., *Anal. Biochem.*, **224**, 302-308 (1995).
113. Samet,J.M., Friedman,M. and Henke,D.C., *Anal. Biochem.*, **182**, 32-36 (1989).
114. Seewald,M. and Eichinger,H.M., *J. Chromatogr.*, **469**, 271-280 (1989).
115. Shaffiq-Ur-Rehman, *J. Chromatogr.*, **567**, 29-37 (1991).
116. Sheeley,R.M., Hurst,W.J., Sheeley,D.M. and Martin,R.A., *J. Liqu. Chromatogr.*, **10**, 3173-3182 (1987).
117. Shimbo,K., *Agric. Biol. Chem.*, **50**, 2643-2645 (1986).
118. Smith,J.D., Mello,C.M. and O-Reilly,D.J., *J. Chromatogr.*, **431**, 395-399 (1988).
119. Snyder,L.R., *J. Chromatogr. Sci.*, **16**, 223-234 (1978).
120. Snyder,L.R., Carr,P.W. and Rutan,S.C., *J. Chromatogr.*, **656**, 537-547 (1993).
121. Snyder,L.R. and Kirkland,J.J., *Introduction to Modern Liquid Chromatography* (2nd edition) (1979) (John Wiley & Sons, New York).
122. Stolyhwo,A., Privett,O.S. and Erdahl,W.L., *J. Chromatogr. Sci.*, **11**, 263-267 (1973).
123. Trumbach,B., Rogler,G., Lackner,K.J. and Schmitz,G., *J. Chromatogr. B*, **656**, 73-76 (1994).
124. Valeur,A., Michelsen,P. and Odham,G., *Lipids*, **28**, 225-229 (1993).
125. Van der Meeren,P., Vanderdeelen,J., Huys,M. and Baert,L., *J. Am. Oil Chem. Soc.*, **67**, 815-820 (1990).
126. Wardlow,M.L., *J. Chromatogr.*, **342**, 380-384 (1985).
127. Wells,G.B. and Lester,R.L., *J. Biol. Chem.*, **258**, 10200-10203 (1983).
128. Yandrasitz,J.R., Berry,G. and Segal,S., *J. Chromatogr.*, **225**, 319-328 (1981).
129. Yeo,Y.K. and Horrocks,L.A., *Food Chem.*, **28**, 197-205 (1988).

Chapter 4

ANALYSIS OF LONG-CHAIN ACYL-COENZYME A ESTERS

Tine Bækdal, Charlotte Karlskov Schjerling, Jan Krogh Hansen and Jens Knudsen

Odense Universitet, Biokemisk Institut, Campusvej 55, DK-5320 Odense, Denmark

A. INTRODUCTION

Long-chain acyl-CoA esters (LCA) play a key role in cellular regulation and metabolism, both in eukaryotes and prokaryotes. In addition LCA play a central role as intermediates in glycerolipid synthesis, β-oxidation, fatty acid elongation and desaturation and in protein acylation.

LCA have been shown to regulate the activity of acetyl-CoA carboxylase (K_i 1-5 nM) [41] and protein kinase C [40]. LCA have also been shown to regulate the mitochondrial adenine nucleotide transporter ($K_i < 1.0$ mM) [60] and intracellular Ca^{2+} fluxes [15,19,23]. Furthermore it has been shown that LCA bind to the nuclear thyroid hormone receptor [34,35]. Finally it has been shown that LCA directly participate in transcriptional regulation in *E. coli* by binding to the transcription factor FadR [8,21]. In order to study these diverse functions of LCA, accurate and quantitative methods for determination of LCA levels and composition are greatly needed.

A large number of methods for determination of LCA levels have been developed. These methods can be divided roughly into two groups. Group one relies on isolation and separation of LCA from short-chain acyl-CoA by acid precipitation followed by alkaline hydrolysis and determination of the released CoASH by different enzymatic or chromatographic assays. Group two relies on extraction of LCA from tissue samples mainly by chloroform/methanol-based methods, followed by separation of individual LCA by reversed-phase high-performance liquid chromatography (HPLC), and calculation of the amount of individual acyl-CoA/g tissue from added internal standard. Alternatively, the extracted acyl-CoA esters can be hydrolysed, derivatized, reduced or transesterified and the amount of acyl chain and composition determined by gas-liquid chromatography (GLC).

In the present review we do not intend to give a detailed description of published methods, but rather to give an overview of existing methods with the main emphasis on HPLC-based methods. Finally we present data which compare results obtained by enzymatic methods based on determination of CoA released by alkaline hydrolysis of acid-precipitated LCA with results obtained with a newly developed HPLC method.

B. METHODS FOR DETERMINATION OF LCA LEVELS BASED ON ISOLATION AND SEPARATION FROM SHORT-CHAIN ACYL-COA BY ACID PRECIPITATION

1. Acid Precipitation and Hydrolysis

LCA determination based on measurement of released CoASH from isolated LCA demands separation of LCA from short-chain acyl-CoA and free CoA. This separation can be obtained by acid precipitation because LCA with acyl chain-length longer than ten carbons are essentially insoluble at low pH. Addition of either cold trichloroacetic acid (0.6 M) or perchloric acid (0.6 M), followed by cell disruption either by homogenisation or sonication, will precipitate all LCA and most proteins, which then can be harvested by centrifugation [55]. By repetitive washing of the precipitate with 0.6 M acid, short-chain acyl-CoA and CoASH are efficiently removed. It is important to realize that this distinction between long- and short-chain acyl-CoA esters is entirely chemical, and therefore may disguise some important metabolic changes. Neither does the method distinguish between CoASH bound in LCA and other CoASH derivatives which might be trapped in the acid precipitate or CoASH bound through disulphide bonds to proteins or other molecules in the precipitate. After removal of short-chain acyl-CoA esters, the acid precipitate is re-suspended in water or a suitable buffer, and the pH is raised to 11.5 with 1 M KOH, followed by incubation at 60°C for 60 minutes to hydrolyse the acyl-CoA esters. If the acyl-CoA esters are to be stored until hydrolysis, residual acid on the sides of the tubes should be removed with filter paper and the precipitate re-suspended in 100 mM MES buffer (pH 6.25) and the pH readjusted to 6.25 with KOH. It is not advisable to wash the precipitate with buffer or water because it might dissolve precipitated LCA. The choice of temperature, pH and time of hydrolysis are important. It is a balance between obtaining complete hydrolysis of LCA without destroying CoASH. Ingebretsen *et al.* [29] found that complete hydrolysis of LCA with a full recovery of released CoASH could be obtained by incubating at 55°C for 60 minutes at pH 12. Variation in pH from 11.5 to 12.5 did not affect CoASH recovery whereas recovery was reduced at temperatures above 60°C. The authors found an absolute requirement of a minimum of 5 mM mercaptoethanol during hydrolysis in order to obtain satisfactory recovery of CoASH. We find that 2.5 mM dithiothreitol (DTT) gives equally good results (results not shown). After hydrolysis pH is adjusted to 6.5 by 1 M perchloric acid, and precipitated $KClO_4$ is removed by centrifugation. It has been reported that the determination of CoASH should be carried out immediately in order to prevent loss of CoASH during storage [24]. However, we find no change in CoASH content upon storage at -80°C for several months (results not shown).

2. CoASH Determination

A large number of enzymatic and chromatographic methods for determination of CoASH released by hydrolysis of LCA have been developed. It is not our intention to give a comprehensive overall review of methods published but rather to give an introduction to the principles of the methods available. The overall problems in determination of CoASH have been sensitivity and reproducibility. The methods divide into two groups:

a. Methods with low sensitivity but high reproducibility, based on direct detection of released CoASH by enzymatic or chromatographic assays without an amplification step (end point assays).

b. Highly sensitive enzymatic methods containing an amplification step (cycling assay). These methods are, due to the high degree of amplification (5-15000 times), very sensitive to experimental errors in sample handling, and are therefore less reproducible.

2.1 End-point assays: The most common end-point assays used are enzymatic determination of CoASH based on the use of one of the following enzymes: 3-Hydroxyacyl-CoA dehydrogenase (3-HADH) [7], phosphate acetyltransferase (PAT) [7], α-ketoglutarate dehydrogenase (αKGDH) [24,25,26,27,56] and acyl-CoA synthetase [43,55,56,57]. The chromatographic method for determination of CoASH or derived CoASH makes use of either ion-exchange [29] or reversed-phase HPLC [5,6,20,28].

2.1.1. The 3-Hydroxyacyl-CoA dehydrogenase method: This method is based on a quantitative conversion of CoASH into acetoacetyl-CoA according to the following reaction with diketene.

$$\begin{matrix} CH_2{=}C{-}O \\ \quad\; | \quad\;\; | \\ \quad H_2C{-}C{=}O \end{matrix} \quad + \quad \text{CoASH} \longrightarrow \text{acetoacetyl-CoA}$$

The acetoacetyl-CoA is then converted into 3-hydroxybutyryl-CoA by 3-HADH.

$$\text{acetoacetyl-CoA} + \text{NADH} + H^+ \rightleftharpoons \beta\text{-hydroxybutyryl-CoA} + \text{NAD}^+$$

The oxidation of NADH is determined by spectrophotometry or fluorometry and the CoASH content calculated. The detection limit in the solution to be assayed is reported to be 30 μM and 0.3 μM by spectrophotometry and fluorometry, respectively. The method is very reproducible. The relative standard deviation is 1.5%. The method is strongly dependent upon 3-HADH not containing other dehydrogenases and on correct preparation of the diketene solution (for experimental details see [7]).

2.1.2. The phosphate acetyltransferase assay: The CoASH in the hydrolysis mixture is converted into acetyl-CoA using phosphate acetyltransferase (PAT). PAT catalyses the reversible transfer of acetyl groups between acetyl phosphate and CoASH according to the following reaction

$$\text{CoASH} + CH_3CO\text{-}OPO_3H_2 \rightleftharpoons CH_3CO\text{-SCoA} + H_3PO_4$$

The thioester bond of acetyl-CoA has an absorption peak at 233 nm and the CoASH content can be calculated directly from the absorbance change at this wavelength. The method has a limited value for determination of CoASH in biological material, because of a high content of UV-absorbing compounds and because of its low sensitivity (2.2 μM CoASH in the assay mixture). Furthermore sodium ions at 10 mM and higher concentrations interfere with the assay. High acetyl-CoA and phosphate should be avoided because they might displace the equilibrium and prevent the reaction from running to completion. The method should not be used with samples containing high DTT or glutathione because of non-enzymatic transfer of acetate from acetyl-CoA to these compounds [50]. The method is very reproducible with a relative standard deviation of 1% and is suitable for determination of CoASH in standard solutions (for experimental details see Michal and Bergmeyer [38]).

To increase the sensitivity of this method, a modified version was developed, where the acetyl-CoA formed was converted into ^{14}C-labelled citrate, using citrate synthase (CS) and [U-^{14}C]-oxaloacetate [42,47]. Originally the method was not designed to measure LCA, but in 1994 the assay was coupled to hydrolysis of the CoA esters, making it a sensitive and easy method for determination of LCA [33]. The PAT reaction was carried out as described above. The enzyme was then denatured at 95°C to avoid cycling, and the acetyl-CoA was converted into [^{14}C]-citrate using citrate synthase as follows

$$\text{acetyl-CoA} + [\text{U-}^{14}\text{C}]\text{-oxaloacetate} \rightleftharpoons [^{14}\text{C}]\text{-citrate} + \text{CoASH}$$

Excess [U-^{14}C]-oxaloacetate was converted into aspartate by glutamate-oxaloacetate transaminase as shown below and the [^{14}C]-aspartate produced was removed by adsorption to a cation exchange resin (AG 50W-X4).

$$[\text{U-}^{14}\text{C}]\text{-oxaloacetate} + \text{glutamate} \rightleftharpoons [^{14}\text{C}]\text{-aspartate} + \alpha\text{-ketoglutarate}$$

The [^{14}C]-citrate formed in the supernatant was determined by liquid scintillation counting, and the CoA content calculated using a CoA-standard curve. The authors report the detection range to be 10-200 pmoles.

2.1.3. The α-ketoglutarate dehydrogenase assay: This method is based on the determination of NADH formed during the conversion of CoASH and α-ketoglutarate into succinyl-CoA by α-ketoglutarate dehydrogenase as follows

$$\text{CoASH} + \alpha\text{-ketoglutarate} + \text{NAD}^+ \rightleftharpoons \text{succinyl-CoA} + \text{CO}_2 + \text{NADH} + \text{H}^+$$

A number of different versions of this method have been described [24,25,26,27,56]. A detailed description of the method is given by Michal and Bergmeyer [38]. The formation of NADH can be measured either by spectrophotometry (340 nm) or by fluorometry (excitation 340 nm; emission 420 nm). The detection limit using fluorometry is 50 pmoles CoA in the fluorometer cuvette. No substances should be present which absorb at the excitation wavelength, or which act as electron acceptors.

2.1.4. The acyl-CoA synthetase assay: Finally, CoASH can be quantified by conversion into sorboyl-CoA using acyl-CoA synthetase [43,55,56,57].

$$\text{CoASH} + \text{sorbate} + \text{ATP} \longrightarrow \text{sorboyl-CoA} + \text{AMP} + \text{PP}_i$$

Sorboyl-CoA can be measured spectrophotometrically at 300 nm. However, acyl-CoA synthetase also absorbs at this wavelength, so it is important to subtract the contribution from the synthetase in the calculation of the final CoASH concentration. The procedure can be used with CoASH concentrations as low as 1 μM in the assay solution [56].

2.1.5. HPLC methods: Ingebretsen *et al.* [29] determined LCA in rat liver mitochondria using acid precipitation followed by alkaline hydrolysis as described above. However, the authors introduced ion-exchange HPLC as a method for determination of released CoASH. CoASH could be separated completely from

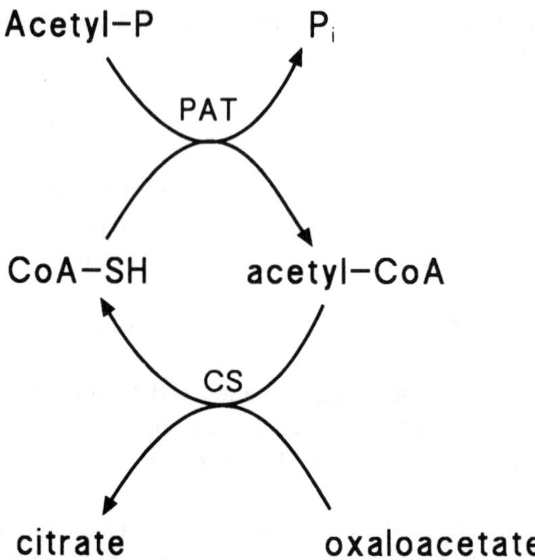

Fig. 4.1. Principle of CoA cycling involving phosphate acetyltransferase (PAT) and citrate synthase (CS) [30].

other compounds in the hydrolysis mixture on a Partisil-10 SAX anion exchange column. The column was equilibrated and CoASH eluted with 196 mM potassium phosphate buffer (pH 3.9), 2.0% (v/v) isopropanol and 0.05% (v/v) thiodiglycol. CoASH content was calculated using an external standard curve. The reproducibility was high (SD < 1.5%) and the method was found to be as sensitive as the enzymatic end-point methods based on fluorometric determination of NADH described above. The method was later modified using a Spherisorb 5 ODS reversed-phase column instead of an ion-exchange column and 220 mM KH_2PO_4 buffer, 11-12% methanol and 0.05% thiodiglycol as elution buffer [5,6,28]. The sensitivity of the method could be increased by derivatizing CoASH with monobromobimane before reversed-phase HPLC [20]. These authors report the lower detection limit to be 3 pmoles. However, the reproducibility was lower than with underivatized CoA; the authors reported the variation to be less than 10% for the same sample analysed on three individual days.

2.2. Cycling methods: All the above methods have limited sensitivity because they all rely on mole to mole conversion of CoASH into a measurable amount of product. The sensitivity of the cycling method is greatly increased by the introduction of an amplification step. However, this step demands very high precision in the preparation of sample and standard solutions. Isolation by acid precipitation and hydrolysis of the LCA are performed as described above. The majority of cycling assays are based on the use of PAT and CS in the cycling step as shown in Figure 4.1. In the presence of excess of acetyl phosphate and oxaloacetate, CoASH is converted into acetyl-CoA by PAT. Subsequently acetyl-CoA is

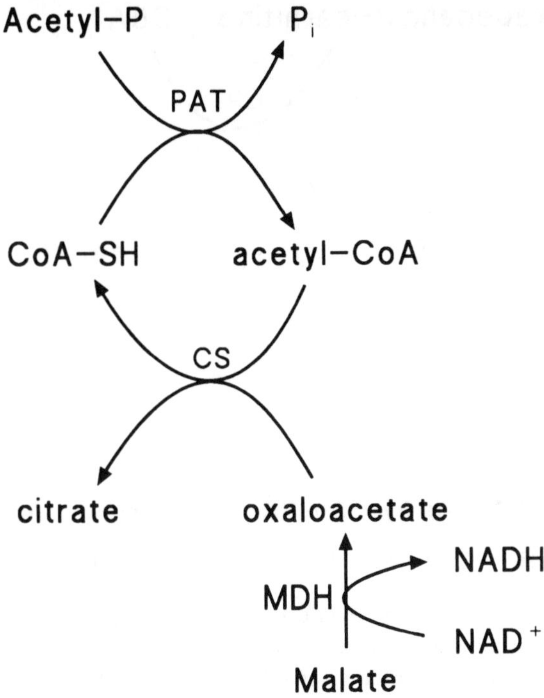

Fig. 4.2. CoA cycling with continuous assay of citrate formed phosphate by acetyltransferase (PAT), citrate synthase (CS) and malate dehydrogenase (MDH) [7].

deacetylated by CS, to form citrate and free CoASH. One cycle of this reaction generates the same amount of citrate and phosphate as the amount of CoASH present in the reaction solution.

The amount of citrate formed can be measured either continuously in a linked assay with malate-dehydrogenase (Figure 4.2) or the reaction can be stopped after an appropriate time and the amount of citrate formed measured in one of the two series of indicator reactions as shown below

I

$$\text{citrate} \xrightarrow{\text{aconitase}} \textit{cis}\text{-aconitase} \xrightarrow{\text{aconitase}} \text{isocitrate}$$

$$\text{isocitrate} + \text{NADP}^+ \xrightarrow[\text{dehydrogenase}]{\text{isocitrate}} \alpha\text{-ketoglutarate} + \text{CO}_2 + \text{NADPH} + \text{H}^+$$

II

$$\text{citrate} \xrightarrow{\text{citrate lyase}} \text{oxaloacetate} + \text{acetate}$$

$$\text{oxaloacetate} + \text{NADH} + \text{H}^+ \xrightarrow[\text{dehydrogenase}]{\text{malate}} \text{malate} + \text{NAD}^+$$

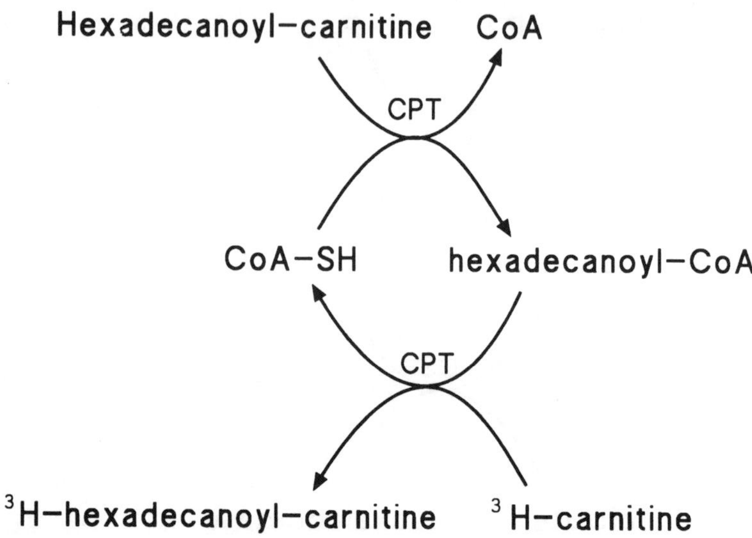

Fig. 4.3. CoA cycling involving carnitine palmitoyltransferase (CPT) [9,11].

The continuous citrate assay (Figure 4.2) was originally invented by Allred and Guy [1] and is described in great detail by Michal and Bergmeyer [7]. The method is highly reproducible; the relative standard deviation (RSD) is reported to be 1.8%. The detection limit is reported to be 40 fmoles in the measuring cuvette when NADH + H^+ formation is measured fluorometrically. The disadvantage of the method is that only a limited number of samples can be measured at a time.

The discontinuous citrate assay, where the cycling reaction is stopped by heating to 100°C for 3 minutes prior to determination of the citrate formed, was introduced by Kato [30]. This method is very sensitive, the reported lower detection limit is 70 fmoles, and this detection limit can be extended to 30 amoles by introducing a second amplification step where NADPH formed by the indicator reaction I is further amplified. The advantage of this method is that it can be used for assaying a large number of samples simultaneously. For detailed experimental procedures see Kato [30].

Finally an alternative cycling or exchange assay for determination of free CoASH based on the use of carnitine palmitoyltransferase (CPT), palmitoylcarnitine and [Me-^{3}H]-carnitine was developed by Skrede and Bremer [49]. The procedure was later modified to measure CoASH liberated from LCA by hydrolysis [11]. The principle of the method is based on the reactions shown in Figure 4.3. These reactions are completely dependent on CoASH. By using [Me-^{3}H]-carnitine the palmitoylcarnitine pool will be labelled radioactively at a rate proportional to the CoASH content of the reaction mixture. The reaction was stopped by addition of concentrated hydrochloric acid and the palmitoylcarnitine was extracted with butanol and quantified by liquid scintillation counting. The sensitivity was reported to be 0.2 nmoles in the assay solution.

C. METHODS BASED ON EXTRACTION, PURIFICATION AND QUANTIFICATION OF TOTAL AND INDIVIDUAL LCA BY GLC OR REVERSED-PHASE HPLC

A number of different methods have been developed for extraction, separation and quantification of total and individual LCA. The chromatographic methods used for the separation and quantification of LCA have included ion-exchange chromatography and reversed-phase HPLC or GLC following hydrolysis, transesterification or reduction of the isolated LCA.

1. GLC-Based Methods

Measurement of LCA by measuring free fatty acids released by hydrolysis of isolated LCA has been done for decades [10]. The original method was, however, rather insensitive and demanded 2-5 rat livers to get measurable amounts of LCA.

The GLC method requires that the LCA is separated completely from all other lipids during purification. A number of different extraction and purification procedures have been developed in order to obtain pure LCA. Most methods use either TLC or alumina chromatography.

1.1. LCA purification by TLC: Stymne and coworkers [51,52] have developed a GLC-method for determination of LCA in rat lung microsomal fractions. Heptadecanoyl-CoA was added as internal standard and the microsomal fraction was extracted with chloroform/methanol as described by Bligh and Dyer [9]. The LCA in the upper methanol/water phase were pre-purified by adsorption on a C_{18} reversed-phase cartridge. The adsorbed LCA were eluted with 0.4 M NH_4OH in methanol/water (4:1) and further purified on silica TLC plates. The plates were developed with butanol/acetic acid/water (50:20:30). The LCA band was scraped off and the esters transmethylated with sodium methoxide to form fatty acid methyl esters. The methyl esters were analysed by GLC and the amount of the individual methyl esters was calculated from the area of the added internal heptadecanoyl-CoA standard [52]. The method was later modified by replacing the pre-purification step with a C_{18} reversed-phase cartridge by a double chloroform wash of the methanol phase [53].

1.2. LCA purification by alumina adsorption chromatography: A different procedure for extraction and purification of acyl-CoA esters followed by transesterification and GLC analysis was developed by Prasad *et al.* [46]. Lyophilized rat liver tissue powder (50 mg) was resuspended in 0.15 ml 0.4 M sodium acetate (pH 4). Then 3.0 nmoles heptadecanoyl-CoA was added as internal standard along with 20 µl glacial acetic acid, 0.4 ml of water and 2.25 ml chloroform/methanol (1:2), and tissue was homogenized for 30-60 sec in a Polytron™ unit. The chloroform phase was removed and the water/methanol phase washed twice with chloroform, and proteins were precipitated by dropwise addition of 2 ml acetonitrile. After removal of precipitated protein by centrifugation, the LCA in the supernatant was adsorbed on alumina. The alumina was washed twice with chloroform/methanol (1:2) to remove glycerolipids and dried with 1.0 ml acetone. After removal of remaining acetone under nitrogen the LCA, still bound to the alumina, were reduced to long-chain alcohols and free CoA by sodium borohydride. The alcohols were extracted with pentane and converted into *t*-butyldimethylsilyl ethers with *t*-butyldimethylchlorosilane prior to separation by GLC. The recovery of added ^{14}C-labelled LCA internal standard was reported to be 47-50%. The authors suggest addition of excess medium-chain acyl-CoA esters as carriers in order to minimize losses of LCA.

The sensitivity and specificity of this method can be improved if the long-chain alcohols produced from reduction of LCA are converted into pentafluorobenzoyl derivatives and analysed by GLC-mass spectrometry in the negative-ion chemical ionization mode [61]. The same group also describes an alternative extraction and TLC purification procedure for LCA prior to the sodium borohydride reduction step.

A major problem with these GLC and GLC-MS methods is errors introduced by contaminating oxygen esters extracted together with LCA. In order to minimize this problem Tamvakopoulos [54] introduced a glycine aminolysis of the LCA prior to esterification of the *N*-acylglycinates formed with pentafluorobenzyl bromide and negative chemical-ionization GLC-MS analysis of the pentafluorobenzyl-*N*-acyl glycinates. The advantage of this procedure is that glycine aminolysis provides a 100 fold discrimination against oxygen esters compared to thioesters [54]. Furthermore, the method increased the sensitivity; the detection limit for hexadecanoyl-CoA was reported to be 300 fmoles. In this investigation LCA were extracted from rat liver by the method reported by Mancha *et al.* [36].

The GLC and GLC-MS methods are laborious but highly sensitive. However, information on the reproducibility and a systematic comparison of the values obtained with those produced by other methods are not available.

2. HPLC Methods

Separation of acyl-CoA esters by reversed-phase paired-ion liquid chromatography was first described by Baker and Schooley [2,3]. They used tetrabutylammonium phosphate as anion-pair agent and a gradient of methanol in water. Using this method the authors were able to obtain complete separation of CoASH and acyl-CoA esters from $C_{2:0}$ to $C_{20:0}$. However, their method was not applied to biological samples, and it did not become popular because it uses a corrosive and expensive ion-pair agent, which produces deterioration of column performance. Subsequently, a number of methods for separation of long- and short-chain acyl-CoA esters by reversed-phase HPLC have been developed. The major problem with the HPLC-based methods is, as for GLC methods, the extraction procedure. The amphipathic nature of LCA makes them difficult to extract without a large contamination of lipid and pigment, especially from plant tissue. Therefore, the extraction procedure must include steps which remove lipids and pigments, for example in the form of a separate lipid extraction step, most often a two-phase chloroform/methanol/water extraction. Furthermore water-soluble impurities need to be removed in a pre-purification step by ion-exchange, reversed-phase or adsorption chromatography or in a selective extraction step in order to obtain satisfactory chromatograms and column life.

The first reported attempt to extract, purify, and characterize LCA in biological material was published by Mancha *et al.* [36]. This method was slightly modified and adapted for quantification and separation by reversed-phase HPLC by Woldegiorgis *et al.* [59]. In the procedure reported by these authors, rat liver tissue was freeze-clamped and powdered. The powdered tissue was suspended in 2 ml isopropanol plus 2 ml 50 mM phosphate buffer (pH 7.2). The suspension was then extracted with petroleum ether 50% saturated with isopropanol to remove fatty acids and non-polar lipids. A chloroform/methanol (1:2) extraction of acyl-CoA esters preceded pre-purification on a neutral alumina AG7 column. The mixture of LCA was separated by C_{18}-reversed-phase HPLC, using a gradient of acetonitrile in 25 mM KH_2PO_4 (pH 5.3) (see analytical scheme in Figure 4.4). Individual peaks were identified using authentic standards and total amounts of acyl-CoA esters in the samples were calculated from peak areas and peak areas of external standards run in parallel. The recovery of acyl-CoA esters added directly on frozen tissue was reported to be 80%. However, actual data on recovery of an added radioactive standard were not shown. The authors found that livers from fed and fasted rats contained 108 ± 11 nmoles/g protein and 248 ± 19 nmoles/g protein, respectively. Using a slightly modified version of the above method, where the alumina column was replaced by a Waters C_{18}-Sep-Pak™ column, the levels of acyl-CoA esters in brown adipose tissue [22], swine heart and skeletal muscle [39] were found to be 235±40 nmoles/g protein, 11.34±1.48 nmoles/g wet weight and 4.35±0.71 nmoles/g wet weight, respectively. For comparison, Molaparast-Saless *et al.* [38] also measured the total amount of LCA extracted, using a method by Ingebretsen *et al.* [29], where the acyl-CoA esters were hydrolysed with KOH and the amount of free CoA measured by HPLC. This comparison shows a good agreement between the two methods. Later, in a report on the effect of fat feeding on tissue levels and composition of acyl-CoA, this group introduced the use of an internal standard to quantify the CoA-esters [17]. However, the authors did not report if the use of an internal versus an external standard improved the method.

In order to adapt the method of Woldegiorgis *et al.* [59] for studies of β-oxidation intermediates in mitochondria, a number of modifications have been introduced [4,31,44,58]. The major differences include an initial acidification in order to measure acid-soluble β-oxidation products, replacement of isopropanol with diethylether for lipid and fatty acid extraction and replacement of the alumina AG7 column with a DEAE-Sephacel™ column in the pre-purification step in order to separate acyl-carnitine from the acyl-CoA esters (see analytical scheme in Figure 4.4). The authors also changed the HPLC elution buffer to 50 mM

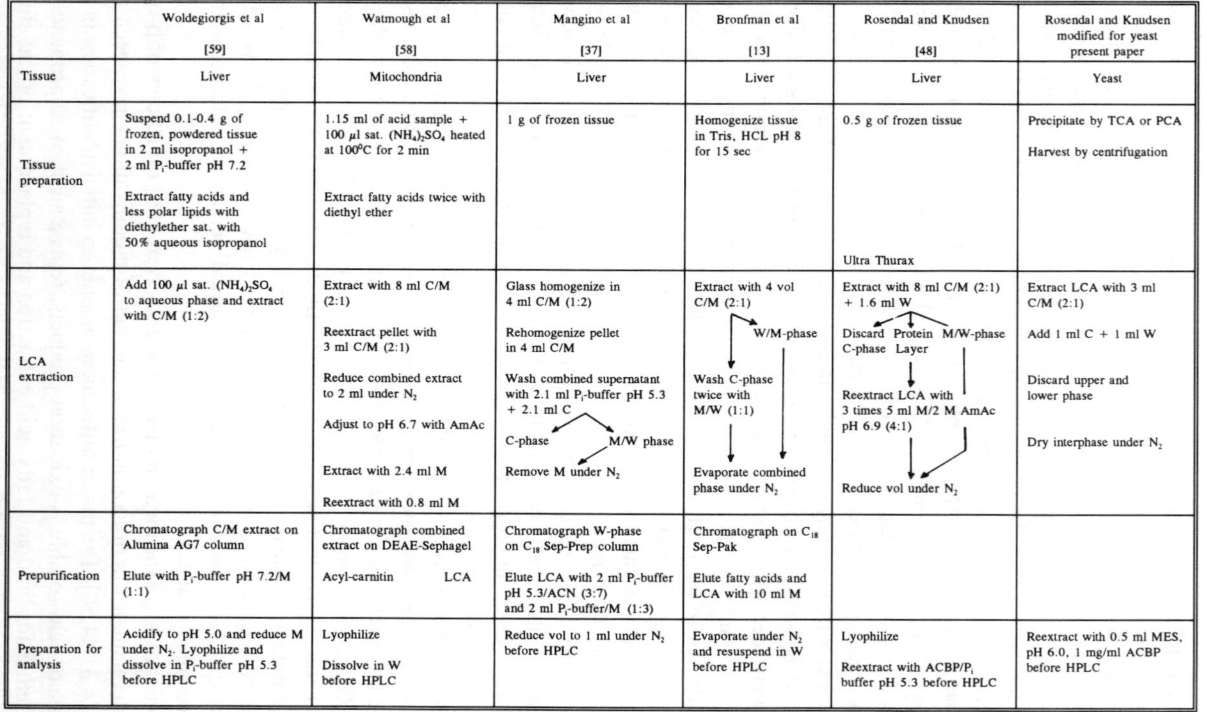

	Woldegiorgis et al [59]	Watmough et al [58]	Mangino et al [37]	Bronfman et al [13]	Rosendal and Knudsen [48]	Rosendal and Knudsen modified for yeast present paper
Tissue	Liver	Mitochondria	Liver	Liver	Liver	Yeast
Tissue preparation	Suspend 0.1-0.4 g of frozen, powdered tissue in 2 ml isopropanol + 2 ml P_i-buffer pH 7.2 Extract fatty acids and less polar lipids with diethylether sat. with 50% aqueous isopropanol	1.15 ml of acid sample + 100 μl sat. $(NH_4)_2SO_4$ heated at 100°C for 2 min Extract fatty acids twice with diethyl ether	1 g of frozen tissue	Homogenize tissue in Tris, HCL pH 8 for 15 sec	0.5 g of frozen tissue Ultra Thurax	Precipitate by TCA or PCA Harvest by centrifugation
LCA extraction	Add 100 μl sat. $(NH_4)_2SO_4$ to aqueous phase and extract with C/M (1:2)	Extract with 8 ml C/M (2:1) Reextract pellet with 3 ml C/M (2:1) Reduce combined extract to 2 ml under N_2 Adjust to pH 6.7 with AmAc Extract with 2.4 ml M Reextract with 0.8 ml M	Glass homogenize in 4 ml C/M (1:2) Rehomogenize pellet in 4 ml C/M Wash combined supernatant with 2.1 ml P_i-buffer pH 5.3 + 2.1 ml C → C-phase / M/W phase Remove M under N_2	Extract with 4 vol C/M (2:1) → W/M-phase Wash C-phase twice with M/W (1:1) Evaporate combined phase under N_2	Extract with 8 ml C/M (2:1) + 1.6 ml W → Discard C-phase / Protein Layer / M/W-phase Reextract LCA with 3 times 5 ml M/2 M AmAc pH 6.9 (4:1) Reduce vol under N_2	Extract LCA with 3 ml C/M (2:1) Add 1 ml C + 1 ml W Discard upper and lower phase Dry interphase under N_2
Prepurification	Chromatograph C/M extract on Alumina AG7 column Elute with P_i-buffer pH 7.2/M (1:1)	Chromatograph combined extract on DEAE-Sephagel Acyl-carnitin LCA	Chromatograph W-phase on C_{18} Sep-Prep column Elute LCA with 2 ml P_i-buffer pH 5.3/ACN (3:7) and 2 ml P_i-buffer/M (1:3)	Chromatograph on C_{18} Sep-Pak Elute fatty acids and LCA with 10 ml M		
Preparation for analysis	Acidify to pH 5.0 and reduce M under N_2. Lyophilize and dissolve in P_i-buffer pH 5.3 before HPLC	Lyophilize Dissolve in W before HPLC	Reduce vol to 1 ml under N_2 before HPLC	Evaporate under N_2 and resuspend in W before HPLC	Lyophilize Reextract with ACBP/P_i buffer pH 5.3 before HPLC	Reextract with 0.5 ml MES, pH 6.0, 1 mg/ml ACBP before HPLC

Fig. 4.4. Procedures for analysis of LCA in biological material by HPLC. Chloroform (C); methanol (M); water (W); acetonitrile (ACN); Acyl CoA-Binding Protein (ACBP); ammonium acetate (AmAC).

KH_2PO_4 buffer (pH 5.3) and extended the acetonitrile gradient to obtain complete separation of both short- and long-chain acyl-CoA esters. This gradient system was first introduced by Causey *et al.* [16]. The same group [31,58] introduced on-line radioactive determination of ^{14}C-labelled LCA and mitochondrial and peroxisomal β-oxidation products. A detailed description of synthesis and separation of dicarboxylic-mono-CoA and monocarnitine esters was presented by Pourfarzam and Bartlett [45].

A very simple method for LCA extraction for HPLC analysis was introduced by Bronfman and coworkers to study activation of the hypolipidaemic drugs ciprofibrate, nofenopin and clofibrate by rat liver and rat liver mitochondria and microsomes [12,13]. After incubation of microsomes with the hypolipidaemic drugs, ATP, Mg^{2+} and CoA, the incubation was stopped by adding chloroform/methanol (2:1), to give a final chloroform/methanol/water ratio of 2.67:1.33:1. The LCA content and composition in the upper water/methanol phase of this extraction were analysed directly by reversed-phase HPLC without further treatment. The authors reported the recovery of dodecanoyl-CoA to be 97% in the upper phase. The HPLC chromatograms showed five peaks, and one was identified to be the CoA ester of the hypolipidaemic drug. For rat liver and rat hepatocytes, the chloroform/methanol extraction was performed as above, but pre-purification of the LCA in the upper phase on a C_{18}-Sep-Pak™ column was introduced. The acyl-CoA esters were eluted from the Sep-Pak™ column with methanol. The eluate was evaporated to dryness with nitrogen, and the LCA were dissolved in water and separated on reversed-phase HPLC (see analytical scheme in Figure 4.4). Using chemically synthesised acyl-CoA esters of the drugs as internal standards added to the initial homogenate, the authors reported recoveries of 70-75% and 90-95%, when rat liver [13] and cultured hepatocytes [13,14], respectively, were analysed. The method is very simple, but does not give good chromatograms.

Rosendal and Knudsen [48] described a modified chloroform/methanol/water method for extracting acyl-CoA esters from rat liver before separation on reversed-phase HPLC. After decapitation of rats, the livers were removed and freeze-clamped. [1-C^{14}]-octadecenoyl-CoA and heptadecanoyl-CoA were added on top of the frozen tissue for recovery calculation and as internal standard respectively. The tissue was homogenized in chloroform/methanol/water (2.5:1.3:3.1) to give a two-phase system with lipids in the lower chloroform phase and short- and some long-chain acyl-CoA esters in the upper methanol/water phase. The interphase containing protein and most of the long-chain acyl-CoA esters were extracted three times with methanol/2 M ammonium acetate (pH 6.9, 4:1) to recover the esters. The three extracts were combined with the upper phase from the chloroform/methanol/water extraction and lyophilized. The dry residue was re-extracted with 0.7 ml 25 mM KH_2PO_4 (pH 5.3), containing 1 mg/ml of Acyl CoA Binding Protein (ACBP) [32] (see analytical scheme in Figure 4.4). Addition of this protein, which has a very high binding affinity for LCA, in the re-extraction step increased the overall recovery of LCA from 20 to 55%. LCA were separated by HPLC using a gradient similar to the one reported by Woldegiorgis *et al.* [59] consisting of acetonitrile in 25 mM KH_2PO_4 (pH 5.3). Identification of individual LCA was made by use of authentic standards and by alkaline hydrolysis, as described by Corkey *et al.* [18]. The amounts of the individual long-chain acyl-CoA esters were calculated from the area of the peaks of individual acyl-CoA esters and that of heptadecanoyl-CoA. A major disadvantage of this method is that more than 80% of the added [1-^{14}]-octadecenoyl-CoA internal standard was bound in the protein interphase layer after phase separation which made it necessary to re-extract the interphase protein layer to obtain sufficient recovery. This group also attempted to introduce a pre-purification step using C_{18}-Sep-Pak™ or BondElut™ reversed-phase pre-purification columns. However, the method was abandoned because of incomplete recovery of added [1-^{14}C]-octadecenoyl-CoA internal standard.

An important modification of the above method was introduced by Mangino *et al.* [37]. The modifications include use of a dual one-phase chloroform/methanol/water extraction followed by removal of tissue precipitate before forming a two-phase system by addition of more chloroform/water. Internal standard [1-^{14}C]-tetradecanoyl-CoA was added directly on top of 1 g of frozen tissue. The tissue was extracted twice with 4 ml chloroform/methanol (1:2). In order to obtain phase separation 2.1 ml of 10 mM KH_2PO_4 (pH 5.3) and 2.1 ml chloroform were added to the combined extracts. The lower lipid containing chloroform phase was removed and the upper methanol/water phase containing the acyl-CoA esters was washed with 2.1 ml chloroform. The upper phase was removed and the methanol evaporated under nitrogen. Further purification of the acyl-CoA esters in the resulting water phase was achieved on a C_{18}-Prep-Sep™ column. The LCA were eluted with 2 ml acetonitrile 10 mM KH_2PO_4 (7:3) and 2 ml of methanol/10 mM KH_2PO_4 (3:1). The combined eluates were concentrated to 1 ml under nitrogen to remove organic solvents and subjected to reversed-phase HPLC on a 5 mm Nucleosil™ C_{18} column using a gradient system of acetonitrile/10 mM KH_2PO_4 (pH 5.3) (see analytical scheme in Figure 4.4). The overall recovery of [1-^{14}C]-tetradecanoyl-CoA after the HPLC step was 61% and the chromatogram obtained was of excellent quality. The authors did not test the recovery of longer-chain acyl-CoA esters and the recovery of [1-^{14}C]-tetradecanoyl-CoA can not be compared directly with the recovery of longer-chain acyl-CoA esters because of a large difference in hydrophobicity. The high recovery (84.4%) of [1-^{14}C]-tetradecanoyl-CoA after the Prep-Sep™ pre-purification step by Mangino and coworkers [37] compared with low recovery of [1-^{14}C]-octadecanoyl-CoA on Sep-Pak™ reported by Rosendal and Knudsen [48] might be due to the difference in chain-length of the acyl-CoA used. The internal standard used for recovery test and quantification should be representative of the average chain-length of acyl-CoA found in the tissue in question and in most cases this would be C_{18} or C_{16} LCA. The Mangino method is clearly an improvement of the extraction procedure for LCA and should be further explored on different tissues and with internal standards of different chain-lengths.

Table 4.1.

Recovery of added [1-^{14}C]-hexadecanoyl-CoA in acyl-CoA extracts from yeast. Cells grown to OD_{550} 2.0 were acidified by addition of 6.6 M perchloric acid to a final concentration of 0.6 M and the cells were harvested by centrifugation. [1-^{14}C]-hexadecanoyl-CoA was added directly to the re-suspended and neutralized cell suspension before extraction with chloroform/methanol/water (3:1:1.8). For experimental details see the text.

	Experiment 1			Experiment 2			
	1	2	3	1	2	3	average ± SD
Water/methanol phase (%)	2.0	1.3	0.5	0.1	0.1	0.1	0.7 ± 0.8
Organic phase (%)	1.2	1.4	1.3	0.9	0.8	0.7	1.1 ± 0.3
ACBP extract. (%)	80.3	82.0	79.6	42.9	69.3	68.7	70.5 ± 14.7
	nmoles LCA/2 x 10^9 cells						
Total acyl-CoA	3.32	3.41	3.29	4.17	3.72	3.96	3.63 ± 0.35

An alternative to the method described by Mangino *et al.* [37], where proteins are removed from the chloroform/methanol extract before phase separation and removal of lipids, would be a method where all LCA are trapped in the interphase layer. By such a method both lipids and water-soluble impurities could be removed in one step. In order to create such conditions we attempted to extract LCA from *Saccharomyces cerevisiae* using different ratios of chloroform/methanol/water. Almost all added [1-^{14}C]-hexadecanoyl-CoA internal standard was trapped in the interphase layer, when a chloroform/methanol/water final ratio of 3:1:1.8 was used for extraction of LCA. In the experiments, yeast cells (200 ml) grown to OD_{550} 2.0 were acidified with perchloric acid or trichloroacetic acid. Cells were harvested by centrifugation and resuspended in 300 ml 100 mM MES buffer (pH 6.0). The pH was readjusted to 6.0 with KOH and [1-^{14}C]-hexadecanoyl-CoA and 1-2 nmoles heptadecanoyl-CoA were added as internal standards to the re-suspended, neutralized cell pellet. Water was added to give a total water volume of 800 ml including cell pellet water, and the cells were extracted with 3.0 ml chloroform/methanol (2:1) for 45 min at 4°C under continuous shaking. Additional chloroform and water, 1 ml of each, were added. After centrifugation for 45 min at 3,000 g at 4°C, the top and bottom phases were carefully removed and discarded. The interphase layer was dried under nitrogen and 0.5 ml 20 mM MES (pH 6.0) with 1.0 mg/ml ACBP was added and the LCA extracted by shaking for 30 min at 4°C. An aliquot of the extract was taken directly for separation on C_{18}-reversed-phase HPLC (see analytical procedure in Figure 4.4).

The results in Table 4.1 show that more than 97% of the added [1-^{14}C]-hexadecanoyl-CoA was trapped in the interphase layer. From 43% to 82% (average 71%) of the [1-^{14}C]-hexadecanoyl-CoA could be re-extracted from the interphase with 0.5 ml 100 mM MES buffer (pH 6.0) containing 1 mg/ml ACBP. The extracted LCA were separated on reversed-phase HPLC and excellent chromatograms (Figure 4.5) and reproducible amounts of acyl-CoA esters were obtained (Table

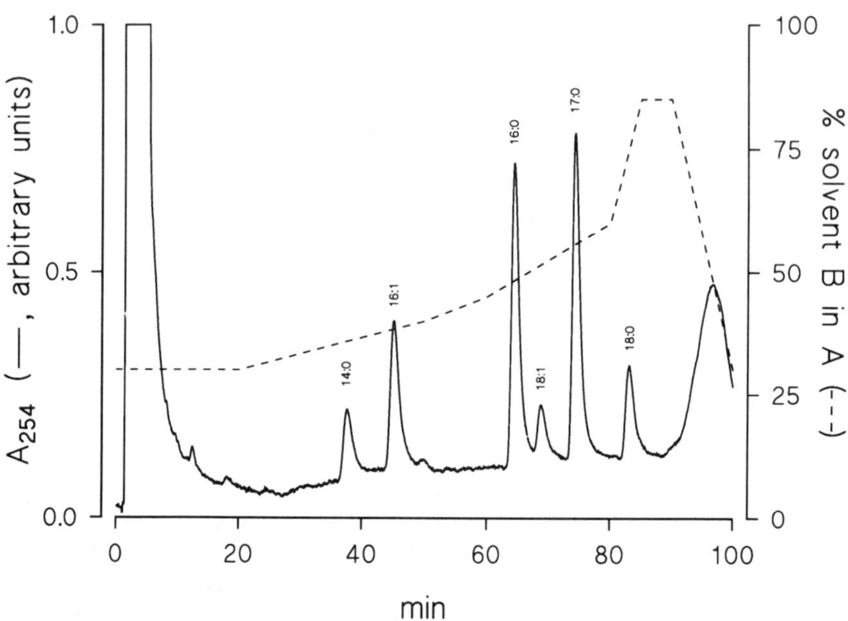

Fig. 4.5. Separation of acyl-CoA esters extracted from *S. cerevisiae*. The cells were harvested and extracted as described in the text. The HPLC column was equilibrated in 30% solvent B (70% acetonitrile in 20 mM KH_2PO_4, pH 5.3) in solvent A (20% acetonitrile in 20 mM KH_2PO_4, pH 5.3) and eluted with a gradient of B in A as indicated (----).

14:0, tetradecanoyl-CoA; 16:0, hexadecanoyl-CoA; 16:1, hexadecenoyl-CoA, 17:0, heptadecanoyl-CoA, 18:0, octadecanoyl-CoA; 18:1, octadecenoyl-CoA.

4.1). This method is now being used routinely for LCA determination in yeast.

We have also applied this method for extraction of LCA from 3T3-L1 cells grown in 56.7 cm^2 wells. Growth was stopped by addition of 7 ml 0.6 M trichloroacetic acid. Cells were harvested by scraping, transferred to a test tube and sonicated. [1-^{14}C]-hexadecanoyl-CoA and heptadecanoyl-CoA were added as internal standards and extraction were carried out as described for yeast cells. The results show that 0% and 2% of the radioactivity were lost in the upper and lower phase respectively, with more than 98% of the radioactivity trapped in the interphase layer (results not shown). However, the re-extraction efficiency of the [1-^{14}C]-hexadecanoyl-CoA from the interphase layer by ACBP-buffer was less efficient. Only 33 to 37% could be recovered in the re-extraction step. This step therefore needs to be optimized.

Whether this method can be extended to liver and other tissues is unknown for the moment. We hope so because the method is simple, gives beautiful chromatograms and has the advantage that metabolism is stopped instantaneously upon addition of 0.6 M trichloroacetic acid to tissue or cell cultures. This is especially important when LCA is to be determined in dilute cell cultures, where a harvesting step is needed before the chloroform/methanol/water extraction can be performed.

The literature reviewed above and results presented from acyl-CoA extractions of yeast clearly show that the direct determination of tissue LCA by extraction and reversed-phase HPLC is still in its infancy. No uniform method for LCA extraction, like the conventional extraction method for lipids, is available. The latest development with the methods by Mangino *et al.* [37] and Rosendal and Knudsen [48] using perchloric acid precipitation prior to extraction with modified chloroform/methanol/water ratios are interesting. The disadvantage of the method of Mangino [37] is that it uses a Prep-Sep™ pre-purification step, which in our hands gives low recovery. In order to evaluate the different methods properly, a systematic comparison of results obtained with the individual methods on the same tissue preparations is strongly needed. At present such a comparison is complicated by the fact that results presented in the literature have been obtained on different strains and animals at different nutritional states. Also a systematic comparison of results obtained by the indirect enzymatic and the direct chromatographic methods is strongly needed. At present no such comparison has been reported (preliminary results concerning this problem are presented below).

Finally, Mangino *et al.* [37] have raised the problem concerning the need to use response factors, when the amount of individual and total LCA is to be calculated from internal standards. Although these authors find a strict linearity with standard curves performed with all authentic standards, the absolute relationship between mass of the individual standards and peak intensity was dramatically different. This is surprising since it is not expected that acylation of the 4-phosphopantetheine SH group of CoASH with different chain-length carboxylic acids would influence the molar extinction coefficient of the adenine group. However, it is possible that the recovery of different acyl-CoA esters on the reversed-phase column is variable.

In order to investigate this problem we have determined response factors by running a mixture of varying amounts of acyl-CoA standards and constant amounts of internal standard heptadecanoyl-CoA. The standard mixtures were made from the same standard stock solutions (5-10 mM) on three individual days and run on reversed-phase HPLC essentially as described above for analysing LCA from yeast, and response factors calculated relative to heptadecanoyl-CoA. The results obtained are shown in Figure 4.6.

The data confirm the results obtained by Mangino *et al.* [37]. The response between injected mass and obtained area is strictly linear for individual standards. We also find a different response between injected mass of individual standards and the obtained peak area (Figure 4.6).

However, the difference obtained in response factors was not reproducible or systematic. The response factor, *i.e.* for hexadecanoyl-CoA, varied from 0.79 to 1.23 on individual days. We do not believe that this is due to a real difference in response factors. It rather focuses on another serious problem concerning handling of LCA solutions. We have known for a long time that it is extremely difficult to take aliquots out from LCA stock solutions in a reproducible manner even though we have worked out a standard procedure for how to thaw, mix and handle

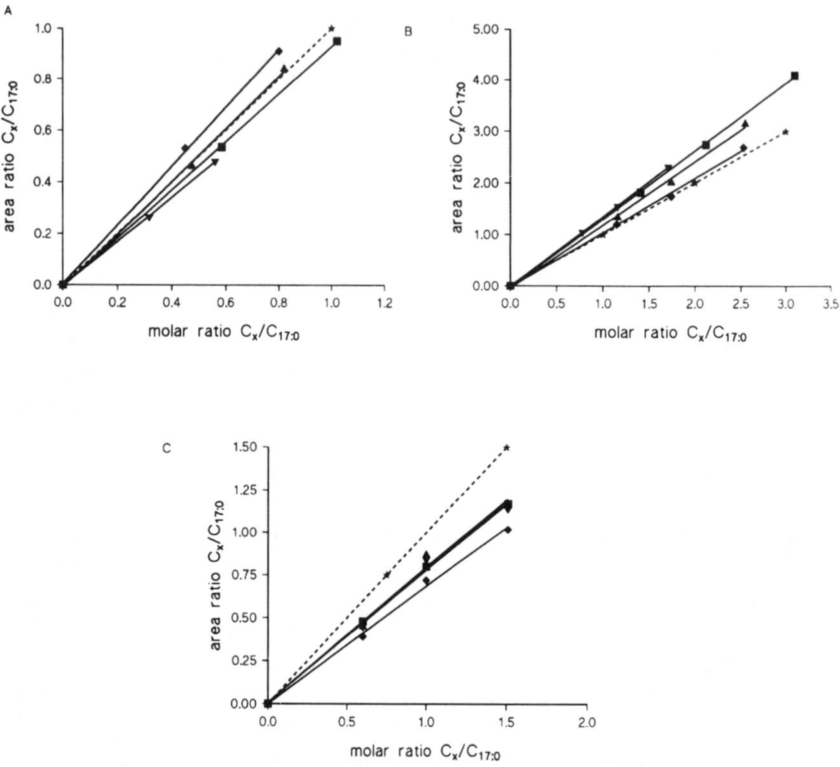

Fig. 4.6. Determination of response factors on three individual days (A, B and C). Mixtures of varying amounts of acyl-CoAs and constant amounts of C_{17}-CoA prepared from the same stock solutions were separated on HPLC as described in the legend to Figure 4.5. ■ pentadecanoyl-CoA, ▲ hexadecanoyl-CoA, ▼ octadecenoyl-CoA, ◆ octadecanoyl-CoA and ★ response factor 1.0.

LCA solutions. We therefore believe that all LCA have the same response factor and that the variation in response factors seen in Figure 4.6 represents variation in the actual concentration of individual standards on different days caused by irreproducible handling rather than differences in response factors.

This is an extremely important factor regarding use of an internal standard because pipetting this CoA ester will be prone to the same errors. We find that this problem can be minimized by addition of ACBP to solutions of acyl-CoA esters. For this reason we always make up large solutions of internal standard with ACBP added at a concentration slightly higher than the internal standard.

D. COMPARISON OF THE INDIRECT ENZYMATIC AND THE DIRECT CHROMATOGRAPHIC METHODS FOR LCA DETERMINATION

The LCA content in fed rat liver determined with the enzymatic cycling assay

Table 4.2.

Comparison of levels of total LCA in fed rat liver determined using the enzymatic cycling assay described by Kato [30] and the HPLC method described by Rosendal and Knudsen [48]. Three different livers a-c were used. In the enzymatic method LCA were precipitated with either trichloroacetic acid or perchloric acid as indicated. The ratio between enzymatic and HPLC determined LCA was calculated for experiments 1, 3 and 6 only. Butyryl-CoA (500 nmoles) was added to one of the samples of liver b, to test if short-chain acyl-CoA esters were precipitated along with the LCA. Values are nmoles LCA/g tissue ± SD.

	Enzymatic (Kato [30])	HPLC (Rosendal and Knudsen [48])	Ratio (Enz./HPLC)
1. Liver (a) trichloroacetic acid	92.3 ± 25.5	58.3 ± 4.4	1.6
2. Liver (a) perchloric acid	82.9 ± 9.6		
3. Liver (b) trichloroacetic acid	105.9 ± 5.4	43.6 ± 0.8	2.4
4. Liver (b) perchloric acid	103.5 ± 6.2		
5. Liver (b) trichloroacetic acid + butyryl-CoA	105.1 ± 2.0		
6. Liver (c) trichloroacetic acid	94.08 ± 16.0	42.0 ± 0.8	2.2

Table 4.3.

Comparison of levels of total LCA in $2x10^9$ yeast cells determined by the enzymatic cycling assay described by Kato [30] and the HPLC method modified for yeast as described in the text. Cells grown to an OD_{550} of 2 were harvested on 3 individual days (1-3) in aliquots of 200 ml or 10 ml for HPLC and enzymatic assay, respectively and LCA precipitated with either trichloroacetic acid or perchloric acid as indicated. The ratio between values obtained by the enzymatic and the HPLC methods was calculated. For experimental details see the text. Values are nmoles LCA/$2x10^9$ cells ± SD for 3 individual determinations.

	Enzymatic (Kato [30])	HPLC (Rosendal and Knudsen [47])	Ratio (Enz./HPLC)
Day 1. trichloroacetic acid	7.7 ± 1.2	3.68 ± 0.19	2.1
Day 1. perchloric acid	7.0 ± 0.5	3.36 ± 0.07	2.1
Day 2. trichloroacetic acid	6.5 ± 0.5	3.85 ± 0.18	1.7
Day 2. perchloric acid	7.4 ± 0.4	3.91 ± 0.23	1.9
Day 3. perchloric acid	6.7 ± 0.1	2.76 ± 0.10	2.4

Table 4.4.

Calculation of concentration of CoASH and heptadecanoyl-CoA stock solutions determined by spectrophotometer at A_{260} =14.7 mM^{-1} cm^{-1} before dilution and enzymatically [30] after dilution. Values are in mM ± SD.

	Spectrophotometric determination (mM)	Enzymatic (mM) (±SD(n-1))	
		Day 1	Day 2
CoA	1050	994 ± 46.4	1031.2 ± 16.4
C_{17}-CoA	40	41.6 ± 4.2	46.4 ± 3.3

described by Kato [30] and the HPLC method described by Rosendal and Knudsen [48] is shown in Table 4.2. A similar comparison of the enzymatic method [30] and the Rosendal and Knudsen method [48] modified for yeast as described in the text, is shown in Table 4.3.

In both comparisons the enzymatic method gave values from 1.6 to 2.4 times that obtained by HPLC methods. We were very surprised to see these results and have no immediate explanation for this difference. The results in Table 4.2, experiment 5, show that the higher value is not due to contamination by short-chain acyl-CoA esters in the enzymatic method. Addition of 500 nmoles of butyryl-CoA/g tissue sample before acid precipitation, did not affect the LCA levels obtained. Similar results were obtained with 3T3-L1 cells (results not shown).

To make sure that this difference was not due to errors in determining the concentration of heptadecanoyl-CoA internal standard used in the HPLC method, or the CoA standard solution used in the enzymatic method, we checked the concentration of heptadecanoyl-CoA and a CoASH stock solution both by spectrophotometry and by the enzymatic assay (see Table 4.4).

The data in Table 4.4 show that the determined difference in tissue levels of LCA by the enzymatic and the HPLC methods is not due to experimental errors introduced by incorrect internal and external standard solutions.

We must therefore conclude that the two methods indeed give different values of LCA in liver and yeast samples (Tables 4.2 and 4.3). It needs to be clarified if this difference is due to incomplete extraction in the HPLC-based methods or to non-LCA CoASH trapped in the acid-precipitated pellet. In the yeast experiments the harvesting step for the two methods is identical; therefore the difference can not be caused by different recovery of LCA in the harvesting step. The fact that the recovery of [1-^{14}C]-hexadecanoyl-CoA internal standard is high in the HPLC method indicates to us that the values obtained by the enzymatic method include a contribution of non-LCA CoASH.

We have been unable to find similar comparisons in the literature. It will be important to clarify which of the values, the enzymatic or the chromatographic, that represents the true value for LCA in the tissues tested. This is especially important because of the accumulating evidence that LCA play a key role in regulation of metabolism, signal transduction and transcriptional regulation. Exact values of intracellular LCA are also important in order to evaluate the role of ACBP and other acyl-CoA binding proteins in intracellular acyl-CoA metabolism and signal transduction.

E. CONCLUSION

The indirect methods for determining cellular LCA levels based on acid precipitation and determination of free CoASH following alkaline hydrolysis of the precipitated pellet are well established. The main choice to be made is between sensitivity and reproducibility. The non-cycling method has lower sensitivity but high reproducibility. The cycling method for CoASH determination is extremely sensitive but less reproducible due to the high amplification factor.

The chromatographic methods are less well established. The GLC-based methods, which rely on determination of the composition and amount of acyl chains released from purified LCA by hydrolysis, reduction or transesterification, are very sensitive. The methods for extraction of LCA are often laborious and great attention should be paid not to introduce errors by contaminating lipids and fatty acids. These lipids can be very difficult to separate from LCA during purification.

The HPLC-based methods have been undergoing a large development over the last few years. The key problem has been to develop efficient extraction procedures, which give separation of LCA from both lipid- and water-soluble contaminants. Both these goals have to be attained in order to get high quality chromatograms and acceptable column life. No standard method applicable to all tissues has yet been developed. Another important aspect is simplification in order to make the methods useful for a large number of samples.

The two most promising approaches seem to be - a chloroform/methanol-based extraction of lipids and acyl-CoA esters with removal of proteins prior to removal of lipid by phase separation [37]. However, this method requires a pre-purification step of the LCA in the methanol/water phase before the HPLC step in order to obtain good chromatograms and column life. The second approach is the modified Rosendal and Knudsen [48] method presented in this paper. In this method, extraction and pre-purification are obtained in one step by creating extraction conditions which facilitate adsorption of LCA in the protein-containing interphase layer. In this way the interphase layer works as a pre-purification medium. This method has been successfully applied to yeast and 3T3-L1 cells. However, it needs further development of the procedure for re-extraction of LCA from the interphase layer. Furthermore the method should be tested on other tissues. The advantage of the method is its simplicity. High recovery of highly purified LCA is obtained in a three-step method which includes a) an acid precipitation, b) a two phase chloroform/methanol/water extraction and c) a re-extraction step.

A major task to be solved concerning LCA determinations is a systematic comparison and evaluation of values obtained by different methods. Most important the cause of the large difference in values obtained by the HPLC methods and the indirect enzymatic method needs to be determined. These investigations are urgent because with the current state of art no method can be pointed out as *the* definitive one which gives the true intracellular concentration of LCA.

Abbreviations:

ACBP: Acyl-CoA binding protein, CoA:Coenzyme A, CoASH: Free Coenzyme A, CPT: Carnitine palmitoyltransferase, CS: Citrate synthase, DTT: Dithiothreitol, GLC: Gas liquid chromatography, GLC-MS: Gas liquid chromatography-mass spectrophotometry, 3-HADH: 3-hydroxyacyl-CoA dehydrogenase, HPLC: High-performance liquid chromatography, a-KGDH: a-ketoglutarate dehydrogenase, LCA: Long-chain acyl-CoA esters, MES:

2-[*N*-Morpholino]ethanesulphonic acid, PAT: Phosphate acyltransferase, RSD: Relative standard deviation, SD: Standard deviation, TLC: Thin-layer chromatography.

REFERENCES

1. Allred,J.B. and Guy,D.G., *Anal. Biochem.*, **29**, 293-299 (1969).
2. Baker,F.C. and Schooley,D.A., *Methods in Enzymology*, 72, 41-52 (1981).
3. Baker,F.C. and Schooley,D.A., Anal Biochem., 94, 417-424 (1979).
4. Bartlett,K., Hovik,R., Eaton,S., Watmough,N.J. and Osmundsen,H., *Biochem. J.*, **270**, 175-180 (1990).
5. Berge,R.K., Aarsland,A., Bakke,O.M. and Farstad,M., *Int. J. Biochem.*, **15**, 191-204 (1983).
6. Berge,R.K., Osmundsen,H., Aarsland,A. and Farstad,M., *Int. J. Biochem.*, **15**, 205-209 (1983).
7. Bergmeyer,H.U., *Methods of Enzymatic Analysis* **7**, 156-177 (1985).
8. Black,P.N. and DiRusso,C.C., *Biochim. Biophys. Acta*, **1210**, 123-145 (1994).
9. Bligh,E.G. and Dyer,W.J., *Can. J. Biochem. Physiol.*, **37**, 911-917 (1959).
10. Bortz,W.M. and Lynen,F., *Biochem. Z.*, **339**, 77-82 (1963).
11. Bremer,J. and Wojtczak,A.B., *Biochim. Biophys. Acta*, **280**, 515-530 (1972).
12. Bronfman,M., Amigo,L. and Morales,M.N., *Biochem. J.*, **239**, 781-784 (1986).
13. Bronfman,M., Morales,M.N., Amigo,L., Orellana,A., Nunez,L., Cardenas,L. and Hidalgo,P.C., *Biochem. J.*, **284**, 289-295 (1992).
14. Bronfman,M., Orellana,A., Morales,M.N., Bieri,F., Waechter,F., Staubli,W. and Bentley,P., *Biochim. Biophys. Res. Commun.*, **159**, 1026-1031 (1989).
15. Cardoso,C.M. and Meis,L.de, *Biochem. J.*, **296**, 49-52 (1993).
16. Causey,A.G., Middleton,B. and Bartlett,K., *Biochem. J.*, **235**, 343-350 (1986).
17. Chen,M.T., Kaufman,L.N., Spennetta,T. and Shrago,E., *Metabolism*, **41**, 564-569 (1992).
18. Corkey,B.E., Brandt,M., Williams,R.J. and Williamson,J.R., *Anal. Biochem.*, **118**, 30-41 (1981).
19. Deeney,J.T., Tornheim,K., Korchak,H.M., Prentki,M. and Corkey,B.E. *J. Biol. Chem.*, **267**, 19840-19845 (1992).
20. Demoz,A., Netteland,B., Svardal,A., Mansoor,M.A. and Berge,R.K., *J. Chromatogr.*, **635**, 251-256 (1993).
21. DiRusso,C.C., Heimert,T.L. and Metzger,A.K., *J. Biol. Chem.*, **267**, 8685-8691 (1992).
22. Donatello,S., Spennetta,T., Strieleman,P., Woldegiorgis,G. and Shrago,E., *Am. J. Physiol.*, **254**, 181-186 (1988).
23. Fulceri,R., Nori,A., Gamberucci,A., Volpe,P., Giunti,R. and Benedetti,A. *Cell Calcium*, **15**, 109-116 (1994).
24. Garland,P.B., *Methods of Enzymatic Analysis*, **7**, 207-211 (1985).
25. Garland,P.B., *Biochem. J.*, **92,** 10c-12c (1964).
26. Garland,P.B., Shepherd,D. and Yates,D.W., *Biochem. J.*, **97**, 587-594 (1965).
27. Guynn,R.W., Veloso,D. and Veech,R.L., *J. Biol. Chem.* **247**, 7325-7331 (1972).
28. Ingebretsen,O.C. and Farstad,M., *J. Chromatogr.*, **202**, 439-445 (1980).
29. Ingebretsen,O.C., Normann,P.T. and Flatmark,T., *Anal. Biochem.*, **96**, 181-188 (1979).
30. Kato,T., *Anal. Biochem.*, **66**, 372-392 (1975).
31. Kler,R.S., Jackson,S., Bartlett,K., Bindoff,L.A., Eaton,S., Pourfarzam,M., Frerman,F.E., Goodman,S.I., Watmough,N.J. and Turnbull,D.M., *J. Biol. Chem.*, **266**, 22932-22938 (1991).
32. Knudsen,J., Mandrup,S., Rasmussen,J.T., Andreasen,P.H., Poulsen,F. and Kristiansen,K., *Mol. Cell. Biochem.*, **123**, 129-138 (1993).
33. Kobayashi,A. and Fujisawa,S., *J. Mol. Cell Cardiol.*, **26**, 499-508 (1994).
34. Li,Q., Yamamoto,N., Inoue,A. and Morisawa,S., *J. Biochem. (Tokyo)*, **107**, 699-702 (1990).
35. Li,Q., Yamamoto,N., Morisawa,S. and Inoue,A., *J. Cell. Biochem.*, **51**, 458-464 (1993).
36. Mancha,M., Stokes,G.B. and Stumpf,P.K., *Anal. Biochem.*, **68**, 600-608 (1975).
37. Mangino,M.J., Zografakis,J., Murphy,M.K. and Anderson,C.B., *J. Chromatogr.*, **577**, 157-162 (1992).
38. Michal,G. and Bergmeyer,H.U., *Methods of Enzymatic Analysis*, **4**, 1967-1972 (1974).
39. Molaparast-Saless,F., Shrago,E., Spennetta,T.L., Donatello,S., Kneeland,L.M., Nellis,S.H. and Liedtke,A.J., *Lipids*, **23**, 490-492 (1988).
40. Nesher,M. and Boneh,A., *Biochim. Biophys. Acta,* **1221**, 66-72 (1994).
41. Nikawa,J., Tanabe,T., Ogiwara,H., Shiba,T. and Numa,S., *FEBS Lett.*, **102**, 223-226 (1979).
42. Pande,S.V. and Caramancion,M.N., *Anal. Biochem.*, **112**, 30-38 (1981).
43. Pearson,D.J. and Tubbs,P.K., *Biochem. J.*, **105**, 953-963 (1967).
44. Pourfarzam,M. and Bartlett,K., *Eur. J. Biochem.*, **208**, 301-307 (1992).
45. Pourfarzam,M. and Bartlett,K., *J. Chromatogr.*, **570**, 253-276 (1991).

46. Prasad,M.R., Sauter,J. and Lands,W.E.M., *Anal. Biochem.*, **162**, 202-212 (1987).
47. Rabier,A., Braiand,P., Petit,P., Kamoun,P. and Cathelineau,L., *Anal. Biochem.*, **134**, 325-329 (1983).
48. Rosendal,J. and Knudsen,J., *Anal. Biochem.*, **207**, 63-67 (1992).
49. Skrede,S. and Bremer,J., *Eur. J. Biochem.*, **14**, 465-472 (1970).
50. Stadtman,E.R., *J. Biol. Chem.*, **196**, 535-546 (1952).
51. Stymne,S. and Glad,G., *Lipids*, **16**, 298-305 (1981).
52. Stymne,S. and Stobart,A.K., *Biochim. Biophys. Acta*, **837**, 239-250 (1985).
53. Sugiura,T., Masuzawa,Y. and Waku,K., *J. Biol. Chem.*, **263**, 17490-17498 (1988).
54. Tamvakopoulos,C.S. and Anderson,V.E., *Anal. Biochem.*, **200,** 381-387 (1992).
55. Tubbs,P.K. and Garland,P.B., *Biochem. J.*, **93**, 550-557 (1964).
56. Tubbs,P.K. and Garland,P.B., *Methods in Enzymology*, **13**, 535-551 (1969).
57. Wakil,S.J. and Hübscher,G., *J. Biol. Chem.*, **235**, 1554-1558 (1960).
58. Watmough,N.J., Turnbull,D.M., Sherratt,H.S.A. and Bartlett,K., *Biochem. J.*, **262**, 261-269 (1989).
59. Woldegiorgis,G., Spennetta,T., Corkey,B.E., Williamson,J.R. and Shrago,E., *Anal. Biochem.*, **150**, 8-12 (1985).
60. Woldegiorgis,G., Yousufzai,S.Y.K. and Shrago,E., *J. Biol. Chem.*, **257**, 14783-14787 (1982).
61. Wolf,B.A., Conrad-Kessel,W. and Turk,J., *J. Chromatogr.*, **509**, 325-332 (1990).

Chapter 5

NUCLEAR MAGNETIC RESONANCE SPECTROSCOPY AND LIPID PHASE BEHAVIOUR AND LIPID DIFFUSION

Göran Lindblom

Department of Physical Chemistry, Umeå University, S-90187 Umeå, Sweden

A. INTRODUCTION

Lipid/water systems exhibit a very rich polymorphism and at least ten different phase structures are known. The phase behaviour of these systems has therefore been investigated extensively in the past and several comprehensive reviews have been published [57,74,139,153,154,215,245]. The ability of lipids to form liquid crystalline phases is of great interest in a variety of disciplines, above all in surface physical chemistry, material science and molecular biology. Lipids are major components of cellular membranes and are increasingly recognized as active participants in the functioning of the living cell. One example is diacylglycerol working as a second messenger in certain signal transduction pathways [181]; other examples will be presented later in this review. In contrast to the long-standing previous view that lipids served the subordinate role of an "inert matrix" or a "slick of oil", it is now clear that lipids play a much more important role for the functioning of biological membranes. Support for the view that lipids are essential for activity of membrane proteins is steadily being published [106,114,154].

Interesting technological applications of lipids are the possibility to polymerize these self-assembling materials in different structures [8,22,135,219] or make use of such lipid microstructures as templates to convert them into stable zeolite-like materials for use in composites [211,212]. There has been a recent stress of activity concerned with the prospect of using lipid liquid crystalline phases as biocompatible encapsulating and controlled-release media [59,138]. The explosive interest in this aspect of lipidology is evident not only from an increase in the number of publications in the area but also by the emergence of commercial enterprises aiming to capitalize on the pharmaceutical and cosmetic applicability of these most versatile materials.

The determination of the phase diagram is commonly the most important characterization of the macroscopic properties of a colloidal system. In general, determination of a complete phase diagram is very tedious and takes a lot of time. The basic principle is to mix the components and observe the number and nature of the phases. Fortunately, this can be done in many different ways. The most apparent and often most versatile method is direct visual observation by the naked eye or in a polarization microscope. Then, only when the phases readily separate macroscopically is it easy to determine the phase diagram. However, for surfactant systems liquid crystalline and other very viscous phases form and these require a long time for achieving macroscopic separation that can be observed easily in distinct regions. For these systems it is very common that phases in equilibrium are dispersed in one another with domain sizes ranging from tens to hundreds of a micrometer. For each prepared sample we face the following questions:

(i) How many phases are present? (ii) What is the nature of the phases (solid, liquid crystalline or liquid) and what is the symmetry of the present solid or liquid crystalline phase structures? (iii) What are the equilibrium compositions of the phases?

If macroscopic phase separation is difficult to obtain during a reasonable period of time, the measurements must be performed on the whole sample. The three most common methods used for such measurements are calorimetry, X-ray scattering and spectroscopic methods like nuclear magnetic resonance (NMR), electron spin resonance (ESR), Fourier-transform infrared (FT-IR) and fluorescence spectroscopy.

Calorimetric measurements are useful only for making a temperature-composition (T-X) diagram. Differential scanning calorimetry (DSC) readily measures the heat capacity as a function of temperature. Small angle X-ray scattering (SAXS) is used for the determination of the gross structure of the aggregates building up the phase. Although difficult, it is still possible to use this method even if several phases are present.

The spectroscopic methods rely on the fact that the measured properties are often sensitive to the local molecular environment. Thus, a spectroscopic investigation of the same molecule in different phases will give rise to different responses. In a multiphase sample this behaviour can be very useful making it possible to count the number of phases and most often also tell something about the nature of the phase.

In this chapter I will draw attention to how different NMR methods can be utilized for determinations of phase equilibria. In particular, ^{2}H and ^{31}P NMR have proven to be very effective, but there are also other nuclei like ^{1}H, ^{14}N, ^{19}F, ^{23}Na, ^{133}Cs, ^{35}Cl etc. that are used on certain occasions. This chapter contains a brief section about the theoretical aspects of NMR of quadrupole nuclei and of NMR of nuclei exhibiting chemical shift anisotropy, and a section regarding the measurements of translational diffusion coefficients of lipids and water with pulsed field gradient NMR. This article is by no means intended to provide an exhaustive compilation of all the literature on NMR studies of phase equilibria published, but is rather a relatively thorough discussion of how phase diagrams are constructed from NMR data with illustrations of some typical examples. There exists a vast literature on this subject dealing with membrane phospholipids, although most of the published reports only deal with systems in excess water. Since these systems have been extensively reviewed by others [44-47], I will only briefly touch on these studies when appropriate. Other kinds of information about the lipid systems that sometimes can be obtained also, like for instance molecular ordering or ion and water association, will be described in some depth. Some useful theoretical models making it possible to extract molecular information from relaxation data are also described. Concerning the NMR diffusion method for determinations of lipid translational diffusion coefficients there exist recent extensive reviews [130,151,231]. Therefore, only a brief discussion of this method will be covered here. A large number of general reviews on phase behaviour of lipids and surfactants have already been published; among these it is worth mentioning the extensive database collected and organized by Hogan and Caffrey, and the handbooks in refs. [169] and [30].

B. GIBBS PHASE RULE AND DESCRIPTION OF PHASE BEHAVIOUR

Amphiphiles and lipids dispersed in water show both thermotropic and lyotropic polymorphism. The most important thermodynamic principle used to describe the phase equilibria is the Gibbs phase rule which simply states that:

$$F + p = c + 2 \tag{1}$$

In this equation F is the number of degrees of freedom, which represents the number of independent intensive variables that remain after we have taken all possible constraints into account, and p is the number of coexisting phases in equilibrium. c is the number of components, the determination of which is not always trivial. Note that the term 'component' has a more technical meaning and should be carefully distinguished from the chemical species or compounds that are present (see *e.g.* [7]). At constant pressure $F = c - p + 1$. For a binary system composed of one lipid and water, $c = 2$ (the minimum number of independent species necessary to define the composition of all the phases present in the system), $F = 3 - p$. Frequently, a membrane lipid forms a lamellar (L_α) phase over the whole concentration region, and $F = 2$. Thus, both the temperature and the lipid/water content can be varied in this single phase region. However, in excess water, where the L_α phase is in equilibrium with 'pure' water, *i.e.* $p = 2$, only the temperature can be varied since $F = 1$. At the gel to L_α phase transition in excess water $p = 3$, $F = 0$, and the system is invariant. Thus, the gel and L_α phases can coexist only at a fixed temperature, often called the main transition, T_m. Note, that a substance containing membrane lipids with acyl chains of varying length, and unsaturation, does not constitute a true binary system with water. However, in many cases this does not create any large difficulties, but we should always be aware that such a "pseudo-binary" system may present "unexpected" results.

For three components we have $F = 4 - p$ and it is therefore necessary to fix the temperature also in order to be able to present the phase diagram in two dimensions. For such a system we thus utilize a triangular diagram with the pure compounds in each corner of the triangle (see below). The maximum number of phases that may be in equilibrium is now three, and a typical characteristic of the ternary phase diagram is the areas of three-phase triangles that occur, compared with the three-phase lines present in the two-component systems.

In the construction of phase diagrams the so-called *lever rule* is very important. A point in a two-phase region of a phase diagram (binary or ternary) indicates not only qualitatively that two phases are present but represents quantitatively the relative amounts of each one. The relative amounts of the two phases that are in equilibrium are determined by the relative distances of the particular point on its tie-line from the respective phase boundaries - this constitutes *the lever rule*. For a binary system tie-lines are always horizontal, but for a ternary system their directions are not that easily predicted and not seldom they have to be determined experimentally. Since the area under a peak in the NMR spectrum is proportional to the number, or fraction, of nuclei giving rise to the signal, this can be used to determine the proportion between different phases in the sample under investigation. From a study of a series of samples with different compositions a phase diagram can be constructed.

The following nomenclature will be used. The upper-case Latin letters characterize the type of long range order (one-, two-, or three-dimensional lattice), the subscript Greek letters stand for ordered (β) or disordered (α) acyl chains, and the subscripts I and II denote normal and reversed for liquid crystalline structures, while 1 and 2 are utilized for micellar solutions:

liquid crystalline phases

L_α = one-dimensional, lamellar (disordered, fluid)
P = two-dimensional oblique or centred (rippled)
H = two-dimensional hexagonal
H_I = normal (oil-in-water)
H_{II} = reversed (water-in-oil)
I = isotropic, cubic
I_I = normal (oil-in-water)
I_{II} = reversed (water-in-oil)

C stands for *crystalline*

isotropic solutions

L_1 = normal micellar solution phase
L_2 = reversed micellar solution phase

C. NMR METHODS FOR STUDIES OF PHASE BEHAVIOUR AND PHASE DIAGRAMS

Any lipid molecule contains at least one nucleus that can be studied by NMR, but some nuclei are more efficient and convenient to use than others. The nucleus used also depends on what kind of information one would like to get about the phase behaviour. If the request is to get a phase transition temperature only the choice of nucleus is more or less arbitrary. In the determination of border- and tie-lines in a phase diagram of a lipid system the NMR methods are all based on the static interactions occurring between spins in NMR, generally observed in solid state NMR. Thus, for ^{1}H NMR the static dipole-dipole couplings, observed through the linewidths of the NMR peaks, are utilized, while for ^{31}P, ^{19}F and ^{13}C the chemical shift anisotropy (CSA), observed in the line shape of the NMR signal, is the appropriate choice of NMR parameter. For nuclei like ^{2}H, ^{14}N, alkali and halide nuclei so-called quadrupole splittings can be used conveniently and these are generally observed in anisotropic phases. Among the appropriate nuclei for phase studies the most popular are ^{31}P and ^{2}H. For phospholipids, or other lipids or surfactants containing phosphorus, it is very convenient to study phase behaviour by ^{31}P NMR, since the phosphorus nucleus has a very high sensitivity in NMR. However, concerning the application of ^{2}H NMR in phase studies, we are faced with the problem that there are very few lipids with sufficient natural abundance of deuterium. This problem may be circumvented either by labelling the lipid (specifically or by perdeuteration) with deuterium or by preparing the samples under study with heavy water. In this article I will discuss these latter methods mainly and give some examples from my own laboratory. I will also briefly relate what kind of molecular information one may be able to obtain from such studies. A short theoretical discussion will therefore be given. Since ^{2}H NMR is the most utilized and generally applied method I will start with a discussion of such studies for determining phase diagrams, followed by ^{31}P NMR and then finally report briefly on how ^{1}H NMR investigations are practised in studies of phase behaviour in lipid-water systems.

The NMR spectrum for a general spin system is determined by the spin Hamiltonian, H, which consists of a number of interaction terms, of which only the following four terms are of interest for us [225]:

$$H = H_{\mathrm{Z}} + H_{\mathrm{CS}} + H_{\mathrm{Q}} + H_{\mathrm{D}} \tag{2}$$

where H_{Z} represents the interaction of the nuclear magnetic moment, μ_{N}, with the magnetic field, $\mathbf{B}_0$ [1]

$$H_{\mathrm{Z}} = - \mu_{\mathrm{N}}\mathbf{B}_0 \tag{3}$$

H_{CS} represents the effect of induced magnetic fields due to orbital electronic motions, *i.e.* the chemical shift, and H_{Q} and H_{D} are the quadrupolar and the dipolar Hamiltonians, respectively.

D. NMR OF QUADRUPOLAR NUCLEI IN LIPID-WATER SYSTEMS

Deuterium, and all the alkali and halogen nuclei (except fluorine) have spin quantum numbers I $\geq$ 1 and consequently possess quadrupole moments. The quadrupolar Hamiltonian, H_Q, arises from the interaction between the nuclear quadrupole moment, eQ, with the electric field gradient (EFG), $\nabla\mathbf{E} = V_{\mathrm{ik}}$, at the nucleus. H_Q in frequency units is [40,271]

$$H_{\mathrm{Q}} = \beta_{\mathrm{Q}} \sum_{\mathrm{q}=-2}^{2} (-1)^{\mathrm{q}} V_{-\mathrm{q}} A_{\mathrm{q}} \tag{4}$$

where V_{-q} is a component of the irreducible electric field gradient tensor:

$$V_{\pm 2} = \frac{1}{2}\left\{\left(\frac{\partial^2 V}{\partial x^2}\right) - \left(\frac{\partial^2 V}{\partial y^2}\right) \pm 2\mathrm{i}\left(\frac{\partial^2 V}{\partial x \partial y}\right)\right\} \tag{5a}$$

$$V_{\pm 1} = \mp\left\{\left(\frac{\partial^2 V}{\partial x \partial z}\right) \pm \mathrm{i}\left(\frac{\partial^2 V}{\partial y \partial z}\right)\right\} \tag{5b}$$

$$V_0 = \sqrt{\frac{3}{2}}\left(\frac{\partial^2 V}{\partial z^2}\right) \tag{5c}$$

A_q is a component of a second rank irreducible tensor operator working on nuclear spin functions:

$$A_{\pm 2} = \frac{1}{2} I_{\pm}^2 \tag{6a}$$

$$A_{\pm 1} = \mp\frac{1}{2}\left(I_{\pm} I_z + I_z I_{\pm}\right) \tag{6b}$$

$$A_0 = \frac{1}{\sqrt{6}}\left(3I_z^2 - \boldsymbol{I}^2\right) \tag{6c}$$

$\beta_Q = \frac{eQ}{2I(2I-1)h}$ is a reduced matrix element as defined in connection with the Wigner-Eckart theorem, and the different I_i's are the common spin operators. The principal axis system of the EFG tensors does not have to coincide with the laboratory reference frame. In that case we must make a transformation of the EFG tensor from the laboratory frame to its principal axis system or, equivalently, to transform the spin operators from the molecular to the laboratory system (the spin operators are defined in the laboratory system). This is conveniently accomplished with the Wigner rotation matrix elements, $D^{(2)}_{q'q}(\alpha\beta\gamma)$, where $(\alpha\beta\gamma)$ are the Eulerian angles defining the transformation [17]. The quadrupolar Hamiltonian can then be written

$$H_Q = \beta_Q \sum_{qq'} (-1)^q V^M_{-q} A^L_{q'} D^{(2)}_{q'q}(\Omega_{LM}) \tag{7}$$

where Ω_{LM} represents the Eulerian angles that specify the transformation between the coordinate system in the molecule (M) and the laboratory system (L). If molecular motion occur in the system under study the Hamiltonian in Equation (7) becomes time-dependent, since there is a time-dependence of Ω_{LM}. For an isotropic solution, like a diluted micellar solution, the mean value of the Wigner rotation matrix elements are zero and the quadrupole interaction manifests itself in relaxation effects only. For an anisotropic liquid crystalline phase, like the L_α or H_{II} phases, the mean value of $D^{(2)}_{q'q}(\Omega_{LM})$ is not necessarily vanishing and quadrupole splittings may be observed in the NMR spectrum. The majority of the lyotropic liquid crystals are uniaxial phases, although there exist exceptions [41]. Here only uniaxial phases will be treated, and the symmetry axis is usually called the director (D). The different coordinate systems are illustrated for a macroscopically aligned lamellar phase in Figure 5.1. The expression for the electric field gradient tensor in the laboratory frame now involves two transformations and two sets of Euler angles: Ω_{DM}, which describe the transformation from the principal-axis system in the molecule to the director system; and Ω_{LD} leading from the director to the laboratory frame the z-axis of which is along the static magnetic field (Figure 5.1). The Hamiltonian now becomes

$$H_Q = \beta_Q \sum_{qq'q''} (-1)^q V^M_{-q} A^L_{q''} D^{(2)}_{q'q}(\Omega_{DM}) D^{(2)}_{q''q'}(\Omega_{LD}) \tag{8}$$

Provided that the magnitude of the electric field gradient is constant, all quantities except $D^{(2)}_{q'q}(\Omega_{DM})$ in Equation (8) remain constant over the molecular motion. Since we have cylindrical symmetry around the director (uniaxial system) the time average $\overline{D^{(2)}_{q'q}(\Omega_{DM})}$ is zero unless q′= 0, giving the following Hamiltonian

$$H_Q = \beta_Q \sum_{qq''} (-1)^q V^M_{-q} A^L_{q''} \overline{D^{(2)}_{0q}(\Omega_{DM})} D^{(2)}_{q''0}(\Omega_{LD}) \tag{9}$$

In an anisotropic liquid crystalline phase the quadrupole interaction does not average to zero and we get a static term which is small compared to the Zeeman term and only the secular part of the Hamiltonian needs to be considered. Furthermore, the electric field gradient tensor is nearly cylindrical symmetric (*e.g.* for ^{2}H it is located along the C-^{2}H bonds) leading to that we only have to consider the V^M_0 term (*i.e.* the asymmetry parameter η is zero) and we get

$$H_Q = \frac{\nu_Q S}{6}(3\cos^2\theta_{LD} - 1)(3I_z^2 - \mathbf{I}^2) \tag{10}$$

where

$$\nu_Q = 3\beta_Q V^M_0 = \frac{3}{2}\beta_Q eq$$

and

$$S = \overline{D^{(2)}_{00}(\Omega_{DM})} = \frac{1}{2}\overline{(3\cos^2\theta_{DM} - 1)}$$

S is an order parameter characterizing the ordering of the electric field gradient tensor in the molecule, *i.e.* it is a measure of the ordering of C-^{2}H or O-^{2}H bonds or EFG of ions associated to an interfacial surface formed by amphiphiles[1]. Thus, to first order, the quadrupolar interaction shifts the Zeeman energy levels by the amount

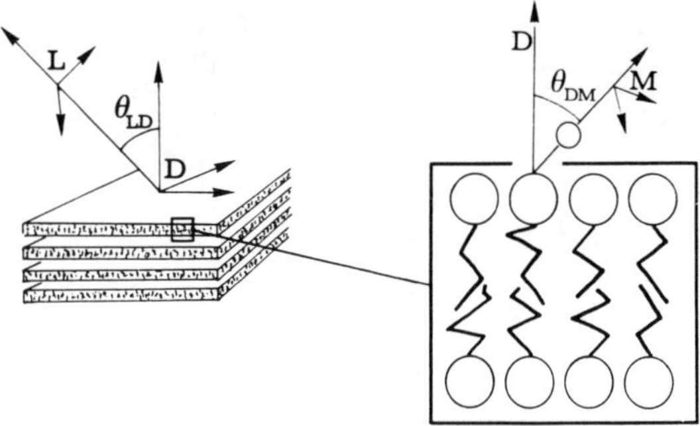

Fig. 5.1. A schematic drawing of an L_α phase, where the three different coordinate systems are defined; the laboratory frame is denoted (L), the director coordinate system (D), and the molecular frame (M). θ_{LD} and θ_{DM} are the angles between the z-axis of the laboratory and director systems and between the z-axis of the director and the molecular systems, respectively.

$$E_m = \frac{\nu_Q S}{6}(3\cos^2\theta_{LD} - 1)[3m^2 - I(I+1)] \tag{11}$$

For deuterium the spin quantum number I = 1 and there are three energy levels

$$\begin{aligned} E_{+1} &= -\gamma B_0 + \frac{\nu_Q S}{6}(3\cos^2\theta_{LD} - 1) \\ E_0 &= \quad\quad - \frac{\nu_Q S}{3}(3\cos^2\theta_{LD} - 1) \\ E_{-1} &= \gamma B_0 + \frac{\nu_Q S}{6}(3\cos^2\theta_{LD} - 1) \end{aligned} \tag{12}$$

where γ is the gyromagnetic ratio. The selection rule for allowed transitions is $\Delta m = \pm 1$ and therefore the two resonances are

$$\begin{aligned} h\nu_+ &= E_{-1} - E_0 = \gamma B_0 + \frac{\nu_Q S}{2}(3\cos^2\theta_{LD} - 1) \\ h\nu_- &= E_0 - E_{+1} = \gamma B_0 - \frac{\nu_Q S}{2}(3\cos^2\theta_{LD} - 1) \end{aligned} \tag{13}$$

yielding two peaks of equal intensity (Figure 5.2) in the ^{2}H NMR spectrum with a quadrupole splitting, $\Delta\nu_Q$, of

$$\Delta\upsilon_Q = \nu_Q S(3\cos^2\theta_{LD} - 1) \tag{14}$$

For a sodium or chloride ion I = 3/2 and there will be four energy levels

[1] More generally, the fluctuations in the molecular ordering is quantified by a traceless order parameter tensor, S_{ii} [209];

$$S_{xx} = \tfrac{1}{2}(\overline{3\cos^2\gamma\sin^2\beta - 1}); S_{yy} = \tfrac{1}{2}(\overline{3\sin^2\gamma\sin^2\beta - 1}); S_{zz}\ \tfrac{1}{2}(\overline{3\cos^2\beta} - 1)$$

These order parameters are elements of a 3x3 ordering matrix which describes the fluctuations of second rank tensors. For a uniaxial system and a cylindrical symmetric EFG only one order parameter S_{zz} is needed to specify the matrix.

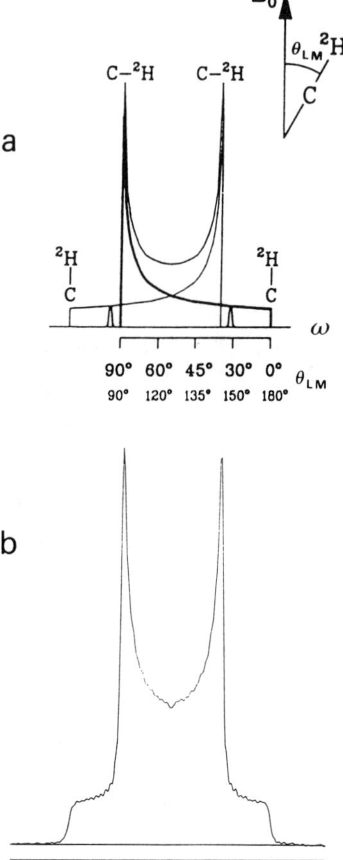

Fig. 5.2. a) Dependence of the NMR peak on the orientation between the molecular segment (a C-^{2}H bond) and the external magnetic field ($\mathbf{B}_0$). Note that the *spin orientation* does not affect the frequency of the NMR signal. For a "powder" sample all possible orientations are present with equal probabilities, generating the typical "powder pattern", which here is shown for a system with axial symmetry.
b) ^{2}H NMR quadrupole splitting of ^2H_2O in an L_α phase composed of DDAO and water.

and the NMR spectrum for a sample with uniform director orientation consists of three peaks with integrated intensity ratio between the peaks being 3:4:3 (Figure 5.3). Generally, there will be 2I equally spaced peaks centred at the Larmor frequency, $\boldsymbol{\omega}_0 = \gamma \mathbf{B}_0$.

Since water molecules or counterions in a liquid crystalline phase may reside in different sites, Equation (14) has to be modified according to

$$\Delta\nu_Q = |\sum_i p_i \nu_Q^i S_i (3\cos^2\theta_{LD} - 1)| = |(\overline{\nu_Q S})(3\cos^2\theta_{LD} - 1)| \qquad (15)$$

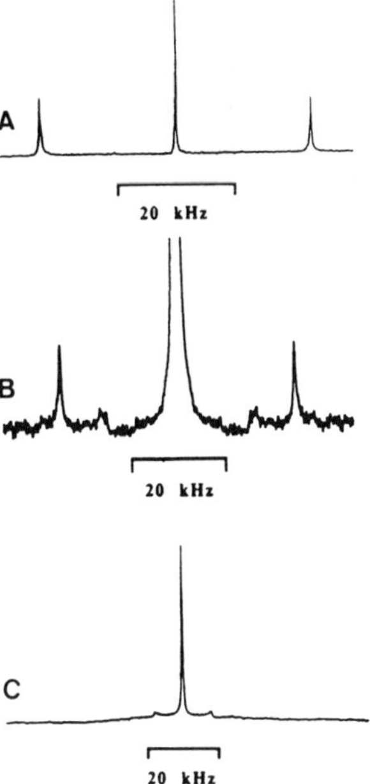

Fig. 5.3. ^{23}Na NMR spectra from

a) a macroscopically aligned L_α phase composed of DOPA and 95% (w/w) water at 25°C.

b) a sample composed of DOPA and 45 % (w/w) water forming a mixture of H_{II} and L_α phases.

c) "powder" sample of DOPG and 30% (w/w) water. Adapted from [155].

where the p_i represent the fraction of ions or water molecules in site i.

For a powder sample (with a random distribution of the director axes) [225] where all values of $\cos\theta_{LD}$ are equally probable the quadrupole splitting in the NMR spectrum corresponds to that for $\theta_{LD} = 90°$ in Equation (14) or (15) and gives

$$\Delta\nu_Q = |\sum_i p_i \nu_Q^i S_i| = |(\overline{\nu_Q S}))| \tag{16}$$

From equations (14) and (15), it is obvious that the $\theta_{LD} = 90°$ orientation of the director axis with respect to the main magnetic field would have a splitting of one-half that obtained at the $\theta_{LD} = 0°$ orientation. Furthermore, the intensity of the $\theta_{LD} = 90°$ peak would be the greatest and would decline steadily to the $\theta_{LD} = 0°$ peak which would have the least intensity. This can be visualized by imagining all of the possible orientations of vectors starting *in origo* and with their arrows on a sphere. The amount of orientations parallel to the magnetic field will be limited to only two, either up or down, whereas the perpendicular orientations can exist anywhere in a plane perpendicular to the magnetic field. Thus, there are more possible orientations perpendicular to the magnetic field.

According to Equation (16) it can be expected that the measured NMR

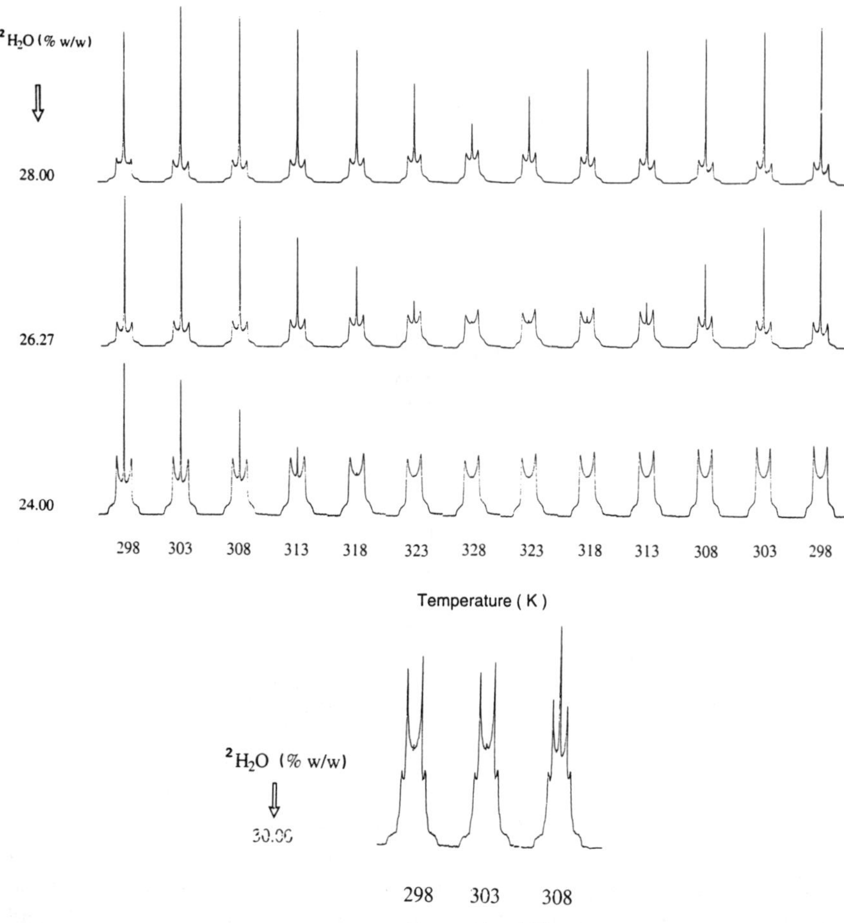

Fig. 5.4. ^{2}H NMR quadrupole splitting as a function of temperature and composition in the system DDAO/gramicidin/water. The molar ratio, X_G, between gramicidin and DDAO is 3:97.

a) The single middle peak originates from a cubic liquid crystalline phase and the splitting is due to an L_α phase. Note the metastable cubic phase at 24% ^2H_2O between 298 and 313K.

b) At 308 K the three phases observed are a cubic, an H_I and an L_α phase. (Courteously provided by G. Orädd).

quadrupole splitting depends on the phase structure, since at least one of the parameters S, p_i or ν_Q should be sensitive to a change in the mesophase structure. Thus, it should be possible to investigate phase equilibria through measurements of quadrupole splittings. For lyotropic liquid crystals it has been shown that measurements on either heavy water or on deuterated amphiphiles is a very convenient and rapid method to obtain almost complete phase diagrams. In most cases the largest effect on the observed quadrupole splitting is due to a change in the order parameter between phases. The magnitude of S is determined by two factors: i) the effect of the degree of anisotropy of the liquid crystalline phase, and ii) the dependence on the angle θ_{DM}. For a lyotropic liquid crystalline phase the value of S can vary between -1/2 and 1 depending on the value of θ_{DM}, and at the "magic angle" $\theta_{DM} = 54.7°$ S is equal to zero. This means that even if the measured quadrupole splitting is small, the anisotropy of the phase itself may be high, and therefore the sample may not consist of a cubic phase or a micellar solution, both of which are isotropic and have $S = 0$. This is, however, usually not a problem in the determinations of phase equilibria, if one also considers the viscosity or optical properties of the sample under study. A further situation where the observed quadrupole splitting may be small, in spite of the fact that the anisotropy is large, occurs when either of the factors in the term $\nu_Q^i S_i$ have opposite signs in different sites causing a partial cancellation of the terms in Equation (16). Information about whether such a case is present or not can be obtained from an examination of the temperature dependence of the splitting. Finally, in studies of phase equilibria it is very important that a single phase gives rise to a characteristic quadrupole splitting and when two or more phases are present the NMR spectrum will consist of a superposition of quadrupole splittings originating from the different phases in the sample, i.e. there is a slow exchange between phase regions (Figure 5.4). This is also most often the case.

Let us look a bit closer at how the geometry of the aggregates building up the mesophase affects the value of the order parameter and the quadrupole splittings. Figure 5.5 shows some examples of the structure of different common liquid crystalline phases. The S values of lamellar and hexagonal phases of the same amphiphile system can easily be related to each other provided the *microscopic structure of the aggregates is identical* in the two phases (*i.e.* the microscopic ordering of lipid acyl chains or the structure at the water-lipid interface. However, if this is not the case information about microstructure may be obtained). For the hexagonal phase the quadrupole Hamiltonian can be written

$$H_Q^H = \beta_Q \sum_{qq'q''q'''} (-1)^q V_{-q}^M A_{q'''}^L D_{q'q}^{(2)}(\Omega_{SM}) D_{q''q'}^{(2)}(\Omega_{DS}) D_{q'''q''}^{(2)}(\Omega_{LD}) \qquad (17)$$

where a transformation between the director coordinate system and the coordinate system fixed at the surface (S) of the rod-like aggregates building up the hexagonal phase has been applied. Since $\overline{D_{q''0}^{(2)}(\Omega_{DS})} = \delta_{q''0} D_{00}^{(2)}(\Omega_{DS})$, averaging over the molecular motion around the rod-like aggregate yields

$$\overline{H_Q^H} = \beta_Q \sum_{qq'''} (-1)^q V_{-q}^M A_{q'''}^L \overline{D_{0q}^{(2)}(\Omega_{SM})} D_{00}^{(2)}(\Omega_{DS}) D_{q'''0}^{(2)}(\Omega_{LD}) \qquad (18)$$

The coordinate system (S) at the surface of the rod is coincident with the director system (D) of the lamellae. This is easily visualized by a simple picture, where a rod in the hexagonal phase has been created from a bilayer in the L_α phase by just curling the monolayers into rods. It is thus obvious that the two coordinate systems (S) and (D) coincide. However, the z-axis of the

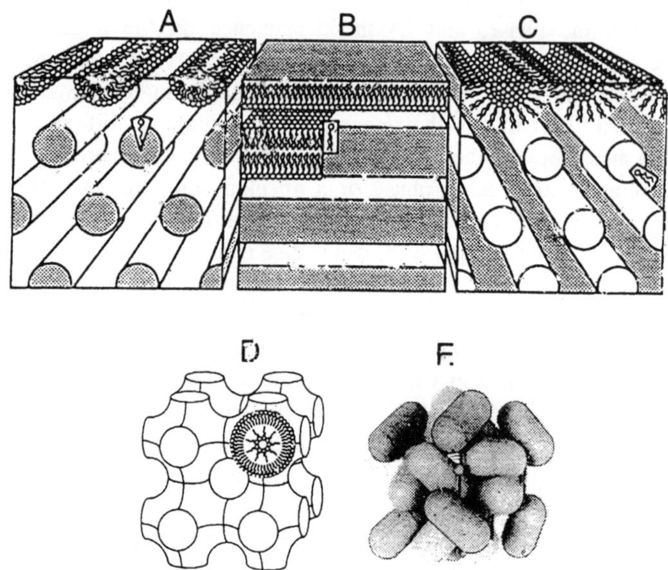

Fig. 5.5. Structure of the most common anisotropic liquid crystalline phases that membrane lipids can form. (A) Normal hexagonal (H_I) phase; (B) lamellar (L_α) phase; and (C) reversed hexagonal (H_{II}) phase [200]. (D) Schematic of a bicontinuous cubic phase belonging to the space group *Im3m*. Lipid monolayers are draped on each side over this so-called minimal surface [154]. (E) A scaled model of a unit cell of a cubic phase structure built up of short rod-like micelles [65].

director system of the hexagonal phase is along the rod-axis, *i.e.* it is perpendicular to the coordinate system (S) giving $D^{(2)}_{00}(\Omega_{DS}) = -\frac{1}{2}$ From Equations (9) and (18) it is then found that

$$\overline{H_Q^H} = -\frac{1}{2}\overline{H_Q^L} \tag{19}$$

where $\overline{H_Q^L}$ is the time-independent quadrupolar Hamiltonian for the L_α phase. Thus, if the microscopic properties (molecular ordering in the different aggregates) are the same in two phases, then the quadrupole splitting will be twice as large for the L_α phase than for the hexagonal phase (H_I or H_{II}).

1. Phase Diagrams Determined by ^2H NMR of ^2H_2O

A convenient and rapid method to establish phase equilibria of lyotropic amphiphile or lipid systems is to use ^2H NMR of ^2H_2O, where rather detailed information about the phase diagram may be obtained. A great advantage with this method is that practically any lipid system can be studied without the requirement that the lipid has to contain a specific NMR-active nucleus, or that a labelling of the lipid itself has to be accomplished. Almost 20 years ago, the first 'complete' phase diagram (Figure 5.6) was constructed from data obtained by ^2H NMR (Figure 5.7), and this was performed for the system dipalmitoylphosphatidylcholine (DPPC) and water in the temperature range 25 to 60°C and the composition range 4 to 15 mol ^2H_2O/mol DPPC [258]. The observation of two independent quadrupole splittings for some DPPC/water compositions (Figure 5.7) over a well-defined temperature range shows that there must exist physically distinguishable water sites with lifetimes of at least 10^{-4} s. Two explanations can be given for these water sites; either there are different binding sites within a homogenous single phase (*e.g.* on the same bilayer) or they can result from water in different phase structures. Since the translational diffusion coefficient is sufficiently large [159] (D_{water} = 4 - 7 10^{-10} m^2s^{-1} in the L_α phase) a fast exchange occurs between different water-binding sites in the homogenous phase, excluding a situation with two independent binding sites in a single phase region. Thus, the observed quadrupole splittings are due to different phases. To get the relative intensity (which is proportional to the amount of a particular phase) of the different NMR quadrupole splittings, computer simulations of the spectra are performed. Based on Gibbs phase rule and the intensities of the quadrupole splittings obtained a phase diagram can be constructed.

As can be inferred from the phase diagram in Figure 5.6 lowering of the water content can trigger phase changes between the liquid crystal and the gel. This is due to the fact that phospholipid bilayers interact through a short-range repulsive force [191], and the lower the water content, the stronger this interaction becomes and this inter-aggregate interaction may cause phase changes. A calculation of the phase diagram of the DDPC/water system (Figure 5.8) has been performed based on derived expressions for the concentration and temperature dependences of the chemical potentials of DPPC and water [84]. The agreement between the calculated and experimentally determined phase diagrams is excellent, demonstrating the crucial role played by the repulsive short-range force in the phase transitions. Thus, there is a close relation between interbilayer forces and the phase behaviour. With decreasing water content the conditions become more advantageous for the bilayer to transform to a condition in which the repulsive interaction is weaker. At the phase transition the interbilayer interaction compensates for the difference in the intrabilayer interactions.

A relative large number of phase diagrams of both surfactants [98,99,109-112,115-119,121,132,168,205-207,269] and membrane lipids [6,16,19,86, 120,143,145,155,174-176,187,188,198,199,236,264] has been determined by this NMR method either used alone or as a complement to other conventional methods.

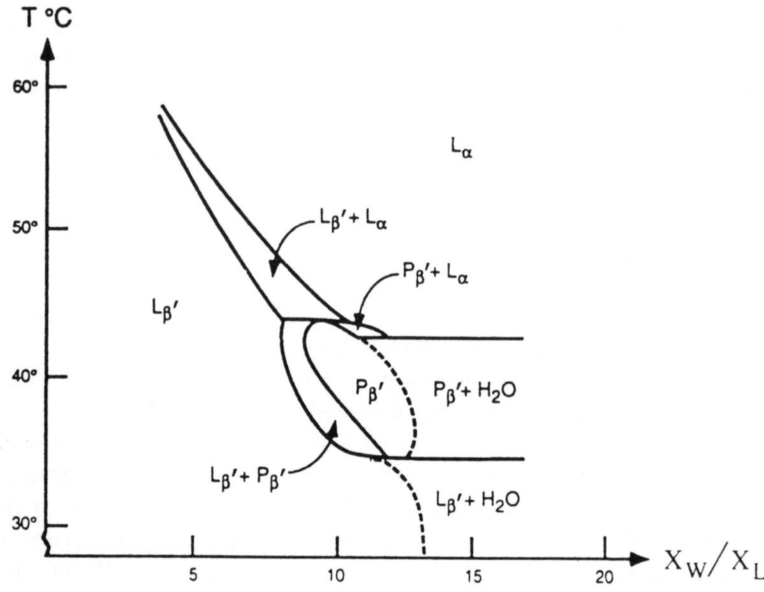

Fig. 5.6. Partial phase diagram of the binary lipid system DPPC-water determined by ^2H NMR of ^2H_2O (see Figure 5.7) [258]. The notations used for the different phases are the same as in Figure 5.7. Concentrations along the abscissa are given in molar fractions of water to lipid, X_W/X_L.

2. Phase Equilibria of Lipids from the Membrane of *Acholeplasma laidlawii*

The parasitic organism *Acholeplasma laidlawii* has its plasma membrane in direct contact with the environment and therefore, in order to keep a stable barrier, has to adjust the lipid content in the membrane according to changes in the surrounding milieu [201]. In the cell membrane of *A. laidlawii* strain A-EF22 there are seven classes of lipids (Figure 5.9)[91-93,201]; the two glucolipids 1,2-diacyl-3-*O*-(α-D-glucopyranosyl)-*sn*-glycerol (MGlcDAG) and 1,2-diacyl-3-*O*-[α-D-glucopyranosyl-(1→2)-*O*-α-D-glucopyranosyl]-*sn*-glycerol (DGlcDAG) constitute the major fraction. Glucolipids with *three acyl chains* are also synthesised by this strain and are 1,2-diacyl-3-*O*-[6-*O*-acyl-(α-D-glucopyranosyl)]-*sn*-glycerol (MAMGlcDAG) [92] and 1,2-diacyl-3-*O*-[α-D-glucopyranosyl-(1→2)-*O*-(6-*O*-acyl-α-D-glucopyranosyl)]-*sn*-glycerol MAD-GlcDAG) [93]. Recently, the chemical structures of the six gluco- and glucophospholipids dissolved in DMSO-d_6 have been determined by 1D and 2D-NMR techniques [91-93]. The approach to determine the structure is straight forward and the reader is referred to the references cited for more details. The high resolution NMR spectra obtained are of very good quality as is illustrated in Figure 5.10, where 1D ^1H NMR spectra of MGlcDAG, DGlcDAG and MADGlcDAG are shown; Figure 5.11 shows a nice text-book example of a ^1H DQF COSY (double quantum filtered correlation spectroscopy) NMR spectrum of MAMGlcDAG. As can be inferred from

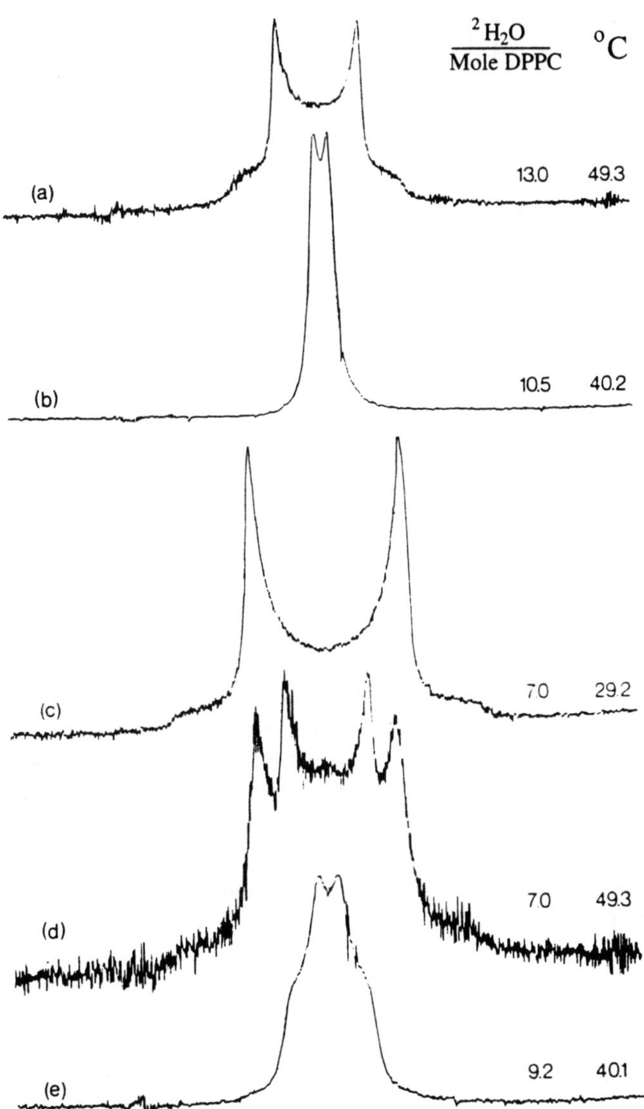

Fig. 5.7. ^2H NMR spectra of ^2H_2O obtained from the DPPC-water system at different compositions and temperatures as indicated on the right hand side of each spectrum.

a) L_α phase; b) $P_{\beta'}$ rippled (the prime on β signifies tilted acyl chains) phase; c) $L_{\beta'}$ gel phase; d) $L_\alpha + L_{\beta'}$ two phase region; e) $L_\alpha + P_{\beta'}$ two phase region. From [258].

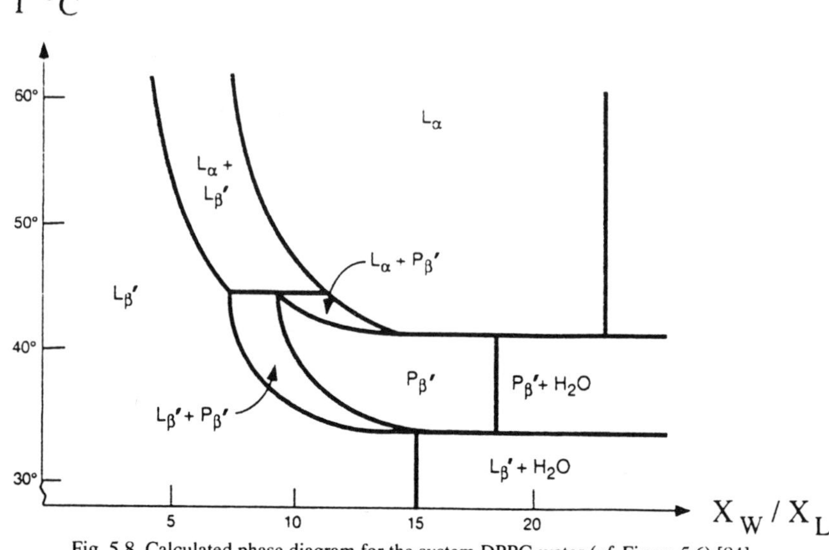

Fig. 5.8. Calculated phase diagram for the system DPPC-water (*cf.* Figure 5.6) [84].

Figure 5.9, three of the seven membrane lipid types, that are synthesised in significant amounts by *A. laidlawii* strain A-EF22, contain a third acyl chain - two of which are the glucolipids mentioned above and the third being a phosphoglucolipid.

In a large number of studies we have shown (see [153,201] for recent reviews) that the physico-chemical properties, or the phase behaviour of the lipids, play a crucial role in the regulation of the membrane lipid composition. In particular, the potential ability of the lipids to form different aggregate structures, or the tendency to form non-bilayer liquid crystalline phases, is of great importance. The three previously known glucolipids in *A. laidlawii* strain A-EF22 play important roles in the regulation of the membrane lipid composition [145,201,276]. The organism strives to maintain a certain balance between bilayer-forming and non-bilayer-forming lipids, resulting in a slightly negative curvature of the lipid monolayers in the membrane [143,154,198,201,281]. DGlcDAG forms only bilayers, while MGlcDAG and MAMGlcDAG are able to form non-bilayer aggregates; MGlcDAG can form I_{II} and H_{II} phases, and MAMGlcDAG forms an L_2 phase when the acyl chains are melted [67,143,145,201]. The newly discovered glucolipid MADGlcDAG must also be considered when discussing the regulation of the lipid composition, since this lipid may constitute up to 15% of the total lipids (A.-S. Andersson, M. Bergqvist, L. Rilfors, and G. Lindblom, to be published). Recently, MADGlc has been shown to form an H_{II} phase above T_m. This suggestion is inferred from a comparison of the phase equilibria of MGlcDAG and its monoacylated derivative MAMGlcDAG; the phase equilibria of MAMGlcDAG are more shifted towards non-bilayer aggregates [145,256].

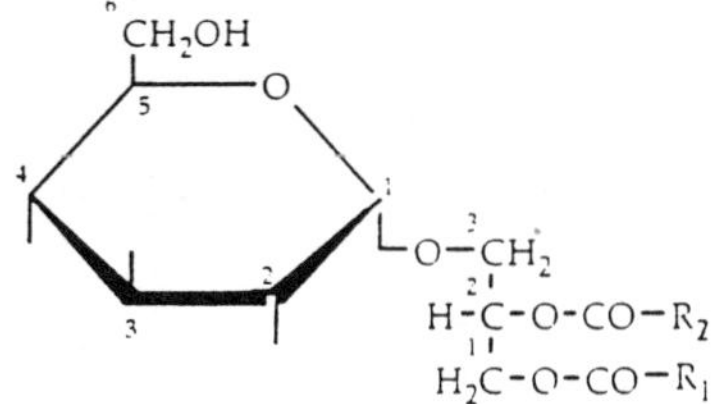

Monoglucosyldiacylglycerol (MGlcDAG)

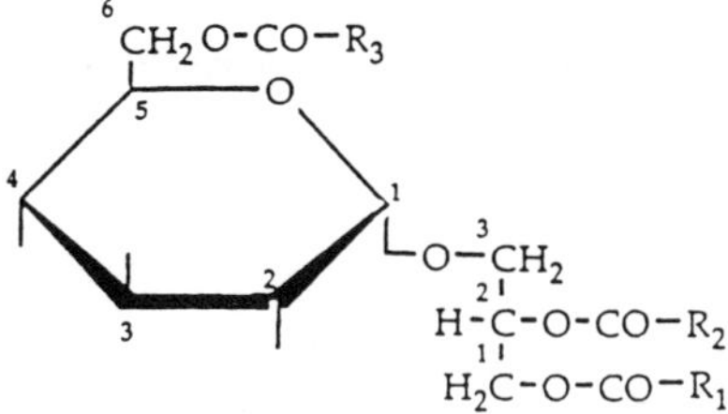

Monoacyl-monoglucosyldiacylglycerol (MAMGlcDAG)

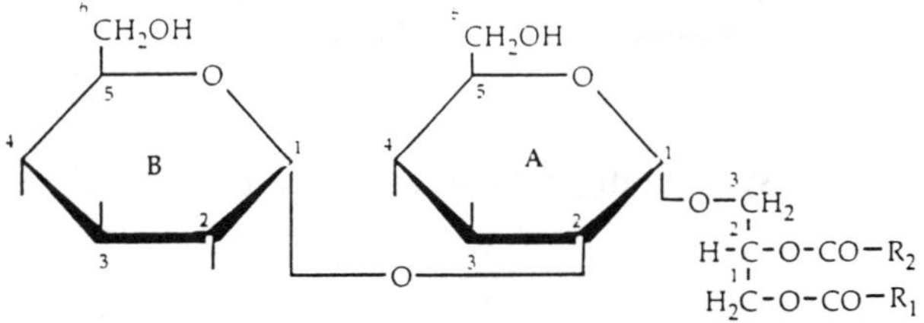

Diglucosyldiacylglycerol (DGlcDAG)

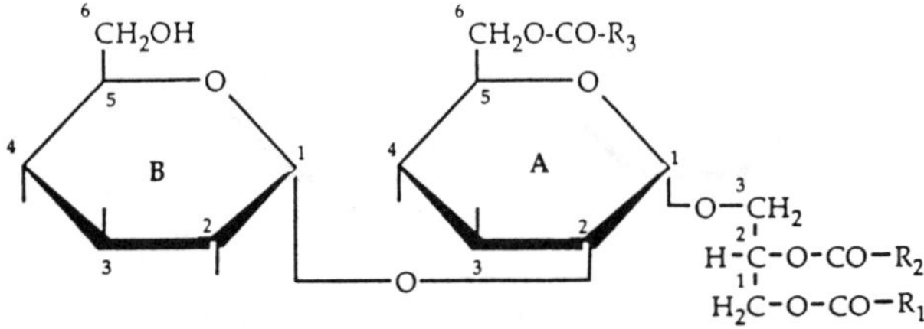

Monoacyl-diglucosyldiacylglycerol (MADGlcDAG)

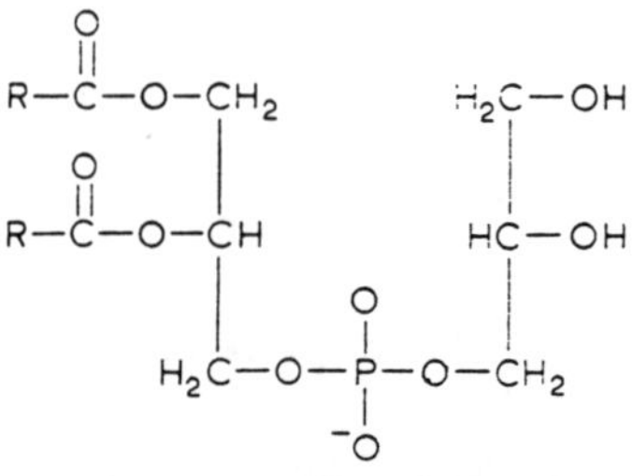

Phosphatidylglycerol (PG)

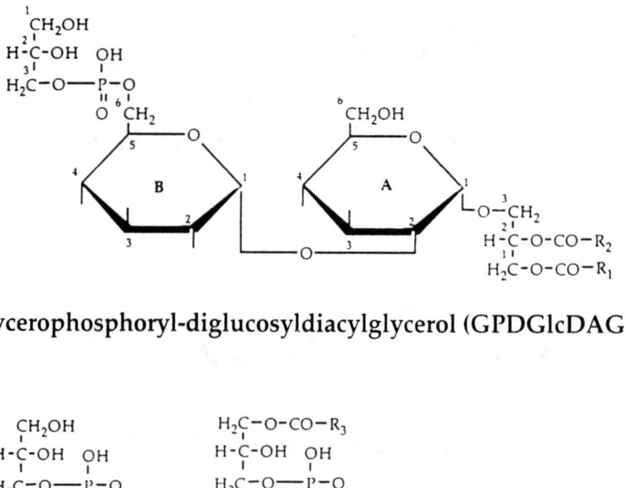

Glycerophosphoryl-diglucosyldiacylglycerol (GPDGlcDAG)

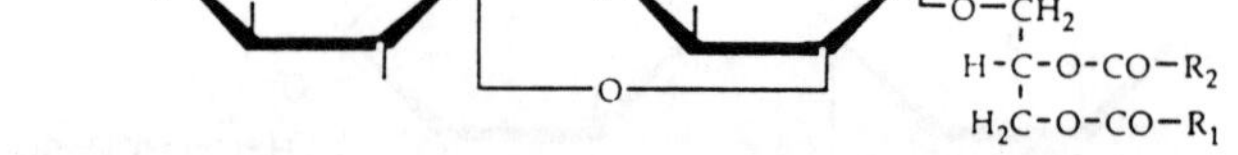

Monoacylbisglycerophosphoryl-diglucosyldiacylglycerol

(MABGPDGlcDAG)

Fig. 5.9. Chemical structures of the seven classes of lipids present in the membrane of *A. laidlawii*. The positions of the OH groups on the sugar rings are indicated.
(Courteously provided by L. Rilfors).

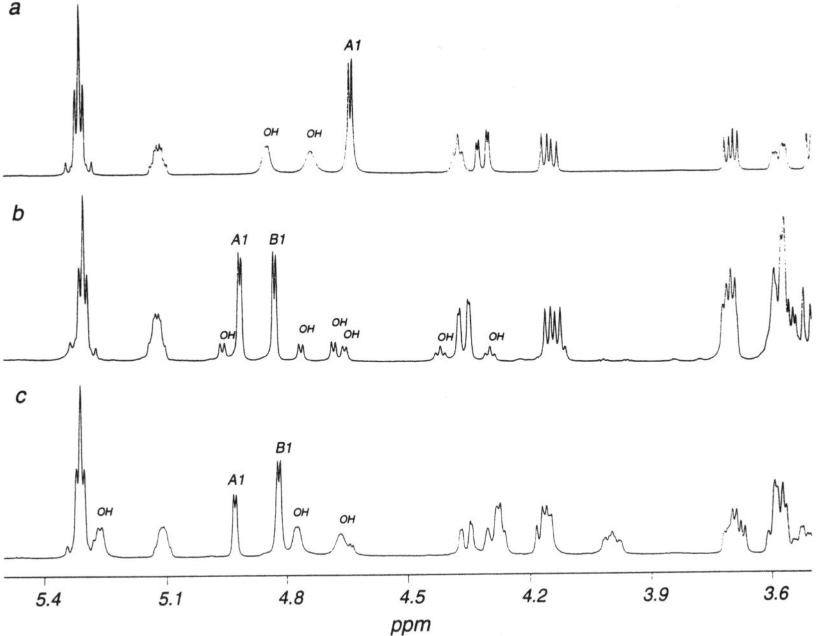

Fig. 5.10. 1D ^{1}H NMR spectrum of the following glucolipids from *A. laidlawii* (a) MAMGlcDAG, (b) DGlcDAG, and (c) MADGlcDAG. The spectra were recorded at a ^{1}H frequency of 500.13 MHz with a spectral width of 5000 Hz and a relaxation delay between successive scans of 10 s. The positions of the anomeric doublets (A1 and B1) are indicated, and hydroxyl peaks not completely removed by chemical exchange are also indicated. From [93].

MADGlcDAG may therefore be able to form a reversed non-lamellar liquid crystalline phase. The phase equilibria of this glucolipid are presently being investigated in our laboratory. Complete phase diagrams do not exist for glycerophosphoryldigluco-syldiacylglycerol (GPDGlcDAG) and monoacyl-bisglycerophosphoryl-diglucosyldiacylglycerol (MABGPDGlcDAG), but the available data from X-ray diffraction and ^{31}P-NMR studies show that both lipids form L_α phases [91,143]. However, investigations of GPDGlcDAG with cryo-transmission electron microscopy have, somewhat surprisingly, revealed that this lipid forms an L_1 phase, which is in equilibrium with the L_α phase at water contents of 99.5 wt% (D. Danino, A. Kaplun, G. Lindblom, L. Rilfors and Y. Talmon, unpublished results). Hence, it is concluded that the membrane lipids synthesised by *A. laidlawii* strain A-EF22 form a remarkable number of phase structures, covering many of the phase structures that can be formed by amphiphiles [154]. DGlcDAG and phosphatidylglycerol form only L_α phases, while MGlcDAG can form L_α, I_{II} and H_{II} phases [201]. MAMGlcDAG represents one extreme since it does not form any (thermodynamically stable) liquid crystalline phases; the lipid transforms from a gel/crystalline phase to an L_2 solution phase when the acyl chains melt [145]. The other extreme is represented by GPDGlcDAG, which can form an L_1 phase, which is in

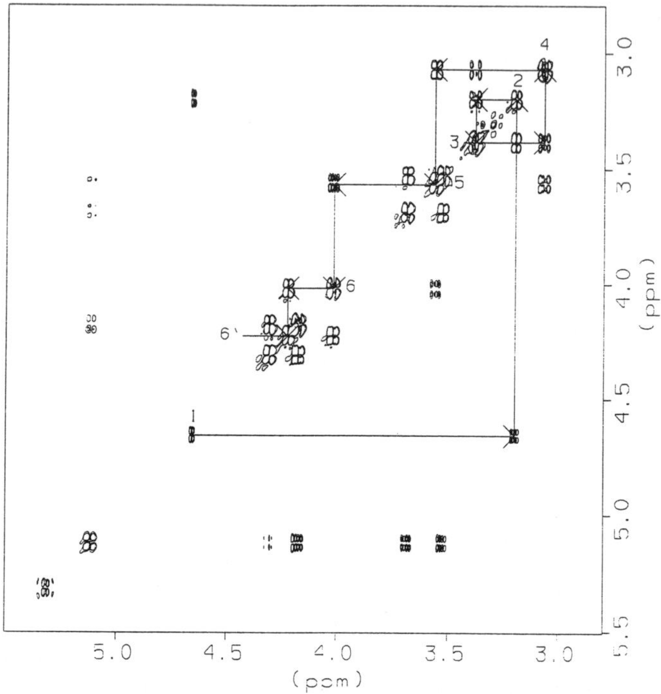

Fig. 5.11. ^{1}H DQF COSY NMR spectrum of MAMGlcDAG recorded at 500.13 MHz. From [92].

equilibrium with the L_α phase. The proportions of all these lipids seem to be carefully balanced by *A. laidlawii* in relation to the prevailing growth conditions.

3. Hydration of Liquid Crystalline Phases

Information about hydration in liquid crystalline phases can also be obtained from ^{2}H NMR quadrupole splittings. A decrease in the ^{2}H NMR quadrupole splitting of $^{2}H_2O$ is nearly always observed with increasing water concentration in liquid crystalline phases. This can be rationalized by applying a simple two-site model (see for example [73,121,152,155]), where a rapid exchange between "free" and "bound" water molecules is assumed. For an L_α phase the physical meaning of these sites can roughly be illustrated as follows; when the bilayers have a maximal amount of water associated to their surfaces ("bound" water), then further addition of water to the L_α phase only results in an increase of "free" water molecules between the lipid lamellas. Thus, an increase in the "free" water results in a linear increase in the distance between the bilayers in the L_α phase. The observed quadrupole splitting within this model then follows

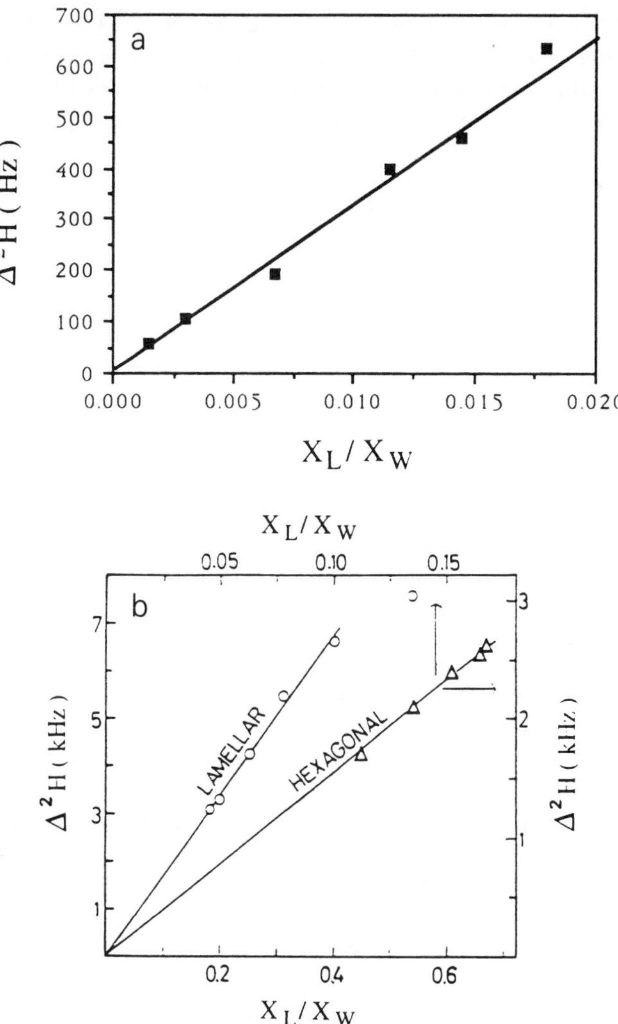

Fig. 5.12. a) The water ^2H NMR quadrupole splitting as a function of the ratio between the mole fractions of lipid (X_L) and water (X_W) in the L_α phase of the DOPA-water system at 25°C. From [155]. b) The same as in a) for the L_α and H_I phases in the system octylammonium fluoride in ^2H_2O at 28°C. From [121].

$$\Delta\nu_Q^{2H} = \frac{nX_L}{X_W}|(\nu_Q S)| \tag{20}$$

where X_L and X_w are the mole fractions of lipid and water, the quadrupole coupling constant ν_Q = 220 kHz for ^2H_2O [80,193] and n is the average number of water molecules "bound" to each head group.

This hydration model is illustrated for two different lipids in Figure 5.12. It can be inferred from this figure, that the observed ^{2}H NMR quadrupole splittings follow equation (20) and give a straight line, passing *through the origin*, over the whole region of water concentrations studied for an L_α phase with the disodium salt of dioleoylphosphatidic acid (DOPA) [155] and for the surfactant octylammonium fluoride. In the latter system both the L_α and the H_I phases showed this behaviour, except at very low water contents for the L_α phase (cf. Figure 5.12b for $X_L/X_W \geq 0.5$) For some membrane lipids at low water contents deviations from a straight line are not uncommon. Thus, for example, it was found that over most of the water concentrations for L_α phases with phosphatidylethanolamine (PE) isolated from *Bacillus megaterium* grown at 20°C and 55°C (PE-20 and PE-55) a straight line is obtained, but not at lower water concentrations [198]. *Iso* and *anteiso* methyl-branched, saturated acyl chains are predominant in *B. megaterium* and the value of the molar ratio of *iso*/*anteiso* acyl chains is about 30-fold higher in PE-55 than in PE-20; the latter lipid is able to take up about 70% more water than the former. Since, for these PE-systems, the deviation from the straight line is observed only at the lowest water content, it is most probably due to two-dimensional swelling [150]. The slopes of the straight lines obtained correspond to 10.5 kHz for PE-20 and 3.4 kHz for PE-55. The different slopes achieved for the two lipids can be due to either that the hydration of the head groups, the *n*-values, or that *S* of the water molecules in the "bound" site, is different for PE-20 and PE-55 (ν_Q in equation (20) is an intramolecular constant). A rough estimate of the *n*-values may be obtained from the deviation of the straight lines, *i.e.* a break or a bend in the straight line is obtained at a point where a further increase in the water content initiates a site of "free" water.

The phase equilibria differ between the two PE preparations: 1) PE-20 is more prone to form reversed non-lamellar phases than PE-55; 2) PE-20 forms both I_{II} and H_{II} phases while PE-55 forms only an H_{II} phase. These differences can be explained by the differences in the acyl chain composition. When the growth temperature is raised, PE molecules with a reduced tendency to form non-lamellar phases are probably synthesised by *B. megaterium* in order to counteract the bilayer destabilizing effect of the temperature (*cf.* the lipid regulation in *A. laidlawii*).

For the glycolipids so far studied [16,143] the quadrupole splittings *versus* X_L/X_W do not give a straight line, indicating that several sites exist. This is expected if water molecules are hydrogen-bonded to different hydroxyl groups and if there is very little "free" water.

4. ^{2}H NMR of Deuterated Lipids

According to Equation (14) the observed quadrupole splittings, $\Delta\nu_Q^{(i)}$, are directly related to the C-^{2}H-bond order parameter, $S_{CD}^{(i)}$, *i.e.*

$$\Delta\nu_Q^{(i)} = \nu_Q S_{CD}^{(i)}(3\cos^2\theta_{LD} - 1) \qquad (21)$$

where ν_Q = 127.5 kHz [21] and where i refers to each C-^{2}H bond in the acyl chain. Similarly, for a powder pattern (Equation 16) we get

$$\Delta\nu_Q^{(i)} = \nu_Q S_{CD}^{(i)} \tag{22}$$

Thus, the maximum splitting possible for a ^{2}H NMR spectrum would be 255 kHz. This corresponds to the θ_{LD} = 0° orientation of a rigid molecule, where S = 1 (complete ordering). If rapid axial rotations exist, as in the case of lipids in the liquid crystalline state, then the maximum splitting is reduced by a factor of two (127.5 kHz). Thus, the quadrupole splittings may be very large and it could be experimentally very difficult to measure the splittings from a Fourier-transformed free induction decay (FID), since the largest splittings will be hidden in the dead time of the spectrometer. Therefore, a clever and simple technique has been developed where a quadrupolar echo-pulse sequence is used for the acquisition of the spectra [12,51].

For a meaningful comparison of the molecular ordering, *i.e.* $S_{CD}^{(i)}$'s, between different phase structures, it should be remembered that the quadrupole splittings of hexagonal phases have to be multiplied by a factor of 2 when compared with the splittings obtained from an L_α phase (*cf.* Equation (19)).

^{2}H NMR spectra of perdeuterated molecules not seldom exhibit a powder pattern that is very difficult to interpret because of the many overlapping spectra. Therefore, it is useful to look at a sample with a single orientation of the director. One way to do this is to macroscopically orient the bilayers between glass slides [141,157]. This technique is time consuming, and a better way is to extract one orientation from the powder-type spectra. By making use of the dePakeing algorithm of Bloom *et al.* [11], the experimental ^{2}H NMR spectra can be numerically deconvoluted to obtain subspectra corresponding to the θ_{LD} = 0° orientation of the bilayer normal relative the main magnetic field in the L_α phase [229]. This can be done by assuming that the quadrupolar splittings scale with the orientation by a factor of $1/2(3\cos^2\theta_{LD}$ -1) = $P_2(\cos\theta_{LD})$ (P_2 is the second Legendre polynomial, *cf.* Equation (14)). This technique greatly improves the resolution of the spectra and makes it possible to determine a large number of order parameters along an acyl chain of a lipid (Figure 5.13).

For lipids the most important factor which modulates the order parameter is the *trans-gauche* isomerizations of the carbon-carbon bonds. In the *all-trans* state, all the C-^{2}H bonds form 90° angle with respect to the director axis and therefore an introduction of *gauche* isomers will cause a change in this angle and a decrease in S. On the NMR time scale this isomerization process is a very rapid one and therefore only a statistical picture of the fraction of *trans* and *gauche* conformations in the chain is obtained. Furthermore, for some systems, where the individual splittings may not be resolved, it can be difficult to get accurate values of the quadrupolar splittings. However, the so-called first and second moments (M_1 and M_2) of the spectra are directly related to the orientational distribution of the quadrupole splittings [1,48]. Therefore, the first two moments give the mean orientational order parameter and its mean square value [49]. Thus, moment analysis of the NMR lineshapes can be of great use in comparing the average molecular order of a system (see also Section F.1).

Often it is observed that the fraction of *gauche* conformations increases dramatically towards the methyl end of the acyl chain of a lipid. The presence of *gauche* isomers leads to a shortening of the lengths of the individual chains. Within the same bilayer this effect differs from chain to chain, therefore causing a broad distribution of the chain ends as a function of the depth in the bilayer. In order to keep the hydrocarbon density constant the spacial distribution of the methyl ends will be broader resulting in a decrease in the molecular ordering. This is taken into account in many of the lattice models for the L_α phase [55,81,178]. This also affects the splittings near the aqueous interface and produces a characteristic plateau region that is not an artefact obtained from several unresolved powder patterns, since for specifically deuterated acyl chains the same plateau region is observed [218].

The most common means to extract geometrical information from the ^2H NMR data is based on the diamond lattice model [210]. The derived order parameters reflect only orientational order, but positional order can be extracted using a simple statistical model [210]. Assuming that the acyl chains are tethered to the aqueous interface, and that rigid body motions can be ignored to a first approximation, the experimental S_{CD}'s can be modelled in terms of rotational isomerizations of the chain segments, that are related to the probabilities of *trans* and *gauche* isomers. It is assumed that only the orientations of the methylene segments falling on a diamond lattice are allowed. A similar model has been presented in which a cubic lattice is assumed instead of a tetrahedral one [53]. However the use of a tetrahedral lattice seems to be preferable [177]. The average projected length, $<L>$, of the acyl chain can be calculated using [249]

$$<L> = l_0 \left[\left(\frac{n-m+1}{2} \right) + \sum_{i=m}^{n-1} \left| S_{CD}^{(i)} \right| + 3 \left| S_{CD}^{(n)} \right| \right] \tag{23}$$

$l_0 = 0.1265$ nm is the normal projection length of one carbon-carbon bond, n is the number of carbons in the acyl chain, m is the number of the carbon segment where the hydrocarbon region is considered to begin ($m = 2$ for the *sn*-1 chain and $m = 3$ for the *sn*-2 chain), $S_{CD}^{(i)}$ is the C-^{2}H bond order parameter of position i in the acyl chain, and $S_{CD}^{(n)}$ the order parameter of the methyl segment.

From the average projected chain length an estimate of the mean molecular area at the lipid/water interface may be obtained. The volume of the acyl chain can be estimated from $v \approx (27.4 + 26.9 \cdot n) \cdot 10^{-3}$ nm^3 [241,275] and the length of the chain from the ^2H NMR data. Using simple geometrical arguments the average cross-sectional area of the acyl chain can be derived [104,177,247-249].

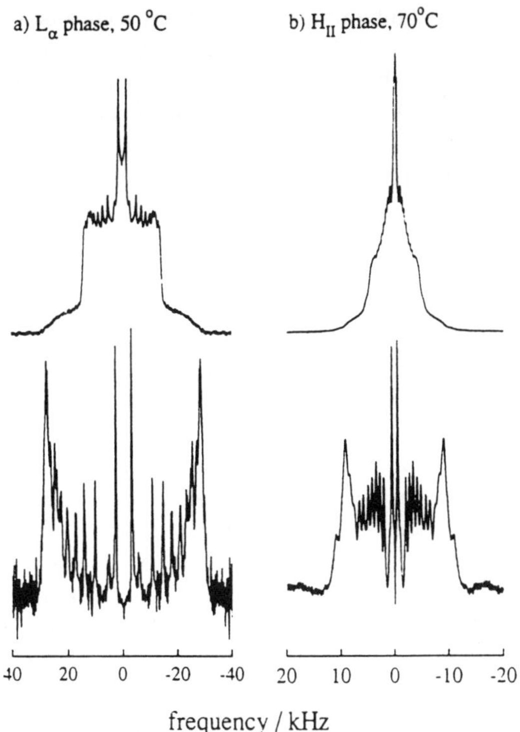

Fig. 5.13. ^{2}H NMR powder spectra (top) of the system PLPE-d_{31} and 50% water. a) The L_{α} phase at 50°C.; b) The H_{II} phase at 70°C. Corresponding dePaked spectra are shown beneath the experimental spectra.

Several studies [105,134,187] have indicated that the *S*-value reflects properties averaged over the entire system and is influenced mainly by *inter*molecular interactions. Therefore, a large number of different probe molecules can be used to report the S_{CD}'s. An example of a probe that may be commonly used is 1-perdeuteriopalmitoyl-2-oleoyl-*sn*-glycero-3-phosphocholine (POPC-d_{31}) because it is relatively easy to synthesise and is predicted to have only small effects on a large variety of systems [134]. It follows that if a small amount of a deuterated lipid is added to a non-deuterated system in such a way that it has a minimal impact on the properties of the original system, the order parameters measured for this deuterated molecule will reflect the molecular ordering of the entire system. Indeed, some very early ^{2}H NMR work involved the use of perdeuterated free fatty acids to report the order profiles of bilayer systems [234]. This technique has several advantages. Only a small amount of deuterated lipid needs to be synthesised, since it is not the major component of the system, and furthermore, the deuterated lipid could be used to report the ordering of systems in which it is difficult to introduce deuterium labelling. A third advantage is that this technique could be used to measure the average properties of complicated mixtures of lipids and even native membranes. Recently, this method was adopted to measure the order parameter of membranes from *A. laidlawii* by incorporating perdeuterated acyl chains biosynthetically [173,251].

Information about the packing of lipids in a bilayer can also be obtained by ^{2}H NMR of deuterated acyl chains, since a change in the packing is influencing the molecular ordering. Recently, we reported the effect of an increasing fraction of MGlcDAG in a DGlcDAG bilayer [67]. In this study lipids from *A. laidlawii* grown on perdeuterated palmitic acid were investigated. Representative ^{2}H NMR spectra of DGlcDAG with varying amounts of MGlcDAG are shown in Figure 5.14. The methylene group assignments were made in comparison with phospholipids, since there are no data on specifically deuterated glucolipids. From the *S*-profiles obtained it is concluded that the packing of the acyl chains is very similar to that observed for phospholipids and that the ordering of MGlcDAG is greater than that of DGlcDAG. The increase in the molecular ordering (or packing) with increasing MGlcDAG content is concluded to be due to the smaller head group of MGlcDAG compared to that of DGlcDAG (*cf.* Figure 5.9). A similar behaviour has been observed for mixtures of POPE-d_{31} and POPC-d_{31} [133].

Several groups have looked at the segmental ordering in intact membranes [50,77,78,113,173,185,195,196,228,232,233,251]. In this case bacteria such as *E. coli* or *A. laidlawii* are grown on perdeuterated fatty acids. ^{2}H NMR spectra can be recorded from either the whole cells, the intact membranes, or the extracted lipids. In general it has been found that the spectra from either whole cells or intact membranes are identical and that they yield ^{2}H NMR spectra very similar to those of synthetic lipids in the L_α phase [173,251]. Furthermore, there were only slight differences found between intact membranes and the extracted lipids indicating that other membrane components do not have a significant influence on the molecular ordering in the lipid bilayers.

Recently, the response of *A. laidlawii* to incorporation of fatty acids of different lengths into the acyl chains of its membranes was reported [251]. As a general trend it was found that when the organism was forced to grow on fatty acids of increasing length the average molecular ordering of the membranes *de*creased. This decrease was found to be correlated with the *de*crease in the amount of MGlcDAG present in the membrane. Since both the *S*-values and the thickness of the membranes were found to vary over a wide range, and were still able to support cell growth, it was concluded that *A. laidlawii* changes the MGlcDAG content to optimize either the membrane permeability or the curvature (packing) stress in the bilayers (see Section D.2).

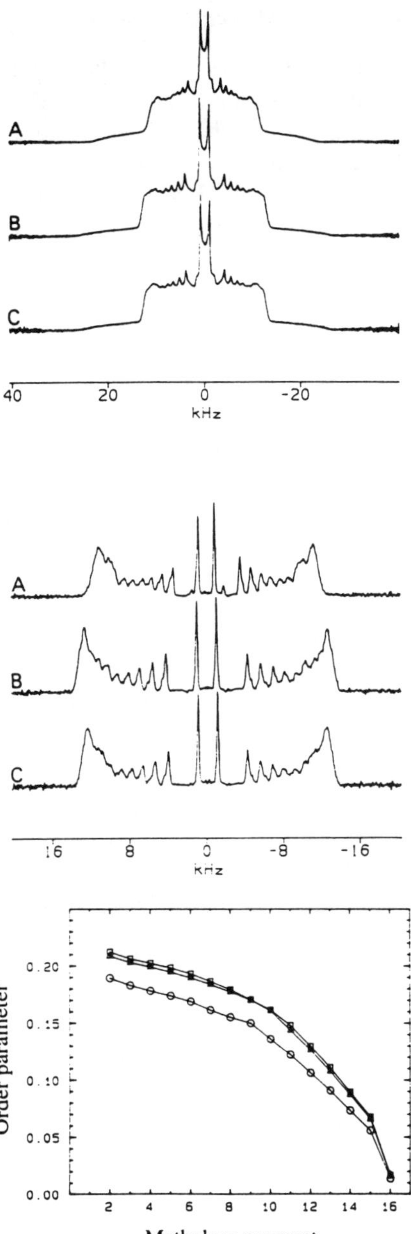

Fig. 5.14. The ^{2}H NMR spectra (top), the corresponding dePaked spectra (middle), and the order parameter profile (bottom) of DGlcDAG-d_{31}, with varying amount of MGlcDAG in the bilayer. (A and ○) pure DGlcDAG-d_{31}; (B and □) DGlcDAG-d_{31}/MGlcDAG (71:29 mol/mol); and (C and Δ) DGlcDAG-d_{31}/MGlcDAG (24:76 mol/mol). From [67].

5. Molecular Ordering in L_α, H_I and H_{II} Phases

The H_{II} phase: Figure 5.13 shows representative ^{2}H NMR spectra of 1-perdeuteriopalmitoyl-2-linoleoyl-*sn*-glycero-3-phosphoethanolamine (PLPE-d_{31}) in both the L_α and H_{II} phases. The major difference in the spectra is that the quadrupolar splittings are reduced in the H_{II} phase compared to the L_α phase. The differences observed by ^{2}H NMR between the H_{II} and L_α phases can be accounted for by variations in the acyl chain packing arising from the difference in geometry between the phases. When going from a planar to a cylindrical geometry there is a change of the symmetry axis along which motions are averaged (Equation 19). However, the reduction observed is *greater* than what can be expected from the change in symmetry only. It should be remarked that the chemical shift anisotropy (see Section E below) for the ^{31}P spectra [249] is scaled by almost exactly -1/2 (-40 ppm for L_α phase and 19 ppm for the H_{II} phase) as seen also for other H_{II} phases [44,155,171]. Figure 5.15 shows that the segmental order in the H_{II} phase is less than that of the L_α phase. It is also evident that the order decreases more uniformly down the chain for the H_{II} phase than for the L_α phase. The same behaviour was observed in mixtures of perdeuterated tetradecanol and 1-palmitoyl-2-oleoyl-*sn*-glycero-3-phosphoethanolamine (POPE) in the H_{II} phase [230] and 1-perdeuteriopalmitoyl-2-oleoyl-*sn*-glycero-3-phosphoethanolamine (POPE-d_{31}) in excess water [133].

It has been helpful to analyse the ^{2}H NMR data by thinking in terms of characteristic shapes, such as cones, truncated cones and cylinders, adopted by the lipids in various phases [101,241,244]. Each lipid in a bilayer occupies a space that on average can be approximated by a cylinder, in which the area at the lipid/water interface is nearly the same as the cross-sectional area along the hydrocarbon chain. For lipids packed in a cylindrical aggregate, with their polar head groups inside and acyl chains out (H_{II} phase), the lipids could be well-approximated by truncated inverted cones with the area at the lipid/water interface being smaller than the area further down the chain (*cf.* Figure 5.5C). It is the difference in the average molecular shape as governed by the aggregate geometry that is reflected in the differences in molecular ordering between the various phases.

Given that the differences in the order parameters between the H_{II} and the L_α phases reflect a change in the cross-sectional area of the acyl chains, it is possible to use the order profiles to determine some of the structural features of the H_{II} phase relative to the L_α phase [247,249]. In this analysis the shape of an individual molecule in the H_{II} phase is assumed to be approximated by a truncated cone. The volume can be estimated from density measurements and x-ray studies [179,180,204,275,278] and the length can be determined from Equation 23, using the ^{2}H NMR-measured order profile *of the H_{II} phase*. Then straight forward geometry can be used to calculate the distance from the centre of the water core to the lipid/water interface, and for PLPE-d_{31} a radius of curvature of 25.4 Å is obtained in the H_{II} phase at 60°C [249].

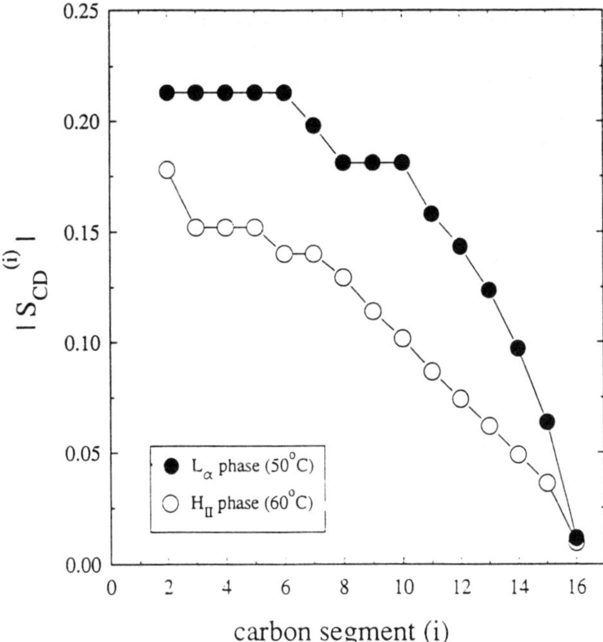

Fig. 5.15. Comparison of the order parameter profiles between the L_α (○) and H_{II} (●) phases containing PLPE-d_{31} derived from the dePaked spectra in Figure 5.13.

The H_I phase: Many single chain amphiphiles such as lysophospholipids and soaps can form H_I phases (Figure 5.5A). For these phases the water volume is continuous and able to fill any polar area not occupied by lipid head groups, while for the H_{II} phases the water is restricted to the rod-like aggregates. Furthermore, the H_I phase can, in principle, take up large amounts of water without significant change in the size of the cylinders, whereas uptake of water in the H_{II} phase causes the cylinder size to increase. As in the analysis of the ^{2}H NMR data for the H_{II} phase, the idea that lipids adopt on average a particular geometric shape in a phase structure can be used also for the H_I phase. In this respect the H_I phase can be seen as just the mirror structure of the H_{II} phase with the area at the lipid/water interface being greater than the area at the ends of the chains.

Figure 5.16 shows the S_{CD}'s of the acyl chain of perdeuterated potassium laurate for both the L_α and the H_I phases. The absolute S-values at the same temperature are greater for the H_I phase (42 wt% H_2O) than for the L_α phase (38 wt% H_2O). As opposed to the H_{II} phase, the average cross-sectional area available for the acyl chain in the H_I phase is less than the area at the lipid-water interface. If it is assumed that the area at the lipid-water interface is the same in the L_α and in the H_I phases, the area would decrease towards the end of the chains for the H_I phase. This causes the chains to be more constrained conformationally in the H_I phase than in the L_α phase, yielding an increase in the S_{CD}'s, as is observed experimentally. It might be expected that the increase in the water content when going from the L_α to the H_I phase could play a role in changing the molecular ordering. However, the change in water content is small and should, if anything, increase the area per molecule in the H_I phase [164].

6. Spin-Relaxation and the Two-Step Model of Molecular Motion

Measurements of NMR relaxation times are very useful in studies of various dynamic processes occurring in biomembranes [18,50,88,89,128,265,271]. ^{2}H NMR is particularly useful since the spin relaxation is dominated by an *intra*molecular quadrupolar interaction (except for ions) which in most cases is relatively easy to interpret. It is beyond the scope of this chapter to present a comprehensive view of ^{2}H NMR relaxation methods and their application to liquid crystalline systems, and I will only touch upon this powerful technique. The measured different spin relaxation times (*e.g.* T_1 and T_2) reflect the rates and amplitudes of the fluctuating motions exerted by the molecules, and in ^{2}H NMR it is the motion of the EFG tensor with respect to $\mathbf{B}_0$ that leads to relaxation. The molecular motions that can be detected by ^{2}H NMR cover a wide dynamic range, which can vary from internal motions of the molecule, such as *trans-gauche* isomerizations and rotational diffusion of an amino acid in the molecule to whole body motions, such as lateral and rotational diffusion of an entire aggregate or macromolecule. Different relaxation rates are usually easy to measure, but their interpretation in terms of the existing particular motions requires extensive mathematical and physical modelling (see *e.g.* ref. [88]).

NMR spin relaxation times may be dependent on the size and geometry of the lipid aggregate in lyotropic liquid crystals. Therefore, it is often possible to get an estimate of such aggregate properties from measurements of spin relaxation times (often in combination with determinations of translational diffusion coefficients [64]) in non-lamellar liquid crystalline phases. Here, only quadrupolar nuclei are considered, where intramolecular interactions dominate the relaxation mechanism, and where molecular ordering can be obtained from quadrupole splittings of the anisotropic liquid crystalline phase. Since only one correlation time is not sufficient to describe the experimental data obtained from lipid aggregates [64,88], where the molecular motion is *locally* anisotropic, even if the system is optically isotropic, a suitable physical model has been developed, which takes this particular property into account. In the simplest model only two modes of motion occurring at different time scales are considered - hence the name *two-step model* [271,273]. There is one fast motion consisting of conformational and orientational fluctuations at a time scale of about 10^{-10} s, and a slow motion described by a correlation time, τ_c^s, which is mainly determined by the lipid lateral diffusion, D_L, around the

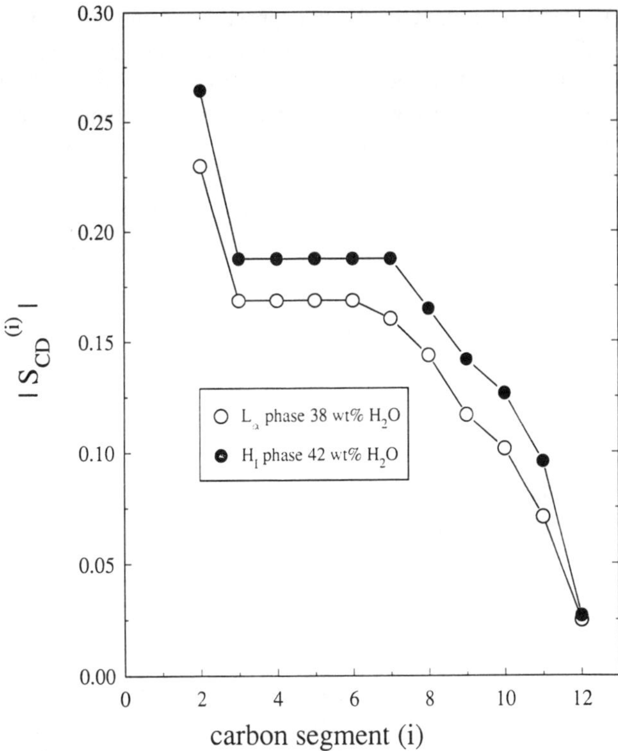

Fig. 5.16. Segmental order profiles, $S_{CD}^{(i)}$, derived from ^{2}H NMR quadrupole splittings, as a function of the chain segment position, i, for potassium laurate-d_{23} in the L_α (○) and H_I (●) phases. Adapted from [249].

curved aggregate surface or by the rotational motion of the whole aggregate. It follows [64] that from measurements of the relaxation times T_1 and T_2, from which τ_c^s is determined, together with the ^{2}H quadrupole splittings and lateral diffusion coefficient of the amphiphile, it is possible to get an estimate of the radius of a micellar aggregate. By assuming that the lateral diffusion coefficient is the same in an L_α phase and a micelle, the radius of the aggregate can be calculated from

$$\tau_c^s = R^2(6D_L)^{-1} \tag{24}$$

where R is the radius of a sphere. Thus for a cubic phase consisting of close-packed spheres an estimate of the radius of the micelles has been obtained and is found to be in excellent agreement with the total length of the amphiphile [64]. The two-step motional model has been utilized in a large number of studies and works surprisingly well [26,27,31-35,62-65,70].

Another example of how relaxation rates can be utilized to get structural information is illustrated in a study, where the transverse relaxation rate ($1/T_{2e}$) in the L_α and H_{II} phases were compared [249]. It was found that the relaxation rate was significantly different in the two phases and was highest in the H_{II} phase. This difference in relaxation rates could be accounted for completely by the diffusion of the lipid molecules around the water cylinders of the H_{II} phase. The translational diffusion changes the orientation of the lipid molecule with respect to the main magnetic field and leads to an additional relaxation mechanism not present in the L_α phase. This also implies that besides the diffusional motion, all other relaxation mechanisms, possibly present in the lipid aggregates, must be similar in the two liquid crystalline phases. This finding also indicates strongly that the *local* molecular dynamics in the liquid crystalline phases are the same or very similar. The translational diffusion in lipid aggregates is discussed further in Section G.

7. Effect of Cholesterol, Bile Salt, and Gramicidin on lipid phase properties

Cholesterol: For many years the effects of cholesterol on the properties of lipid membranes have been studied extensively by a number of spectroscopic methods. From ^{2}H NMR investigations of perdeuterated lipids it has been carefully confirmed that cholesterol increases strongly the molecular ordering in fluid bilayers [146,152,182,186,235]. It is also well known that cholesterol tends to have a condensing effect on many lipids in the L_α phase, but it was not until recently that the ternary phase diagram of DPPC/cholesterol/water was presented [264]. For the determination of the phase diagram a very useful spectral subtraction method was developed in this study. The phase diagram shows many interesting features. As the cholesterol concentration increases in the lipid bilayer two L_α phases with different molecular ordering are in equilibrium with each other. This unusual behaviour has been explained completely by a theoretical model developed by Ipsen and colleagues [100]. Similar experimental results concerning the phase behaviour have been reported in a study of the effect of cholesterol on the acyl chain order of three phosphatidylcholines with 14, 16 and 18 carbons/acyl chain [208]. A molecular mechanism was proposed in which increasing the concentration of cholesterol has the effect of stretching the acyl chains of phospholipids by increasing the population of *trans* conformers up to a stage where the hydrophobic thickness is considerably larger than the cholesterol molecule. Beyond this concentration, a partially interdigitated phase is formed. The influence of cholesterol on the lipid lateral diffusion in various L_α phases has been investigated also (see below) by pulsed field gradient (PFG) NMR [129,146] and it is interesting to note that although cholesterol may strongly increase the ordering in a bilayer, the lipid translational diffusion is more or less unaffected [146].

Bile salt: Another steroid, sodium cholate, has a strong effect on the physicochemical properties and the phase diagram of phospholipid systems [250,257]. Bile salts in binary, ternary and quaternary systems have been extensively studied by Small and coworkers [226,227]. The bile salts are surface-active compounds with several properties that differ from those of common amphiphiles, like soaps or lipids. Bile salts aggregate in aqueous solution, but they do not form typical micelles with a well-defined critical micelle concentration, nor do they form liquid crystalline phases with water. The bile salt molecule consists of a stiff and roughly planar steroid fused-ring system, in which one side is hydrophobic and the other one is hydrophilic. The latter property is due to the presence of up to three hydroxyl groups located on the same side of the ring system. These properties seem to be the key to interpretation of the atypical properties of the bile salts as surfactants.

The molecular organization of the lipid aggregates formed in the ternary system PC/sodium cholate/water has been thoroughly studied for the liquid crystalline phases [250,257] and for the L_1 phase [97,144,183,213]. The phase diagram of this ternary system consists of L_α, H_I, I_{II} and L_1 phases [257]. Figure 5.17 shows ^{2}H NMR spectra of the L_α and the H_I phases, where the order profiles differ significantly, and it is concluded that cholate influences the molecular ordering in the bilayer in an opposite way compared to cholesterol. Unlike the cholesterol molecule, cholate cannot penetrate the bilayer readily, but a substantial fraction of these molecules lies flat on the bilayer surface as a result of the two-sided hydrophilic-hydrophobic nature of the molecule. Thus, cholate induces a major perturbation in the acyl chains, resulting in a drastic decrease in the S-values, and as the cholate content increases the plateau region of the order profile disappears. An important conclusion from these studies is that the hydrocarbon core of the rod-like aggregates in the H_I phase is continuous, in contrast to a previous suggestion. This also suggests that the aggregates at low water content in the L_1 phase consist of long, rod-like micelles, as later confirmed by neutron- and light-scattering investigations [96,97].

Further support for the suggestion that the effect of cholate is to alter the monolayer curvature of the lipid aggregates has been obtained in the system dioleoylphosphatidylethanolamine (DOPE)-sodium cholate-water, where the following sequence of phase transitions with increasing cholate concentration was observed: $H_{II} \rightarrow L_\alpha \rightarrow$ cubic $\rightarrow L_1$ phase [250].

Gramicidin: There is a vast literature on gramicidin solubilized in lipid/water system (see the excellent review by Killian [122] and references therein). Here I will only discuss some of our own recent work on gramicidin solubilized in the system *N,N*-dimethyldodecylamine oxide (DDAO) and water. The ternary phase diagram is shown in Figure 5.18, where it can be inferred that DDAO is unable to incorporate gramicidin into any other phase but the L_α phase. At gramicidin concentrations as low as 0.5 mol%, an L_α phase is formed at all water contents. This is in agreement with investigations performed on mixtures of gramicidin and lysophosphatidylcholine (LPC), where lamellar aggregates were induced upon addition of gramicidin [123]. Gramicidin has also been solubilized in micelles of sodium dodecylsulphate (SDS). In order to obtain small micelles suitable for high-resolution NMR studies, it was necessary to add small amounts of trifluoroethanol (TFE) to the solutions [5]. However, with a recently developed lyophilization technique it is possible to solubilize gramicidin into detergent micelles without the use of TFE [124].

The phase diagram tells us that for the aggregates formed at a water content above 40.5% (w/w) the molar ratio between DDAO and gramicidin is 9:1 and that this ratio increases at lower water contents. These stoichiometric values are comparable with the value of 7:1 for LPC-gramicidin bilayers reported previously [123]. Raman and ESR studies on phospholipids indicate ratios of about 5:1 to 10:1, corresponding to 10 to 20 acyl chains per gramicidin molecule [36].

In the L_α phase at 22.4 % (w/w) water the quadrupole splittings are much larger in the presence of gramicidin. This is most probably because the DDAO molecule has to stretch in order to incorporate gramicidin into the bilayer. It has been shown that the ordering of the acyl chains in PC bilayers increases with increasing gramicidin content for chain lengths with less than 18 carbons, whereas above 18 carbons no change is observed upon addition of gramicidin to the bilayer [267]. The stretching of the hydrocarbon chain of DDAO results in a longer plateau region of the *S*-profile. Only five individual signals are resolved; the methyls of the head group, the terminal methyl group of the hydrocarbon chain, and the three methylene groups at the farther end of the chain.

From an equation corresponding to Equation 23, involving the first moment, M_1, the chain length, <L>, of the DDAO molecule is calculated [187] (Table 5.1). <L> increases from 13.1 Å for $X_G = 0$ to 13.6 Å for $X_G \approx 5$, where X_G is the molar fraction of gramicidin solubilized within the DDAO bilayer. This is in agreement with the idea that there is an "ideal" chain length of 15 Å for solubilization of gramicidin. The DDAO molecule is shorter than the ideal chain length and therefore has to stretch when gramicidin is solubilized in the bilayer. A calculation of half the bilayer thickness, which corresponds to the length of the DDAO molecule, was performed using data from X-ray investigations of DDAO bilayers (Table 5.1). Although the values from the two methods differ by approximately 0.1 nm, they both show an increase in the chain length as the gramicidin content is increased in the bilayer. The difference obtained is not surprising considering the assumptions made in both methods.

Computer simulations of the order parameter profile have been performed for lipids in bilayers with embedded integral proteins or peptides of both positive and negative mismatch (*i.e.* the bilayer-spanning peptide is either longer or shorter than the bilayer thickness, respectively) [72]. For a positive mismatch of 3.5 Å, which would correspond to the case of gramicidin in a DDAO bilayer, a significant increase in the *S*-values of the segments near the polar head group is obtained. This situation corresponds to the experiments

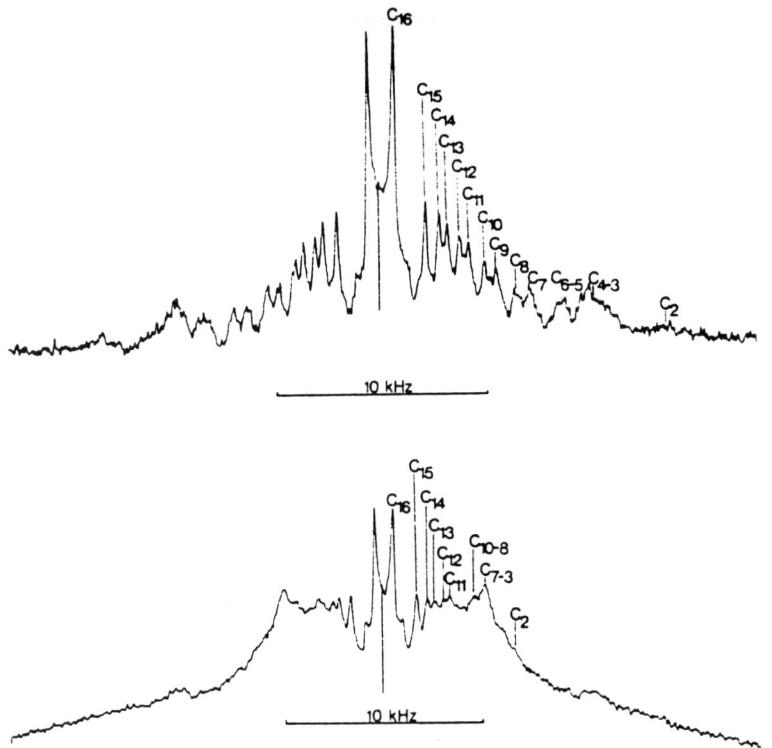

Fig. 5.17. ^{2}H NMR spectra of L_α (top) and H_I (bottom) phases of the ternary system POPC-d_{31}, sodium cholate and water. Quadrupole splittings from most of the C-^{2}H bonds are resolved in these powder spectra. The numbering refers to the carbon atom of C-^{2}H beginning with the carboxylic end of the palmitoyl acyl chain. Adapted from [257].

Table 5.1

Comparison between the chain lengths of DDAO determined with ^{2}H NMR and X-ray diffraction. From [187].

X_G (mol-%)	Alkyl chain length (nm) from ^{2}H NMR (Eq. (23))	Average molecular length (nm) from X-ray diffraction
0	1.31	1.20
1.5	1.29	—
3	1.35	1.24
4	1.32	—
5	1.35	1.28
10	1.38	1.31
15	1.36	1.31
20	1.36	1.31

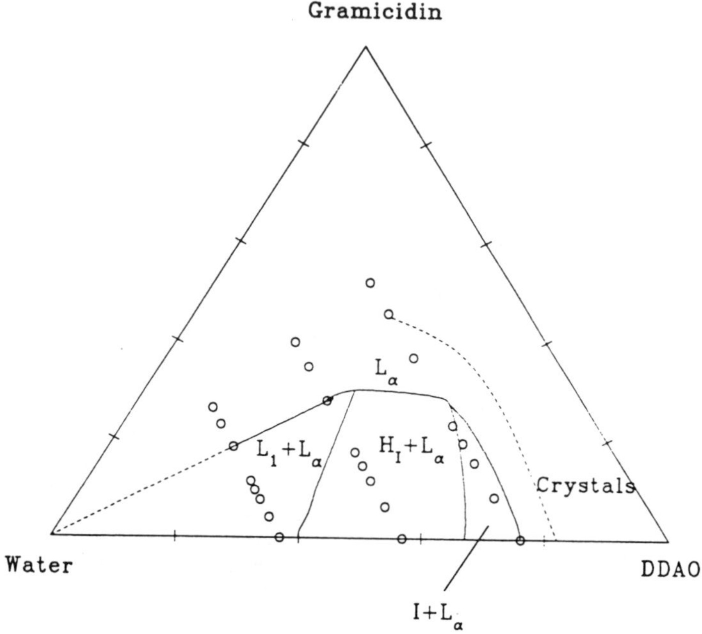

Fig. 5.18. Tentative partial phase diagram of the DDAO/gramicidin/H_2O system. The various phases are indicated in the diagram; L_I, the micellar solution phase, H_I, the normal hexagonal phase, I, the cubic phase, L_α, the lamellar phase. From [187].

with $X_G \geq 5$, where all the DDAO molecules are adjacent to gramicidin in an L_α phase. A comparison between the experimental *S*-values and those calculated theoretically shows that the two *S*-profiles obtained agree reasonable well. However, as we go down the hydrocarbon chain a slight deviation is observed, but the trends are the same in both cases.

Addition of gramicidin above a molar ratio between gramicidin and DDAO equal to 1:9 results in further changes in the *S*-profile of DDAO, which can be understood by considering the shape of the gramicidin molecule. The gramicidin molecule has a shape of an inverted cone which occupies a larger area at the polar-apolar interface than further down in the interior of the bilayer [14]. Thereby, the molecular packing at the polar head group methyls will be tighter than in the interior of the bilayer. The packing pressure arising in the polar head group region cannot be reduced by decreasing the bilayer thickness, since the alkyl chains are forced to stretch in order to cover the hydrophobic surface of a gramicidin dimer. Therefore, an increase in the packing at the interface of the bilayer and a corresponding decrease in the packing of the hydrocarbon chains in the interior of the bilayer will occur in order to compensate for this mismatch. This is reflected in the measured increase of the *S*-value of the head group and the corresponding decrease in the terminal part of the chain. Similar arguments have been used to explain data obtained from investigations of ^2H quadrupolar splittings and CSA's of ^{13}C and ^{31}P on gramicidin-DMPC bilayers [13].

8. ^{23}Na NMR in Studies of Counterion Binding

The experimental discovery of NMR quadrupole splittings of Na^+ counterions, about 25 years ago, for an L_α phase composed of egg phosphatidylcholine, sodium cholate and water [140] resulted in the development of a convenient technique for studies of ion binding of quadrupolar nuclei in anisotropic liquid crystalline phases. It should be noted that this NMR method is probably still the only one with which ion association in lyotropic liquid crystals can be investigated directly. A non-zero electric field gradient arises when the ion is close to a charged surface so that the solvation shell of the ion at an interface will be asymmetric [271]. In concert with the treatment of the water quadrupole splittings discussed earlier, counterions close to the aggregate charged surface produce a contribution to the observed quadrupole splitting, while those in the centre region between two aggregate surfaces show no splitting [150,152,155,161,253,254,271,272]. There is a fast chemical exchange between the two sites, "free" and "bound" ions, and according to Equation (16) the observed quadrupole splitting can be written

$$\Delta \upsilon_Q^{Na} = | p_f(\overline{\nu_Q S})_f + p_b(\overline{\nu_Q S})_b | = | p_b(\overline{\nu_Q S})_b | \quad (25)$$

where the subscripts f and b refer to free and bound ions, and it is assumed that the splitting in the free site is equal to zero. If there are several different bound sites the observed splitting is the average over all the sites,

$$\Delta \upsilon_Q^{Na} = | \sum_i p_{bi}(\overline{\nu_Q S})_{bi} | \quad (26)$$

Figure 5.19 shows that at water concentrations between about 50 and 98 wt % the ^{23}Na quadrupole splitting is constant for some different anionic lipids, implying that the fraction of bound ions, p_b, is constant over a large range of water concentrations. This kind of ion association is also observed for L_α phases composed of simple surfactants and water and it can be described by a physical model, called the "ion condensation" model [155,272].

For a comprehension of the electrostatic properties of charged bilayers, let us briefly look into the counterion distribution between two similarly charged planar surfaces in pure water. In this consideration the counterions are treated as point charges embedded in a dielectric continuum of relative permittivity, ε, bounded by two planar negatively charged surfaces with surface charge density σ. It can be shown that the ion concentration at the planar surface is given by

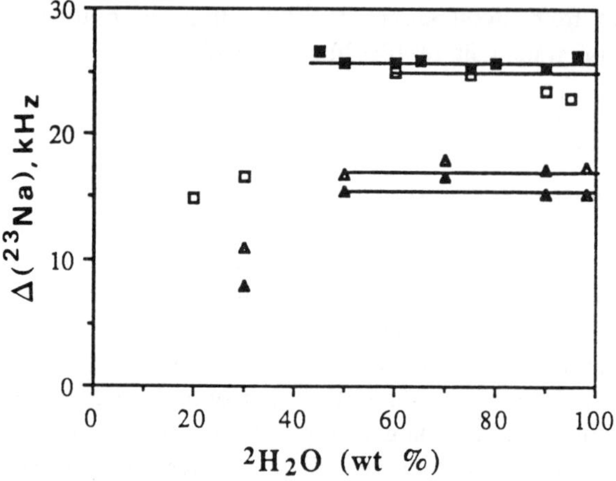

Fig. 5.19. ^{23}Na NMR quadrupole splittings as a function of the water content at 25 °C for the following lipid systems; sodium salt of DOPA-^2H_2O (■), DOPS-^2H_2O (Δ), DOPG-^2H_2O (▲), and DPG-^2H_2O (□). From [155].

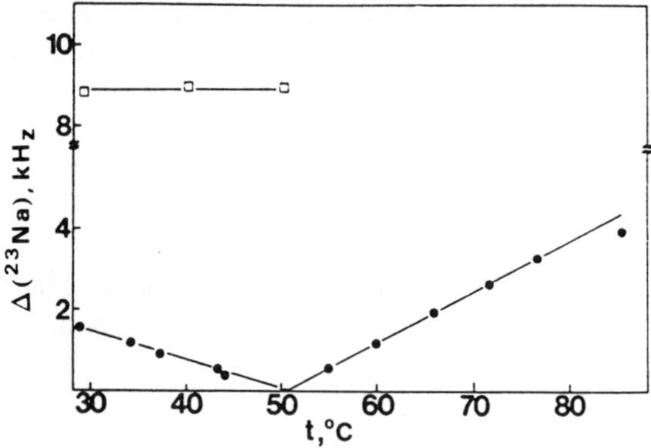

Fig. 5.20. ^{23}Na quadrupole splitting $\Delta(^{23}Na)$ as a function of the temperature for a) (●) an L_α phase composed of sodium octanoate (24.3% w/w), decanol (40.4% w/w), and water (24.3% w/w); b) (□) an H_I phase composed of sodium octylsulphate (58.3% w/w) and water (41.7% w/w). From [161].

$$n_{surface} = n_o + \sigma^2/2\varepsilon\varepsilon_o kT \tag{27}$$

where n_o is the concentration of ions at the centre between the surfaces and the other symbols have their usual meaning. Equation (27) is called the contact value theorem [71]. It can be concluded from this equation that the counterion concentration at the surface is independent of the ion valency, z, and that for two surfaces that are far apart, n_o will approach zero, and we have

$$n_{surface} = \sigma^2/2\varepsilon\varepsilon_o kT \tag{28}$$

It can be concluded that almost all of the counterions are located close to the bilayer surfaces. Furthermore, according to equation (28), the counterion concentration close to the bilayer surface *will not depend on the water content* between the bilayers (at sufficiently large distances between the bilayers). This is in accordance with the experimental observation for a rather large number of systems (*cf.* Figure 5.19). Similarly, for the H_{II} phase of the sodium salt of DOPA the counterion association to the aggregate surface is high, but somewhat smaller than for the L_α phase (see Figure 5.27). It was found that for a two-phase sample of DOPA-water the ^{23}Na quadrupole splitting was equal to 26.8 kHz for the L_α phase and 16.2 kHz for the H_{II} phase [155]. Thus, the ratio between the observed splittings for the two phases is about 1.7, sufficiently close to the theoretical value of two (*cf.* Equation 19), considering the number of assumptions involved.

Note that the result of the ion condensation model is in sharp contrast to that expected from a simple chemical equilibrium model having a binding site characterized by an equilibrium binding constant. In such a model a dilution of the system would lead to a decrease in the number of bound ions, which would be reflected in a decrease in the measured ^{23}Na quadrupole splitting. At lower water contents the observed quadrupole splittings may be dependent on the water concentration and/or temperature [149,150,161]. Such a behaviour of the splitting might arise from a change in the p_{bi}'s describing the fractions of ions in several different binding sites, which happen to have an opposite sign of the order parameter S_i. A possible explanation could then be that at high water contents the sodium ions reside on the edge of the polar head groups, causing the electric field gradient to be more or less perpendicular to the aggregate surface. When the water concentration is reduced the counterions may replace some of the water molecules between the polar head groups and S_i may change sign upon such a process. Thus, it has been shown that quadrupole splittings may pass through zero upon a variation in temperature or water concentration [149] (Figure 5.20).

E. NMR THEORY OF CSA IN LIPID-WATER SYSTEMS

The presence of electrons in the molecule containing the nucleus in question gives rise to chemical shielding. The strong, external magnetic field, $\mathbf{B}_0$, induces electronic currents that produce, in turn, an induced magnetic field $\mathbf{B}'$ that modifies the local magnetic field $\mathbf{B}_{loc}$ at the site of the nucleus in question. $\mathbf{B}'$ will be proportional to $\mathbf{B}_0$, and generally diamagnetic; for molecules executing rapid and random rotation in liquids $\mathbf{B}' = -\sigma_0\mathbf{B}_0$, where σ_0 is the isotropic chemical shielding parameter. In solids and liquid crystals it may be necessary to write $\mathbf{B}' = -\tilde{\sigma}.\mathbf{B}_0$, where $\tilde{\sigma}$ is a second rank tensor, which is represented by a 3x3 matrix:

$$\mathbf{B}' = (B_x', B_y', B_z') = -\begin{pmatrix} \sigma_{xx} & \sigma_{xy} & \sigma_{xz} \\ \sigma_{yx} & \sigma_{yy} & \sigma_{yz} \\ \sigma_{zx} & \sigma_{zy} & \sigma_{zz} \end{pmatrix}\begin{pmatrix} 0 \\ 0 \\ B_z \end{pmatrix} = (\sigma_{xz}B_z, \sigma_{yz}B_z, \sigma_{zz}B_z)$$

where the induced field is written as a row vector and the external magnetic field as a column vector. Thus, the induced field is not necessarily directed antiparallel with the external magnetic field, but is determined by the components σ_{xz}, σ_{yz} and σ_{zz} of the chemical shift tensor.

Thus, the Zeeman interaction between the magnetic field and the nuclear spins is influenced by the chemical shift $\tilde{\sigma}$, which has both a scalar part $\sigma^{(0)}$ and a second rank tensor part $\sigma^{(2)}$. The latter part is orientation dependent (*cf.* the quadrupole interaction) and carries the information about the molecular arrangement. The Hamiltonian for the orientation dependent part of the chemical shift, H_{CS}, represented in the irreducible spherical tensor notation, is

$$H_{\mathrm{CS}} = -\sum_{iqq'q''} \gamma_i \sqrt{5}\, \sigma^{(2)}_{-q}(i) B_{q'} I_{q''}(i) \begin{pmatrix} 1 & 1 & 2 \\ q' & q'' & -q \end{pmatrix} \tag{29}$$

where γ_i is the gyromagnetic ratio of nucleus i and B_q is a component of the magnetic field. The different components of the nuclear spin angular momentum operator are

$$I_{\pm 1} = \mp \frac{1}{\sqrt{2}} I_{\pm} = \mp \frac{1}{\sqrt{2}} \left(I_x \pm i I_y \right);\; I_0 = I_z \tag{30}$$

and the components of the anisotropic part of the chemical shift tensor are

$$\begin{aligned} \sigma^{(0)} &= \frac{1}{3}\mathrm{Tr}\tilde{\sigma} = \frac{1}{3}(\sigma_{xx} + \sigma_{yy} + \sigma_{zz}), \\ \sigma^{(2)}_{\pm 2} &= \frac{1}{2}(\sigma_{xx} - \sigma_{yy} \pm 2i\sigma_{xy}), \\ \sigma^{(2)}_{\pm 1} &= \mp(\sigma_{zx} \pm i\sigma_{zy}), \\ \sigma^{(2)}_{0} &= \frac{1}{\sqrt{6}}(3\sigma_{zz} - \mathrm{Tr}\tilde{\sigma}) \end{aligned} \tag{31}$$

Finally

$$\begin{pmatrix} 1 & 1 & 2 \\ q' & q'' & -q \end{pmatrix}$$

is a Wigner 3-*j* symbol [17]. Equation (29) is valid for an arbitrary Cartesian coordinate system. One convenient choice of coordinate system is to make $B_0 = B_z$ (the magnetic field along the z-axis as denoted above), which then is the only non-zero component of the magnetic field. However, the tensor $\sigma^{(2)}$ is a molecular quantity and is thus best defined in a coordinate system fixed in the molecule. Usually one chooses a principal coordinate system where σ_{xx}, σ_{yy} and σ_{zz} give the only non-zero contributions. Equation (29) then becomes

$$H_{\mathrm{CS}} = -\sqrt{5}B_z\sum_{iqq'}\gamma_i\sigma_{q'}^{(2)}(i)I_q(i)D_{q'-q}^{(2)}(\Omega_{\mathrm{LM}})\begin{pmatrix}1 & 1 & 2\\ 0 & q & -q\end{pmatrix} \tag{32}$$

As the molecule rotates the Eulerian angles change, while the remaining factors are time-independent for rigid molecules. In analogy with the treatment of the quadrupolar interaction, the anisotropic part of the chemical shielding of the Hamiltonian can be treated as a perturbation to the scalar part of the Zeeman interaction. In first order it is only the secular terms, which have q = 0, that give any contributions. Time averaging then yields

$$H_{\mathrm{CS}} = -\sqrt{\frac{2}{3}}B_z\sum_{iq}\gamma_i\sigma_q^{(2)}(i)I_z(i)\overline{D_{q0}^{(2)}(\Omega_{\mathrm{LM}})} \tag{33}$$

This equation is quite general and is valid for a liquid as well as for a liquid crystalline phase. For the latter phase we now transform the tensors in Equation (31) in two steps (see Figure 5.1). Firstly, a transformation from the molecular to the director system through the Eulerian angles Ω_{DM}, followed by a transformation to the laboratory system through Ω_{LD}. This leads to

$$H_{\mathrm{CS}} = -(3\cos^2\theta_{\mathrm{LD}} - 1)\frac{1}{\sqrt{6}}B_z\sum_{iq}\gamma_i I_z(i)(-1)^q\overline{\sigma_q^{(2)}(i)D_{q0}^{(2)}(\Omega_{\mathrm{DM}})} \tag{34}$$

Thus, also for the chemical shift the anisotropic part is multiplied by $(3\cos^2\theta_{\mathrm{LD}} - 1)$, and the time averaged term contains the order parameters. If the chemical shift tensor in the molecular coordinate system (or the principal axes system) has cylindrical symmetry we can extract the order parameter from a measurement of the chemical shift anisotropy (CSA) in the NMR spectrum, analogously to how S is obtained from the ^{2}H NMR spectrum. Unfortunately, most often the molecular chemical shift tensor does not fulfil the criterion of cylindrical symmetry and S-values cannot be obtained. This situation pertains, for instance, to one of the efficiently applied nuclei in studies of the phase behaviour of lipids, namely ^{31}P [217]. However, since CSA measurements are obtained with the *sign* in the NMR spectrum, information about phase structure is quite easy to get as will be discussed below.

From the Hamiltonian in Equation (34) it follows that the *effective* chemical shift tensor is cylindrically symmetric, reflecting the cylindrical symmetry about the normal to the lipid aggregate in the liquid crystalline phase (*i.e.* in the director coordinate system). Therefore, the observed peaks in the NMR spectrum will depend on the orientation of the cylindrically symmetric tensor with respect to the magnetic field. An NMR signal may appear in a limited region of frequencies determined by θ_{LD}, which can vary between 0° and 90° (Figure 5.21). The distance between the peaks at these two extreme values of θ_{LD} can be measured and is usually called the CSA, and is defined as $\Delta\sigma = (\sigma_{\parallel} - \sigma_{\perp})$. A powder sample consists of microcrystallites exhibiting a random orientation of the directors and the

NMR spectrum will consist of a superposition of signals from all the different directions θ_{LD}. The characteristic NMR line shape of CSA with axial symmetry is shown in Figure 5.21 which may be compared with a quadrupolar NMR spectrum (Figure 5.2; note that the quadrupolar spectrum can be seen as two mirror CSA line shapes). The line shape from a sample with axial symmetry can be understood qualitatively as follows. Here, the orientation of only one axis relative to the magnetic field is of relevance, namely the one for which the chemical shift is σ. The probability is larger that this axis is oriented perpendicular to, rather than parallel with, the magnetic field, since the number of vectors with arrows pointing to the equator on a sphere will be larger than those pointing towards the south or north poles. Thus, the intensities of the NMR signals in the spectrum are distributed so that $\sigma_\perp$ has the highest intensity and $\sigma_\parallel$ the lowest. It is then a simple task to determine the values of $\sigma_\parallel$ and $\sigma_\perp$ from the shoulder and high peak in the spectrum, respectively. The chemical shift anisotropy for a powder sample can be obtained from Equation (34) yielding

$$\Delta\sigma = (\sigma_{||} - \sigma_\perp) = \frac{3}{2}(\sigma_{zz} - \sigma_0)\overline{D^{(2)}_{00}(\theta_{DM})} + \sqrt{\frac{2}{3}}(\sigma_{xx} - \sigma_{yy})[\overline{D^{(2)}_{02}(\theta_{DM}, \varphi_{DM})} + \overline{D^{(2)}_{0\text{-}2}(\theta_{DM}, \varphi_{DM})}] = \frac{3}{4}(\sigma_{zz} - \sigma_0)\overline{(3\cos^2\theta_{DM} - 1)} + (\sigma_{xx} - \sigma_{yy})\overline{\sin^2\theta_{DM}\cos 2\varphi_{DM}} = \frac{3}{2}(\sigma_{zz} - \sigma_0)S + \frac{2}{3}(\sigma_{xx} - \sigma_{yy})(S_{xx} - S_{yy}) \tag{35}$$

The averaged terms are recognized as the order parameters (S=S_{zz}, S_{xx} and S_{yy} according to the notation used previously). Notice that the second rank S matrix is traceless

$$\sum_i S_{ii} = 0 \tag{36}$$

For a cylindrically symmetric shielding tensor of the molecule (*i.e.* $\sigma_{xx} = \sigma_{yy}$) we get

$$\Delta\sigma = (\sigma_{||} - \sigma_\perp) = \frac{3}{2}(\sigma_{zz} - \sigma_0)S = (\sigma_{zz} - \sigma_{xx})S \tag{37}$$

where we have used $\sigma_0 = \frac{1}{3}(\sigma_{xx} + \sigma_{yy} + \sigma_{zz})$. Thus, S can be determined from Equation (37), provided the molecular shift tensor is known and cylindrically symmetric. If this is not the case, two order parameters are necessary to describe the orientation of the molecule in the liquid crystalline phase and Equation (35) may be written as

$$\Delta\sigma = (\sigma_{||} - \sigma_\perp) = (\sigma_{zz} - \sigma_{yy})S_{zz} + (\sigma_{xx} - \sigma_{yy})S_{xx} \tag{38}$$

If ions, like ^{19}F [116,238] or ^{133}Cs [192,274], are investigated the equations (37) and (38) have to be multiplied with the relative population of ions, p, in the

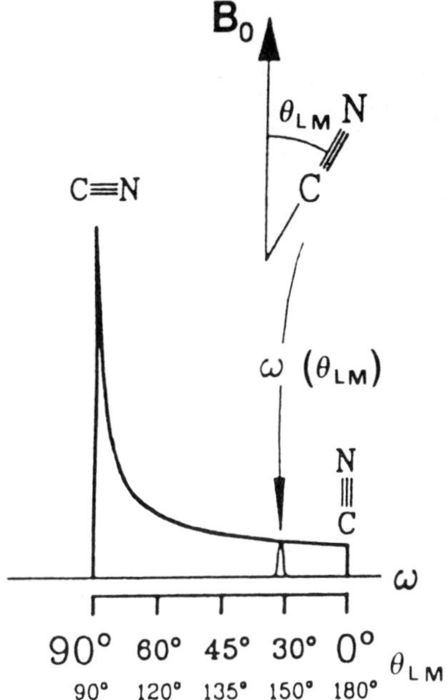

Fig. 5.21. Dependence of the NMR peak on the orientation between the molecular segment described by the shift tensor and $\mathbf{B}_0$ (*cf.* Figure 5.2).

anisotropic site (assuming that there is a fast exchange between the different sites occurring in the liquid crystal).

In general it is found that for an L_α phase $\Delta\sigma < 0$. Thus, the spectra of L_α phases exhibit a high field peak and a low field shoulder (Figure 5.22). In parallel with the quadrupole splittings for an H_{II} phase the CSA is twice as large as for an L_α phase and with opposite sign, which, however in contrast to quadrupole splittings, is observed in the NMR spectrum with CSA, *i.e.* $\Delta\sigma^{L_\alpha} = -2\Delta\sigma^{H_{II}}$.

1. ^{31}P NMR in the Determination of Phase Diagrams

A large number of investigations of phase transitions of membrane lipids have been published [52,217], but reports of determinations of complete phase diagrams are considerably less. Although the observation of a low field shoulder in the NMR spectrum most often represents an L_α phase, exceptions have been reported [126,127]. Thus, it is usually advisable to use a complementary method to get conclusive results. Another obstacle that may arise, in particular if the studies are performed at high magnetic fields, is that frequently the lipid aggregates align macroscopically in the external field, leading to the fact that the shoulders, both at low and high field, are removed from the spectrum. This dilemma is, however, relatively easy to solve by a careful study of the changes in the chemical shift

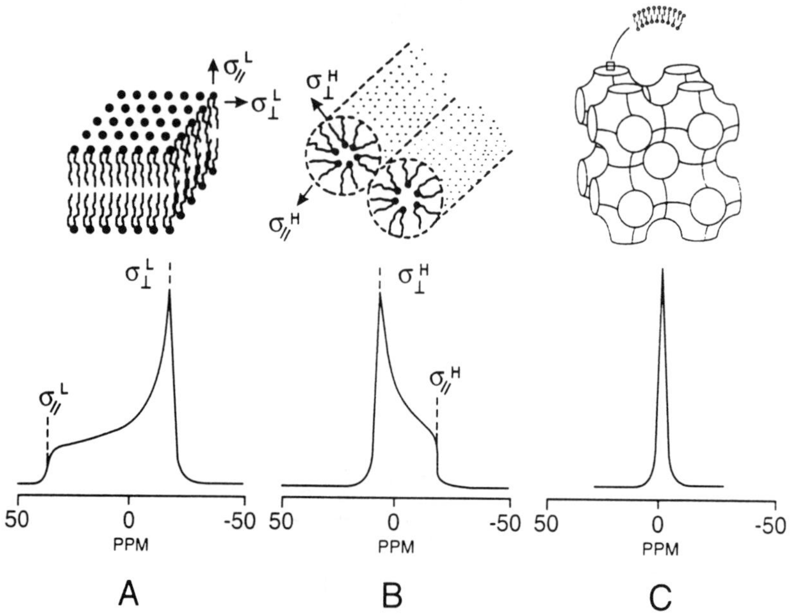

Fig. 5.22. Typical ^{31}P NMR line shapes for some different liquid crystalline phase structures: A) L_α phase, B) H_{II} phase, and C) cubic phase.

with time in the spectrometer, and a possibly wrong conclusion about the presence of an isotropic phase can be circumvented. It should be noted that the molecular chemical shift tensors are similar in most phospholipids [95,125,126,217]. Obviously, also the simulations of spectra, in order to get the fractions between different phases, rely on the recording of true powder spectra [66]. It is necessary to avoid effects of distorsions due to technical insufficiencies inherent in the spectrometer and this is accomplished by phase cycling and spin echo detection [194].

One of the first phase diagrams (Figure 5.23) determined with mainly ^{31}P NMR was for the system DOPC-monooleoylglycerol (MO)-water [86]. As can be inferred from this figure a cubic phase occupies a large area of the ternary phase diagram. Previously, it had been shown that in the binary system MO-water, *i.e.* the base of the phase triangle in Figure 5.23, there are two cubic phase structures, namely an *Ia3d* at low water contents and a *Pn3m* at high water contents [162]. It is interesting to note that there are two different L_α phases and an H_{II} phase in the ternary system. Thus, by mixing samples of these two L_α phases, an H_{II} phase may be formed. Obviously, the simple molecular shape argument does not apply here and it has been concluded that there is an interaction between the polar head groups of MO and DOPC resulting in an effective shape that favour the H_{II} or cubic phase [184]. This phase diagram may also be taken as an excellent example of how phase diagrams can be utilized in pharmaceutical technology. The L_2 phase is isotropic with low viscosity and can be readily utilized to dissolve

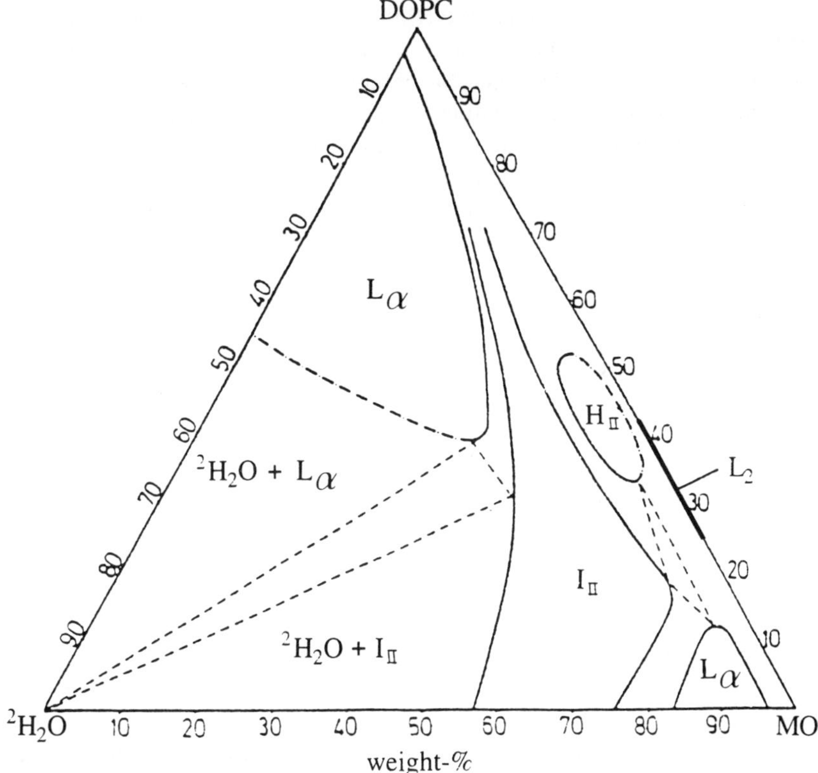

Fig. 5.23. Ternary phase diagram for the system DOPC/MO/^2H_2O at 28°C. For the notation of the different phases see text. From [86].

molecules, like hydrophobic drugs and peptides. The phase obtained can be easily applied *e.g.* under the skin covering the teeth or at any other place subcutaneous. At this location the L_2 phase will take up water from the surroundings in the skin. This process is represented in the phase diagram by going on a straight line from the L_2 phase towards the water corner. As the water content is increasing in the "drug"-sample the composition is approaching the water corner through the cubic phase, which in turn is in equilibrium with water. Thus, a very high viscosity typical for a cubic phase will come up in the "drug"-sample. Two advantageous effects will be obtained in this way, first the sample becomes liquid crystalline, exhibiting high viscosity, and second the drug is slowly released from the cubic structure to avoid large side-effects of the drug (an example of slow-release distribution of drugs).

A partial phase diagram has been determined for the ternary system DOPC-dioleoylphosphatidylethanolamine (DOPE)-water [66]. In this article, some methodological aspects are discussed also. The CSA is greater for DOPC than for DOPE (6-9 ppm in the L_α phase), a difference that was ascribed to different *S*-values for the two lipid head groups. ^{31}P NMR spectra from an L_α phase of DOPC-DOPE-water below maximum hydration exhibit two resolved, superimposed powder spectra (Figure 5.24). The CSA of both phospholipids is greater at excess water than below maximum hydration, showing that the molecular ordering of the head groups depends on the degree of hydration. The spectral resolution between DOPC and DOPE in the L_α phase is strikingly diminished at excess water concentration, due to an increased broadening of the spectral line shape. Furthermore, it was concluded that both head groups have a more upright orientation in relation to the bilayer surface above maximum hydration [152,246]. From the ^{31}P NMR spectra it was possible to determine the relative differences in the DOPC/DOPE content between the various lipid phases in equilibrium. As expected the proportion of DOPE is decreased in the L_α phase, while it increases in the H_{II} phase, when these phases are in equilibrium.

Since we have a particular interest in cubic phases (see also Section G on lipid diffusion) partial or complete phase diagrams of systems containing such phases have been investigated. Diagrams have been determined for lipids containing both a single acyl chain, like the lysolipids in water [6] or a lysolipid and a fatty acid in water [15], as well as for ternary systems containing PCs [154, 155, 88, 223, 224, 250].

A change in the molecular shape does not explain why an H_{II} phase is formed at increasing water contents in PC-alkane-water systems [156,223,224]. This problem was solved neatly by Gruner [82], who used a phenomenological concept, the *spontaneous curvature*, H_o, of the lipid monolayer[2]. This concept was first introduced by Helfrich [94] for PC bilayer systems. Gruner and coworkers have shown that the energy of curvature plays a dominant role in the formation of H_{II} phases. The total curvature free energy, G_{curve}, per area, A, is given by [2]

$$G_{curve} = K_m \, A \, \langle (H-H_o)^2 \rangle \tag{39}$$

where H is the local mean curvature of the monolayers, and K_m is the elastic bending constant. It can thus be seen that it is advantageous to have H close to H_o to minimize the free energy of curvature. However, for a phospholipid monolayer to form a cylinder of radius $R_o = 1/H_o$ there will also be a cost of non-zero packing energy. In particular, the acyl chains of the PC molecules must stretch to fill the interstitial regions between the cylindrical aggregates building up the H_{II} phase (Figure 5.25). It can be concluded that the smaller the radius of the water cylinder is, the smaller are the hydrophobic interstices and the easier it is for the lipid acyl chains to elongate to fill the volume. This explains why the H_{II} phase tends to form at low water contents. Similarly, it can be understood why an increasing acyl

[2]There are two fundamental types of curvature used in differential geometry that characterize each point on a surface: the Gaussian curvature $K=R_1^{-1} \cdot R_2^{-1}$ and the mean curvature $H=\frac{1}{2}(R_1^{-1}+R_2^{-1})$, where R_1 and R_2 are the principal radii. The surfaces of so called minimal surface satisfy $H=0$, so that every point is a balanced saddle point with $R_1=-R_2$. The convention often used is that for a lipid monolayer $H>0$ when the layer curves towards the hydrocarbon chain region and $H<0$ when the layer curves towards the aqueous region.

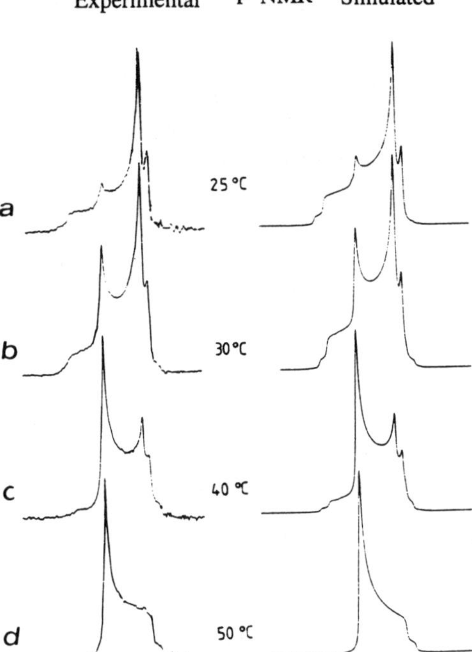

Fig. 5.24. ^{31}P NMR spectra recorded at different temperatures from a system composed of DOPC/DOPE/^2H_2O. The molar ratio between DOPC/DOPE was 1:2 and the water content 10% (w/w). From [66].

chain length also favours the H_{II} phase formation. It has been shown in several studies [223,224,245] that the void volumes appearing between the cylinders in the H_{II} phase can be eliminated by addition of hydrophobic molecules, like alkanes. NMR studies strongly indicate that these molecules preferentially partition into these interstitial regions between the cylinders in the H_{II} phase structure [223]. It can be expected that the more saturated the acyl chains of the phospholipids are, the greater the packing constraints for formation of the H_{II} phase are, and the larger the water cylinders in the H_{II} phase must be. It has also been observed that for a phospholipid-dodecane-water system the size of the water cylinders in the H_{II} phase increases in the following order: DPPC > 1-palmitoyl-2-oleoyl-PC > DOPC ≈ dilinoleoyl-PC (Figure 5.26).

Obviously, the size of the polar head group plays an important role for the formation of an H_{II} phase; smaller head groups favour H_{II} phases. The ability of a lipid monolayer to adopt negative radii of curvatures is determined in part by the ability of the head groups to be packed closely together: the area per lipid molecule is smaller at the head group than in the hydrocarbon chain region for a monolayer with negative curvature. It can be expected that lipids with a charged

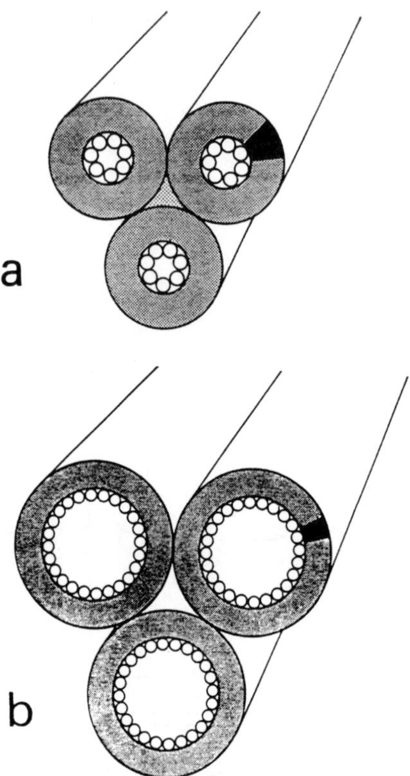

Fig. 5.25. Schematic illustration of an H_{II} phase having acyl chains with (A) a high degree of unsaturation, and (B) a low degree of unsaturation. From [224].

head group are less prone to pack closely to each other because of the electrostatic repulsion between the head groups. Hence, such lipids should have a smaller tendency to adopt reversed-phase structures than uncharged lipids with otherwise similar chemical structures. Steric interactions between head groups must also affect their ability to pack closely together. Therefore, lipids with bulkier head groups should be less prone to form non-lamellar reversed-phase structures.

As can be inferred from the phase diagram in Figure 5.27 an H_{II} phase is formed for the DOPA-water system up to a water content of about 35 wt %, and an L_α phase can swell up to a water concentration of at least 98 wt %. When a large fraction of DOPA molecules is neutralized by association of counterions, the head groups can be packed quite closely together. Therefore, the lipid monolayer is able to adopt a negative curvature, leading to the formation of an H_{II} phase. With increasing water concentration in the DOPA-water system, the interstitial regions between the growing lipid-water cylinders, building up the H_{II} phase, will increase. The creation of such void volumes is energetically unfavourable, and will eventually lead to an induction of an L_α phase at a certain water content (Figure 5.27). When the water concentration is further increased the L_α phase

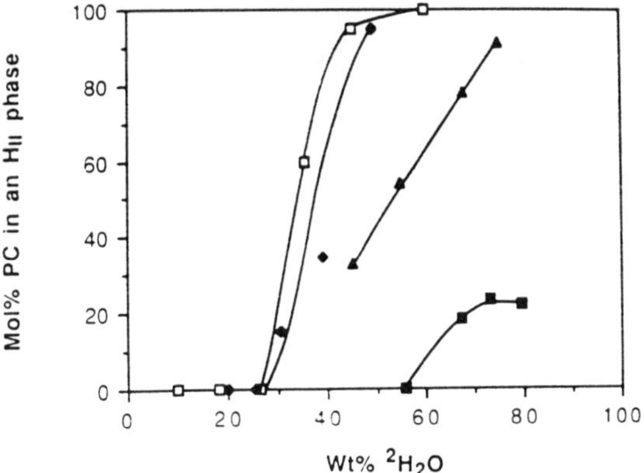

Fig. 5.26. Onset of the formation of an H_{II} phase at 45°C in different PC-*n*-dodecane-water systems. The molar ratio of dodecane/PC = 2. (◆) DLiPC; (❑) DOPC; (▲) POPC; (■) DPPC. From [224].

swells, resulting in large distances between the bilayers, due to the long-range coulombic forces between the charged bilayers. It is well known that even a small increase in the surface charge density may drastically increase the tendency for water to be taken up by an L_α phase [15,85]. The thickness of the water layer at 98 wt % of water for the DOPA-water system has been estimated to be approximately 200 nm.

2. NMR of Counterions and Phase Diagrams

Counterions utilized for determinations of phase behaviour are relatively rare. Only two examples will be mentioned here. The CSA of ^{19}F is quite large and it is therefore very convenient to use this counterion in NMR studies. Typical spectra are shown in Figure 5.28, and the corresponding phase diagram is shown in Figure 5.29 [121]. Similarly ^{133}Cs has been used for studies of counterion binding and phase behaviour [192].

F. ^1H NMR IN STUDIES OF LIQUID CRYSTALLINE PHASES

Because of strong dipolar couplings in anisotropic liquid crystalline phases ^1H NMR exhibits broad featureless spectra. The interpretation of such NMR spectra poses problems, but at the same time there is a lot of information to be gained. In general, however, proton NMR does not constitute the method of first choice in determinations of phase diagrams of lipid-water systems. On the other hand, in diffusion studies protons are advantageous, having a large gyromagnetic ratio and high sensitivity, were it not beclouded by the strong dipolar couplings. Fortunately, this problem can be circumvented, as we will see later, in studies of lipid diffusion of L_α phases. A general applied procedure that can be conveniently utilized to get detailed information about phase equilibria is not available, and we

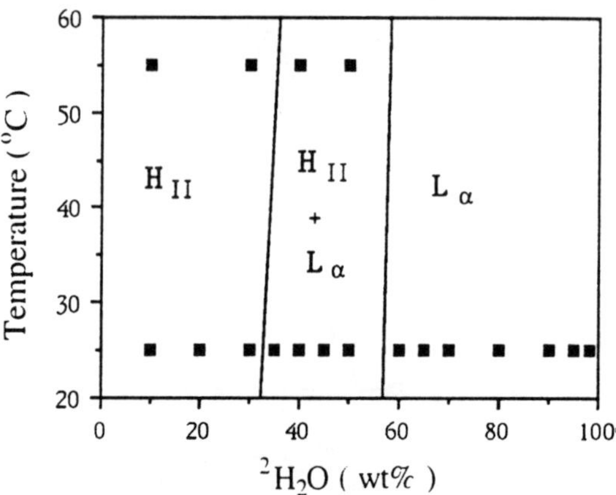

Fig. 5.27. Tentative phase diagram for the system DOPA-^2H_2O. The phase equilibria were deduced by ^{31}P NMR and X-ray diffraction. From [155].

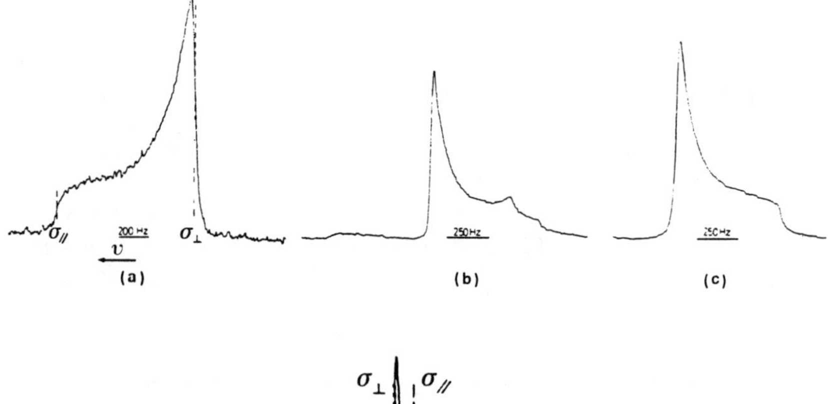

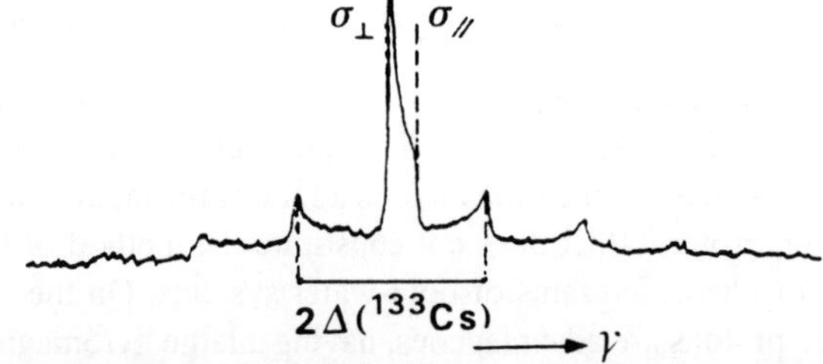

Fig. 5.28. Top) ^{19}F NMR spectra from liquid crystalline phases of the system octylammonium fluoride and water. The phases are from left to right; L_α, L_α + H_I, and H_I. From [116], and bottom) Typical powder spectrum of ^{133}Cs NMR (I = 7/2) for an L_α phase. From [192].

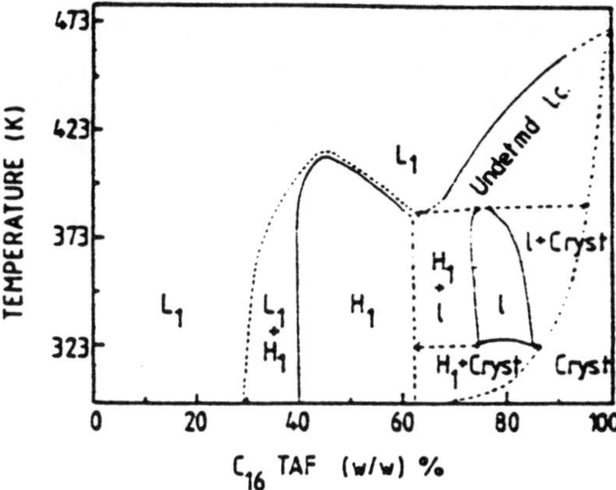

Fig. 5.29. Phase diagram of the binary system hexadecyltrimethylammonium fluoride (CTAF) and water. From [116].

are therefore confined to the information we get about phase transitions and estimates of the molecular dynamics in the system.

The dipolar part of the Hamiltonian can be written

$$H_D = \sum_{i>j,q} (-1)^q F_{-q}^{(2)}(ij) A_p^{(2)}(ij) \tag{40}$$

where the sum i > j is over all spin pairs. The components of the tensor $F_{-q}^{(2)}$ depend on the distance between, and the orientation of, the two nuclear magnetic dipoles i and j. For lipid bilayers in strong magnetic fields, provided that the rotational and translational diffusion is fast so that a zero average for all the intermolecular dipolar couplings is attained, H_D in angular frequency units is given by

$$H_D = \sum_{i>j} F_0^{(2)}(ij) A_p^{(2)}(ij) \tag{41}$$

Here $F_0^{(2)}(ij)$ is the irreducible component along $\mathbf{B}_0$ of the dipolar tensor. The component $F_0^{(2)}(ij)$ is related to $F'^{(2)}_0(ij)$, a component along the internuclear vector, r_{ij}, by

$$F_0^{(2)}(ij) = \tfrac{1}{2}(3\cos^2\theta_{LD} - 1) F'^{(2)}_0(ij)$$

and

$$F'^{(2)}_0(ij) = 2\gamma_i \gamma_j r_{ij}^{-3} \overline{D_{00}^{(2)}(\Omega_{ij})}$$

The dipolar Hamiltonian for all the protons in the hydrocarbon chain can then be written as

$$H_D = \tfrac{1}{2}(3\cos^2\theta_{LD} - 1) \sum_{i>j} F'^{(2)}_0(ij) A_0^{(2)}(ij) \tag{42}$$

Now the sum i > j is only over intramolecular pairs. Thus, again we see that the single factor $\frac{1}{2}(3\cos^2\theta_{LD} - 1)$ multiplies all the anisotropic terms. This fact results in that a remarkable and characteristic bandshape, called super-Lorentzian, is observed for powder samples of liquid crystalline phases in ^{1}H NMR. This spectrum is then a superposition of spectra from microcrystallites with random orientations of their directors. Wennerström [268] succeeded to reproduce such a spectrum almost quantitatively for a surfactant-water system. It was also possible to simulate the ^{1}H NMR spectrum of a non-oriented L_α phase composed of phosphatidylcholine and water [259]. There is a dramatic change in the proton NMR spectrum at the gel to L_α phase transition. The gel phase shows a signal typical of an organic solid, although some motional freedom is present as indicated by the relatively small second moment of the signal (see below). The basic reason for the difference between the spectra in the gel and liquid crystalline phases can be found by a consideration of H_D in the Equations (40) and (42). In the gel phase the lipid lateral diffusion is slow and neither cylindrical symmetry around the normal to the aggregate nor the intermolecular dipolar interactions are averaged to zero. Thus, a gel-type of spectrum can be taken as an indication of slow lateral diffusion, $D_L \leq 10^{-13}$ m^{2}s^{-1}, while the observation of a super-Lorentzian bandshape for a liquid crystal implies a faster translational diffusion, $D_L \geq 10^{-13}$ m^{2}s^{-1}. From a study of the effect of cholesterol on the proton NMR bandshape above and below T_m it was concluded that cholesterol induces an increase in the lipid lateral diffusion below T_m resulting in a change of the state of the gel phase, and the NMR spectrum gradually exhibits a super-Lorentzian bandshape indicative of a liquid crystalline phase behaviour [259].

1. Measurements of Moments of the NMR Signal

For a lipid in an anisotropic liquid crystalline or solid system the *intra*- and *inter*-methylene proton dipolar interactions are of comparable magnitude, that in turn are much larger than the range of proton chemical shifts, precluding the observation of individual ^{1}H NMR resonances as discussed above [10,102]. The dipole-dipole interaction results in homogeneously broadened NMR resonances, *i.e.* lines may overlap, be degenerate, and produce a band of absorptions that is in general difficult to associate with any particular transition. However, it is possible to obtain some information about the average value of the dipolar interactions in the sample from such a band of frequencies. To accomplish this, it is necessary to use an expansion technique originally suggested in 1948 by van Vleck [1,225]. The expansion is in terms of moments of the NMR line. The n'th moment of a spectrum is given by [225]. The expansion is in terms of moments of the NMR line. The n'th moment of a spectrum is given by [225]

$$M_n = \frac{\int x^n F(x) dx}{\int F(x) dx} \tag{43}$$

where F(x) is the spectral lineshape, $x = \omega - \omega_o$, and ω_o is the Larmor frequency. The second moment, M_2, is particularly useful for monitoring molecular motion. For example a transition from a gel to an L_α phase will result in a large decrease in the second moment M_2.

Bloom et al. [10] have used moments, calculated from ^1H NMR spectra, to study acyl chain order of PC in model membranes. Likewise, we have used moments to draw conclusions about aggregate structures and phase transitions.

In the gel/crystalline phase the linewidth of the ^1H NMR signal may be as large as 86 kHz [102]. Thus, the corresponding M_2 will be large. A transition to a liquid crystalline phase reduces the contributions from the intermolecular dipolar interactions to M_2 to almost zero because of the rapid lateral diffusion and fast rotation about the long axis of the molecule. The intramolecular dipolar interactions are significantly lowered due to the increased flexibility of the acyl chains in the liquid crystalline phase. Furthermore, the dipolar interactions in an H_{II} phase are averaged to a greater extent than in an L_α phase due to lateral diffusion around the cylindrical aggregates (*cf.* Equation (19)) and to an increased acyl chain flexibility. For an isotropic phase, like the I_{II} phase, the dipolar interactions are averaged out completely. The M_2 of the three liquid crystalline phases is thus expected to decrease in the order $L_\alpha > H_{II} > I_{II}$. Applications of M_2 measurements in the study of phase behaviour are also rather rare, but they constituted an invaluable complementary method in the determination of the L_2 phases in systems containing glycolipids [145,256].

The hydration properties and the phase structure of 1,2-di-*O*-tetradecyl-3-*O*-(3-*O*-methyl-β-D-glucopyranosyl)-*sn*-glycerol (3-*O*-Me-β-D-GlcDAlG) in water have been studied via differential scanning calorimetry, ^1H NMR and ^2H NMR spectroscopy, and X-ray scattering. The apparent second moment M_2 of ^1H NMR spectra is related to the amount of orientational order in lipid molecules and has been measured for 3-*O*-Me-β-D-GlcDAlG at various temperatures and water concentrations [256]. Values of M_2 from the ^1H NMR spectra of 3-*O*-Me-β-D-GlcDAlG versus temperature are given in Table 5.2. The M_2 for 3-*O*-Me-β-D-GlcDAlG decreases from $5.8 \cdot 10^9 \, s^{-2}$ at 30°C to $3.8 \cdot 10^9 \, s^{-2}$ at 60°C and to a value of $0.043 \cdot 10^9 \, s^{-2}$ at 70°C. The large change in the moment with increasing temperature is indicative of a transition from a well-ordered state (rigid acyl chains and head group) to a fluid state with significant orientational freedom.

For comparison, ^1H NMR spectra of DPPC in 35 wt% ^2H_2O at 30, 40 and 50°C are shown in Figure 5.30 along with spectra from 3-*O*-Me-β-D-GlcDAlG with the same hydration at temperatures of 30 and 80°C. At 30°C, DPPC is in the $L_{\beta'}$ gel phase characterized by mostly rigid acyl chains and the presence of whole-molecule rotation about the lipid long axis [49,166,258]. The spectrum from the low temperature phase of 3-*O*-Me-β-D-GlcDAlG is much broader than that from the $L_{\beta'}$ phase of DPPC, indicating that 3-*O*-Me-β-D-GlcDAlG in its crystalline phase is experiencing much less motional averaging than DPPC in its $L_{\beta'}$ phase. A smaller amount of rotation about a preferred motional axis for 3-*O*-Me-β-D-GlcDAlG is the most reasonable explanation for the observed spectral difference.

Table 5.2.

Apparent second moments M_2 from high power ^{1}H NMR spectra of 1,2-di-*O*-tetradecyl-3-*O*-(3-*O*-methyl-ß-D-glucopyranosyl)-*sn*-glycerol (3-*O*-Me-ß-D-GlcDAlG) and DPPC at various temperatures. From [256].

	$M_2 \cdot 10^9$ s^{-2}	
T/°C	3-*O*-Me-ß-D-GlcDAlG	DPPC
30	5.8	2.9
40	5.6	2.3
50	5.6	0.33
60	3.8	0.28
70	0.04	

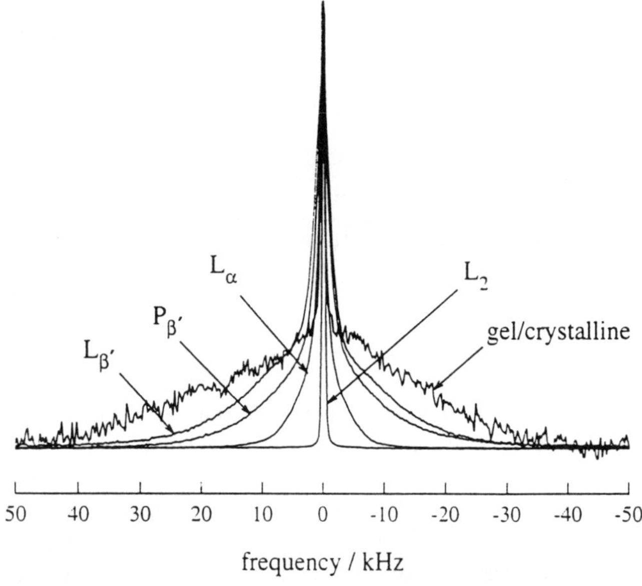

Fig. 5.30. High power ^{1}H NMR spectra from various lipid aggregate structures. The crystalline and L_2 structures are 1,2-di-*O*-tetradecyl-3-*O*-(3-*O*-methyl-ß-D-glucopyranosyl)-*sn*-glycerol in 35 wt % ^2H_2O at 30 and 80°C respectively; the $L_{\beta'}$, $P_{\beta'}$ and L_α structures are DPPC in 35 wt% ^2H_2O at 30, 40 and 50°C, respectively. From [256].

At 40°C, DPPC is in the $P_{\beta'}$ phase that is characterized by rippled bilayers, semi-rigid acyl chains and some long axis rotation [103,258]. The increased rotation of lipids about their long axis is observed in the narrowing of the spectra. The acyl chains in both $L_{\beta'}$ and $P_{\beta'}$ gel phases experience only a small amount of *trans-gauche* isomerization [172]. At 50°C, DPPC is in the L_α phase characterized by melted, highly flexible acyl chains, fast rotational diffusion about the lipid long axis and fast lateral diffusion along the bilayer surface [259]. The spectrum corresponding to this phase displays the typical super-Lorentzian lineshape due to fast motional averaging about the bilayer normal. The spectrum from the melted L_2 phase of 3-*O*-Me-β-D-GlcDAlG is much narrower than that of the super-

Lorentzian lineshape from DPPC in the L_α phase due to the increased motional averaging of 3-*O*-Me-β-D-GlcDAlG in this phase.

Apparent second moments from the DPPC spectra are included in Table 5.2 and quantify the observed spectral differences seen in Figure 5.30. At the lowest temperatures studied (30°C), the M_2 of DPPC is nearly half that of 3-*O*-Me-β-D-GlcDAlG. The M_2 of DPPC at 50°C is significantly larger than that of 3-*O*-Me-β-D-GlcDAlG in the high temperature phase (70°C). At 50°C DPPC is in the L_α phase and intermolecular dipolar interactions are effectively removed by fast lateral diffusion and do not contribute to M_2. The difference in M_2 is due, then, to a smaller amount of intramolecular interactions in 3-*O*-Me-β-D-GlcDAlG because of increased flexibility or fast reorientational averaging.

G. NMR STUDIES OF LIPID TRANSLATIONAL DIFFUSION

The NMR methods with pulsed magnetic field gradients (PFG) provide one of the most attractive techniques for studies of molecular transport in lipid systems. One of the most successful applications of PFG NMR is its use in extracting structural information about heterogeneous systems such as complex liquids and liquid crystals. In particular, it has been an invaluable tool in the determinations of the structure of cubic liquid crystalline phases [154,155]. PFG NMR also provides a method with which lipid lateral diffusion coefficients in an L_α phase can be measured directly [157]. In recent years the applicability of the NMR diffusion techniques has been growing rapidly due to the great improvements in the NMR equipment for diffusion and NMR microscopy [23]. Some extensive reviews have been published [130,151,231].

1. Theoretical Aspects of NMR Diffusion

The basic spin-echo experiment for diffusion measurements is illustrated in Figure 5.31. A preparation period is usually applied first, where the spin system is allowed to evolve to the initial state of the spin-echo sequence. This state must have some transverse magnetization, which in most cases is accomplished by a $\pi/2$-pulse, which flips the equilibrium magnetization, $\mathbf{M}_0$, into the xy-plane. Nuclear spins with different precession rates are then allowed to dephase in the xy-plane during a time τ. At the time τ the dephasing process is reversed and the nuclear spins begin to rephase and eventually they meet again to form a spin-echo. In order to extend the diffusion time, the magnetization can be stored along the z-axis for time periods comparable with the relaxation time T_1 between the dephasing and rephasing durations. The most commonly practised spin-echo experiment is the so-called "Hahn echo" first used by Carr and Purcell [28], consisting of a $\pi/2$ - π - acquire pulse sequence (Figure 5.31). However, any two pulses in sequence will create a refocussed signal [1,279], and in fact the first recorded spin-echo was actually created by a $\pi/2$ - $\pi/2$-sequence [87].

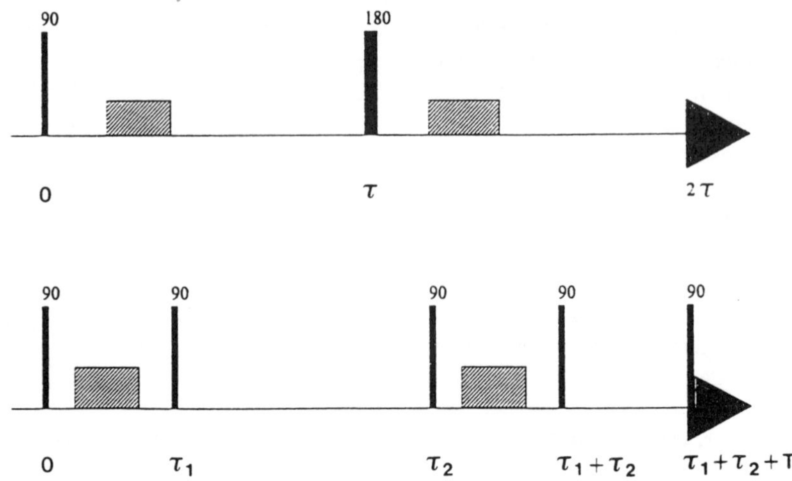

Fig. 5.31. The pulse sequences used in the NMR diffusion experiments. The radio frequency pulses are shown as filled vertical rectangles with the flip angles denoted above each pulse. The magnetic field gradient pulses are shown as hatched rectangles and the data acquisition is indicated with a black triangle.
Top: the SE experiment. Bottom: the LED sequence.

The monitoring of self diffusion of molecules in a sample is accomplished by the application of magnetic field gradients during the dephasing and rephasing periods. These gradients cause the nuclear spins in different local positions in the sample to precess at different Larmour frequencies, thereby enhancing the dephasing process. If the spins maintain their positions in the space throughout the experiment, they will refocus completely into a spin-echo by the $\pi/2$ - π pulse sequence. On the other hand, if they change their positions in space during the experiment, their precession rates will also change, and the refocussing will be incomplete, resulting in a decrease in the intensity of the spin-echo. The attenuation of the spin-echo due to translational diffusion, in terms of the macroscopic nuclear magnetization [255], $\mathbf{M}(\mathbf{r},t) = M_x\mathbf{e}_x + M_y\mathbf{e}_y + M_z\mathbf{e}_z$, is described by a combination of the Bloch's and Fick's equations and can be obtained from the equation

$$\frac{\partial \mathbf{M}(\mathbf{r},t)}{\partial t} = \gamma \mathbf{M} \times \mathbf{B} - \frac{M_x\mathbf{e}_x + M_y\mathbf{e}_y}{T_2} - \frac{M_z - M_0}{T_1}\mathbf{e}_z + D\nabla \cdot \nabla \mathbf{M} \quad (44)$$

The extent of attenuation of the spin-echo signal depends on the set-up of the experiment. Customarily, there are two different pulse sequences that are utilized in the diffusion experiments, namely the Hahn spin echo sequence (SE) [28] and a modification of the so-called stimulated spin echo sequence (LED) [79] (Figure 5.31). In both sequences, two gradient pulses of strength, g, and duration, δ, are applied during the defocussing and refocussing intervals, and the time between the gradient pulses is called Δ. The height or intensity of the recorded echo-signal, S(t), is described by Equation (45)

$$S(t) = S(0)e^{-(g\gamma\delta)^2 D(\Delta - \frac{\delta}{3})}e^{-R} \quad (45)$$

where $t = \tau_1 + \tau_2 + T$ or $t = 2\tau$ depending on how the experiment is performed (Figure 5.31). R is the attenuation of the echo due to relaxation, and for the SE experiment $R = 2\tau/T_2$ and for the LED experiment $R = (2\tau_1/T_2) + ((\tau_2 - \tau_1 + T)/T_1)$, where τ, τ_1, τ_2, and T are defined in Figure 5.31. Several of the parameters in Equation (45) may be varied in a diffusion experiment, t, but generally, all parameters except δ are kept constant so that the second exponent in Equation (45) is constant during an experiment. Note, that the attenuation of the echo signal caused by spin relaxation (T_1 and T_2) will be different for all the possibly present resonance peaks in an NMR spectrum. The relative intensities of the NMR peaks are therefore not directly proportional to the masses of the molecules or groups having different chemical shifts.

2. NMR Measurements of Lipid Lateral Diffusion in Oriented Bilayers

The translational motion of lipids can be measured with the conventional NMR diffusion PFG-SE method for anisotropic liquid crystalline systems, if all static proton dipole-dipole interactions can be removed in some way. For measurements of the lateral diffusion of lipids in an L_α phase this can be accomplished by a macroscopical alignment of the bilayers between glass plates for example [157,158,203]. From Equation (42) it is obvious that if the bilayer is macroscopically aligned and oriented at the magic angle $\theta_{LD} = 54.7°$, $H_D = 0$ and the static dipolar couplings are reduced to zero. In the NMR experiment the translational diffusion is measured in the direction of the magnetic field gradient. For a lipid membrane or an L_α phase the diffusion coefficient depends on the orientation and two diffusion constants D_L and $D_\perp$ describe the diffusion. D_L is the lateral diffusion coefficient for motion parallel with the membrane and $D_\perp$ describes the diffusional motion perpendicular to the bilayer. The measured diffusion coefficient, D, for an oriented bilayer system is then

$$D = D_L \sin^2\theta_{LD} + D_\perp \cos^2\theta_{LD} \quad (46)$$

For a lamella oriented at the magic angle $\sin^2\theta_{LD} = \frac{2}{3}$, and in lipid bilayers it is reasonable to assume that $D_\perp$ is orders of magnitude smaller than D_L and therefore the second term in Equation (46) can be neglected.

Table 5.3
Amphiphile lateral diffusion coefficients, D_L, in some macroscopically aligned L_α phases.

Amphiphile system	Composition (wt %)	Temperature (°c)	$D_L \cdot 10^{11}$ (m^2 s^{-1})	References
Surfactants				
Octylammonium chloride	72	24	33	[157]
Water	28			
Aerosol OT	71	24	1.0	[148]
Water	29			
		35	1.7	[157]
Sodium octanoate	22	24	21	
Decanol	40			
Water	38			
Lithium perfluoro-		25	3	[252]
octanoate	72			
Water	28			
Biological lipids				
MO	82	22	1.1	[148]
Water	12			
		35	1.6	[277]
DGlcDAG	87	45	3.9	
Water				
	13			
DOPC	80	35	0.5	[146]
Water	20			
		45	0.9	[146]
POPC	80	35	0.6	
Water	20			
DOPC	50	35	0.7	[146]
Cholesterol	30			
Water	20			
DOPC	70	35	0.5	[257]
Sodium cholate	10			
Water	20			
Lysooleoyl-PC	84	26	0.2	[197]
Water	16			
Dimyristoyl-PC	75	50	1.2	[270]
Water	25			
DPPC	75	52	0.9	[197]
Water	25			
DOPC	46.2	36	0.3	
DOPE	43.8	46	0.4	
Water	10.0	56	0.6	

In the earliest experiments this method was used for the determination of the lateral diffusion coefficient of some surfactants [157,158] and PC's [157,158,203] in L_α phases. A most annoying difficulty in these early measurements of the lateral diffusion coefficient of lipids was the contribution from water protons to the spin echo signal, which resulted in a too large diffusion coefficient of the lipids. This problem can be solved by a careful exchange of all free protons for deuterium. As an example the lateral diffusion coefficient of a typical amphiphile like sodium octanoate was measured to be equal to $2.1 \cdot 10^{-10}$ m^2s^{-1} at 24°C (Table 5.3) [157].

The lateral diffusion coefficient is also affected by a change in the water content in the L_α phase (Table 5.3) [129]. It was found that the lateral diffusion of DPPC increased by a factor of 2 for a change in the ^2H_2O content between 15 to 40 wt%. Wu & Huang [280] have calculated the change in the lateral diffusion as a function of the water content according to the theory of Brownian motion in thin sheets. They found that the dependence of phospholipid diffusion on hydration can be attributed primarily to the change in the bulk water mobility in the multilamellar phase. Recently, we reported on the effect of glycerol on translational and rotational motions in aqueous lipid bilayers [190]. A continuous decrease in the lipid diffusion coefficient was observed upon an increasing glycerol concentration; thus, it was found that at room temperature $D = 12.6 \cdot 10^{-12}$ m^2s^{-1} for an aqueous L_α phase, while $D = 1.9 \cdot 10^{-12}$ m^2s^{-1} when all water was displaced by glycerol. It was concluded that glycerol between the lipid bilayers affects the lipid diffusion because of changes in the viscosity in the interbilayer regions. This is also in accordance with the findings of Wu & Huang [280].

As expected intuitively D_L depends on the hydrocarbon chain length, but the dependence is rather weak [129,157]. Different polar head groups, on the other hand, affect the translational motion of amphiphiles strongly [157]. Thus, D_L of lyso-PC is somewhat smaller than for the corresponding PC, although the latter has two acyl chains and the lyso-compound only one [70,197]. Most probably the bulkier head group of lyso-PC plays an important role for the lipid lateral diffusion. In studies [143,277] of the lateral diffusion of the two dominant glucolipids in the bacterium *A. laidlawii* it was found that the translational motion of DGlcDAG was about ten times faster than for DOPC [146] (Table 5.3). Probably this difference in D_L can be ascribed to differences in the interactions in the polar head group region and the packing of the lipids in the bilayer [143,157]. Addition of a monoglucolipid resulted [277] in a decrease in D_L, which was suggested to be due to hydrogen bonding between the polar head groups [143]. The theoretical models used in molecular dynamics simulations of the lateral diffusion in lipid bilayers disregard the effect of the polar head group and therefore the calculated D_L's are much larger than the measured ones. A value of the diffusion coefficient is obtained that is more than an order of magnitude too large [56,260,262]. The reason for the discrepancy is that head group-head group interactions are essential for slowing down the diffusion process.

It can be inferred from Table 5.3 that cholesterol has a very small effect on the PC lateral diffusion in lipid bilayers. This is in accordance with recent molecular dynamics simulations, where it was found that cholesterol does not have any significant effect on the lateral diffusion of the chains [56]. On the other hand, cholesterol has a great influence on the molecular ordering in bilayers [146,152,186,235]. Thus, cholesterol increases the molecular ordering in an L_α phase, but the diffusional motion is unaffected. In a study [197] of the lipid lateral diffusion in bilayers with a mixture of DOPC, DOPE and water, both lipids were found to have about the same D_L [146].

The diffusional motion for liquid crystalline powder samples has also been measured [202,222,257]. In order to reduce the static dipole-dipole couplings a combined multiple pulse homonuclear decoupling and multiple PFG technique has been used to determine lipid diffusion coefficients [222]. By this method the diffusion has been determined for DPPC at 25°C with 15 wt-% ^2H_2O in the $L_{\beta'}$ phase and for potassium oleate with 30 wt-% ^2H_2O in the L_α phase, where the D-values were found to be $1.6 \cdot 10^{-14}$ and $1.3 \cdot 10^{-12}$ $m^2\,s^{-1}$, respectively. The lateral diffusion coefficient is not obtained directly, but an average value over the randomly oriented microcrystallites [159]. In such a measurement a too-small diffusion coefficient is obtained compared to the local lateral diffusion. In an early study Roberts [202] used the stimulated SE technique to determine the surfactant diffusion in the H_I and L_α phases of potassium laurate/^2H_2O at 80°C. Somewhat surprisingly he got about the same diffusion coefficient in both phases, namely $D_{lamellar} = 2.4 \cdot 10^{-10}$ $m^2\,s^{-1}$ and $D_{hexagonal} = 2.3 \cdot 10^{-10}$ $m^2\,s^{-1}$, *i.e.* independent of phase structure.

An interesting application of NMR diffusion is the use of PFG double quantum SE decays for studies of the translational diffusion, parallel and perpendicular to the magnetic field, of dichloromethane dissolved in two thermotropic liquid crystals [170].

A method has been presented where the diffusion coefficient can be extracted from a simulation of ^{31}P NMR spectra from a lipid vesicle dispersion. Experimental spectra are obtained by use of a DANTE pulse sequence [136] and the spectral simulation was based on molecules with axially symmetric lineshapes undergoing lateral diffusion on a curved surface. By fitting a set of spectra recorded from a system which is allowed to evolve for a variable time interval after the DANTE excitation, a lateral diffusion coefficient could be calculated from the measured correlation time. Köchy and Bayerl [131] measured ^2H quadrupolar transverse relaxation times obtained by the Carr-Purcell-Meiboom-Gill (CPMG) pulse sequence on spherical phospholipid bilayers on a solid support. They took advantage of the fact that the CPMG sequence can progressively filter out contributions to the T_2 relaxation arising from motions like lateral diffusion, which are slow on the NMR time scale. For supported bilayers of POPC they got $D = 4 \cdot 10^{-12}$ $m^2\,s^{-1}$ at 30°C in good agreement with PFG NMR [146]. An attempt to estimate the lipid lateral diffusion in phospholipid vesicles from ^{13}C spin-spin relaxation was less successful [58]. The authors used a very simple model to describe the spin relaxation, a model which is similar to the so called 'two-step model' [90,271,273]. The lateral diffusion coefficient required to fit the values of T_2 is larger by almost two orders of magnitude than the values obtained with PFG NMR.

3. NMR Measurements of Water and Ion Diffusion in Liquid Crystalline Phases

Most of the investigations of water diffusion in lyotropic liquid crystalline systems have been conducted on powder samples. The intermolecular exchange of protons in the water molecule is rapid on the NMR time scale and therefore the water signal is relatively sharp even for a non-oriented sample. When the diffusion coefficient is determined for a powder sample with the NMR technique [159], the resulting echo has contributions from lamellas of all orientations θ_{LD} and

$$S(2\tau) = \int_{-1}^{1} \exp\left\{-2\tau/T_{2(\theta_{LD})} - \sin^2\theta_{LD}(\gamma g\delta)^2 D(\Delta - \delta/3)\right\} d\cos\theta_{LD} \qquad (47)$$

The first measurements, on powder samples containing 70% sodium palmitate and 30% water, of water diffusion coefficients in L_α phases were performed by Blinc and coworkers [9]. The complications described by Equation (47) were not considered in their determination of the water diffusion. Assuming that the water molecules in the lamellas oriented close to the magic angle make the major contribution to the spin echo, the measured diffusion coefficient should be multiplied by 3/2 to give $D_L = 8 \cdot 10^{-10}\ m^2\ s^{-1}$ [159]. Water diffusion coefficients have, however, also been determined for oriented L_α phases using both proton [252] and deuterium PFG NMR [24]. Callaghan and coworkers measured the diffusion coefficients of ^2H_2O parallel and perpendicular to the bilayer surface to $D_L = 1.7 \cdot 10^{-9}\ m^2\ s^{-1}$ and $D_\perp = 5 \cdot 10^{-11}\ m^2\ s^{-1}$. They also reported the first solid-echo PFG ^2H NMR experiment performed on an L_α phase of potassium palmitate and water in the presence of quadrupole splittings.

In the lithium perfluoro-octanoate/water system, self-diffusion coefficients are reported for water, lithium and perfluoro-octanoate ions in the L_α phase, as a function of temperature [252]. The difference between the diffusional motion along and across the perfluoro-octanoate bilayers was small for both water and lithium ions. The Li^+ diffusion coefficient was only a factor of three larger than that of perfluoro-octanoate. The Li^+ diffusion coefficient has also been determined for an L_α phase composed of octanoate, water and decanol [68]. The Li^+ diffusion coefficient of this L_α phase was independent of the composition, which is due to a constant fraction of bound ions in accordance with the ion condensation model [155,272]. Recently, the PFG NMR method was applied to estimate the rates of ionophore-mediated transmembrane exchange of Li^+ in liposomes [266]. There was a good agreement between the transmembrane exchange rate estimated from PFG and magnetization-transfer ^7Li NMR experiments.

Table 5.4
Lipid diffusion coefficients, D, for some cubic liquid crystalline phases.

Lipid system	Composition (wt %)	Temperature (°C)	$D_L \cdot 10^{11}$ ($m^2\ s^{-1}$)	References
Aerosol OT	72.4	24	0.36	[148]
Water	27.6	35	0.59	
Potassium octanoate	60	24	8.8	[157]
Water	40			
Sodium octanoate	39.4	24	<0.1	[157]
Octane	4.3			
Water	56.3			
Dodecyltrimethyl ammonium chloride	84.0	26	0.8	[64]
Water	16.0			
MO	88	43	1.5	[148]
Water	12			
		57	2.6	
Lyso-lauroyl-PC	40	29	0.006	[70]
Water	60			
Lyso-oleoyl-PC	79	26	0.14	[70]
Water	21			
DOPC	97	60		[142]
Water	3		0.5	
DOPC	45	35	0.1	[257]
Sodium cholate	30			
Water	25	35		
MGlcDAG	97	35	0.1	[143]
Water	3			
DOPC	26.5	59	0.03	[188]
Dioleoylglycerol	61.3		1.5	
Water	12.3		3.0	
DOPC	46.23	66	0.37	[197]
DOPE	43.8	73	0.50	
Water	10.0	83	0.64	
		90	0.78	

Water diffusion coefficients have been measured on powder samples in the binary system of DPPC and water [159]. The diffusion coefficients were found to be 4-10 times smaller than the translational diffusion coefficient in bulk water and were essentially unaffected by changes in the phase structure.

Some recent investigations of the effect of the liquid crystalline phase structure on the measured water diffusion coefficients have been published [25,29,39,43].

4. Lipid Translational Diffusion in Cubic Liquid Crystalline Phases

Lipid translational diffusion coefficients have been determined for a large number of different lyotropic cubic phases (Table 5.4). There is a great interest in the physico-chemical properties of membrane lipids, in particular the formation of cubic and H_{II} phases, since these lipid phase structures are believed to be involved in many biologically important membrane processes, like fusion [38,220,221] and membrane lipid regulation [143,145,154,198,201]. With X-ray diffraction methods [83,164,167,245] the space group of the cubic structure most often can be obtained, but seldom is it possible to distinguish a bicontinuous structure from a structure consisting of closed micellar aggregates. The strength of the PFG NMR method is that it can discriminate between these two structures, since restricted motions in space can be easily observed [242]. The first PFG-NMR study of amphiphile diffusion in a cubic phase was done by Charvolin and Rigny [37] for a soap system of potassium laurate that formed a cubic phase at 80°C ($D = 2 \cdot 10^{-10}\ m^2\ s^{-1}$). Such a large diffusion coefficient is not compatible with closed aggregates. Conclusions about whether a cubic phase consists of closed aggregates or not are based on a comparison between lipid diffusion coefficients measured by PFG-NMR for neighbouring phases, with which the cubic phase is in equilibrium [20,157,158,160].

The apparent diffusion coefficients of cubic and L_α phases can be compared qualitatively by using a simple, rough model [148,157,160]. The restricted lipid diffusion within aggregates building up the cubic lattice can be described by a local diffusion tensor, $\tilde{D}$, where the component representing restricted diffusion in a certain direction is set to zero. The measured diffusion coefficient, D, is then an average value of the local diffusion tensor

$$D = \frac{1}{3}\overline{\mathrm{Tr}\tilde{\mathbf{D}}} \tag{48}$$

where the bar denotes an average over all the sites in a unit cell of the liquid crystalline phase. Depending on how many directions that are effectively available for diffusional motion of the lipid, as measured by PFG NMR, three possibilities can be distinguished:

i) For a cubic phase with closed aggregates, like micelles, the lipid diffusion is restricted in all three directions and all the diagonal elements in the tensor are equal to zero and D is ideally equal to zero.

ii) A cubic structure of connected, long, rod-like aggregates permits lipid diffusion only in one direction (along the rod) giving

$$D = \frac{1}{3}\overline{\mathrm{Tr}\mathbf{D}_{rod}} = \frac{1}{3}D_{\parallel} \tag{49}$$

iii) For a cubic phase built up of lipid bilayers (Figure 5.5D) the diffusion, D_L^{cub}, can occur in two directions and

$$D = \frac{2}{3}D_L^{cub} \tag{50}$$

Thus, it can be expected that the aggregate structure in the cubic phase will influence appreciably the magnitude of the measured diffusion coefficients. In particular, the PFG-NMR provides a convenient means to distinguish between bicontinuous cubic phases from those having closed aggregates. According to the above equations a crude estimate of the geometry of the aggregates building up the cubic phase should also be obtainable, by comparing diffusion coefficients measured for lipids in the L_α and cubic phases. Such a comparison assumes that the local environment of the molecules in the diffusing process is the same in the two phases.

For the comparison of diffusion coefficients of lipids and water in cubic phases, an attractive mathematical model based on a description of the cubic structure by periodic minimal surfaces can be used [2-4,154,214]. Recently, Anderson and Wennerström [3] presented numerical results for the geometric obstruction factor of self-diffusion in different bicontinuous, cubic phases. The lipid diffusion was modelled as a diffusion of a particle, confined to the polar-apolar dividing surface, with a constant diffusion coefficient D_o. It was proven analytically that the effective diffusion coefficient for a particle diffusing over any minimal surface of cubic symmetry is exactly equal to $2/3D_o$.

4a. ***Bicontinuous phases*****:** Early PFG-NMR and X-ray diffraction investigations of the bicontinuous cubic phases in the MO/water system suggested that the phase structures were based on lamellar bilayer units [148] (*cf.* Figure 5.5D). The NMR diffusion results of the MO-system were compared with those of the Aerosol OT/water system, and the simple model of lipid diffusion in different cubic structures described previously was applied. The cubic phase of the latter system was suggested to be built up of rod-like aggregates. By using equations (49) and (50) surprisingly good predictions were made about the aggregate shape in the cubic phases.

Table 5.5

Water and lipid diffusion in cubic phases of monooleoylglycerol and water [69]. D_{cub}^{water} / D_0 is the measured water diffusion coefficient relative to that of bulk water. Temperature 25°C.

^1H_2O (wt %)	Phase	D_{cub}^{water} / D_0	$D_{cub}^{lipid} \cdot 10^{12}$ ($m^2 s^{-1}$)	p^b	n
28.0	*Ia3d*	0.201	14.4	0.56	4.3
30.4	*Ia3d*	0.237	15.1	0.49	4.2
33.4	*Ia3d/Pn3m*	0.276	15.3	0.41	4.1
36.6	*Pn3m*	0.292	15.8	0.39	4.4
39.5	*Pn3m*	0.300	16.9	0.38	5.0

The cubic phases of several monoacylglycerols exist over extended water and temperature regions [163]. These phases can dissolve substantial amounts of hydrophobic and amphiphilic molecules [60,61,86,137] as well as glycerol [190]. The water diffusion coefficient in cubic phases *versus* the MO and water content has been investigated. The water diffusion coefficient, D_{cub}^{water}, in a simple two-site model can be written [69] as

$$D_{cub}^{water} = p_b D_{cub}^{lipid} + (1 - p_b)\beta D_o \tag{51}$$

where p_b is the fraction of bound water, D_{cub}^{lipid} is the lipid diffusion coefficient in the cubic phase, β is the obstruction factor for the diffusion of unbound water in the cubic phase, and D_o is the diffusion of bulk water. The association of water obtained from the application of this simple model (Table 5.5) should be taken as a quantification of the extent to which the surface of the aggregates affects the dynamic properties of water. The diffusion coefficient of water showed [69] a discontinuous increase around 34% water, which harmonizes with the two-phase region between the *Ia3d* and *Pn3m* cubic structures [162]. The lipid diffusion was found to increase with increasing unsaturation of the acyl chains of the lipid, and it decreases when molecules like cholesterol, gramicidin A, and lyso-oleoyl-PC are solubilized in the cubic phase. In a cubic phase of MO, DOPC, and water, the individual lipid diffusion coefficients could be determined *simultaneously*. The results gave arguments for the belief that, when probe techniques are used, the choice of the probe itself may influence the measured lipid diffusion coefficient. Recently, we have investigated the cubic phase of DDAO having a molar ratio of 97:3 between DDAO and gramicidin and 27 wt% water. At 25°C the diffusion coefficient of water, $D_{water} = 6.3 \cdot 10^{-11}$ m^2s^{-1} and the diffusion coefficients of the methyl groups of the polar head group and the terminal methyl group were measured to be $7 \cdot 10^{-12}$ m^2s^{-1} (G. Orädd and G. Lindblom, to be published). The latter coefficient should be compared with the diffusion of DDAO in a cubic phase in

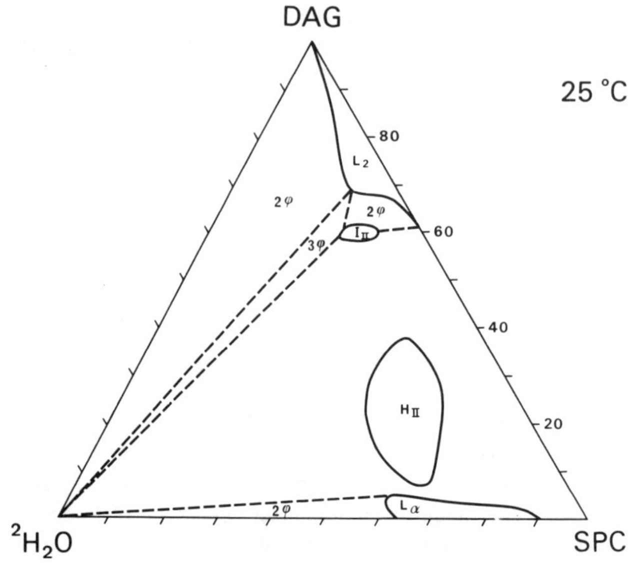

Fig. 5.32. The phase diagram of the system soya phosphatidylcholine (SPC), diacylglycerol (DAG) and ^2H_2O at 25°C. The various phase regions are denoted by: L_2; reversed micellar solution phase, I_{II}; cubic liquid crystalline phase, H_{II}; reversed hexagonal phase and L_α; lamellar liquid crystalline phase. The number of phases are indicated in the various two- and three-phase regions by 2^φ and 3^φ, respectively. Adapted from [188].

the absence of gramicidin which is equal to $1{\cdot}10^{-11}$ m^2s^{-1}, *i.e.* gramicidin causes quite a small effect of obstruction.

Several membrane lipid-water systems give rise to freeze-fracture replicas of three-dimensional regular arrays of closely packed globular elements, which were called "lipidic particles" [263]. These structural pictures have often been poorly classified and they have been suggested to show closed lipid aggregates, such as reversed micelles. PFG NMR and X-ray diffraction studies of some of these liquid crystalline phases, giving rise to such freeze-fracture replicas, indicated that they are (*i*) cubic and (*ii*) bicontinuous cubic phases (with one exception) [197]. There is a rather large accumulation of diffusion coefficients to verify the structure of bicontinuous cubic phases (Table 5.4) [16,42,64,70,143,151,157,160,189,198, 199,223,257,277].

4b. ***Discontinuous phases:*** Commonly, the discontinuous cubic structure (I_I) are located between the L_1 and H_I phases in the phase diagrams [75]. X-ray diffraction showed that this structure belongs to the space group *P43n* or *Pm3n*. It was long debated whether it consisted of closed aggregates [20,64,65,70,76,147] or of a network of connected rod-like micelles with enclosed spherical micelles [167,243]. This debate is now settled [54,261] and it is conclusively shown by PFG NMR (Table 5.4) [65,66,147] that this cubic structure is discontinuous. From

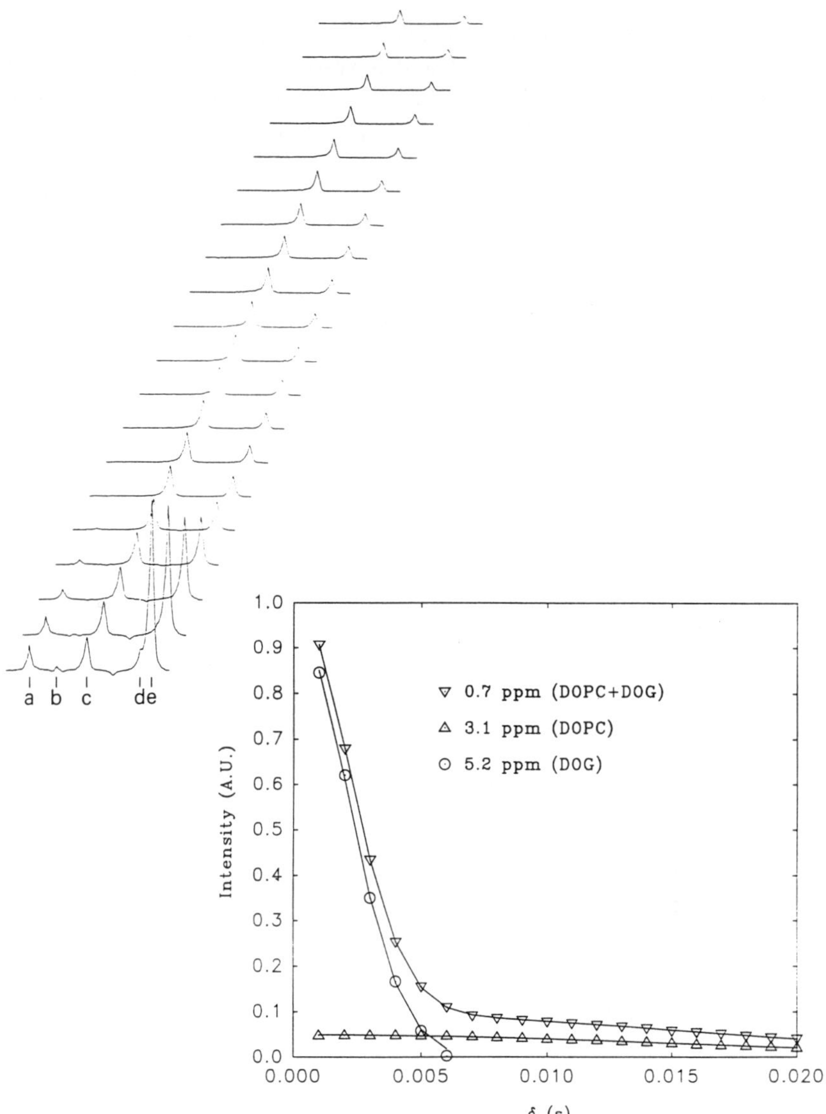

Fig. 5.33. a) Stacked plot from a SE ^{1}H NMR diffusion experiment of a cubic phase of the DOPC/DOG/$^{2}H_2O$ system at 59°C. The ^{1}H NMR peaks in the spectrum are from the left to the right: (a) DOG hydroxyl group (5.2 ppm); (b) water (4.6 ppm); (c) DOPC choline head group (3.1 ppm); (d) acyl chain methylenes of DOPC and DOG (1.1 ppm) and (e) the terminal methyl groups of the hydrocarbon chains of DOPC and DOG (0.7 ppm). The following settings were used: $\tau = \Delta = 300$ ms, g = 0.5734 Tm^{-1} and $\delta = 1$ ms for the spectrum at the bottom in the firgure and δ is then increased by 1 ms between successive NMR spectra from the bottom to the top.

(b) Fit of the intensities of the peaks in the NMR spectra shown in a). The amplitudes of the different peaks have been arbitrarily chosen. The results obtained from the fits are: 5.2 ppm: $D = 1.5 \cdot 10^{-11}\ m^2s^{-1}$; 3.1 ppm: $D = 3 \cdot 10^{-13}\ m^2s^{-1}$; and 0.7 ppm: $D_1 = 1.5 \cdot 10^{-11}\ m^2s^{-1}$ (90%) and $D_2 = 3 \cdot 10^{-13}\ m^2s^{-1}$ (10%). From [188].

NMR line shapes [65], spin relaxation [237,240] and time-resolved quenched fluorescence [107,147] it is concluded that it consists of slightly elongated, closed micelles (Figure 5.5E). Recently, a cubic phase consisting of *reversed* micelles was established [216]. The structure of this phase, which belongs to the space group *Fd3m*, and which may be composed of either DOPC and dioleoylglycerol (DOG) or oleic acid and sodium oleate, was recently solved by low-resolution crystallography [165]. Figure 5.32 shows the phase diagram of the ternary system PC-DOG-water at 28°C. From X-ray data it was suggested that the structure of this cubic phase consists of a complex packing of two types of quasi-spherical reversed micelles, and by PFG-NMR it was shown conclusively that closed aggregates build up the structure [188]. The diffusion coefficient of DOPC in such a cubic phase is very small, close to 10^{-13} m^2s^{-1}, while the diffusion coefficients of water and DOG are much larger, in the order of 10^{-11} m^2s^{-1} (Figure 5.33 and Table 5.4) It was concluded that the reversed micelles contain mainly DOPC and that the DOG merely plays the role of a solvent surrounding the micelles [188].

5. *Translational Diffusion in Other Non-Lamellar Liquid Crystalline Phases*

The translational diffusion of all three components has been measured in the ternary system of DOPC, water and dodecane, where dodecane induces an H_{II} phase [223]. From a comparison between diffusion coefficients it was concluded that dodecane resides between the water cylinders in the H_{II} phase.

Some lyotropic liquid crystals, called nematics, orient spontaneously in a strong magnetic field. Utilizing UV linear dichroism on a solubilized chromophore and PFG NMR on such a nematic phase containing potassium dodecanoate, it was shown that this phase is built up of long rod-like micelles [108,239]. The translational diffusion coefficient of the surfactant was obtained from a macroscopically aligned sample oriented at the magic angle. The surfactant diffusion along the rod was $D_L = 5 \cdot 10^{-11}$ $m^2 s^{-1}$ for potassium laurate and $D_L = 8 \cdot 10^{-11}$ $m^2 s^{-1}$ for sodium decyl sulfate [108]. These values compare well with the lateral diffusion coefficients for the corresponding lamellar mesophases (Table 5.3). By means of the relation $<x^2> = 2Dt$ the length of the rods can be estimated to be longer than 2000 nm.

AKNOWLEDGEMENT

I would like to thank Mrs Eva Vikström for an excellent work with the figures. I am also grateful to Drs. Lennart B.-Å. Johansson, Leif Rilfors, Maria Maddalena Sperotto and Per-Olof Westlund for valuable comments on the manuscript.

ABBREVIATIONS

CPMG, Carr-Purcell-Meiboom-Gill pulse sequence; CSA, chemical shift anisotropy; DDAO, N,N-dimethyldodecylamine oxide; DGlcDAG, 1,2-diacyl-3-

O-[a-D-glucopyranosyl-(1→2)-*O*-a-D-glucopyranosyl]-*sn*-glycerol; DOG, dioleoylglycerol; DOPA, disodium salt of dioleoylphosphatidic acid; DOPE, dioleoylphosphatidylethanolamine; DPG, diphosphatidylglycerol; DPPC, dipalmitoylphosphatidylcholine; DSC, differential scanning calorimetry; EFG, electric field gradient; eQ, nuclear quadrupole moment; ESR, electron spin resonance; FID, free induction decay; M_1 and M_2, first and second moments,respectively; FT-IR, Fourier transform infrared spectroscopy; GPDGlcDAG, glycerophosphoryldiglucosyldiacylglycerol; Hamiltonians: H, Hamiltonian: H_z, Hamiltonian of the Zeeman energy: H_{CS}, Hamiltonian of the chemical shielding: H_Q, quadrupolar Hamiltonian: H_D, dipolar Hamiltonian: ^{1}H DQF COSY, double quantum filtered correlation spectroscopy; H_0, spontaneous curvature; LED, stimulated spin echo sequence; LPC, lysophosphatidylcholine; MABGPDGlcDAG, monoacylbisglycerophosphoryldiglucosyldiacylglycerol; MADGlcDAG, 1,2-diacyl-3-*O*-[α-D-glucopyranosyl-(1→2)-*O*-(6-*O*-acyl-α-D-glucopyranosyl)]-*sn*-glycerol; MAMGlcDAG, 1,2-diacyl-3-*O*-[6-*O*-acyl-(α-D-glucopyranosyl)]-*sn*-glycerol; 3-*O*-Me-ß-D-GlcDAlG, 1,2-di-*O*-tetradecyl-3-*O*-(3-*O*-methyl-ß-D-glucopyranosyl)-*sn*-glycerol; MGlcDAG, 1,2-diacyl-3-*O*-(a-D-glucopyranosyl)-*sn*-glycerol; MO, monooleoylglycerol; NMR, nuclear magnetic resonance; PE, phosphatidylethanolamine; PFG, pulsed field gradient; Phases: L_α, lamellar phase: $L_{\beta'}$, gel phase: $P_{\beta'}$, gel rippled phase: H_I, normal hexagonal phase (oil-in-water): H_{II}, reversed hexagonal phase (water-in-oil): I_I, normal cubic phase (oil-in-water, isotropic): I_{II}, reversed cubic phase (water-in-oil, isotropic): C, crystalline: L_1, normal micellar solution phase: L_2, reversed micellar solution phase; PLPE-d_{31}, 1-perdeuteriopalmitoyl-2-linoleoyl-*sn*-glycero-3-phosphoethanolamine; POPC-d_{31}, 1-perdeuteriopalmitoyl-2-oleoyl-*sn*-glycero-3-phosphocholine; POPE-d_{31}, 1-perdeuteriopalmitoyl-2-oleoyl-*sn*-glycero-3-phosphoethanolamine; S, order parameter; SAXS, small angle X-ray scattering; SDS, sodium dodecylsulphate; SE, Hahn spin echo sequence; TFE, trifluoroethanol.

REFERENCES

1. Abragam,A., *The Principles of Nuclear Magnetism*, Oxford University Press, London, 1961.
2. Anderson,D.M., Gruner,S.M. and Leibler,S. *Proc. Natl. Acad. Sci. USA*, **85**, 5364-5368 (1988).
3. Anderson,D.M. and Wennerström,H. *J. Phys. Chem.*, **94**, 8683-8694 (1990).
4. Andersson,S., Hyde,S.T., Larsson,K. and Lidin,S. *Chem. Rev.*, **88**, 221-242 (1988).
5. Arseniev,A.S., Barsukov,I.L., Bystrov,V.F., Lomize,A.L. and Ovchinnikov,Y.A. *FEBS Lett.*, **186**, 168-174 (1985).
6. Arvidson,G., Brentel,I., Khan,A., Lindblom,G. and Fontell,K. *Eur. J. Biochem.*, **152**, 753-759 (1985).
7. Atkins,P.W., *Physical Chemistry*, Oxford University Press, Oxford, 1994.
8. Bennet,D.E. and O´Brien,D.F. *J. Am. Chem. Soc.*, **116**, 7933-7934 (1994).
9. Blinc,R., Easwaran,K., Pirs,J., Volfan,M. and Zupancic,I. *Phys. Rev. Letters*, **25**, 1327-1330 (1970).
10. Bloom,M., Burnell,E.E., Mackay,A.L., Nichol,C.P., Valic,M.I. and Weeks,G. *Biochemistry*, **17**, 5758-5762 (1978).
11. Bloom,M., Davis,J.H. and Mackay,A.L. *Chem. Phys. Lett.*, **80**, 198-202 (1981).
12. Bloom,M., Davis,J.H. and Valic,M.I. *Can. J. Phys.*, **58**, 1510-1517 (1980).
13. Braach-Maksvytis,V.L.B. and Cornell,B.A. *Biophys. J.*, **53**, 839-843 (1988).

14. Brasseur,R. and Ruysschaert,J.M. *Biochem. J.*, **238**, 1-11 (1986).
15. Brentel,I., Lindblom,G. and Arvidson,G. *Biochim. Biophys. Acta*, **904**, 401-404 (1987).
16. Brentel,I., Selstam,E. and Lindblom,G. *Biochim. Biophys. Acta*, **812**, 816-826 (1985).
17. Brink,D.M. and Satchler,G.R., *Angular Momentum*, University Press, Oxford, 1968.
18. Brown,M.F. *J. Chem. Phys.*, **77**, 1576-1599 (1982).
19. Bryant,G., Pope,J.M. and Wolfe,J. *Eur. Biophys. J.*, **21**, 223-232 (1992).
20. Bull,T. and Lindman,B. *Mol. Cryst. Liq. Cryst.*, **28**, 155-160 (1974).
21. Burnett,L.J. and Muller,B.H. *J. Chem. Phys.*, **55**, 5829-5831 (1971).
22. Caffrey,M., Hogan,J. and Rudolph,A.S. *Biochemistry*, **30**, 2134-2146 (1991).
23. Callaghan,P.T., *Principles of Nuclear Magnetic Resonance Microscopy*, Clarendon Press, Oxford, 1991.
24. Callaghan,P.T., Le Gros,M.A. and Pinder,D.N. *J. Chem. Phys.*, **79**, 6372-6381 (1983).
25. Callaghan,P.T. and Söderman,O. *J. Phys. Chem.*, **87**, 1737-1744 (1983).
26. Canet,D., Marchal,J.P., Nery,H., Robin-Lherbier,B. and Cases,J.M. *J. Colloid Interface Sci.*, **93**, 241-251 (1983).
27. Canet,D., Turki,T., Belmajdoub,A. and Diter,B. *J. Phys. Chem.*, **92**, 1219-1222 (1988).
28. Carr,H.Y. and Purcell,E.M. *Phys. Rev.*, **94**, 630-638 (1954).
29. Celebre,G., Coppola,L. and Ranieri,G.A. *J. Chem. Phys.*, **97**, 7781-7785 (1992).
30. Cevc,G. (Editor). *Phospholipids Handbook*, Marcel Dekker, Inc., New York, N.Y., (1993).
31. Chachaty,C. *Prog. Nucl. Magn. Reson. Spectrosc.*, **19**, 183-222 (1987).
32. Chachaty,C. and Bredel,T. *J. Chem. Soc., Faraday Trans.*, **88**, 1893-1900 (1992).
33. Chachaty,C. and Bredel,T. *J. Phys. Chem.*, **95**, 5335-5344 (1991).
34. Chachaty,C., Caniparoli,J.-P., Faure,A. and Tistchenko,A.M. *J. Phys. Chem.*, **92**, 6330-6339 (1988).
35. Chachaty,C., Warr,G.G., Jansson,M. and Li,P. *J. Phys. Chem.*, **95**, 3830-3836 (1991).
36. Chapman,D., Cornell,B.A., Eliasz,A.W. and Perry,A. *J. Mol. Biol.*, **113**, 517-538 (1977).
37. Charvolin,J. and Rigny,P. *J. Chem. Phys.*, **58**, 3999-4008 (1973).
38. Chernomoridik,L.V., Melikyan,G.B. and Chizmadzhev,Y.A. *Biochim. Biophys. Acta*, **906**, 309-352 (1987).
39. Chung,J. and Prestegard,J.H. *J. Phys. Chem.*, **97**, 9837-9843 (1993).
40. Cohen,M.H. and Reif,F. in *Solid State Physics*, vol. 5, pp. 321-438 (1957) (ed. D. Turnbull, McGraw-Hill Book Company, N.Y.).
41. Collings,P.J., Photinos,D.J., Bos,P.J., Ukleja,P. and Doane,J.W. *Phys. Rev. Lett.*, **42**, 996-999 (1979).
42. Conroy,J.P., Hall,C., Leng,C.A., Rendall,K., Tiddy,G.J.T., Walsh,J. and Lindblom,G. *Progr. Colloid Polym. Sci.*, **82**, 253-262 (1990).
43. Coppola,L., La Mesa,C., Ranieri,G.A. and Terenzi,M. *J. Chem. Phys.*, **98**, 5087-5090 (1993).
44. Cullis,P.R. and De Kruijff,B. *Biochim. Biophys. Acta*, **559**, 399-420 (1979).
45. Cullis,P.R. and de Kruijff,B. *Biochim. Biophys. Acta*, **507**, 207-218 (1978).
46. Cullis,P.R. and de Kruijff,B. *Biochim. Biophys. Acta*, **513**, 31-42 (1978).
47. Cullis,P.R., Hope,M.J., de Kruijff,B., Verkleij,A.J. and Tilcock,C.P.S., in *Phospholipids and Cellular Regulation*, vol. 1, pp. 1-59 (1985) (ed. J.F. Kuo, CRC Press, Boca Raton, Florida).
48. Davis,J.H. *Biochim. Biophys. Acta*, **737**, 117-171 (1983).
49. Davis,J.H. *Biophys. J.*, **27**, 339-358 (1979).
50. Davis,J.H., Bloom,M., Butler,K.W. and Smith,I.C.P. *Biochim. Biophys. Acta*, **597**, 477-491 (1980).
51. Davis,J.H., Jeffrey,K.R., Bloom,M., Valic,M.I. and Higgs,T.P. *Chem. Phys. Lett.*, **44**, 390-394 (1976).
52. de Kruijff,B., Cullis,P.R., Verkleij,A.J., Hope,M.J., van Echteld,C.J.A. and Taraschi,T.F. in *The Enzymes of Biological Membranes*, pp. 131-204 (1984) (ed. A.N. Martinosi, Plenum, N.Y.).
53. De Young,L.R. and Dill,K.A. *Biochemistry*, **27**, 5281-5289 (1988).
54. Delacroix,H., Gulik-Krzywicki,T., Mariani,P. and Luzzati,V. *J. Mol. Biol.*, **229**, 526-539 (1993).
55. Dill,K.A. and Flory,P.J. *Proc. Natl. Acad. Sci. USA*, **77**, 3115-3119 (1980).
56. Edholm,O. and Nyberg,A.M. *Biophys. J.*, **63**, 1081-1089 (1992).
57. Ekwall,P. *Adv. Liq. Cryst.*, **1**, 1-142 (1975).
58. Ellena,J.F., Lepore,L.S. and Cafiso,D.S. *J. Phys. Chem.*, **97**, 2952-2957 (1993).
59. Engström,S. *Polymer Prep.*, **31**, 157-158 (1990).
60. Ericsson,B., Eriksson,P.O., Löfroth,J.E. and Engström,S. *ACS Symp. Ser.*, **469**, 251-265 (1991).

61. Ericsson,B., Larsson,K. and Fontell,K. *Biochim. Biophys. Acta*, **729**, 23-27 (1983).
62. Eriksson,P.-O., Arvidson,G. and Lindblom,G. *Israel J. Chem.*, **23**, 353-355 (1983).
63. Eriksson,P.-O., Johansson,L.B.-Å. and Lindblom,G., in *Surfactants in Solution*, vol. 1, pp. 219-236 (1984) (eds. K.L. Mittel and B. Lindman, Plenum Press, New York).
64. Eriksson,P.-O., Khan,A. and Lindblom,G. *J. Phys. Chem.*, **86**, 387-393 (1982).
65. Eriksson,P.-O., Lindblom,G. and Arvidson G. *J. Phys. Chem.*, **89**, 1050-1053 (1985).
66. Eriksson,P.-O., Rilfors,L., Lindblom,G. and Arvidson,G. *Chem. Phys. Lipids*, **37**, 357-371 (1985).
67. Eriksson,P.-O., Rilfors,L., Wieslander,Å., Lundberg,A. and Lindblom,G. *Biochemistry*, **30**, 4916-4924 (1991).
68. Eriksson,P.O. and Lindblom,G. *Springer Series in Chemical Physics* **11**, 290-295 (1980).
69. Eriksson,P.O. and Lindblom,G. *Biophys. J.*, **64**, 129-136 (1993).
70. Eriksson,P.O., Lindblom,G. and Arvidson,G. *J. Phys. Chem.*, **91**, 846-853 (1987).
71. Evans,D.F. and Wennerström,H., *The Colloidal Domain*, VCH Publishers, Inc., New York, 1994.
72. Fattal,D.R. and Ben-Shaul,A. *Biophys. J.*, **65**, 1795-1809 (1993).
73. Finer,E.G. and Darke,A. *Chem. Phys. Lipids*, **12**, 1-16 (1974).
74. Fontell,K. *Mol. Cryst. Liq. Cryst.*, **63**, 59-82 (1981).
75. Fontell,K. *Adv. Colloid Interface Sci.*, **41**, 127-147 (1992).
76. Fontell,K., Fox,K.K. and Hansson,E. *Mol. Cryst. Liquid Cryst.*, **1**, 9-17 (1985).
77. Gally,H.U., Pluschke,G., Overath,P. and Seelig,J. *Biochemistry*, **19**, 1638-1643 (1980).
78. Gally,H.U., Pluschke,G., Overath,P. and Seelig,J. *Biochemistry*, **18**, 5605-5610 (1979).
79. Gibbs,S.J. and Johnson,C.S., *J. Magn. Reson.*, **93**, 395-402 (1991).
80. Glasel,J.A. in *Water, a Comprehensive Treatise*, vol. 1, pp. 215-254 (1972) (ed. F. Franks, Plenum Press, New York).
81. Gruen,D.W.R. *Chem. Phys. Lipids*, **30**, 105-120 (1982).
82. Gruner,S.M. *Proc. Natl. Acad. Sci. USA*, **82**, 3665-3669 (1985).
83. Gruner,S.M., Cullis,P.R., Hope,M.J. and Tilcock,C.P.S. *Ann. Rev. Biophys. Biophys. Chem.*, **14**, 211-238 (1985).
84. Guldbrand,L., Jönsson,B. and Wennerström,H. *J. Colloid Interface Sci.*, **89**, 532-541 (1982).
85. Gulik-Krzywicki,T., Tardieu,A. and Luzzati,V. *Mol. Cryst. Liq. Cryst.*, **8**, 285-291 (1969).
86. Gutman,H., Arvidson,G., Fontell,K. and Lindblom,G., in *Surfactants in Solutions*, vol. 1, pp. 143-152 (1984) (Eds. K.L. Mittal and B. Lindman, Plenum Press, New York).
87. Hahn,E.L. *Phys. Rev.*, **80**, 580-594 (1950).
88. Halle,B. *J. Phys. Chem.*, **95**, 6724-6733 (1991).
89. Halle,B., Ljunggren,S. and Lidin,S. *J. Chem. Phys.*, **97**, 1401-1415 (1992).
90. Halle,B. and Wennerström,H. *J. Chem. Phys.*, **75**, 1928-1943 (1981).
91. Hauksson,J.B., Lindblom,G. and Rilfors,L. *Biochim. Biophys. Acta*, **1214**, 124-130 (1994).
92. Hauksson,J.B., Lindblom,G. and Rilfors,L. *Biochim. Biophys. Acta*, **1215**, 341-345 (1994).
93. Hauksson,J.B., Rilfors,L., Lindblom,G. and Arvidson,G. *Biochim. Biophys. Acta*, **1258**, 1–9 (1995).
94. Helfrich,W. *Z. Naturforsch.*, **28**, 693-703 (1973).
95. Herzfeld,J., Griffin,R.G. and Haberkorn,R.A. *Biochemistry*, **17**, 2711-2718 (1978).
96. Hjelm,R.P., Thiyagarajan,P. and Alkan,H. *J. Appl. Cryst.*, **21**, 858-863 (1988).
97. Hjelm,R.P., Thiyagarajan,P., Sivia,D.S., Lindner,P., Alkan,H. and Schwahn,D. *Prog. Colloid Polym. Sci.*, **81**, 225-231 (1990).
98. Ikeda,K., Khan,A., Meguro,K. and Lindman,B. *J.Colloid Interface Sci.*, **133**, 192-199 (1989).
99. Ikeda,K., Khan,A. and Meguro,K.L., *J. Colloid Interface Sci.*, **133**, 192-199 (1989).
100. Ipsen,J.H., Karlström,G., Mouritsen,O.G., Wennerström,H.W. and Zuckermann,M.J. *Biochim. Biophys. Acta*, **905**, 162-172 (1987).
101. Israelachvili,J.N., Mitchell,D.J. and Ninham,B.W. *J. Chem. Soc. Faraday Trans. II*, **72**, 1525-1568 (1976).
102. Janes,N., Rubin,E. and Taraschi,T.F. *Biochemistry*, **29**, 8385-8388 (1990).
103. Janiak,M.J., Small,D.M. and Shipley,G.G. *J. Biol. Chem.*, **254**, 6068-6078 (1979).
104. Jansson,M., Thurmond,R.L., Barry,J.A. and Brown,M.F. *J. Phys. Chem.*, **96**, 9532-9544 (1992).
105. Jansson,M., Thurmond,R.L., Trouard,T.P. and Brown,M.F. *Chem. Phys. Lipids*, **54**, 157-170 (1990).
106. Jensen,J.W. and Schutzbach,J.S. *Biochemistry*, **27**, 6315-6320 (1988).
107. Johansson,L.B.-Å. and Söderman,O. *J. Phys. Chem.*, **91**, 7575-7578 (1987).

108. Johansson,L.B.-Å., Söderman,O., Fontell,K. and Lindblom,G. *J. Phys. Chem.*, **85**, 3694-3697 (1981).
109. Jokela,P., Jönsson,B. and Khan,A. *J. Phys. Chem.*, **91**, 3291-3298 (1987).
110. Jönsson,B., Jokela,P., Khan,A., Lindman,B. and Sadaghiani,A. *Langmuir*, **7**, 889-895 (1991).
111. Jönsson,B., Jokela,P., Khan,A., Lindman,B. and Sadaghiani,A. *Langmuir*, **7**, 889-895 (1991).
112. Kang,C. and Khan,A. *J. Colloid Interface Sci.*, **156**, 218-228 (1993).
113. Kang,S.Y., Gutowsky,H.S., Huang,J.C., Jacobs,R., King,T.E., Rice,D. and Oldfield,E. *Biochemistry*, **18**, 3257-3267 (1979).
114. Keller,S.L., Besrukov,S.M., Gruner,S.M., Tate,M.W., Vodyanoy,I. and Parsegian,V.A. *Biophys. J.*, **65**, 23-27 (1993).
115. Khan,A., Das,K.P., Eberson,L. and Lindman,B. *J. Colloid Interface Sci.*, **125**, 129-138 (1988).
116. Khan,A., Fontell,K. and Lindblom,G. *J. Phys. Chem.*, **86**, 383-386 (1982).
117. Khan,A., Fontell,K., Lindblom,G. and Lindman,B. *J. Phys. Chem.*, **86**, 4266-4271 (1982).
118. Khan,A., Jönsson,B. and Wennerström,H. *J. Phys. Chem.*, **89**, 5180-5184 (1985).
119. Khan,A., Lindman,B. and Shinoda,K. *J. Colloid Interface Sci.*, **128**, 396-406 (1989).
120. Khan,A., Rilfors, L., Wieslander,Å. and Lindblom,G. *Eur. J. Biochem.*, **116**, 215-220 (1981).
121. Khan,A., Söderman,O. and Lindblom,G. *J. Colloid Interface Sci.*, **78**, 217-224 (1980).
122. Killian,J.A. *Biochim. Biophys. Acta*, **1113**, 391-425 (1992).
123. Killian,J.A., de Kruijff,B., van Echteld,C.J.A., Verkleij,A.J., Leunissen-Bijvelt,J. and De Gier,J. *Biochim. Biophys. Acta*, **728**, 141-144 (1983).
124. Killian,J.A., Trouard,T.P., Greathouse,D., Chupin,V. and Lindblom,G. *FEBS Letters*, **348**, 161-165 (1994).
125. Kohler,S.J. and Klein,M.P. *Biochemistry*, **15**, 967-973 (1976).
126. Kohler,S.J. and Klein,M.P. *Biochemistry*, **16**, 519-526 (1977).
127. Kohler,S.J. and Klein,M.P. *J. Am. Chem. Soc.*, **99**, 8290-8293 (1977).
128. Kothe,G. and Stohrer,J., in *The Molecular Dynamics of Liquid Crystals*, vol. C, p. 431 (1994) (Ed. G.R. Luckhurst and C.A. Veracini, Kluwer Academic Publishers, Dordrecht).
129. Kuo,A.L. and Wade,C.G. *Biochemistry*, **18**, 2300-2308 (1979).
130. Kärger,J., Pfeifer,H. and Heink,W., in *Advances in Magnetic and Optical Resonance*, vol. 13, pp. 1-89 (1988) (ed. W.S. Warren, Academic Press, Inc., San Diego, CA).
131. Köchy,T. and Bayerl,T.M. *Phys. Rev. E*, **47**, 2109-2116 (1993).
132. La Mesa,C., Khan,A., Fontell,K. and Lindman,B. *J. Colloid Interface Sci.*, **103**, 373-391 (1985).
133. Lafleur,M., Cullis,P., Fine,B. and Bloom,M. *Biochemistry*, **29**, 8325-8333 (1990).
134. Lafleur,M., Cullis,P.R. and Bloom,M. *Eur. Biophys. J.*, **19**, 55-62 (1990).
135. Lamparski,H., Lee,Y.-S., Sells,T.D. and O´Brien,D.F. *J. Am. Chem. Soc.*, **115**, 8096-8102 (1993).
136. Larsen,D.W., Boylan,J.G. and Cole,B.R. *J. Phys. Chem.*, **91**, 5631-5634 (1987).
137. Larsson,K., Gabrielsson,K. and Lundberg,B. *J. Sci. Food Agric.*, **29**, 909-914 (1978).
138. Lasic,D.D., *Liposomes from Physics to Applications* (1993) (Elsevier, Amsterdam).
139. Laughlin,R.G., in *Adv. Liq. Cryst.*, vol. 3, pp. 41-148 (1978) (Ed. G.H. Brown, Academic Press, New York).
140. Lindblom,G. *Acta Chem. Scand.*, **25**, 2767-2768 (1971).
141. Lindblom,G. *Acta Chem. Scand.*, **26**, 1745-1748 (1972).
142. Lindblom,G. *Acta Chem. Scand.*, **B35**, 61-62 (1981).
143. Lindblom,G., Brentel,I., Sjölund,M., Wikander,G. and Wieslander,Å. *Biochemistry*, **25**, 7502-7510 (1986).
144. Lindblom,G., Eriksson,P.-O. and Arvidson,G. *Hepatology*, **4**, 1295-1335 (1984).
145. Lindblom,G., Hauksson,J.B., Rilfors,L., Bergenståhl,B., Wieslander,Å. and Eriksson,P.-O. *J. Biol. Chem.*, **268**, 16198-16207 (1993).
146. Lindblom,G., Johansson,L.B.-Å. and Arvidson,G. *Biochemistry*, **20**, 2204-2207 (1981).
147. Lindblom,G., Johansson,L.B.-Å., Wikander,G., Eriksson,P.-O. and Arvidson,G. *Biophys. J.*, **63**, 723-729 (1992).
148. Lindblom,G., Larsson,K., Johansson,L., Fontell,K. and Forsén,S. *J. Am. Chem. Soc.*, **101**, 5465-5470 (1979).
149. Lindblom,G., Lindman,B. and Tiddy,G.J.T. *Acta Chem. Scand.*, **A 29**, 876-878 (1975).
150. Lindblom,G., Lindman,B. and Tiddy,G.J.T. *J. Am. Chem. Soc.*, **100**, 2299-2303 (1978).
151. Lindblom,G. and Orädd,G. *Progr. Nucl. Magn. Reson. Spectrosc.*, **26**, 483-516 (1994).
152. Lindblom,G., Persson,N.-O. and Arvidson,G. *Adv. Chem. Ser.*, **152**, 121-141 (1976).

153. Lindblom,G. and Rilfors,L., in *Structural and Dynamic Properties of Lipids and Membranes*, pp. 51-76 (1992) (Eds. P.J. Quinn and R.J. Cherry, Portland Press, London).
154. Lindblom,G. and Rilfors,L. *Biochim. Biophys. Acta*, **988**, 221-256 (1989).
155. Lindblom,G., Rilfors,L., Hauksson,J.B., Brentel,I., Sjölund,M. and Bergenståhl,B. *Biochemistry*, **30**, 10938-10948 (1991).
156. Lindblom,G., Sjölund, M. and Rilfors,L. *Liq. Cryst.*, **3**, 783-790 (1988).
157. Lindblom,G. and Wennerström,H. *Biophys. Chem.*, **6**, 167-171 (1977).
158. Lindblom,G. and Wennerström,H. in *VIIth International Conference on Magnetic Resonance in Biological Systems.*, St Jovite, Quebec, Canada, (1976).
159. Lindblom,G., Wennerström,H. and Arvidson,G. *Int. J. Quant. Chem.*, **XII, suppl. 2**, 153-158 (1977).
160. Lindblom,G., Wennerström,H., Arvidson,G. and Lindman,B. *Biophys. J.*, **16**, 1287-1295 (1976).
161. Lindblom,G., Wennerström,H. and Lindman,B. *Amer. Chem. Soc. Symp.*, **34**, 372-396 (1976).
162. Longley, W. and McIntosh,T.J. *Nature*, **303**, 612-614 (1983).
163. Lutton,E.S. *J. Am. Oil Chem. Soc.*, **42**, 1068-1070 (1965).
164. Luzzati,V., in *Biological Membranes*, vol. 1, pp. 71-123 (1968) (Ed. D. Chapman, Academic Press, New York).
165. Luzzati,V., Vargas,R., Gulik,A., Mariani,P., Seddon,J. and Rivas,E. *Biochemistry*, **31**, 279-285 (1992).
166. Mackay,A.L. *Biophys. J.*, **35**, 301-313 (1981).
167. Mariani,P., Luzzati,V. and Delacroix,H. *J. Mol. Biol.*, **204**, 165-189 (1988).
168. Marques,E., Khan,A., da Graca Miguel,M. and Lindman,B. *J. Phys. Chem.*, **97**, 4729-4736 (1993).
169. Marsh,D., *Handbook of Lipid Bilayers* (CRC Press, Inc, Boca Raton, Fl, 1990).
170. Martin,J.F., Selwyn,L.S., Vold,R.R. and Vold,R.L. *J. Chem. Phys.*, **76**, 2632-2634 (1982).
171. McLaughlin,A.C., Cullis,P.R., Berden,J.A. and Richards,R.E. *J. Magn. Reson.*, **20**, 146-165 (1975).
172. Mendelsohn,R., Davies,M.A., Brauner,J.W., Schuster,H.F. and Dluhy,R.A. *Biochemistry*, **28**, 8934-8939 (1989).
173. Monck,M.A., Bloom,M., Lafleur,M., Lewis,R.N.A.H., McElhaney,R.N. and Cullis,P.R. *Biochemistry*, **31**, 10037-10043 (1992).
174. Morrow,M.R. and Daves,J.H. *Biochim. Biophys. Acta*, **904**, 61-70 (1987).
175. Morrow,M.R., Singh,D., Lu,D. and Grant,C.W.M. *Biochim. Biophys. Acta*, **1106**, 85-93 (1992).
176. Morrow,M.R., Srinivasan,R. and Grandal,N. *Chem. Phys. Lipids*, **58**, 63-72 (1991).
177. Nagle,J.F. *Biophys. J.*, **64**, 1476-1481 (1993).
178. Nagle,J.F. *J. Chem. Phys.*, **63**, 1255-1261 (1975).
179. Nagle,J.F. and Wiener,M.C. *Biochim. Biophys. Acta*, **942**, 1-10 (1988).
180. Nagle,J.F. and Wilkinson,D.A. *Biophys. J.*, **23**, 159-175 (1978).
181. Newton,A.C. *Annu. Rev. Biophys. Biomol. Structure*, **22**, 1-25 (1993).
182. Nezil,F.A. and Bloom,M. *Biophys. J.*, **61**, 1176-1183 (1992).
183. Nichols,J.W. and Ozarowski,J. *Biochemistry*, **29**, 4600-4606 (1990).
184. Nilsson,A., Holmgren,A. and Lindblom,G. *Biochemistry*, **30**, 2126-2133 (1991).
185. Oldfield,E., Meadows,M. and Glaser,M. *J. Biol. Chem.*, **251**, 6147-6149 (1976).
186. Oldfield,E., Meadows,M., Rice,D. and Jacobs,R. *Biochemistry*, **17**, 2727-2740. (1978).
187. Orädd,G., Lindblom,G., Arvidson,G. and Gunnarsson,K. *Biophys. J.*, **68**, 547-557 (1995).
188. Orädd,G., Lindblom,G., Fontell,K. and Ljusberg-Wahren,H. *Biophys. J.*, **68**, 1856-1863 (1995).
189. Orädd,G., Lindblom,G., Johansson,L.B.-Å. and Wikander,G. *J. Phys. Chem.*, **96**, 5170-5174 (1992).
190. Orädd,G., Wikander,G., Lindblom,G. and Johansson,L.B.-Å. *J. Chem. Soc. Faraday Trans. I.*, **90**, 305-309 (1994).
191. Parsegian,V.A., Fuller,N. and Rand,P.R. *Proc. Natl. Acad. Sci. USA*, **76**, 2750-2759. (1979).
192. Persson,N.-O. and Lindblom,G. *J. Phys. Chem.*, **83**, 3015-3019 (1979).
193. Persson,N.-O. and Lindman,B. *J. Phys. Chem.*, **79**, 1410-1418 (1975).
194. Rance,M. and Byrd,R.A. *J. Magn. Reson.*, **52**, 221-240 (1983).
195. Rance,M., Jeffrey,K.R., Tulloch,A.P., Butler,K.W. and Smith,I.C.P. *Biochim. Biophys. Acta*, **688**, 191-200 (1982).
196. Rance,M., Jeffrey,K.R., Tulloch,A.P., Butler,K.W. and Smith,I.C.P. *Biochim. Biophys. Acta*, **600**, 245-262 (1980).

197. Rilfors,L., Eriksson,P.-O., Arvidson,G. and Lindblom,G. *Biochemistry*, **25**, 7702-7711 (1986).
198. Rilfors,L., Hauksson,J.B. and Lindblom,G. *Biochemistry*, **33**, 6110-6120 (1994).
199. Rilfors,L., Khan,A., Brentel,I., Wieslander,Å. and Lindblom,G. *FEBS Letters*, **149**, 293-298 (1982).
200. Rilfors,L., Lindblom,G., Wieslander,Å. and Christiansson,A., in *Biomembranes*, vol. 12, pp. 205-245 (1984) (Eds. L.A. Manson and M. Kates, Plenum Press, New York).
201. Rilfors,L., Wieslander,Å. and Lindblom,G., in *Subcellular Biochemistry: Mycoplasma Cell Membranes.*, vol. 20, pp. 109-166 (1993) (Eds. S. Rottem and I. Kahane, Plenum Press, New York).
202. Roberts,R.T. *Nature (London)*, **242**, 348 (1973).
203. Roeder,S.B.W., Burnell,E.E., Kuo,A.-L. and Wade,C.G. *J. Chem. Phys.*, **64**, 1848-1849 (1976).
204. Rossini,F.D., Pitzer,K.S., Arnett,R.L., Braun,R.M. and Pimentel,G.C., *Selected Values of Physical Properties of Hydrocarbons and Related Compounds*, Carnegie Press, Pittsburg, 1953.
205. Sadaghiani,A.S. and Khan,A. *J.Colloid Interface Sci.*, **146**, 69-78 (1991).
206. Sadaghiani,A.S. and Khan,A. *Langmuir*, **7**, 898-904 (1991).
207. Sadaghiani,A.S., Khan,A. and Lindman,B. *J. Colloid Interface Sci.*, **132**, 352-362 (1989).
208. Sankaram,M.B. and Thompson,T.E. *Biochemistry*, **29**, 10676-10684 (1990).
209. Saupe,A. *Z. Naturforschung*, **19 A**, 161-171 (1964).
210. Schindler,H. and Seelig,J. *Biochemistry*, **14**, 2283-2287 (1975).
211. Schnur,J.M. *Science*, **262**, 1669-1676 (1993).
212. Schnur,J.M., Price,R., Schoen,P., Yager,P., Calvert,J.M., Georger,J. and Singh,A. *Thin Solid Films*, **152**, 181-206 (1987).
213. Schurtenberger,P. and Lindman,B. *Biochemistry*, **24**, 7161-7165 (1985).
214. Scriven,L.E. *Nature*, **263**, 123-125 (1976).
215. Seddon,J.M. *Biochim. Biophys. Acta*, **1031**, 1-69 (1990).
216. Seddon,J.M., Bartle,E.A. and Mingins,J. *J. Phys. Condens. Matter.*, **2**, 285-290 (1990).
217. Seelig,J. *Biochim. Biophys. Acta*, **515**, 105-140 (1978).
218. Seelig,J. and Seelig,A. *Biochem. Biophys. Res. Commun.*, **57**, 406-411 (1974).
219. Sells,T.D. and O´Brien,D.F. *Macromolecules*, **27**, 226-233 (1994).
220. Siegel,D.P. *Biophys. J.*, **65**, 2124-2140 (1993).
221. Siegel,D.P. *Chem. Phys. Lipids*, **42**, 279-301 (1986).
222. Silva-Crawford,M., Gerstein,B.C., Kuo,A.-L. and Wade,C.G. *J. Am. Chem. Soc.*, **102**, 3728-3732 (1980).
223. Sjölund,M., Lindblom,G., Rilfors,L. and Arvidson,G. *Biophys. J.*, **52**, 145-153 (1987).
224. Sjölund,M., Rilfors,L. and Lindblom,G. *Biochemistry*, **28**, 1323-1329 (1989).
225. Slichter,C.P., *Principles of Magnetic Resonance*, Springer-Verlag, New York, 1990.
226. Small,D.M., *Handbook of Lipid Research*, Boca Raton, Fla., 1990.
227. Small,D.M., Bourges,M. and Dervichian,D.G. *Biochim. Biophys. Acta*, **125**, 563- (1966).
228. Smith,I.C.P., Butler,K.W., Tulloch,A.P., Davis,J.H. and Bloom,M. *FEBS Letters*, **100**, 57-61 (1979).
229. Sternin,E., Bloom,M. and Mackay,A.L. *J. Magn. Reson.*, **55**, 274-282 (1983).
230. Sternin,E., Fine,B., Bloom,M., Tilcock,C.P.S., Wong,K.F. and Cullis,P.R. *Biophys. J.*, **54**, 689-694 (1988).
231. Stilbs,P. *Prog. Nucl. Magn. Reson. Spectrosc.*, **19**, 1-45 (1987).
232. Stockton,G.W., Johnson,K.G., Butler,K.W., Polnaszek,C.F., Cyr,R. and Smith,I.C.P. *Biochim. Biophys. Acta*, **401**, 535-539 (1975).
233. Stockton,G.W., Johnson,K.G., Butler,K.W., Tulloch,A.P., Boulanger,Y., Smith,I.C.P., Davis,J.H. and Bloom,M. *Nature*, **269**, 267-268 (1977).
234. Stockton,G.W., Polnaszek,C.F., Tulloch,A.P., Hasan,F. and Smith,I.C.P. *Biochemistry*, **15**, 954-966 (1976).
235. Stockton,G.W. and Smith,I.C.P. *Chem. Phys. Lipids*, **17**, 251-263 (1976).
236. Söderman,O., Arvidson,G., Lindblom,G. and Fontell,K. *Eur. J. Biochem.*, **134**, 309-314 (1983).
237. Söderman,O. and Henriksson,U. *J. Chem. Soc. Faraday Trans. I.*, **83**, 1515-1529 (1987).
238. Söderman,O., Khan,A. and Lindblom,G. *J. Magn. Reson.*, **36**, 141-146 (1979).
239. Söderman,O., Lindblom,G., Johansson,L.B.-Å. and Fontell,K. *Mol. Cryst. Liq. Cryst.*, **59**, 121-136 (1980).
240. Söderman,O., Walderhaug,H., Henriksson,U. and Stilbs,P. *J. Phys. Chem.*, **89**, 3693-3701 (1985).

241. Tanford,C., *The Hydrophobic Effect*, John Wiley, New York, 1980.
242. Tanner,J.E. and Stejskal,E.O. *J. Chem. Phys.*, **49**, 1768-1777 (1968).
243. Tardieu,A. and Luzzati,V. *Biochim. Biophys. Acta*, **219**, 11-17 (1970).
244. Tartar,H.V. *J. Phys. Chem.*, **59**, 1195-1199 (1955).
245. Tate,M.W., Eikenberry,E.F., Turner,D.C., Shyamsunder,E. and Gruner,S.M. *Chem. Phys. Lipids*, **57**, 147-164 (1991).
246. Thayer,A.M. and Kohler,S.J. *Biochemistry*, **20**, 6831-6834 (1981).
247. Thurmond,R.L., Dodd,S.W. and Brown,M.F. *Biophys. J.*, **59**, 108-113 (1991).
248. Thurmond,R.L. and Lindblom,G., in *Structural and Biological Roles of Non-bilayer Forming Lipids*, in press (Ed. R. Epand, JAI Press, Greenwich, Connecticut).
249. Thurmond,R.L., Lindblom,G. and Brown,M.F. *Biochemistry*, **32**, 5394-5410 (1993).
250. Thurmond,R.L., Lindblom,G. and Brown,M.F. *Biophys. J.*, **60**, 728-732 (1991).
251. Thurmond,R.L., Niemi,A.R., Lindblom,G., Wieslander,Å. and Rilfors,L. *Biochemistry*, **33**, 13178-13188 (1994).
252. Tiddy,G. *J. Chem. Soc., Faraday Trans. I*, **73**, 1731-1737 (1977).
253. Tiddy,G.J.T., Lindblom,G. and Lindman,B. *J. Chem. Soc., Faraday Trans. I*, **74**, 1290-1300 (1978).
254. Tiddy,G.J.T., Rendall,K. and Trevethan,M.A. *Comun. Jorn. Com. Esp. Deterg.*, **15**, 51-62 (1984).
255. Torrey,H.C. *Phys. Rev.*, **104**, 563-573 (1956).
256. Trouard,T.P., Mannock,D.A., Lindblom,G., Rilfors,L., Akiyama,M. and McElhaney,R.M. *Biophys. J.*, **67**, 1090-1100 (1994).
257. Ulmius,J., Lindblom,G., Wennerström,H., Johansson,L.B.-Å., Fontell,K., Söderman,O. and Arvidson,G. *Biochemistry*, **21**, 1553-1560 (1982).
258. Ulmius,J., Wennerström,H., Lindblom,G. and Arvidson,G. *Biochemistry*, **16**, 5742-5745 (1977).
259. Ulmius,J., Wennerström,H., Lindblom,G. and Arvidson,G. *Biochim. Biophys. Acta*, **389**, 197-202 (1975).
260. van der Ploeg,P. and Berendsen,H.J.C. *Mol. Physiol.*, **49**, 233-248 (1983).
261. Vargas,R., Mariani,P., Gulik,A. and Luzzati,V. *J. Mol. Biol.*, **225**, 137-145 (1992).
262. Venable,R.M., Zhang,Y., Hardy,B.J. and Pastor,R.W. *Science*, **262**, 223-226 (1993).
263. Verkleij,A.J., Mombers,C., Leunissen-Bijvelt,J. and Ververgaert,P.H.J.T. *Nature*, **279**, 162-163 (1979).
264. Vist,M.R. and Davis,J.H. *Biochemistry*, **29**, 451-464 (1990).
265. Vold,R.L. and Vold,R.R., in *The Molecular Dynamics of Liquid Crystals*, vol. C, p. 431 (1994) (Eds. G.R. Luckhurst and C.A. Veracini, Kluwer Academic Publishers, Dordrecht)
266. Waldeck,A.R., Lennon,A.J., Chapman,B.E. and Kuchel,P.W. *J. Chem. Soc. Faraday Trans.*, **89**, 2807-2814 (1993).
267. Watnick,P.A., Dea,P. and Chan,S.J. *Proc. Natn. Acad. Sci. USA*, **87**, 2082-2086 (1990).
268. Wennerström,H. *Chem. Phys. Lett.*, **18**, 41-44 (1973).
269. Wennerström,H., Khan,A. and Lindman,B. *Adv. Colloid Interface Sci.*, **34**, 433-449 (1991).
270. Wennerström,H. and Lindblom,G. *Q. Rev. Biophys.*, **10**, 67-96 (1977).
271. Wennerström,H., Lindblom,G. and Lindman,B. *Chem. Scr.*, **6**, 97-103 (1974).
272. Wennerström,H., Lindman,B., Lindblom,G. and Tiddy,G.J.T. *J. Chem., Faraday Trans. I*, **75**, 663-668 (1979).
273. Wennerström,H., Lindman,B., Söderman,O., Drakenberg,T. and Rosenholm,J.B. *J. Am. Chem. Soc.*, **101**, 6860-6864 (1979).
274. Wennerström,H., Persson,N.-O., Lindblom,G., Lindman,B. and Tiddy,G.J.T. *J. Magn. Reson.*, **30**, 133-136 (1978).
275. Wiener,M.C., Tristram-Nagle,S., Wilkinson,D.A., Campbell,L.E. and Nagle,J.F. *Biochim. Biophys. Acta*, **938**, 135-142 (1988).
276. Wieslander,Å., Nordström,S., Dahlqvist,A., Rilfors,L. and Lindblom,G. *Eur. J. Biochem.*, **227**, 734-744 (1995).
277. Wieslander,Å., Rilfors,L., Johansson,L.B.-Å. and Lindblom,G. *Biochemistry*, **20**, 730-735 (1981).
278. Wilkinson,D.A. and Nagle,J.F. *Biochemistry*, **20**, 187-192 (1981).
279. Woesner,D.E. *J. Chem. Phys.*, **34**, 2057- (1961).
280. Wu,W.-G. and Huang,C.-H. *Lipids*, **16**, 820-822 (1981).
281. Österberg,F., Rilfors,L., Wieslander,Å., Lindblom,G. and Gruner,S.M. *Biochim. Biophys. Acta*, **1257**, 18–24 (1995).

Chapter 6

PLANT GLYCOLIPIDS: STRUCTURE, ISOLATION AND ANALYSIS

Ernst Heinz

Institut für Allgemeine Botanik, Universität Hamburg, Ohnhorststr. 18, 22609 Hamburg, Germany

A. INTRODUCTION

Plant glycolipids comprise compounds, which can be divided into distinct groups with regard to their hydrophobic aglyca. In terms of quantitative abundance, glycosyl diacylglycerols usually rank first followed by glycosylated ceramides and sterols. All these compounds are characteristic lipid constituents of protoplasmic membrane systems, whereas acyl derivatives of glucose and sucrose are extracellular lipids from the surface cover of plants, in which they can exceed all other glycolipids with regard to quantity. In many blue-green algae, glycosyl derivatives of hydroxyalkanes, hydroxy ketoalkanes (or hydroxy fatty acids?) and carotenoids occur, that are not generally present in higher plants.

Compared to the more than 250 different structures elucidated for human and animal glycolipids [1], the number of well-defined plant glycolipids is lower by several factors. Nevertheless, despite many common problems in analysis of plant glycolipids, this field of research actually splits into different areas mainly with regard to the various aglyca. It is in the analysis of these moieties, in which several laboratories have developed their specific expertise, mainly because their interests are concentrated on problems of biosynthesis, which vary with the different aglyca.

Analysis of glycolipids requires methodology for isolation of the different compounds, elucidation of their chemical structures, release and separation of constituent building blocks as well as quantification. Finally, in many cases the chemists' ambitions are satisfied only, when the structures analysed have been confirmed by synthesis. During its development this field of research has been reviewed several times. The last comprehensive survey by Kates [2] dates back about five years and contains a large and diverse collection of relevant data. In view of this valuable source the present summary will place more emphasis on aspects, which have not been covered or which will have continuous interest in the future. Undoubtedly, biochemical studies on the metabolism of glycolipids will depend on their isolation and separation, and additional work on structural elucidation of new compounds will be carried out as for example in the group of cerebrosides and phytoglycolipids. In such cases modern techniques of two-dimensional NMR and mass spectroscopy (MS) will be applied, most likely in cooperation with experts in the field and, therefore, an outline of this methodology cannot be the task of this article. Instead, selected references will be given for the application of these methods with plant glycolipids to demonstrate their potential for comparison with the painstaking work, which was the only alternative for analysis of complex glycolipids before the availability of such methods.

Apart from structural analysis, there is an increasing demand for easy and reliable separation and quantification of various compounds, ideally by high-performance liquid chromatography (HPLC), of the whole spectrum of lipids present in complex mixtures as extracted from various tissues. A routine quantification method in combination with measurement of radioactivity would be very useful for physiological and biochemical aspects of plant lipidology, including the increasing number of analyses required to characterize the performance of transgenic plants expressing introduced genes of lipid metabolism in sense or antisense orientation.

B. ESTERS OF GLUCOSE AND SUCROSE

1. Biological Functions

These compounds have often been neglected in general reviews and therefore, a short introduction to their biological relevance is given. The epidermal surface of plants belonging to the family of *Solanaceae* is often covered by a sticky and viscous exudate, which can account for up to 25% of the plant dry weight [3]. Characteristic components of this extracellular material are short- and medium-chain fatty acid esters of glucose and sucrose, which in some cases may account for more than 80% of the epicuticular exudate [3]. The presence of this exudate on the plant surface is correlated with the resistance of these plants against a wide range of arthropod pests. The viscous character of the sugar esters is believed to cause physical impediments such as entrapment, while their toxicity confers feeding deterrence and may even inhibit larval development of various pests [4-6].

The sugar esters are secreted by characteristic glandular trichomes, which grow out from epidermal cells. They have a multicellular stalk and a secretory glandular head. The cells of a multicellular head have a full complement of organelles and in particular many chloroplasts with well-developed thylakoids [7]. This results in the green colour of the heads, which have the ability to incorporate various labelled precursors into sugar esters [8-10]. Another type of shorter trichomes does not participate in synthesis and secretion of these esters.

In a wild species of tomato (*Lycopersicon pennellii*) quantitative measurements on the productivity of these secretory hairs have been carried out [3]. On leaves glandular trichomes are formed at a density of 200-300/10 mm^2. Their secretory activity results in an epicuticular exudate cover of 150-280 mg/g leaf dry weight corresponding to 480-740 $\mu g/cm^2$. This represents 600-800 nmol glucose ester/cm^2 as compared to at most 400 nmol glycosyl diacylglycerols/cm^2 [11]. These data demonstrate that the sugar esters represent the highest proportion of glycolipids produced by these tissues and the high activity of the secretory heads.

In contrast to the wild *L. pennellii* the cultivated tomato *L. esculentum* does not produce glandular trichomes [3]. The epidermis of this species is not covered by an epicuticular exudate and likewise, glucose esters are missing. This is correlated with the susceptibility of *L. esculentum* to a wide variety of pests and explains the agronomic interest in this field of glycolipid research. The capacity to develop glandular trichomes, which are capable of secreting glucose esters, is solely dependent on the epidermis and not on subepidermal tissue. This was shown by biochemical and morphological studies of periclinal chimeras between *L. pennellii* and *L. esculentum* [5]. The *L. pennellii* epidermis on a body of *L. esculentum* cells was sufficient for the production of glandular trichomes at the same density as found in the wild *L. pennellii.* They secreted a cover of glucose esters resulting in aphid resistance of chimeric plants, which are in fact *L. esculentum* enclosed by a heterologous *L. pennellii* epidermis. Therefore, the expression of this syndrome is confined to the genotype of the epidermal layer, despite the fact that aphids feed from subepidermal tissues.

In tobacco, similar glandular trichomes are responsible for exudate and sucrose esters production, which account for 55% of the exudate [9] and contribute to the economically relevant flavour of dried tobacco. In one genotype the presence of a recessive allele (1406) at a single locus causes a lack of secretory capability of these trichomes [7]. Ultrastructural studies revealed that this phenotype was characterized by unusual inclusions in the cytoplasm of secretory cells, which in contrast to wild type cells also lacked chloroplasts. On the other hand, these studies showed that the proliferation of endomembranes usually related to secretory functions were not detected in wild type cells. So far, it is not known where the 1406-mutation may interfere with the multiplicity of reactions involved in biosynthesis and secretion of glucose and sucrose esters.

Studies on the biosynthesis of the characteristic branched short- and medium-chain fatty acids present in these sugar esters (see below) suggest that precursors are derived from branched-chain amino acids [8-10]. Only a few studies have so far addressed these questions, from which most regarding enzymology and compartmentation of synthesis and secretion cannot be answered at present.

2. CHEMICAL STRUCTURES

i) Glucose Esters. Glucose esters comprise various forms of increasing degree of acylation from di- to penta-*O*-acyl-D-glucopyranoses in both α- and β-anomeric forms (see Table 6.2). The acyl groups include acetyl residues and branched short- or medium-chain acyl residues up to C12 (Table 6.1). General acylation patterns or specific locations of particular fatty acids are hardly discernible; only acetate groups have not so far been found at C3 and C4. A representative example from *Solanum aethiopsicum* [12] is shown in Figure 6.1a: 2-*O*-acetyl-3-*O*-isobutyryl-4-*O*-isocaproyl-α-D-glucopyranose. Rather unique among the glucose esters are those isolated from *Datura metel*, in which a 1,2,3-tri-*O*-hexanoyl-α-D-glucopyranose (Figure 6.1b) predominates [13]. In many cases, where different fatty acids are esterified, the exact location of particular acyl groups was not analysed and only fatty acid mixtures released from these lipids are given (Table 6.1). They typ-

Table 6.1.
Acyl group composition (mol%) of glucose and sucrose esters from surface lipids of *Nicotiana umbratica,* and their R_f-values after separation by TLC in the solvents indicated using HPTLC plates [24]. Acyl group positions on glucose and fructose (primed numbers) residues are identified by ring carbons.

	D-Glucopyranose ester			**Sucrose esters**				
Acid	1,2,3,4,6	2,3,4,6	2,3,4	2,3,4,6,1′,3′	2,3,4,1′,3′	2,4,6,3′	2,3,4,3′	2,3,4
Acetic	40.3	25.4	3.0	48.8	42.3	46.4	16.5	0.8
Propionic	0.2	1.2	tr	0.4	0.5	0.4	tr	0.5
Methylpropionic	9.4	19.0	4.4	3.4	18.9	0.7	4.6	1.5
Methylbutyric	12.3	17.4	27.0	15.8	27.2	19.8	21.2	21.1
3-Methylpentanoic	29.4	25.2	46.1	23.3	5.3	30.8	41.9	51.2
4-Methylpentanoic	6.1	8.9	11.8	4.1	3.9	1.1	10.0	11.9
Methylhexanoic	1.9	2.3	5.3	2.5	1.7	0.8	4.3	7.9
Methylheptanoic	0.4	0.6	2.4	1.7	0.2	tr	1.5	5.1
R_f in Hexane/Acetone 7:3	0.64	0.46	0.31					
R_f in $CHCl_3$/Acetone 1:1				0.77	0.61	0.53	0.38	0.23

ically contain 2-methylpropanoic, 2(or 3)-methylbutanoic, 3-methylpentanoic and so forth up to 8-methyldecanoic and 10-methylundecanoic acids [6]. Methyl branching in the *anteiso* position as found in C_4-C_{10} acids can result in enantiomers, but in most investigations, the stereochemistry of such fatty acids has not been investigated. If they originate from isoleucin, then they should always have the same configuration as for example in the 3*S*-methylpentanoyl residue [14] present in a sucrose ester (see below). Occasionally, glucose esters contain up to 37% of a unique 2-methylbutenoic acid of unknown stereochemistry [15]. n-Decanoic and n-dodecanoic acid were usually minor components as in the 2,3,4-tri-*O*-acyl glucose from *L. pennellii*, where they account for 10.4 and 3.4%, respectively [16]. It had been discussed, whether such medium-chain fatty acids could be due to the activity of a specific plastidial acyl-ACP-thioesterase interfering with chain elongation during fatty acid biosynthesis. In this context, it should be recalled that a putative short-chain acyl ACP-thioesterase was cloned by chance from *Arabidopsis* [17], which is known not to produce significant proportions of such fatty acids. In the meantime, this possibility has been excluded [18]. New labelling experiments with *Lycopersicon*, *Nicotiana* and *Petunia* [19] have provided evidence that straight- and branched-chain C_6-C_{12} acids are synthesised by stepwise 1C-elongation cycles from α-ketoacids in reactions formally identical with part of the mitochondrial citric acid cycle covering the segment from citrate synthase to α-ketoglutarate dehydrogenase. A linear or branched α-ketoacid is condensed with acetyl-CoA followed by isomerization (*cf.* aconitase) and dehydrogenation/decarboxylation (cf. isocitrate dehydrogenase). The resulting linear or branched α-ketoacid (n+1) may be cycled through this sequence for further 1C-elongation, and this is finally completed by a CoASH-dependent α-ketoacid dehydrogenase (*cf.* α-ketoglutarate dehydrogenase) to yield the activated branched or linear acyl-CoA with an odd or even number of carbon atoms.

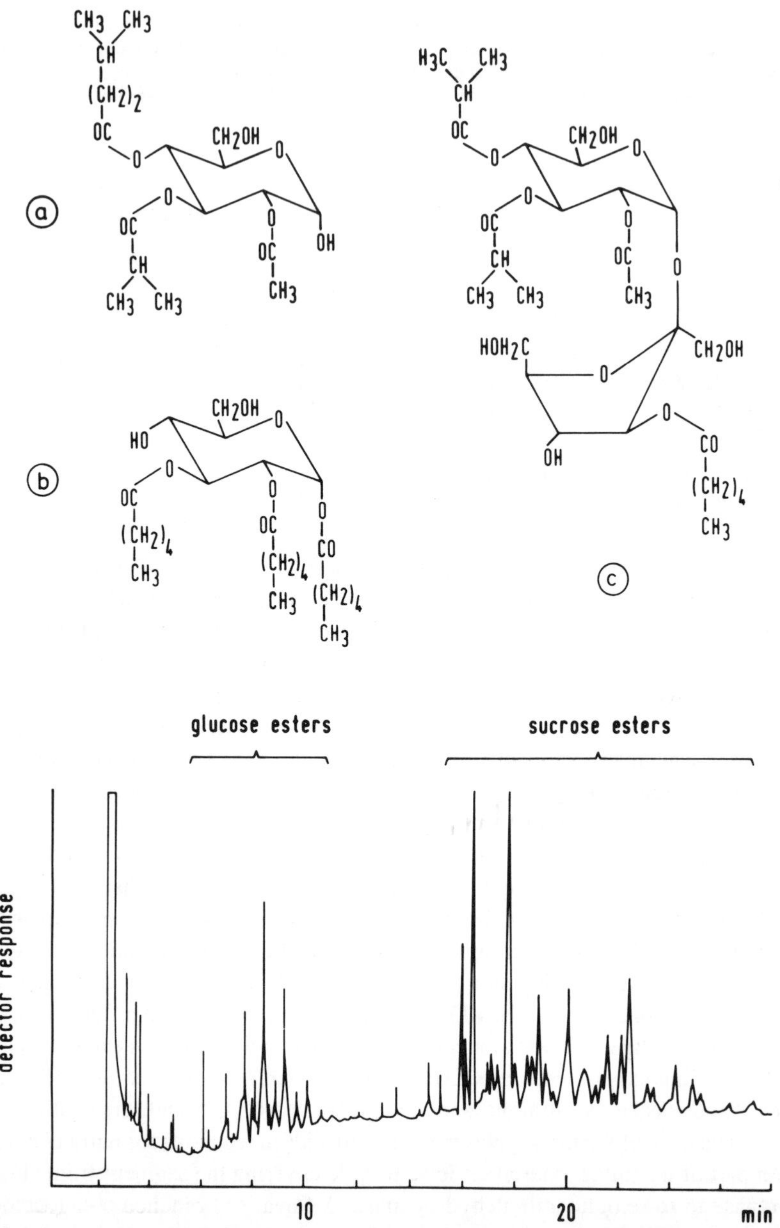

Fig. 6.1. Structures and chromatographic separation of glucose and sucrose esters from surface lipids of *Solanaceae*. As examples 2-*O*-acetyl-3-*O*-isobutyryl-4-*O*-isocaproyl-α-D-glucopyranose (a), 1,2,3-tri-*O*-caproyl-α-D-glucopyranose (b) and 2-*O*-acetyl-3,4-di-*O*-isobutyryl-3′-*O*-caproyl-sucrose (c) are shown. The resolution of TMS ether derivatives of glucose and sucrose esters present in surface lipids from *Nicotiana umbratica* as separated by capillary GLC with temperature programming between 250 and 300°C (5°/min) and FID detection is shown in the lower part of the figure (redrawn from [30]).

Contrary to expectations, these acyl-CoA thioesters are not used for direct acylation of glucose. Apparently, they are first hydrolysed and then linked to the anomeric hydroxyl group of glucose by a UDP-glucose: free fatty acid β-D-glucosyltransferase reaction [20,21]. It is the resultant 1-*O*-acyl-β-D-glucopyranose, which is the actual donor for acyl transfer to non-anomeric hydroxyl groups of partially acylated glucose esters. These reactions are dismutations, each releasing free glucose from the donor and an acceptor with an additional acyl group. The predominance of shorter-chain fatty acids and the solubility of the resultant esters may cause a preferential loss of monoacylated forms in the course of extraction and work-up or even during weathering of plants.

ii) Sucrose Esters. Sucrose esters represent a similarly complex mixture of acylated forms, both regarding acyl groups, esterified positions and degree of acylation (Tables 6.1 and 6.2). In general the glucopyranosyl part is more extensively acylated than the fructofuranosyl moiety. The glucosyl part may carry from one to four acyl groups, *i.e.* it may be fully acylated. The fructosyl group, which often is not acylated at all, carries at most two acyl groups located at C1′ and C3′ or C6′. Also in sucrose esters, acetyl groups can be present, that seem to be excluded from C3 and C4 of glucose, in which positions they have also not been found in free glucose esters. In *L. hirsutum* C_{11} and C_{12} medium-chain fatty acids may be concentrated at C3 of glucose [22]. Together with the specific location of acetyl groups this may point to specificities of the acyltransferases involved in decorating sugars with acyl groups. It has been pointed out that this derivatization very often affects the positions C2, C4 and C1′, which are crucial for the sweet taste of sucrose [22]. The masking of these epitopes may abolish their interaction with taste receptors or even result in an adverse taste.

As with the glucose esters, only a few sucrose esters have been analysed in all details, *i.e.* including the specific position of particular acyl groups. Such an example is given in Figure 6.1c [23]. In general, only mixtures of fatty acid moieties released from these compounds are given as presented in Table 6.1, which compares the acyl residues from various glucose and sucrose esters from one plant [24]. In this case, C2, C6, C1′ and C3′ may carry acetyl groups, and this explains the change in the proportion of this acid in the different compounds, whereas variations in other fatty acids (for example methylpropionic acid) are not easily explained.

So far the genera *Datura*, *Lycopersicon*, *Nicotiana*, *Petunia* and *Solanum* from the *Solanaceae* have been found to produce acylated sugars [25]. *Nicotiana* produces both glucose and sucrose esters, whereas *Datura* may produce only glucose esters. Recently acyl sugars from a wild tomato *L. pennellii* have been investigated with respect to the geographic distribution of this species [6], which grows wild in the Andes of Peru. Since it can be crossed with the cultivated tomato *L. esculentum, L. pennellii*-derived acyl sugars may provide a system to protect the cultivated crop against pests. It was found that both the ratio of glucose to sucrose

Table 6.2.

^{1}H- and ^{13}C-NMR data of glucose and sucrose esters [24]. Only signals from sugar carbons and protons are listed, from which those attributed to the fructosyl residue are indicated by primed numbers. The free glucose esters can occur as α- or β-anomers as indicated, whereas sucrose is α-D-glucopyranosyl-β-D-fructofuranoside. The spectra were recorded in $CDCl_3$, shifts are given in δ ppm downfield from tetramethylsilane as internal standard and coupling constants can be found in the original publications. The [1]H-signals for all acetyl groups in sucrose octaacetate have been assigned in deuterated pyridine/benzene (1:1) [34].

	^{13}C-shifts							^{1}H-shifts					
carbon atoms	α-1,2,3,4,6	α-2,3,4	β-2,3,4	2,3,4,6,1′,3′	2,3,4,3′	2,3,4	**acetate signals**	2,3,4	2,3,4,3′	2,3,4,1′,3′	β-2,3,4	α-2,3,4	α-1,2,3,4,6
1	89.2	90.1	95.7	89.7	89.8	88.8	-	5.66	5.64	5.65	4.71	5.46	6.34
2	69.4	71.5	73.1	70.4	70.8	70.8	2.012	4.84	4.85	4.87	4.91	4.85	5.11
3	69.3	69.1	71.6	69.1	69.1	69.6	1.897	5.50	5.50	5.48	5.31	5.62	5.55
4	67.7	69.3	68.7	67.7	68.9	69.0	1.869	4.90	4.92	5.14	5.03	4.96	5.20
5	70.2	69.5	74.6	68.9	71.6	71.0	-	4.21	4.13	4.21	3.54	4.06	4.09
6	61.8	61.4	61.4	61.6	61.9	61.7	2.000	3.55	3.87	4.15	3.61	3.61	4.11
1′				64.8	64.6	63.0	1.915	3.49	3.58	4.05			
2′				102.6	104.2	104.7	-	-	-	-			
3′				79.2	79.9	77.4	2.048	4.23	5.18	5.22			
4′				71.4	71.9	73.7	1.843	4.12	4.55	4.51			
5′				82.6	82.4	81.5	-	3.59	3.69	3.88			
6′				60.3	60.6	61.3	1.941	3.59	3.79	3.77			

esters and the mixture of acyl groups vary with the individual plants of different origin. In addition, ontogenetic variations may occur.

3. Isolation and Separation

Epicuticular lipids are easily washed off from fresh leaves by immersion in chloroform or dichloromethane for short times (5 sec to 5 min), and the shorter this time, the lower the contamination by intracellular glycolipids. The partitioning of the extract against water as sometimes carried out [6] may bear the risk of losing some compounds of low acylation. After removal of the solvent, the residue is redissolved in acetone and cooled to 0°C to precipitate waxes [23], which are removed by filtration or centrifugation. The resultant lipid mixture is then ready for fractionation either by silicic acid column chromatography or by preparative silica gel thin layer chromatography (TLC).

Larger quantities of lipids are fractionated by silicic acid column chromatography. Chloroform elutes glucose esters with five to three acyl groups, and sucrose esters with six acyl residues are obtained also by prolonged elution with this solvent [24]. Glucose derivatives with two ester groups are eluted with chloroform/acetone (1:1, v/v, as throughout in the following for all solvent ratios), which also elutes sucrose esters with six to three acyl groups. Pure acetone is used to elute sucrose diesters [26]. A stepwise increase of acetone following extensive elution with chloroform may result in increased resolution of groups.

For further separation by silica gel TLC on preparative plates (0.5-1.0 mm thick) 100 to 200 mg of material were handled on a single plate [12,23]. A run in chloroform/methanol (9:1) separates di- and tri-esters of glucose (R_f 0.3 and 0.6, respectively), which can be eluted from the silica gel by acetone or chloroform/methanol (3:1). For localisation they can be detected by spraying with water (for preparative purposes) or with any of the fluorescent reagents used for lipid detection. With chloroform/methanol (6:1) sucrose esters are resolved in the range of R_f 0.3-0.6 depending on their degree of acylation (further data on glucose and sucrose esters in Table 6.1).

The partially purified or enriched fractions obtained by preparative TLC or column chromatography are subsequently subjected to additional separation for isolation of pure compounds. Representative examples are given in the following. Silica gel TLC with chloroform/methanol (9:1) was used to obtain the following compounds in pure and crystalline form:
2-*O*-acetyl-3-*O*-isobutyryl-6-*O*-isocaproyl-α-D-glucopyranose, R_f 0.66, m.p. 92-93°C [12] and 2,6,1′-tri-*O*-(3-methylbutyryl)-3-*O*-(2-methylbutyryl)-sucrose, R_f 0.48, m.p. 120-122°C [22].
Silica gel HPTLC with hexane/acetone (7:3) [27] separates pentaacylglucose (R_f 0.57), tetraacylglucose (R_f 0.41), triacylglucose (R_f 0.25) and pentaacylsucrose (R_f 0.11). Chloroform/acetone (1:1) was suitable for resolution of sucrose esters [24] with 3-6 acyl groups spread between R_f 0.23-0.77 (Table 6.1). Apart from separation according to degree of acylation, there is also some separation accord-

ing to position of acyl groups. A free primary as compared to a secondary hydroxyl group results in stronger retention because of better interaction with the silica gel.

Reversed-phase (RP-C18) TLC with acetone/water (7:3) was used to separate glucose esters as for example 1,2,3-tri-*O*-hexanoyl-α-D-glucopyranose with R_f 0.44 [13] and sucrose esters. Separation of the following collection of sucrose esters in acetone/water (3:1) shows the high resolution obtained by this method [23,28].

acyl groups on sucrose	aliphatic C-atoms	R_f value
2-acetyl-3′-hexanoyl-3,4-di-isobutyryl (mp 115-116°)	12	0.59
6-caproyl-3,3′,4-tri-isobutyryl	14	0.54
2-acetyl-3′,4-di-hexanoyl-3-isobutyryl	14	0.51
6-caproyl-3′-isobutyryl-3,4-di-methylbutyryl	16	0.47
2-acetyl-3′-decanoyl-3,4-di-isobutyryl	16	0.45

HPLC on silicic acid columns is carried out with hexane/acetone [15,24,26,27] or chloroform/acetone mixtures [29] and refractive index (RI) detection. Hexane/acetone (7:3) was used for separation of glucose esters (tetraacylglucose R_t 10 min; triacylglucose R_t 13 min) [15], whereas sucrose esters [24] were eluted at higher acetone proportions (hexane/acetone, 1:1). They can even be separated according to location of various acyl substituents [26] as shown by the resolution of 4,3′- (R_t 11.2 min), 3,4- (R_t 12.8 min) and 6,3′-di-*O*-acyl sucrose (R_t 16.3 min) in hexane/acetone (4:6).

RP-HPLC gives the highest resolution as already indicated by the results of RP-TLC. A RP-C18 column was used [16] to separate nine different 2,3,4-tri-*O*-acyl glucose esters by isocratic elution with acetonitrile/water (3:1) and UV detection at 210 nm. Under these conditions, the 2,3,4-tri-*O*-methylpropanoyl ester (with 9 aliphatic acyl carbons) emerges at 9.5 min, whereas the 2,3,4-tri-*O*-acyl ester with methylpropanoyl, methylbutanoyl and *n*-decanoyl groups (16 aliphatic carbons) elutes at 30.6 min. In addition, β-anomers elute ahead of α-anomers because of the higher hydrophilicity of an equatorially exposed hydroxyl group.

Gas-liquid chromatography (GLC) has been used to separate acetyl [13] and trimethylsilyl (TMS) derivatives [30,31] of glucose and sucrose esters, that are formed by standard techniques with acetic anhydride or *N,O*-bis(trimethylsilyl)trifluoroacetamide, respectively. Separation is achieved by capillary GLC using stationary phases of SE-54, DB-1 or DB-5. For example, with a temperature program of 5°/min and a range of 250-300°C, TMS ethers of glucose esters emerged between 6 and 11 min and sucrose esters between 14 and 26 min [30] (Figure 6.1, lower part). In combination with GLC-MS the resultant fingerprint patterns may give the most useful and detailed data in biochemical and ecological studies.

4. Quantification

Quantitative analyses of sugar esters in surface lipids have been based on the determination of glucose and sucrose. They are released by saponification followed by enzymatic (glucose), chemical (reducing sugar) or colorimetric (total sugar) determination [6,16]. As indicated above, a more sensitive method may be GLC of TMS ethers [30] including a suitable internal standard for glucose and sucrose esters, preferably of synthetic origin and designed not to overlap with naturally occurring compounds. A completely different approach is based on the retention of the water-soluble stain Rhodamine B from an aqueous solution in a dried film of glucose and sucrose esters [32]. The sugar esters are extracted from leaves (2-10 cm^2 of lamina) by dipping for 20 sec in acetonitrile. After evaporation, the residue is covered with an aqueous solution of Rhodamine B, followed by several washings and final redissolution of the sugar ester/stain complex in acetonitrile/water (1:1). Limits of detection by measurement of absorption (550 nm) or fluorescence (emission at 582 nm) are 10 μg and 0.1 μg, respectively, and linearity extends at least up to 200 μg and 2 μg, respectively. This very simple method may be the method of choice for routine measurements in applied screenings.

The synthesis of 2,3,4-tri-*O*-acyl-glucose [16], four different 3,4,6-tri-*O*-acyl sucrose esters [14] and 2,3,4,6-tetra-*O*-acyl sucrose [33] with particular acyl groups as found in exudates from *Solanaceae* has been described [14,33] and may be of relevance for selecting internal standards for quantitative work.

5. Structural Analysis

The general assignment as glucose or sucrose esters can be made by chromatographic resolution of sugars released by alkaline or acid hydrolysis from the esters [13,16,26]. Detailed structural elucidation of intact compounds was based mainly on NMR spectroscopy and mass spectrometric fragmentation patterns. Glucose and sucrose esters are recognized by their proton-decoupled ^{13}C-spectra displaying characteristic signals for glucose and sucrose carbons. Two-dimensional homo- and heteronuclear (^{1}H-^{1}H, ^{1}H-^{13}C) methods allow the resolution and identification of one-dimensionally heavily overlapping signal groups. The positions of the signals for the ring and terminal hydroxymethylene protons allow the immediate identification of acylation patterns of glucose and fructose moieties (Table 6.2), since the deshielding by an ester group results in downfield shifts of the corresponding carbinol protons. Protons of primary hydroxymethylene groups are shifted least (about 0.5 ppm) followed by the anomeric proton of glucose esters (about 0.9 ppm) and the methine ring protons (up to 1.6 ppm, Table 6.2). Peracetylation with acetic anhydride followed by NMR reexamination will identify newly shifted carbinol protons and at the same time the originally free hydroxyl groups [34]. In ^{13}C-spectra acylation at several positions does not result in such easily recognizable changes because of overlapping downfield and upfield shifts in α- and β-positions (Table 6.2). Non-acylated positions with free hydroxyl groups were recognized after shaking of the sample solution in deutero-chloroform with a (1:1) mixture of D_2O/H_2O. The DO-^{13}C coupling results in a small deuterium-induced downfield shift of the corresponding ^{13}C-signal giving rise to a 1:1-doublet instead of the original singlet [29]. Also in ^{13}C-spectra, signals for anomeric carbons are the most downfield and those for hydroxymethylene groups the most upfield ones. In the original references, additional data on the splitting of ^{1}H-signals and their coupling constants are given, that demonstrate their connectivities and the axial position of all ring protons in these sugar esters.

The anomeric configuration is recognized by differences in the position and splitting of the signal for the anomeric proton: large (small) coupling constants $^{3}J_{HH}$ for β(α)-linkage due to a trans-axial (gauche-equatorial) correlation between H1 and H2 (with the exception of mannose). In addition α-H1 is further downfield than β-H1, which is reverse to the ^{13}C1 signals [24] (Table 6.2). Both ^{1}H- and ^{13}C-spectra of α- and β-anomers show additional differences not to be detailed here. When samples of pure anomeric forms of α- and β-2,3,4-tri-*O*-acyl-glucopyranose in deutero-chloroform were repeatedly measured for several days, identical spectra were obtained [15]. This demonstrates the stability of the pure anomeric forms in deutero-chloroform, whereas in water or acetonitrile/water mixtures they equilibrate to an α/β-mixture due to mutarotation [16].

The exact localisation of acetyl groups in the presence of other acyl substituents is possible by peracetylation with deuterated acetic anhydride followed by ^{1}H-NMR analysis of acetate signals. In deuterated pyridine/benzene (1:1), all acetyl signals of sucrose octaacetate are resolved and assigned [34] (Table 6.2). A regiospecific assignment of individual acetyl groups to specific positions of the sugar moieties is also possible by 2D-NMR experiments on long range ^{1}H-^{13}C heteronuclear interactions [23,24,26]. This approach has been extended to other short-chain acyl residues [24] and represents a major advance in structural work. Apart from complete assignments of all glucose and sucrose ^{13}C-signals, many data on NMR (^{13}C, ^{1}H) and mass spectra of short-chain ester groups are documented in the original references [16,25], and just the downfield signals of the atoms forming the double bond of 2-methylbutenoic acid [15] may be mentioned (139.5 and 122.7 ppm for ^{13}C; 6.9 ppm for ^{1}H).

An overall attribution of acyl group mixtures to the glucose and fructose moieties of sucrose esters is possible by mass spectrometry because of fragmentation of the glycosidic bond and charge retention on either fragment. It has also been accomplished by release of acylated glucose and fructose residues from intact sucrose esters by mild acid hydrolysis (2.5 mM hydrochloric acid in hexane/methanol/water) for 48 h at room temperature [31]. The hydrolysis of glycosidic bonds normally requires slightly harsher conditions, but the linkage in sucrose is particularly labile due to the involvement of both anomeric centres. On the other hand, acyl groups may easily change their original positions under such conditions. Induced acyl migration (15 min at 0° in methanol with 5% ammonia) has been used to prepare 2,3,6-tri-*O*-acyl-glucose from the 2,3,4-tri-*O*-acyl isomer [22]. Needless to say, such acyl migrations have to be avoided in the course of the isolation of such compounds.

6. Analysis of Short-Chain Fatty Acids

Qualitative and quantitative analysis of short- and medium-chain fatty acids present in sugar esters was accomplished mainly by GLC/MS. Fatty acids were separated either in free form or as methyl, ethyl, propyl or butyl esters by temperature-programming between 50 and 180°C. During preparation and handling of esters a selective loss of more volatile compounds has to be avoided. Methyl and ethyl esters are released by short treatment (10 min) at room temperature with sodium methoxide or ethoxide [22] in dry methanol or ethanol, respectively, followed by deionization with Amberlite (H^+). Propyl esters [16], which are claimed to be better for critical separations of isomeric branched-chain acids, were prepared by heating in *n*-propanol/benzene/sulphuric acid (19:1:0.5). Butyl esters were obtained from free acids by heating in *n*-butanol with a few drops of concentrated sulphuric acid for 1 h at 110°C and subsequent extraction into hexane [31].

For labelling studies, the free acids were resolved by isocratic HPLC on an Aminex-column in 5 mM sulphuric acid containing 2% acetonitrile with UV detection at 210 nm [10]. Addition of carriers aided in collecting the sample peaks to be used for subsequent scintillation counting. Acetic acid emerges after 11.2 min, and hexanoic acid after about 37 min.

C. HETEROCYST GLYCOSIDES

1. Structure

Certain groups of cyanobacteria reduce molecular nitrogen in specialized cells, so-called heterocysts. As nitrogenase, the key enzyme of this reaction, is sensitive towards oxygen, a prime function of heterocyst cell envelopes is the minimization of oxygen access into the cell interior [35]. It is believed that heterocyst glycolipids contribute to this oxygen shield by forming a thick and non-fluid barrier in which oxygen mobility and penetration are reduced.

After their discovery in 1968 [36], research on heterocyst glycosides has been concentrated on structural elucidation and labelling studies, whereas a closer characterization of enzymes such as glycosyltransferases or hydroxylases has not been carried out yet. So far, two groups of glycosides have been characterized [37-41], which differ with regard to their aglyca. In one group (probably the predominant one) the sugar is linked glycosidically to the primary hydroxyl group of linear polyhydroxyalkanes or polyhydroxyketones, whereas in the other group polyhydroxy-alkanoic acids are esterified to the anomeric hydroxyl group of the sugar residues (this structure most likely has to be abandoned, see below). In both groups, glucose and galactose are linked in α-pyranose form, although glucose is predominant (80-90%). It is not known whether galactose is formed by epimerization of lipid-linked glucose as characteristic for other cyanobacterial galactolipids.

A major task in structural work on these glycosides is the assignment of the stereochemistry at asymmetric carbon atoms, which carry the secondary hydroxyl groups. As a result of such studies [39-41], the following seven aglyca were elucidated in all details (optical rotations in chloroform/methanol 2:1; compounds marked by an asterisk as 1-methoxy derivatives [39-41]):

1,3*R*, 25*R*-hexacosanetriol	$[\alpha]_D$ + 7.7
1,3*S*, 25*R*-hexacosanetriol	$[\alpha]_D$ + 12.0
1,3*R*, 27*R*-octacosanetriol	$[\alpha]_D$ + 8.1
3-keto-1,25*R*-hexacosanediol	$[\alpha]_D$ - 4.8*
3-keto-1,25*S*,27*R*-octacosanetriol	$[\alpha]_D$ - 21.1*
1,3*R*,25*S*,27*R*-octacosanetetrol	$[\alpha]_D$ -166.9
27-keto-1,3*R*-octacosanediol	

The glycosides resulting from attachment of a sugar residue at C1 are shown in Figure 6.2. The analyses show that the ω-1 carbon atoms C25 and C27 have always the *R*-configuration, whereas C3 in hexacosanetriol occurs as an epimeric *R*/*S*-mixture. In addition, oxidation to keto groups can replace a secondary hydroxyl group at C3 (in the C_{26} and the C_{28} compounds) or at C27 (in the C_{28} compound) resulting in the corresponding ketodiols [39-41]. The secondary hydroxyl groups have been suggested to result from alkane hydroxylation. The occurrence of 3*R*/*S*-epimers in hexacosanetriol is remarkable in view of the usual enantioselectivity of alkane hydroxylases (or ketoreductases?). Apart from the C_{26} and C_{28} triol derivatives a 1,3,25,27-tetrahydroxyoctacosane has been released from a glycoside mixture [37] and its stereochemistry elucidated recently [41].

The glycoside esters were found to contain 25-hydroxyhexacosanoic and 25(or 26),27-dihydroxy-octacosanoic acid of unknown stereochemistry, to which glucose and minor proportions of galactose are esterified. These structures were deduced from mass spectrometrical studies [38], but it has been pointed out that the isomeric nature of ketohydroxy and acylhydroxy structures may necessitate a reinvestigation of the glycoside esters [39-41]. The significance of the alkali-lability of the glycosidic bond in these compounds, at first sight a strong argument in favour of a glycoside ester, may lose in weight in view of the proximity of a keto group, which may induce β-elimination under alkaline conditions. Interestingly, treatment of the "glycoside ester" with sodium borohydride (*i.e.* reduction of the β-keto to a β-hydroxy group, see below) prevented its subsequent hydrolysis by alkali [38], which does support the β-keto structure. In a recent reinvestigation the previously proposed structures (the glucose esters of hydroxy fatty acids) have been revised [41].

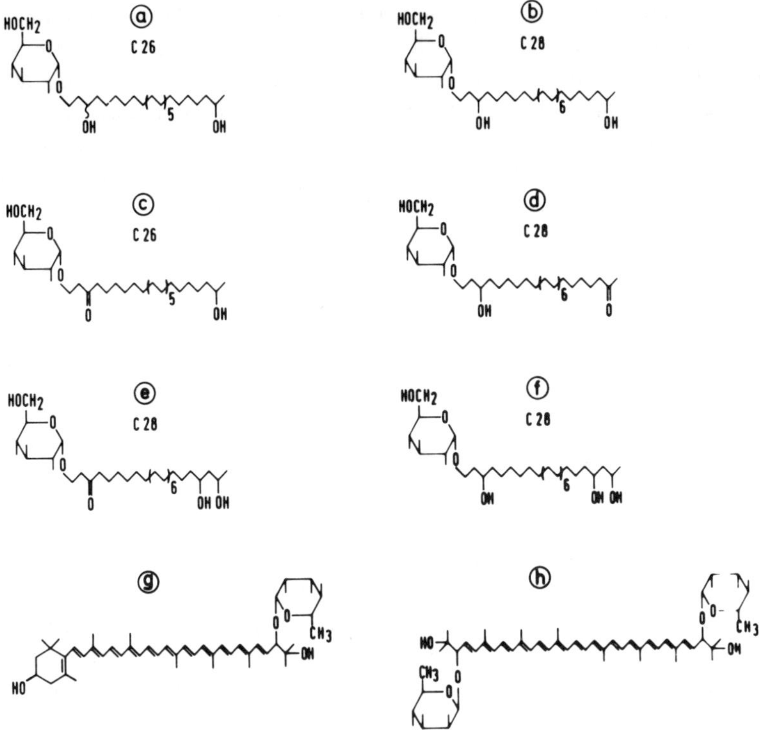

Fig. 6.2. Structures of heterocyst glycosides from cyanobacteria. Only α-D-glucopyranoside derivatives have been shown (a-f), but up to 18% of α-D-galactopyranosides may be present as well. The sugar-carrying C_{26} and C_{28} polyols and ketopolyols (c,d,e) are described in the text, including the stereochemistry at their asymmetric carbons. The β-keto-hydroxyalkane glycosides (c,e) are isomeric with the structures of glucose esters of hydroxy fatty acids originally proposed, that in fact may not exist and have not been included in the figure. The relative ratio of a:c:e:f was 100:30:3:8 in *Anabaena cylindrica* [41]. Several cyanobacteria produce glycosylated carotenoids such as the rhamnoside g = myxoxanthophyll and the dirhamnoside h = oscillaxanthin.

2. *Isolation and Separation*

Cyanobacterial cells, either suspended, sedimented or freeze-dried, are extracted several times with mixtures of chloroform and methanol or isopropanol (2:1 to 1:1). Yields of crude lipids range from 10-28% (dry weight basis) or about 30 times the amount of chlorophyll [42]. It has been recommended that solvent partitioning of crude lipid extracts should be avoided, since heterocyst glycosides may get lost with insoluble material at the interphase [43]. Instead, after redissolution in chloroform/methanol crude lipids may be purified from non-lipid contaminants by passage through columns of Sephadex LH20 or RP-18 cartridges. Subsequent fractionation procedures will depend on the quantities available.

For larger quantities of lipids, silicic acid column chromatography will result in prefractionation and enrichment. Regarding polarity and chromatographic behaviour, heterocyst glycosides range between mono- and digalactosyldiacylglycerols (MGD and DGD, see below), which are the predominant lipids in cyanobacteria (though not in heterocysts) and which have to be removed. MGD was eluted with chloroform/acetone (8:2) [43] followed by subsequent elution of heterocyst glycosides with increasing proportions of methanol in chloroform up to 10% [38], at which concentration DGD starts to elute. Alternatively, elution from DEAE-columns in chloroform/methanol (95:5) has been used [37], but the difficulties in removing galactolipids and separation of glycoside groups remain.

Further purification relies on separation by silica gel TLC, for which purpose three solvent systems have often been used: chloroform/methanol (8:2); chloroform/methanol/acetic acid/water (170:30:20:7.4); acetone/benzene/water (91:30:8). A demonstration of the chromatographic complexity of heterocyst glycolipids and the simplification of this pattern by saponification is shown in Figure 6.3 [44]. Because of the difference in the number of hydroxyl groups, the esters and ketones run ahead of the normal glycosides. On borate-impregnated plates glucosides travel ahead of galactosides [42], as also observed with other lipids having gluco/galacto-epimeric pairs. It should be mentioned that several cyanobacteria produce mono- and diglycosyl derivatives of carotenoids [45] (Figure 6.2g,h) such as for example myxoxanthophyll, which is a myxol rhamnoside [46]. The polarity of these orange-coloured compounds results in chromatographic behaviour very similar to heterocyst glycosides with R_f-values between the two galactolipids (Figure 6.3a). Therefore, their presence may pose additional problems during purification.

Recently, HPLC has been included in purification of heterocyst glycosides (Figure 6.3b,c) and found to achieve the highest resolutions [42,43]. Underivatized fractions (from 5 μg to 1 mg) separated by TLC were subjected separately to isocratic RP-C18 HPLC with methanol/water (91:9) and RI detection. Depending on the cyanobacterial species, each fraction was resolved into various peaks. In addition, the TLC fractions were subjected to perbenzoylation (pyridine/benzoyl chloride 10:2 for 90 min at 60°C) for subsequent isocratic separation by RP-C18 HPLC with methanol/dichloromethane (9:1) and UV detection at 230 nm. Again, separation of each fraction into several compounds was observed. Glucose/galactose-pairs were eluted as doublets, in which the faster moving component was the glucoside. A similar separation of two diastereomers, which differed only by one pair of epimeric hydroxyl groups in the hydrocarbon chain, was observed after acetylation of 1-(*O*-α-D-glucopyranosyl)-3*R*/3*S*-25*R*-hexacosanediol. Silica gel-HPLC with hexane/isopropanol (99:1) separated the two diastereomeric hexaacetates, from which the 3*R*-isomer eluted earlier [39]. The original compounds could be recovered by deacetylation, which was achieved by stirring the hexaacetates with sodium carbonate for 16 h in methanol.

3. *Quantification*

Quantification of heterocyst glycosides has only occasionally been attempted

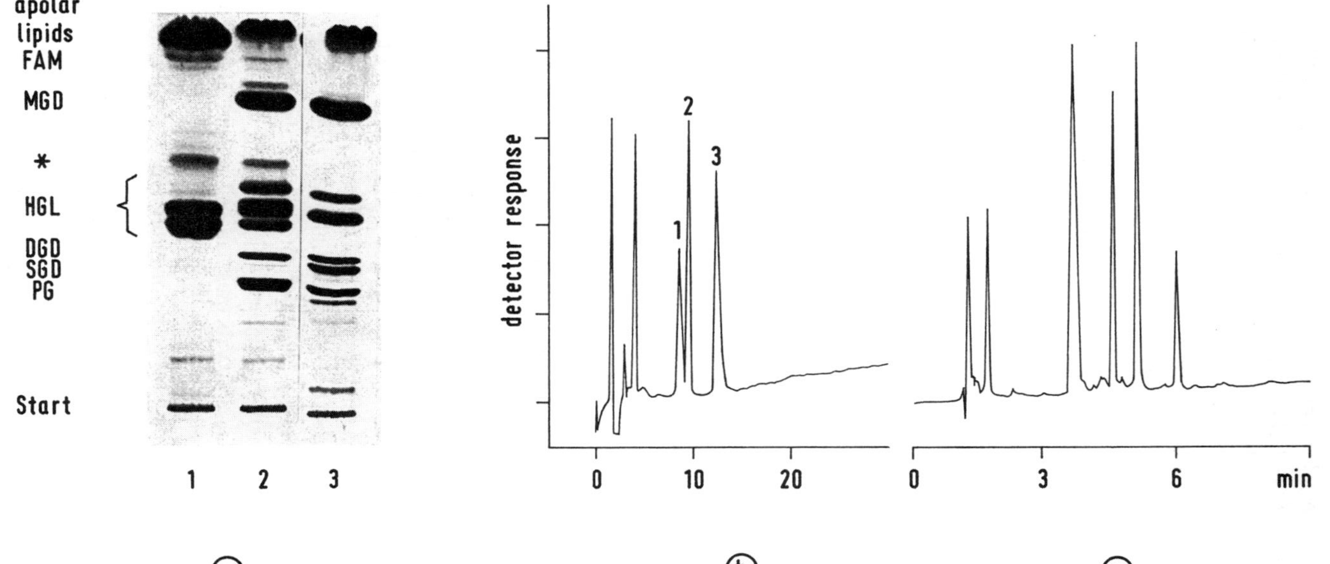

Fig. 6.3. Resolution of cyanobacterial heterocyst glycolipids by TLC (a) and HPLC (b,c)· $^{14}CO_2$-labelled lipids from *Tolypothrix* (a2) and *Nostoc* (a3) were separated in acetone/benzene/water (91:30:8) or treated with sodium methoxide before chromatography (a1 from a2) and subjected to radioautography. From the three heterocyst glycosides HGL in a2 only two are alkali-stable. The compound marked by an asterisk represents a carotenoid glycoside. MGD, DGD, SQD = monogalactosyl-, digalactosyl- and sulfoquinovosyl diacylglycerol, respectively, PG = phosphatidylglycerol, FAM = fatty acid methyl esters resulting from transmethylation (redrawn from [44]). b: heterocyst glycosides isolated by preparative TLC from *Anabaena* and rechromatographed in underivatized form by reversed-phase HPLC isocratically with methanol/water (91:9) and UV detection at 205 nm; c: normal-phase HPLC of heterocyst glycosides in perbenzoylated form with gradient elution (0.5 to 5 % 2-propanol in hexane) and UV detection at 230 nm. The complete identity of the separated peaks in b and c (including 1-3 in b) has yet to be determined (redrawn from [42,43]).

by using colorimetry. Compounds were separated by TLC followed by direct colour development with silica gel scrapings in anthrone/sulphuric acid [37]. The calibration covered a range of 50-200 μg and shows the relative insensitivity of this method [47]. The recently developed HPLC separation of benzoylated derivatives is far more sensitive [42,43]. It can be carried out with 60 μg of total lipid extract corresponding to about 2 μg of chlorophyll. The lower detection limit at 230 nm of individual glycosides is in the range of 20 pmoles = 10 ng. A drawback is at present the difference in the extinction coefficients because of the presence of different numbers of benzoyl groups in the different compounds, but this can be eliminated by calibration. The use of a non-overlapping internal standard may be useful. In this context, the procedure worked out to synthesise 1-*O*-α-D-glucopyranosyl-3*R*,25*R*-hexacosanediol may receive additional relevance [48].

4. Structural Analysis

Identity of constituent sugars and aglyca, anomeric configuration and glycoside attachment site to the polyhydroxy alkane, location of hydroxyl groups in these aglyca and assignment of their stereochemistry were elucidated by spectroscopic studies of intact glycosides either in free or derivatized form as well as of hydrolysis products, and in particular with the polyhydroxy alkanes in various forms of derivatization.

Intact glycosides were converted by standard procedures into acetyl, trimethylsilyl or methyl derivatives to be analysed by NMR or direct-inlet MS. For GLC/MS, TMS ethers were used that eluted at about 300°C from SE-30 columns. The 3′-keto glycosides can be converted to the corresponding alcohols by reduction with sodium borohydride in methanol (15 min at room temperature) [38,39]. After acetylation, separation by HPLC (see above) shows the presence of two 3′-epimeric diastereomers [39].

Acid hydrolysis was carried out by heating to reflux for 16-28 h in acidified (2N hydrochloric or sulphuric acid) methanol or methanol/water (9:1). The aglyca were extracted into chloroform and purified by silica gel TLC in chloroform/methanol (9:1) or diethyl ether/hexane (7:4). The recovered aglyca were converted to acetyl or TMS derivatives for direct-inlet MS or for GLC/MS on SE-30 (elution temperature 250-280°C). These derivatives have also been used for radio-GLC in the course of biochemical studies [49]. GLC/MS was used for the localisation of methoxy groups in methoxyalkanes released by acid hydrolysis from permethylated glycosides. α-Fragmentation and charge retention in either fragment is used to determine the position of the original hydroxy group. The sugar derivative obtained simultaneously was analysed after reduction/acetylation for confirmation of the pyranose ring size (see below).

Regarding the structural ambiguity (glycoside ester versus β-keto glycoside) discussed above, it is interesting to note that acid methanolysis of the 3-keto glucoside (Figure 6.2c) does not yield 1,25-dihydroxy-3-keto-hexacosane, but 25-hydroxy-3-keto-1-methoxy-hexacosane [39-41]. This β-keto methyl ether is isobaric with a methyl ester. In this context, the conditions for alkaline saponification of "glycoside esters" should be mentioned, that do not differ from normal procedures, but likewise may split β-keto glycosides. Hydrolysis was accomplished by 0.5N potassium hydroxide in methanol (30 min at 0°C) [38], by methanol/concentrated ammonia (5:1, 1 h at 100°C) or simply by sodium methoxide in methanol [44]. The formation of 1-methoxy-3-keto derivatives has been explained by β-elimination followed by acid- or base-catalysed addition of methanol to the 1-en-3-keto intermediate [41].

The stereochemical assignment of asymmetric centres in the released triols or keto diols has been accomplished by ^{1}H-NMR and circular dichroism (CD) spectroscopy of suitable derivatives [39-41] representing an alternative to measurement of optical rotation. For ^{1}H-NMR spectroscopy triols were esterified separately with the pure enantiomers of a chiral acyl chloride, in this case *S*(+)- or *R*(-)-α-methoxy-α-trifluoromethylphenylacetyl chloride, MTPA-Cl. The secondary hydroxyl groups at C3 and ω-1 gave diastereomeric pairs of esters (so-called Mosher esters). The NMR-signals of their neighbouring protons, *i.e.* of the terminal methyl and of the C2-methylene group, have slightly different chemical shifts in the different diastereomers. The sign of this difference (positive or negative for $\Delta\delta = \delta_{R\text{-MTPA}} - \delta_{S\text{-MTPA}}$) can be correlated to the configuration of the asymmetric C3 and ω-1 carbon by reference to compounds of known stereochemistry (see below). This method has been applied to tris-(*R*)- or tris-(*S*)-MTPA esters of triols or to the MTPA-ester of the ω-1 hydroxyl group alone after blocking the 1,3-diol system as an isopropylidene group (by conversion to an acetonide with 2,2-dimethoxy propane). MTPA esters are formed by heating the alcohol in pyridine with an excess of *S*(+)- or *R*(-)-MTPA-Cl in the presence of 4-dimethylaminopyridine (24 h at 50°C). The shift differences between diastereomeric pairs are admittedly rather small, 0.05 ppm for methylene and 0.08 ppm for methyl protons, but apparently reproducible. Diastereomeric reference pairs of known configuration were prepared from enantiomerically pure *R*- and *S*-1,3-butanediols, that displayed the following chemical shifts and shift differences for the relevant protons:

	^{1}H-NMR chemical shifts (ppm)					
	(3*R*)-1,3-butanediol			(3*S*)-1,3-butanediol		
protons	MTPA-ester		$\Delta\delta =$	MTPA-ester		$\Delta\delta =$
at	*R*	*S*	$\delta_R - \delta_S$	*R*	*S*	$\delta_R - \delta_S$
C2	2.01	1.97	+0.04	1.97	2.00	-0.03
C4	1.27	1.35	-0.08	1.35	1.26	+0.09

It is obvious that this method, because of the resolution of NMR spectra, allows immediate stereochemical assignment of individual asymmetric centres. MTPA-esters have also been used for configuration analysis of 2-hydroxy fatty acids from cerebrosides by HPLC separation of the *SS*- and *SR*-diastereomeric pairs (see below). Because of the wider spreading of signals, ^{13}C- or ^{19}F-NMR spectra may be of additional use for solving these problems [50]. In fact, the sensitivity of ^{13}C-shifts to stereochemical correlations has been used to assign the relative stereochemistry of the C25,C27-diol system in octacosane-1,3,25,27-tetrol [41]. By reacting with [1,3-$^{13}C_2$]-2,2-dimethoxypropane two ^{13}C-enriched acetonide groups were generated and analysed by ^{13}C-NMR. The signals for methyl groups in acetonides show up at two different shift values (about 20 and 30 ppm), when formed from 1,3-*syn*-diols, but at nearly identical position (25 ppm), when formed from 1,3-*anti*-diols. The *bis*-acetonide from the natural tetrol showed two pairs of ^{13}C-methyl signals at 19.3/19.8 and 30.0/30.4 ppm and, therefore, the C25/C27-diol has the relative *syn*-stereochemistry. In the context of ^{13}C-shift sensitivity it should be mentioned that the ^{13}C-spectra of the C3-epimeric compounds (Figure 6.2) in acetylated form revealed slightly different shifts for the α-anomeric glucose carbons as well as for C1-C3 of the long-chain triol (see Table 6.3 [39]).

A different approach towards the elucidation of the stereochemistry of a 1,3-diol is based on the circular dichroism exciton chirality method [39]. It depends on the interaction of neighbouring chromophores such as *p*-bromobenzoyl groups esterifying 1,3-diols. Depending on the configuration of C3, Cotton effects of opposite signs are observed that allow absolute assignments with reference to compounds of known configuration. Triols were converted to tris-*p*-bromobenzoates in the same way as described above for MTPA esters with *p*-bromobenzoyl chloride.

A large body of NMR- and MS-data is evidence for the structures shown in Figure 6.2. Only a few data are shown in Table 6.3 to demonstrate the changes in ^{1}H- and ^{13}C-spectra as a consequence of substituting a proton in the aliphatic aglycon by oxygen to yield hydroxyl, glycosyl or keto groups. The deshielding effect of a carbonyl oxygen on the bonding carbon (about 209 ppm in ketones as compared to 170 ppm of ester carbonyls), as well as on the neighbouring carbons and protons is particularly evident and supports the structure of the two glycosides with keto groups in different positions. Among others, the presence of the ω-1 keto group results in a collapse of the high-field methyl doublet to a singlet and its down-field shift. The absence of a carboxylate ester ^{13}C-carbonyl signal at 170-175 ppm in the spectrum of the alkali-labile "glucoside ester" and the presence of a ^{13}C-keto signal at 208 ppm (compound e in Table 6.3) is firm evidence for the absence of an ester group in this compound. The presence of three free hydroxyl groups in the deglycosylated triols was evident in ^{1}H-spectra in pyridine, where OH-signals for C1, C3 and ω-1 were detected at δ 6.18, 5.98 and 5.86 ppm, respectively.

D. STERYL GLYCOSIDES

1. Possible Cellular Functions?

Similar to other eucaryotic organisms, plants produce a variety of sterols and sterol derivatives [51]. The variations concern the structure of the tetracyclic sterol

Table 6.3.

^{3}C- and ^{1}H NMR data of heterocyst glycosides, which are α-D-glucopyranosides of different long-chain alcohols with additional hydroxyl and keto groups. The long-chain alcohols n the glucosides a-f are those shown in Figure 6.2, a = 1,3,25*R*-hexacosanetriol; acet a*R* = 1,3*R*,25*R*- and acet a*S* = 1,3*S*,25*R*-octacosanetriol in acetylated form; d = 27-keto-1,3*R*-octa-osanediol; e = 3-keto-1,25*S*,27*R*-octacosanetriol; f = 1,3*R*,25*S*, 27*R*-octacosanetetrol; 1 = nonglucosylated 1,3*R*,25*R*-hexacosanetriol. The data of the acetylated epimers a*R*/a*S* are hown separately to demonstrate the sensitivity of some ^{13}C-shifts towards stereochemical details (see text). For non-equivalent protons in methylene groups the different shifts are iven. The spectra were recorded in deuterated pyridine (P) or chloroform (C), shifts are given in δ ppm downfield from tetramethylsilane as internal standard and coupling constants can be found in original publications.

	^{13}C-shifts										^{1}H-shifts								
	polyol							glucose			polyol						glucose		
arbon umber	1	a	acet a*R*	acet a*S*	f	e	d	a	acet a*R*	acet a*S*	1	a	acet a*R*	f	e	d	a	acet a*R*	acet a*S*
1	60.5	66.3	65.1	65.3	66.2	63.9	66.2	100.7	96.1	95.7	4.32	4.40;4.11	3.71;3.35	4.38;4.10	4.40;4.04	4.39;4.11	5.48	4.99	5.05
2	41.3	38.2	34.1	33.8	38.2	43.0	38.2	73.9	70.7	70.8	2.14	2.09		2.07	2.88	2.07	4.25	4.86	4.86
3	69.8	68.7	71.1	71.5	68.6	208.9	68.7	75.5	70.2	70.1	4.32	4.20	5.00	4.20	-	4.19	4.72	5.47	5.47
4	38.9	38.8	34.6	34.5	38.8	43.4	38.8	72.3	68.8	68.5					2.50		4.30	5.04	5.07
5	26.6	26.7	25.3	25.2	26.4	24.0	26.4	74.5	67.3	67.2							4.46	3.98	4.06
6							63.0	62.0	61.8						4.60;4.49	4.25;4.08	4.29;4.07		
23	26.5	26.6	25.4	25.4	26.2	26.2													
24	40.4	40.4	36.0	35.9	39.1	39.1	29.5												
25	67.2	67.4	71.2	71.8	71.6	71.6	24.2				4.11	4.11	4.90	4.22	4.22				
26	24.5	24.4	20.0	19.9	46.7	46.7	43.6				1.45	1.47	1.19	2.01;1.82	201;1.82	2.43			
27					67.8	67.8	208.3							4.44	4.44	-			
28					24.8	24.8	29.8							1.48	1.48	2.14			
olvent	P	P	C	C	P	P	P	P	C	C	P	P	C	P	P	P	P	C	C
ref	[39]	[39]	[39]	[39]	[41]	[41]	[40]	[39]	[39]	[39]	[39]	[39]	[39]	[41]	[41]	[40]	[39]	[39]	[39]

nucleus, the derivatisation of the 3-β-hydroxyl group of ring A, the side chains attached at C17 to the cyclopentane ring D as well as its involvement in additional ring structures resulting in steroidal saponins or (glyco)alkaloids. Compared to the large number of exactly known chemical structures, the knowledge of biological functions of steryl glycosides is less comprehensive. It is well established that free sterols, including plant sterols, interfere with packing and ordering of acyl groups in lipid bilayers and membranes [52]. Free sterols contribute to the fluidity of membranes, which in turn governs ion permeability as well as insertion and function of membrane proteins. It has been suggested that ER membranes as sites of protein insertion require a less ordered lipid bilayer and accordingly less cholesterol than membranes with barrier functions [53]. This was used to explain the low content of cholesterol in endoplasmic reticulum (ER) membranes and its gradient to higher concentrations going from ER (4%) via Golgi (11%) to plasma membranes (31%) as also observed in plants (values in brackets from barley roots [54]). Plasma membranes and tonoplasts from plants have been found repeatedly to contain 20-30 mole% of free sterols [55,56]. If they are concentrated in the cytoplasmic leaflet of the bilayer as observed in erythrocytes [57], then sterols may well account for about 50 mole% of the lipid constituents in this layer.

The unexpected interference of mutations in sterol biosynthesis with different metabolic functions in yeast [58,59] supports the above mentioned role of sterols in membrane function and biogenesis. But in addition to such more general functions as bulk constituents, so called sparking functions have been detected by studies on sterol auxotrophs. Growth of yeast cells requires sterols of specific structure, such as for example ergosterol, and this requirement can be satisfied by minute proportions added to a bulk sterol such as cholestanol [60].

In plants subcellular membranes contain additional sterol derivatives [51], sometimes in quantities exceeding free sterols [61]. These comprise steryl esters [62], steryl glycosides and acylated steryl glycosides [51]. Steryl glycosides and their acylated form were also found in animals [63,64], and because of their presence in human food of plant origin their metabolism in rats has been studied [65]. Regarding bulk and sparking functions, differences in the various forms of sterol derivatives may be relevant, but little is known in plants. Sterols, steryl glycosides and acyl steryl glycosides had similar effects on the phase transition of bilayers of dipalmitoyl lecithin [66,67], whereas bilayers of unsaturated soybean lecithin responded differently; sitosterol reduced water permeability and increased acyl chain ordering, whereas stigmasterol was far less effective [68]. Therefore, neglecting differences in composition of sterols, sugars or acyl groups in the various steryl derivatives as practised below may represent an unjustified simplification.

It has been shown that the proportions of various sterol derivatives may be adjusted to specific demands depending on the type of subcellular membrane and in response to endogenous or environmental signals [61]. For example, acclimation of rye to low temperature is paralleled by an increase of free sterols at the expense of glycosylated forms in plasma membranes [55]. Several of the enzymes involved in sterol biosynthesis and interconversion have been studied and may become relevant for genetic engineering of sterol derivatives in membranes with respect to stress resistance of plants. The total quantity of sterols in plants seems to be controlled by 3-hydroxy-3-methylglutaryl-CoA reductase, which has been cloned several times [69]. The increased activity of this enzyme in a tobacco mutant [70] resulted in an increased synthesis of sterols, which were deposited in the form of steryl esters in discrete lipid droplets, whereas the proportion of free sterols in membranes was kept constant (sterol glycosides were not investigated). The formation and hydrolysis of steryl esters require two separate enzymes, of which a sterol esterase from yeast has been cloned [71]. Similarly, formation and hydrolysis of steryl glycosides require two different enzymes, which have been extensively studied in plants [72-74]. In particular, the UDP-glucose:sterol β-D-glucopyranosyl transferase has recently been purified to homogeneity and is being sequenced [75]. Further details on the biosynthesis and metabolism of plant sterols and their derivatives may be found in recent reviews [51,76,77].

2. *Chemical Structures*

Phytosterols are found as aglyca in glycosides of varying degree of complexity carrying up to five glycosyl residues (SG-SG_5) in linear sequences [78]. On the other hand, in saponins or other steroidal glycosides with modified sterol nuclei, branched oligosaccharide chains composed of various sugars and attached at different positions to the aglycon occur. In the normal Δ5- or Δ7-sterol glycosides the glycosidic bond involves the single equatorial hydroxyl group at C3 of ring A, which is on the same side (β) as the axial methyl group at C10. The predominating steryl glycoside is a β-D-glucopyranoside, but many additional steryl glycosides with different glycosyl residues in α- or β-linkage have been isolated (see Figure 6.4) including β-D-galactopyranosyl [79], β-D-glucuronopyranosyl [82], α-L-rhamnopyranosyl [80], α-D-riburonofuranosyl [83], β-D-xylopyranosyl [81] and α-D-xyluronopyranosyl [84] residues. Also a mannoside has been detected but not yet analysed with respect to its anomeric structure [85]. Recently, a 3-*O*-α-L-arabinopyranosyl derivative of pregn-5-en-3β,20-diol has been isolated from a tropical plant [86]. In this glycoside, the sterol nucleus carries a shortened 1-hydroxy ethyl side chain (not shown in Figure 6.4).

Acylated derivatives of steryl glucosides (ASG), first isolated from potato tuber lipids [87], usually carry a long-chain fatty acid at C6 of glucose. The normal C_{16} and C_{18} fatty acids known from membrane lipids are esterified in this position. Palmitic acid predominates, but oleic, linoleic and linolenic esters have been isolated also [88]. Occasionally, other acyl derivatives have been identified which differ from the predominating 6-*O*-acyl-glucosyl derivative in the position and nature of acyl groups as well as in the glycosyl residue itself [89,90]: 2-*O*- and 4-*O*-stearoyl-β-D-xylopyranoside, 2-*O*-tetracosanoyl-β-D-xylopyranoside and 2,4-di-*O*-acetyl-6-*O*-stearoyl-β-D-glucopyranoside (Figure 6.4). The sterol profiles present in the various forms of derivatives may vary to some extent, and in this

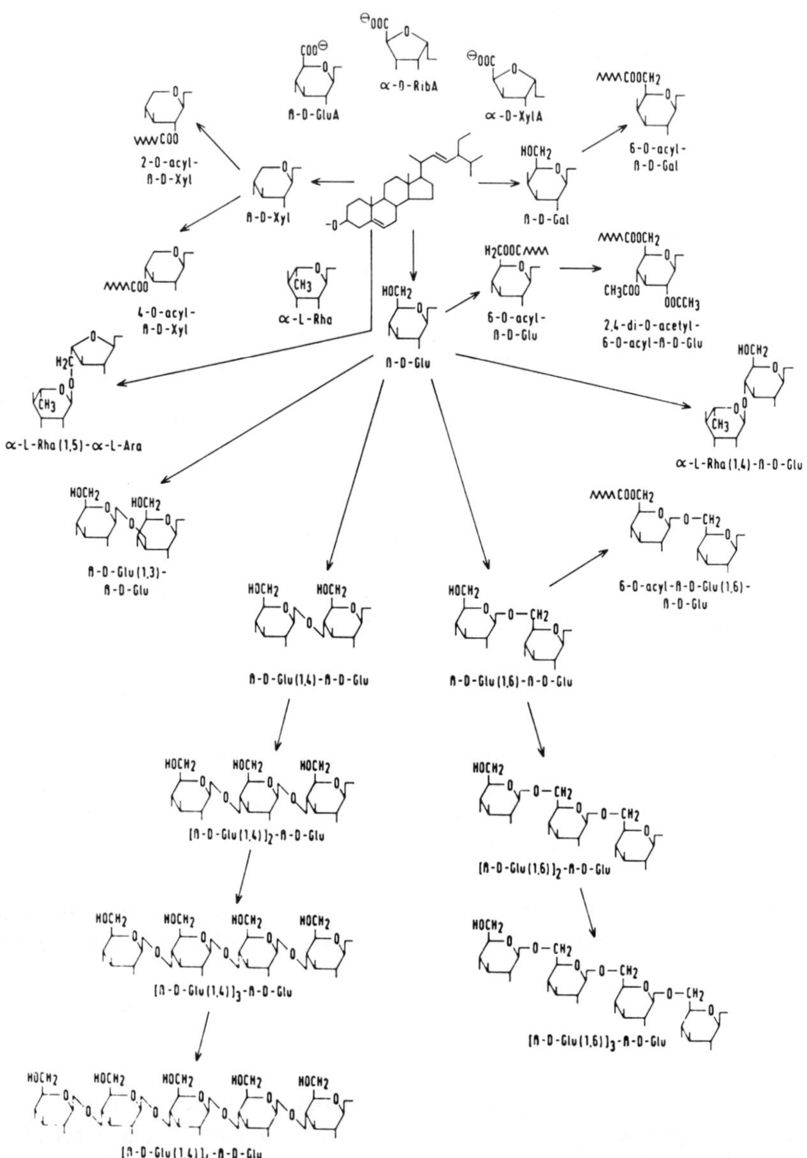

Fig. 6.4. Structures of steryl glycosides. The upper area shows steryl monoglycosides in free or acylated form. Possible biosynthetic correlations are indicated by arrows. The sugar residues are named in abbreviated form, A stands for uronic acid; of the sterols only stigmasterol is shown. The compounds in the next row further down are disaccharide derivatives, from which only the cellobiosides (β-1,4) and gentiobiosides (β-1,6) are acceptors for further elongation to the corresponding tri-, tetra- and even pentasaccharide side chains as shown. A unique inositol derivative of steryl glucoside is shown in Figure 6.7a. The great number of different, but similar structures may indicate the existence of a large family of sugar nucleotide:sterol glycosyltransferases. From these, only the UDP-glucose:sterol glucosyltransferase has been isolated and is being cloned [75].

context, the above mentioned differentiation between bulk and sparking sterols should be recalled. It is not known whether SG, irrespective of their actual constituents (sugar, sterol, acyl group), have sparking function in a bulk phospholipid bilayer, or whether the predominating steryl β-D-glucopyranoside (bulk) has to be supplemented by sparking SG with more exotic sterols, sugars or acyl groups. Whatever their functions are, they seem to be indispensable as judged from their presence in all organisms carrying out oxygenic photosynthesis. In non-green tissues such as roots, their proportions are comparable to phospholipids and they thus have to be considered as major membrane constituents. In cyanobacteria their presence has been described only once [91], whereas more and more eucaryotic algae [79] are being identified as containing SG.

Further glycosidation of steryl monoglycosides results in steryl biosides. Attachment of a second β-D-glucopyranosyl group to C6, C4 or C3 of the first β-D-glucopyranosyl residue results in steryl gentiobiosides (β1→6) [92,93], cellobiosides (β1→4) [78] or laminariobiosides (β1→3) [94], respectively. Furthermore, a terminal α-L-rhamnopyranosyl group may be attached either to C4 of the β-D-glucopyranosyl or the C5 of an internal α-L-arabinofuranosyl residue, respectively [95]. Compared to the widespread occurrence of acylated steryl monoglycosides, acylated steryl biosides have only rarely been detected [96], such as sitosterol[6″-*O*-palmitoyl-β-D-glucopyranosyl(1″→6′)-β-D-glucopyranoside]. A rather unique form of an acylated derivative of a steryl glycoside has been isolated from banana and has been crystallized as the acetate [96]. It contains an acylated inositol (Figure 6.7a) and has the structure of sitosterol [2″-*O*-palmitoyl-*myo*-inosityl(1″→6′)-β-D-glucopyranoside]. This compound occurs also in its non-acylated form. Addition of further β-D-glucopyranosyl residues to normal biosides results in either linear cello(β1→4)- [97] or gentio(β1→6)-triosides [93], whereas tetra- and pentaosides are only found in linear β1→6 linkage, *i.e.* belonging to the gentio-family [78,93].

As mentioned at the beginning, other terpenoid glycosides have been analysed in numbers exceeding those just described as SG. Regarding their structures in terms of amphiphilic character, some are similar to SG [98,99] and accordingly, they may be membrane constituents as well. But with a very few exceptions, the interest in these compounds has so far been satisfied with the elucidation of their chemical structure. It is an open question whether it is justified to consider one group as secondary products and the other as membrane constituents (of yet unknown function). In this context it should be pointed out that in oat leaves terpenoid glycosides such as avenacosides [100] occur in quantities comparable to the predominating membrane galactolipids of chloroplasts and exceeding ASG and SG by a factor of 10.

3. Isolation and Separation

For extraction of steryl derivatives various solvents and solvent mixtures have been used, but in view of the fact that steryl glycosides and their derivatives are extracted together with all the other lipids, a useful mixture regarding solubility and ease of subsequent evaporation is chloroform/methanol (2:1). Whether the plant material should be steam-killed before extraction as often recommended has to be decided in view of a possible acyl migration in acylated steryl glycosides.

Silicic acid column chromatography can be used for prefractionation when starting with larger quantities of total lipids. Regarding the polarity of eluting solvents to be used, it should be pointed out that ASG is more apolar than MGD, whereas DGD may be compared in chromatographic properties to steryl biosides. Elution with hexane/diethyl ether (1:1) will elute apolar lipids including steryl esters and free sterols [87], whereas ASG was eluted with hexane/diethyl ether (3:2) [88,89], dichloromethane/ethyl acetate (2:3) [88] or chloroform/acetone (92:8) [101]. The 2,4-di-*O*-acetyl-6-*O*-stearoyl-SG was eluted with hexane/diethyl ether (3:2) [89]. Increasing proportions of methanol in chloroform were used to recover ASG (2% methanol) [87,90], SG (5% methanol) [87], steryl biosides (6-8% methanol) [90,92], steryl triosides (35% methanol) and steryl tetraosides (above 40% methanol) [93]. SG can also be eluted with chloroform/acetone (6:4) [101]. It should be mentioned that MGD and DGD are obtained with 4% and 10% of methanol in chloroform, respectively. A more severe problem is the separation of SG from ceramide giycosides, which will be referred to below (see sphingoglycolipids).

Silica gel TLC is generally used for final purification. With the exception of ASG, the other SG can be freed from contaminating ester glycerolipids by alkaline saponification. On the other hand, this treatment will not affect cerebrosides, which travel between SG and DGD in most solvents (see Section on cerebrosides). Depending on the compound to be isolated, solvent systems of different polarity have been applied. Chloroform/methanol/water (65:25:4) [93,97] separates steryl esters (R_f 1.0), free sterols (R_f 0.95), ASG (R_f 0.84), SG (R_f 0.66), SG_2 (R_f 0.43), SG_3 (R_f 0.28) and SG_4 (R_f 0.17). The separation obtained by this solvent is shown in Figure 6.5c, which at the same time gives an impression of the proportions and resolution of steryl and ceramide glycosides with increasing numbers of sugars. Steryl esters (R_f 0.84) and free sterols (R_f 0.25) are more reliably separated in pure chloroform or dichloromethane, in which ASG and SG do not move from the start [102]. These two can be separated [102] in chloroform/methanol (9:1) (ASG with R_f 0.56 and SG with R_f 0.16) or in ethyl acetate/methanol (97.5:2.5), in which ASG travels with R_f 0.5 [103]. Free sterols may be further separated by double development in dichloromethane into 4,4-dimethyl, 4-α-methyl and 4-desmethyl sterols (R_f 0.4-0.25) [104]. Acetylation of free sterols and development in chloroform on silver nitrate-impregnated plates separates according to number of double bonds and results in a spreading between R_f 0.33 and 0.84 [105]. A similar separation of intact SG may be expected, but has never been tried. In the absence of other lipids the purified compounds will crystallize on evaporation of solvents and with the exception of ASG, they can be redissolved in only a few solvents such as pyridine or chloroform/methanol mixtures. Therefore, nearly all steryl glycosides and biosides have been obtained in crystalline form. Steryl

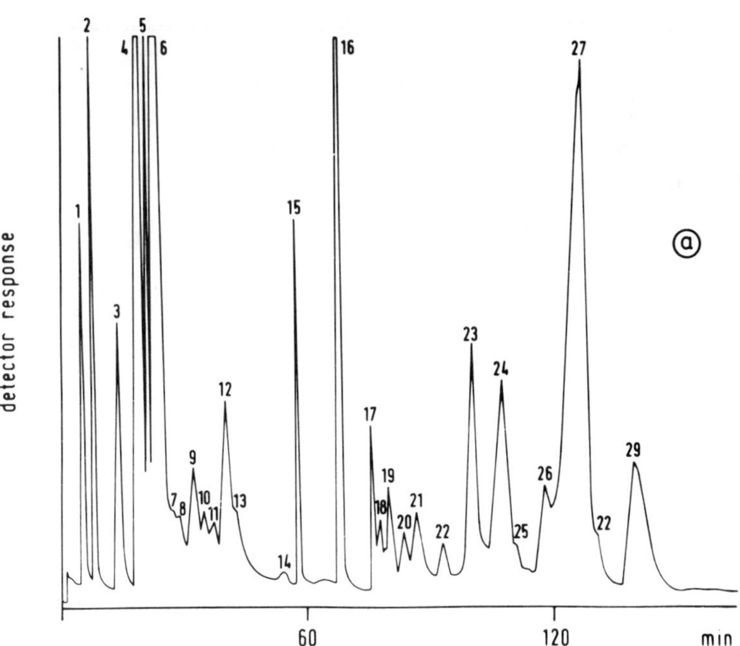

detector response
1
2
3
4
5
6
7
8
9
10
11
12
13
14
15
16
17
18
19
20
21
22
23
24
25
26
27
22
29
60
120
min
a

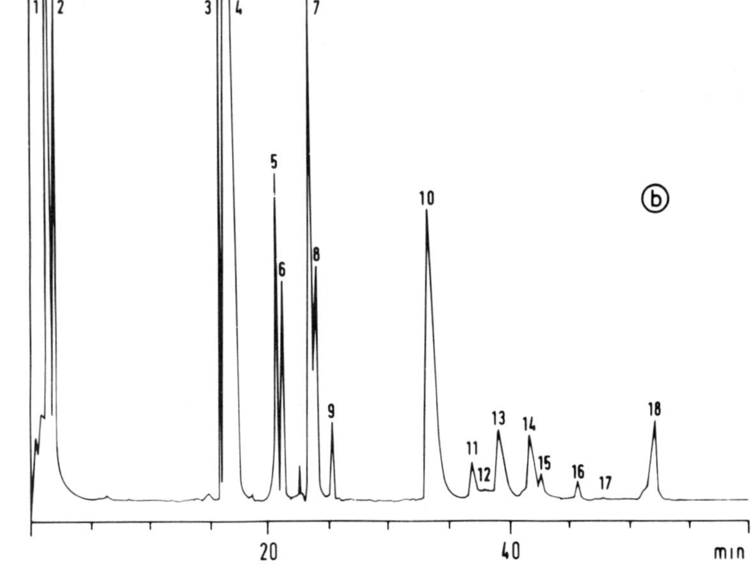

detector response
1
2
3
4
5
6
7
8
9
10
11
12
13
14
15
16
17
18
20
40
min
b

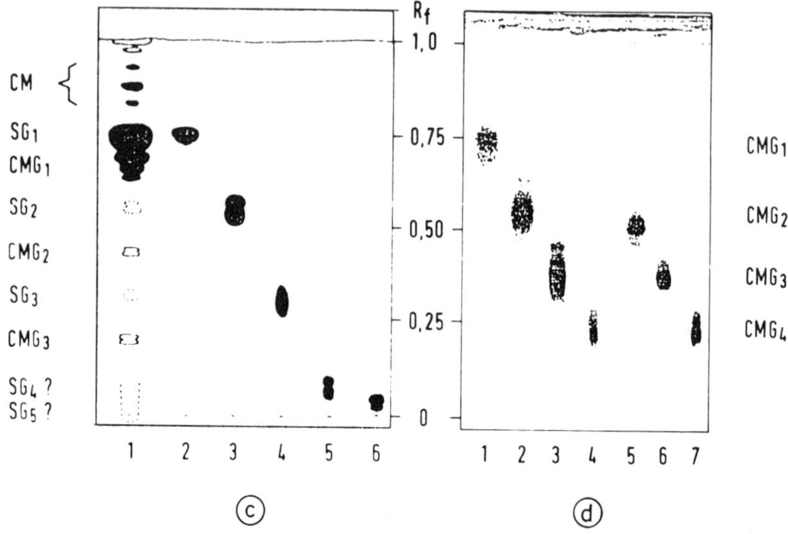

Fig. 6.5. Separation and isolation of sterol and ceramide glycosides from plant lipid mixtures by HPLC and TLC. a: normal-phase HPLC of a lipid extract from developing soybeans. Compounds were eluted with two gradients (diethyl ether in pentane up to peak 14 followed by methanol/8% ammonia in chloroform) and detected by a moving wire/FID combination. The components of interest in the present context are 15 = acylated steryl glycoside, 16 = monogalactosyl diacylglycerol, 17 = steryl glycoside, 19 = cerebroside, 22 = digalactosyl diacylglycerol; 21,23,24,26 = phosphatidyl glycerol, -ethanolamine, -inositol and -choline, respectively; 27 = phosphatidic acid, 29 = lyso-phosphatidyl choline (redrawn from [106]).
b: normal-phase HPLC of a lipid extract from wheat flour. Compounds were eluted with a ternary gradient of hexane/tetrahydrofuran (99:1), 2-propanol and water and detected with an evaporative light-scattering detector. The components of interest in the present context are 6 = acylated steryl glycoside, 7 = monogalactosyl diacylglycerol, 8 = monogalactosyl monoacylglycerol, 9 = steryl glycoside, 10 = digalactosyl diacylglycerol; 13,14,16 = phosphatidyl ethanolamine, phosphatidyl glycerol and phosphatidyl choline, respectively; 18 = lyso-phosphatidyl choline (redrawn from [107]).
c,d: separation of sterol and ceramide glycosides by TLC in chloroform/methanol/water (65:25:4). The resolution of the alkali-stable fraction from rice bran lipids (lane 1 in c) demonstrates the original proportions of free ceramides (CM), sterol and ceramide monoglycosides (SG_1, CMG_1) and the higher glycosylated forms with two, three and even four sugar residues, from which the steryl glycosides run ahead of the corresponding ceramide glycosides always. Lanes 2-6 represent steryl glycosides after isolation as used for structural identification. Compounds were detected by charring after spraying with 50 % H_2SO_4.
d: separation of individual ceramide glycosides from rice bran (lanes 1-4) and wheat grain (5-7). The spreading of spots is ascribed to structural differences in the ceramide parts. Compounds were detected with an anthrone/H_2SO_4 reagent (c and d were redrawn from [150,180]).

glycuronosides can be converted to methyl esters by treatment with diazomethane. The conversion of a carboxylate into a methyl ester group results in increased TLC mobility and helps in separation of otherwise overlapping spots. In dichloromethane/ethyl acetate (4:1) sitosterol xylopyranoside (R_f 0.07), sitosterol xyluronofuranoside methyl ester (R_f 0.32) and sitosterol riburonofuranoside methyl ester (R_f 0.50) are separated [84]. Similarly, sitosterol glucuronopyranoside methyl ester (R_f 0.42) travels ahead of sitosterol glucopyranoside (R_f 0.23) in hexane/ethyl acetate (3:2) [82].

For spectroscopic or further chemical analysis acetylated or permethylated derivatives have been prepared that were subjected to a final purification with chloroform/benzene/acetone (8:2:2) or similar proportions depending on the number of glycosyl residues [93,97].

The superior resolution of HPLC is only gradually being applied for purification of SG and their derivatives. Isocratic HPLC on a preparative column of Lichrosorb-NH_2 with dichloromethane/methanol (41:4) was used to separate the positional isomers of sitosterol-3-*O*-(2′-*O*- and 4′-*O*-stearoyl)-β-D-xylopyranoside [89]. In one of the earliest applications of HPLC to plant lipids [106] a very good separation of many lipids into sharp peaks including ASG and SG was obtained (Figure 6.5a), but unfortunately this system using a flame-ionization detector (FID) has not been developed for routine application with modern instrumentation. A similar separation was obtained on a silica gel HPLC column with gradient elution using mixtures of hexane/tetrahydrofuran/isopropanol/water (Figure 6.5b). Wheat flour lipids were resolved into 18 remarkably sharp peaks including ASG and SG, which were detected by the evaporative light scattering detector [107].

RP-HPLC has been carried out with SG following their isolation from total lipid extracts by TLC. One method separates SG directly without derivatisation [108], whereas the other uses 1-anthroylnitrile derivatives [109], but both methods were confined to steryl β-D-glucopyranosides. Free SG were separated on a C6-RP column [108] by elution with a gradient of acetonitrile in water (50 to 100%) with recording at 200 nm which detects isolated double bonds. The elution sequence is the same as observed with free sterols. Unresolved pairs were Δ7-avenasterol/cholesterol and Δ7-stigmastenol/stigmasterol, which on the other hand are separated as free sterols by GLC [104]. Derivatisation with 1-anthroylnitrile [109] results in hydrophobic derivatives with four additional chromophore groups. Isocratic HPLC on a C18-RP column was carried out with a more apolar solvent of acetonitrile/dichloromethane (68:32) with detection at 254 nm. Also in this case, elution behaviour was exclusively dependent on the sterol component, and retention times decreased with shorter side chains, increased numbers of double bonds or shifting their position from Δ7 to Δ5, or from C24 to C22. Also this method was handicapped by some unresolved pairs such as 7-stigmasterol/sitosterol and campesterol/stigmasterol derivatives, which represent some of the most abundant plant sterols. Both methods were designed primarily for quantification (see below).

Separation of intact SG by GLC was possible in the form of TMS derivatives [110]. Isothermal chromatography at 265°C on a short packed column (60 cm) of SE-30 resolved the TLC-purified SG into five components. The two major ones were identified by GLC/MS (see below) as glucosides of stigmasterol and sitosterol and had retention times of 50 and 55 min, respectively. The use of modern capillary columns may result in improved resolution and sharper peaks.

4. Quantification

Quantitative measurements of all the different steryl derivatives in plants have been carried out only occasionally [61,102,105]. In view of the subcellular compartmentation and its variation with cell types the significance of an overall analysis from intact plants or plant organs may be questioned. Irrespective of the material extracted, quantification requires previous separation by TLC and appropriate quantities to be detected and recovered from the plates. After TLC separation, SG and ASG have been quantified by the anthrone method [56,111]. It is based on the sugar content of these compounds and requires 40-200 nmoles for calibration [56], which indicates its relative insensitivity. TLC spots are scraped off and mixed with a solution (1.5 ml) of anthrone in sulphuric acid for subsequent heating, centrifugation and optical measurement. At this point it may be mentioned that chemical or enzymatic methods [112] for quantification of free sterols are somewhat more sensitive. Calibration of the chemical method was carried out with 15-60 nmoles of sitosterol [56], which were scraped from TLC plates and mixed with the *o*-phthaldialdehyde reagent, but colour yield may vary with different sterols. ASG can also be determined via GLC or HPLC of its constituent fatty acids. Similarly, SG has been determined by GLC of free sterols released by acid hydrolysis after admixture of a suitable internal standard [108]. Compared to the anthrone procedure the sensitivity of this method is better by two orders of magnitude, but it has the severe drawback that several plant sterols such as the Δ5- and Δ7-avenasterols are completely destroyed by acid hydrolysis. This was demonstrated by comparing the profiles of free sterols released by acid or enzymatic hydrolysis with β-glucosidase [108].

The HPLC methods mentioned above circumvent the problems of hydrolysis. The sensitivity obtained with the evaporative light-scattering detector as set up in a study with wheat flour lipids [107] was slightly better than reached with direct colorimetry after TLC as evident from the calibration curves, which covered a range of 10-200 μg for each component. The detection of underivatized SG at 200 nm varies with the number of double bonds and requires calibration, since compounds with two double bonds (stigmasterol) show only 1.5 times the extinction of compounds with one double bond (cholesterol). But it is an advantage that free sterols instead of SG can be used for detector calibration in terms of peak area/nmol [108]. The detection limit was about 0.2 nmol (80 ng) with linearity checked up to 13 nmoles. This sensitivity is similar to GLC procedures and allows several parallel measurements with the lipid extract from 1g of leaf fresh weight.

The conversion of SG to anthroyl derivatives [109] followed by HPLC separation and detection at 254 nm eliminates the dependency on different numbers of double bonds and at the same time increases the sensitivity by an additional two orders of magnitude to a detection limit of 0.5 ng and a linear working range of 1-50 ng. On the other hand, reliable identification after TLC separation requires in the order of 40 nmoles (2 μg) of SG. For calibration a synthetic cholesterol glucoside [109] was used. At this point it may be mentioned that cholesterol [113] and sitosterol β-D-glucopyranosides have been synthesised several times including a ^{14}C-labelled form [65] and the 6-*O*-acyl derivative with a stearoyl, oleoyl or linolenoyl ester group [88,114]. In one case, even sitosteryl α-D-glucopyranoside was isolated from the anomeric glycosidation mixture [88]. In addition coupling of appropriately blocked lactose and maltose derivatives with sitosterol and cholesterol led to the corresponding biosides [115,116]. Such compounds may be useful standards for quantification. Quantitative data on biosides or compounds with even more sugar residues are not available. Their abundance seems to be rather low as judged from a yield of 4-10 mg of these compounds starting from 10 kg of beans [93].

5. Structural Analysis

Isolation and structural analysis of many of the compounds mentioned above were carried out at times when sophisticated 2D-NMR techniques were not yet established routinely. This early work was based on derivatisation and degradation, and methods such as methylation/hydrolysis, acetylation/chromic oxide oxidation, sugar analysis by chromatographic techniques and GLC/MS predominated. These methods are still in use, but they play a more confirmatory and supporting role in a field in which most structural details are determined by ^{1}H- and ^{13}C-NMR spectroscopy of underivatized or easily prepared derivatives of intact molecules. A few prominent features of such spectra will be outlined, but for further information on NMR spectroscopy in structural elucidation of oligosaccharides and glycosides the original papers and a recent review should be consulted [117].

In ^{1}H-NMR spectra of underivatized SG [79-84,88-90,103] several signals for sterol and sugar protons are easily recognized (Table 6.4). Methyl signals from the sterol part show up as singlets (tertiary), doublets (secondary) or triplets (primary), depending on their angular or side-chain position. Their signals are found in the range of 0.6-1.0 ppm and have been fully assigned in most cases. In ASG [88] even the triplet of the terminal methyl group from the acyl residue was identified in addition to the six methyl groups of sitosterol. Methyl groups linked to a double bond in the side chain such as in stigmasta-5,25(27)-diene-3β-ol are shifted to lower field at about 1.6 ppm. Olefinic protons of the B ring (H5 or H7) and the side chain (H22, H23) show up at lower field (between 5.0-5.3 ppm). In ASG they may overlap with signals originating from unsaturated acyl groups [88], whereas the two olefinic protons of an exo (terminal) methylene group are found at 4.5-4.7 ppm [79,103]. H3 is regularly seen at 3.5-3.9 ppm, which is only slightly further downfield compared to non-glycosylated sterols [118].

The sugar protons give rise to signals between 3-4 ppm, and only the anomeric proton H1′ is further downfield and readily recognized as a doublet with a characteristic splitting diagnostic for an α- or β-linkage. In ASG [88-90,103] the carbinol protons from acyl-carrying positions are shifted downfield: the two non-equivalent H6′ protons in 6-*O*-acyl-glucopyranosides to 4.1-4.5 ppm (non-acy-

Table 6.4.

^{1}H-and ^{13}C-NMR data of steryl glucosides and of some acylated derivatives. a = free spinasterol (Δ7, Δ22); b = free stigmasterol (Δ5, Δ22); c = stigmasterol 3-*O*-β-D-glucopyranoside; d = sterol 3-O-6′-*O*-acyl-β-D-glucopyranoside; e = sitosterol 3-O-β-D-glucuronopyranoside methyl ester, not including the signals for the methyl ester group of the glucuronosyl residue at 56.3 ppm (^{13}C) and 3.68 ppm (^{1}H); f = sitosterol 3-*O*-2′,4′-di-*O*-acetyl-6′-*O*-stearoyl-β-D-glucopyranoside; g = peracetylated sitosterol 3-*O*-β-D-glucopyranoside.Only sterol and glucose signals are given, coupling constants can be found in the original publications. The spectra were recorded in deuterated pyridine (P) or chloroform (C), shifts are given in δ ppm downfield from tetramethylsilane as internal standard.

	^{13}C-shifts									^{1}H-shifts					
	sterol			glucose						sterol			glucose		
carbon atom	a	b	c	c	d	d	e	f	g	a	b	c	g	f	g
1	37.1	37.9	37.5	102.6	102.9	101.3	101.6	99.4	99.7				5.05	4.53	4.59
2	31.4	32.3	30.3	75.3	75.2	73.9	74.2	74.4	71.5				4.05	4.80	4.96
3	71.0	71.3	78.1	78.6	78.6	76.3	74.0	79.9	72.9	3.59	3.52	3.97	4.27	3.70	5.21
4	38.0	43.5	39.3	71.7	71.7	70.5	70.1	71.3	68.5			2.63; 2.38	4.30	4.91	5.08
5	40.2	141.9	140.9	78.4	78.4	76.3	76.5	71.7	71.7				3.97	3.61	3.69
6	29.6	121.2	121.9	62.8	64.7	63.4	174.4	62.3	62.1		5.35	5.35	4.56; 4.41	4.13; 4.21	4.13; 4.26
7	117.4	32.6	32.2							5.15					
8	139.5	32.2	32.1												
9	49.4	50.5	50.4												

10	34.2	36.9													
11	21.5	21.4													
12	39.4	39.9													
13	43.3	42.4													
14	55.1	57.1													
15	23.0	24.6													
16	28.5	29.3													
17	55.8	56.1													
18	12.0	12.2								0.55	0.69				
19	13.0	19.6								0.80	1.01				
20	40.8	40.8													
21	21.4	21.5								1.02	1.02				
22	138.7	138.8								5.15	5.15	5.21			
23	129.4	129.5								5.02	5.01	5.11			
24	51.2	51.4													
25	31.9	32.2													
26	21.1	19.2								0.85	0.84				
27	19.0	21.3								0.80	0.79				
28	25.4	25.7													
29	12.3	12.5								0.80	0.80				
solvent	C	P	P	P	P	C	C	C	C	C	C	P	P	C	C
ref	[115]	[116]	[146]	[116]	[118]	[100]	[79]	[86]	[115]	[115]	[115]	[135]	[116]	[86]	[115]

lated at 3.5-3.9 ppm); H2′ and H4′ in acylated xylopyranosyl residues to 4.7 ppm (from 3.6) and 4.9 ppm (from 3.7), *i.e.* even further downfield than H1′ (4.6 ppm).

Peracetylation yields derivatives, which display acetate signals of diagnostic number and position [118]. Missing acetate signals in peracetylated ASG will in turn identify the position of the acyl residue on the sugar ring [64].

Glycuronosides [82-84] were characterized after esterification of the carboxyl group with diazomethane followed by acetylation. This treatment results in characteristic methyl signals at 3.45-3.7 ppm for carboxyl methyl esters and three (glucuronopyranosides) or two (xylurono- and riburonofuranosides) acetate signals spread around 2.0 ppm. Subsequent reduction with lithium aluminium hydride in tetrahydrofuran and acid hydrolysis liberated glucose, xylose or ribose, respectively.

In ^{13}C-NMR spectra a full assignment of all carbon atoms has been achieved and a few characteristic details will be summarized [79-84,88-90,119-121]. The most upfield signals between 11 to 21 ppm are ascribed to the various methyl groups. The angular C18 group is found reproducibly at 11-12 ppm in both Δ5- and Δ7-sterols, whereas the position of the signal of the second angular C19 methyl group varies with the location of the double bond in the B-ring: in Δ5-sterols it is at 19 ppm, but in Δ7-sterols at 13 ppm. In contrast to the ^{1}H-methyl signals, ^{13}C-methyl signals are not shifted by the presence of neighbouring double bonds. On the other hand, the olefinic carbons are amongst the most downfield signals, since desaturation results in a shift of about 100 ppm to lower field. In Δ5-sterols, C5/C6 are at 141/122 ppm and C7/C8 of Δ7-sterols at 117/139 ppm. Olefinic side chain carbons are similarly shifted to 138/129 ppm for C22/C23 and to 147/111 ppm for C25/C27 of the exo methylene group. Even further downfield between 169-174 ppm are carbonyl carbons from acetates, ester groups of ASG or methyl esters from glycuronosides, which at the same time give rise to a methoxy signal at 56 ppm [82-84]. A characteristic signal is C3, which due to its glycosidation and in contrast to the ^{1}H-signals is shifted significantly from about 71 ppm in free sterols to about 79 ppm in the glycosides [79,120] and thus is the only aliphatic carbon from the aglycon found within the range of sugar carbons.

The sugar carbons represent a separate block of signals falling within 62-104 ppm. The β-anomeric carbon is most downfield, followed by the α-anomeric signal and the other carbons in characteristic sequence and spacings, which allow identification of different sugars as well as furanose and pyranose ring forms [117].

The terminal C6- or C5-hydroxymethylene groups result in the most upfield signals of this block at about 62 ppm in hexopyranosides and 64 ppm in pentopyranosides [90] as measured in pyridine. Acyl substitution at C6 results in a downfield shift of C6′ from 62.8 to 64.7 ppm in sitosterol 3-*O*-β-D-glucopyranoside [121]. Acyl substitution at a ring position as for example at C2′ of sitosterol 3-*O*-β-D-xylopyranoside [90] results in a downfield shift of the α-carbon (in this case C2′ by about 1.6 ppm) and in an upfield shift of the β-carbons (in this case C1′ and C3′ by about 3 ppm). Glycosylation results in similar changes [117], which in conjunction with 2D-NMR techniques allow the analysis of various sugars and their sequence in oligosaccharides as separate spin systems and by this permit complete assignment of oligosaccharides. The demanding elucidation of complex steroidal saponins [98-99] with several branched oligosaccharide side-chains can serve as impressive example of the potential of NMR techniques in this field.

Mass spectrometry with various ionization modes was applied to many of the compounds just described, but in most cases the full potential of this method, for example with regard to sequencing of oligosaccharides as possible with permethylated compounds [122], was not used or required. MS supported the overall structure of SG, and often only molecular ions (if present) and a very few fragment ions are tabulated. Underivatized SG yields $[M-H]^+$ by fast-atom bombardment (FAB)-MS (for stigmast α-5,25-dienol galactoside at m/z = 573 [79]), EI-MS gives M^+ (for sito-/stigmasterol glucoside at m/z = 576/574 [117]) and FD-MS produces $[M+Na]^+$ (at m/z = 761/759 and m/z = 923/921 for sito-/stigmasterol biosides and triosides, respectively [93]). Underivatized ASG does not form persistent M^+, apparently because of rapid formation of [M-fatty acid]$^+$. On the other hand, acetylated ASG gives with EI-ionisation M^+ (at m/z = 968 for the acetylated derivative of 6-*O*-stearoyl sitosterol glucoside [101]) at relatively high abundance (20%). Further fragmentations of the glycosidic and 6-*O*-ester bonds give prominent fragments, which are ascribed to the sterol and the acylated sugar part and thus are of diagnostic relevance.

GLC/MS has been applied to TMS derivatives of intact SG [110]. M^+ at m/z = 862/860 were missing because of rapid fragmentation to form [M-trimethylsilanol-CH_3]$^+$ at m/z = 759/757 as heaviest ions. Cleavage on either side of the glycosidic oxygen and mutual charge retention yield strong signals for the steryl fragment at m/z = 397/395 and a less abundant TMS-sugar fragment at m/z = 351. A similar fragmentation of permethylated oligosaccharides provides a means for sequencing [122].

Most of the compounds described above were analysed before the advent of sophisticated NMR and MS techniques by classical carbohydrate methodology such as chromatographic identification of released sugars; permethylation/hydrolysis followed by sodium borohydride-reduction/acetylation for identification of partially methylated alditol acetates, particularly for analysis of biosides up to pentaosides [92,93]; acetylation/chromic oxide oxidation [93,97] followed by methanolysis and GLC analysis of the surviving α-linked sugars. These techniques will not be described here, since they have been documented repeatedly in great detail [122], in particular permethylation with improved formation of methyl sulphinyl carbanion from potassium hydride or butyl lithium [123]. This method can be extended by combination with reductive cleavage followed by final GLC/MS of reduced and acetylated hydrolysis products [124]. Extended lists of fragmentation patterns of partially methylated alditol acetates [125] allow the analysis of linkage types as well as of pyranose and furanose ring forms [124]. In this field experience and routine have decreased the samples of glycolipids required for a complete analysis to quantities (about 20 μg or less) [123], which are at least two orders of magnitude lower than that required for ^{13}C-NMR spectra.

From all these techniques just the hydrolysis conditions will be outlined, since they may be critical for biochemical studies on labelling of the various constituents and the subsequent identification of hydrolysis products. The conversion of ASG to SG is carried out by any kind of saponification, the mildest applying sodium methoxide (5 mM, 15 min, 35°C) [108]. Acid hydrolysis of SG has been carried out with a wide range of concentrations of hydrochloric or sulphuric acid in anhydrous methanol or ethanol (for release of methyl glycosides) or with the inclusion of water (for release of free sugars). It has to be pointed out that all of these methods will destroy some particularly labile sterols [108]. A general procedure may apply 1N hydrochloric acid in methanol and reflux for 5 h, but up to 15% sulphuric acid and 18 h reflux have been used. Even under the normal conditions, glycuronosides will decarboxylate and, therefore, prior to acid hydrolysis they may be converted to normal glycosides by conversion of the terminal carboxyl to a hydroxyl group [82-84]. Furanosides may be split under milder conditions (reflux for 90 min in 0.2M hydrochloric acid) [84]. It may be added that permethylated compounds are hydrolysed in 0.5N sulphuric acid in 90% acetic acid at 80° for 16 h to prevent partial loss of methoxy groups [123,125]. Enzymatic hydrolysis of SG (0.5 mg dissolved in 30 μl of dimethyl sulphoxide) was accomplished with β-glucosidase (10 mg in 0.5 ml buffer) for 15 h [108]. This treatment was used to liberate unaffected sterols and was also carried out with sitosterol gentiobioside [96] On the other hand, release of free sterols from their glycosides may also be accomplished by periodate oxidation/hydrazinolysis (see glycosphingolipids) thus avoiding acid hydrolysis.

Partial acid hydrolysis of biosides has been achieved by reducing reflux time to 10 min [90] or by using milder conditions such as 0.02N sulphuric acid in ethanol for 10 days at room temperature [95]. This treatment released only rhamnose from both α-L-rhamnopyranosyl(1→4)-β-D-glucopyranosyl and from α-L-rhamnopyranosyl(1→5)-α-L-arabinofuranosyl sitosterol, which is remarkable in view of the inner furanosyl residue in the second compound. A useful silica gel TLC system to identify the sugars released from these compounds would be n-butanol/acetone/methanol (4:5:1), in which rhamnose (R_f 0.69), arabinose (R_f 0.47) and glucose (R_f 0.39) are separated [98].

Hydrolysis mixtures are extracted with organic solvents (chloroform, ethyl acetate, petroleum ether) for recovery of free sterols, which are converted directly or after TLC separation into acetates, followed by repurification on TLC plates (see above) and final GLC/MS identification [104,126]. The aqueous acidic phase is neutralized (ion exchange, barium carbonate, silver carbonate) and used for sugar identification either by TLC (see above) or any of the GLC methods after conversion to TMS ethers of methyl glycosides or alditol acetates [125], which are formed from free sugars by sodium borohydride reduction and acetylation. Solid phase extraction Sep Pak-C18 cartridges have become popular for separation of sugars (elution with water/methanol, 9:1) from sterols (elution with methanol/water, 9:1).

A general problem in glycolipid analysis is the assignment of D- or L-configuration to the constituent sugars, and this will be discussed at this point, since steryl glycosides contain sugars of both configurations. The assignment can be made by measuring optical rotations of isolated compounds. In case that only one sugar is present, the sign (+ or -) of optical rotation and not its magnitude measured with the aqueous hydrolysis mixture was sufficient for this purpose [39]. Enzymes of absolute enantioselectivity such as D-glucose and D-galactose oxidase can be used to confirm the presence of these two D-hexoses [38]. A generally applicable method is based on the conversion of the two chemically identical enantiomers (for example D- and L-arabinose) into two chemically different diastereomers by reaction with a chiral reagent (as for example (*R*)-2-butanol). The resulting (*R*)L and (*R*)D diastereomers are separated chromatographically on non-chiral columns as are many sugars which differ from each other as diastereomeric epimers (separation of enantiomers as diastereomers). On the other hand, the enantiomers (*R*)L and (*S*)D will fall together.

In principle the chiral reagent could be attached to any position of the enantiomer, but in most cases the free (non-chiral) carbonyl group is derivatised to bind via glycosidic linkage an aglycon with an additional chiral centre. Heating with acidified (*R*)-2-butanol (18 h) results in (*R*)-2-butyl glycosides, which after acetylation or TMS ether formation are separated by GLC on SE 30 into the various (*R*)L and (*R*)D pairs [127]. Another derivatisation involves Schiff-base formation with (*S*)-α-methylbenzylamine followed by reduction with sodium cyanoborohydride and acetylation to yield 1-[(S)-N-acetyl-α-methylbenzylamino]-1-deoxyalditol acetates [128]. These can be separated into (*S*)L and (*S*)D pairs by HPLC on silica gel columns by elution with hexane/ethanol (19:1) and detection at 230 nm. In all these cases reference derivatives are prepared from commercially available D- and L-sugars.

E. GLYCOSPHINGOLIPIDS

1. Biosynthesis and Functions

N-Acylated sphinganine derivatives are the characteristic and lipophilic aglyca of this group of glycolipids. Sphinganine is formed from L-serine and palmitoyl-CoA by palmitoyl-CoA:L-serine *C*-palmitoyltransferase EC 2.3.1.50. This reaction can be blocked specifically by L-cycloserine or the recently detected antifungal agents sphingofungin B and C, which are effective in the nanomolar range [129]. The resultant 2(*S*)-amino-1-hydroxy-3-keto-octadecane or 3-ketosphinganine is reduced with NADPH to 2(*S*)-amino-1,3(*R*)-dihydroxy-octadecane or sphinganine by D-erythro-sphinganine:NADP$^{\oplus}$3-oxidoreductase EC 1.1.1.102. The natural sphinganine has D-*erythro* configuration, *i.e.* the secondary amino and hydroxyl groups both have D-configuration. The amino group is acylated to yield *N*-acyl sphinganine = ceramide. This reaction is catalysed by two enzymes, from which the acyl-CoA:sphinganine *N*-acyltransferase EC 2.3.1.24 requires acyl-CoA, whereas the ceramide synthase works with free fatty acids [130]. The mycotoxin fumonisin inhibits selectively the *N*-acyltransferase at μM concentrations [131]. The enzymes involved in ceramide biosynthesis are localized in ER membranes and face the cytoplasm [132]. The subsequent addition of glycosyl residues to the primary hydroxyl group of ceramide proceeds in Golgi membranes. The UDP-glucose:ceramide glucosyltransferase EC 2.4.1.80 is exposed to the cytoplasmic face, whereas all subsequent glycosylation reactions are facing the lumen [133] and necessitate translocation of both substrates. Apart from further glycosylation, other reactions of two different types modify the sphinganine part of ceramides. Desaturations are possible at C4 and C8 and result in *trans*(C4,C8)- or *cis*-double bonds (only C8), one desaturation product being sphingosine or 2(*S*)-amino-1,3(*R*)-dihydroxy-4-*trans*-octadecene. In addition, hydroxylation of sphinganine at C4 may introduce a further secondary hydroxyl group of the D-configuration resulting in 2(*S*)-amino-1,3(*S*),4(*R*)-trihydroxyoctadecane or D-*ribo*-phytosphingosine. Several combinations of 4-*trans*, 8-*cis* or *trans* and 4(*R*)-hydroxy modifications are found in addition to the generally present 2-amino-1,3-dihydroxy pattern.

From the enzymes involved in this metabolism, the two subunits of serine palmitoyl transferase [134,135], a ceramide galactosyltransferase [136] and a galactosyl as well as a glucosyl ceramide glycosylhydrolase [137,138] have been cloned from yeast and animal sources. In plants the reactions up to ceramide can be measured enzymatically [130,139], but subsequent glycosylation could not be demonstrated. Very little is known about the enzymes involved in hydroxylation and desaturation reactions, from which the last ones may use ceramide as the actual substrate [140].

In animal cells, sphingosine and its derivatives have been recognized as playing a pivotal role in signal cascades regulating for example the cell cycle [141]. In plants, fumonisin is highly toxic indicating that also in these organisms the accumulating free sphinganine may interfere with signalling at elevated concentrations [130,142]. The phytotoxicity of mycotoxins [142] may further indicate that it is in this part of metabolism where invading fungi break plant resistance. Therefore, methods for quantification of sphingolipids and their biosynthetic intermediates may receive relevance in applied studies on plant pest resistance.

Yeast mutants with a defective serine palmitoyl transferase cannot live without addition of exogenous sphinganine. A second site suppressor mutant of this strain revealed an unexpected function of sphinganine [143]. In normal cells the long-chain base is apparently the only acceptor for a very-long-chain (C_{26}) fatty acid to become membrane-bound in amide linkage as part of ceramide, which then is used for synthesis of mannosylated inositylphosphoryl-ceramide. In the suppressor mutant, the selectivity of an unrelated enzyme, the 1-acyl-glycerol-3-phosphate acyltransferase was changed to introduce a very-long-chain fatty acid into the *sn*-2 position of diacylglycerols, in which normally only C_{16} and C_{18} fatty acids are present. The novel diacylglycerol with the long-chain acyl group at *sn*-2 substitutes for long-chain ceramide in the formation of complex lipids, which in the mutant strain had a diacylglycerol membrane anchor with hydrophilic head groups normally only found in glycosphingophospholipids [144].

Following their biosynthesis, final make-up in Golgi stacks and subcellular sorting and trafficking, a large proportion of glycosphingolipids ends up in the exterior leaflet of plasma membranes to participate in glycocalyx formation [132,133]. In plant cells, tonoplasts and plasma membranes contain glycosphingolipids as major lipid constituents [54-56], and most of the plasma membrane-localized glycosphingolipids are also oriented towards the external medium.

Glucocerebrosides have been shown to induce the formation of fruiting bodies when added to mycelia of the mushroom *Schizophyllum commune* [145,146], although this hormone-like effect is not understood. In addition, glucocerebroside with a 2(*R*)-hydroxy palmitoyl residue linked to 4*E*,8*Z*-D-*erythro*-4,8-sphingadienine was effective as Ca″-ionophore with erythrocyte membranes, in which system the enantiomeric compound (all asymmetric centres, including those of glucose, in inverted configuration) and the galactocerebroside were inactive [147].

In membranes with a significant proportion of these lipids the mobility of the ceramide moiety may play an important role in dealing with variations in hydration and temperature [148]. Desaturation and hydroxylation in the proximal part of sphinganine enable the modulation of acyl chain packing in this membrane segment, which has a high degree of order, since the closest *cis*-double bonds in phospholipid acyl groups are normally located at C9. In addition, the amide-linked fatty acid usually carries at C2 a hydroxyl group in a long aliphatic chain with zero or at most one double bond (see below). The specific requirement for this long-chain acyl group has been referred to above.

2. Chemical Structures

Ceramide residues are found in two types of plant glycolipids. Neutral cerebrosides carry a glycosyl or oligosaccharide unit at the primary hydroxyl group of the sphinganine derivative, whereas the so-called phytoglycolipids are negatively charged derivatives of ceramide-1-phosphate, to which glycosylated inositols are bound via a phosphodiester linkage. Plant cerebrosides (Figure 6.6) contain from 1 to 4 sugar residues in linear chains [78,149-151]. There are two different monoglycosyl ceramides with either a β-D-mannopyranosyl or a β-D-glucopyranosyl residue, of which the glucoside is predominant in photosynthetic tissues [55,56,152]. For addition of further sugar residues, mainly the glucosyl ceramide is used, and glycosylations always involve C4 of the glucosyl residue. The added sugars are either exclusively β-D-mannopyranosides resulting in di-, tri- and tetraglycosyl ceramides with a terminal chain of β(1→4)-linked mannopyranosyl residues attached to C4′ of glucosyl ceramide. These compounds are characteristic for wheat grain [151,153], whereas in other plants such as rice chain-elongation by β(1→4)-linked hexopyranosyl residues can use either glucosyl or mannosyl residues [150]. Addition of a glucosyl residues results in chain termination, and only compounds with terminal mannosyl residues are used for further elongation. This results in a series of di-, tri- and tetraglycosyl ceramides, which are ter-

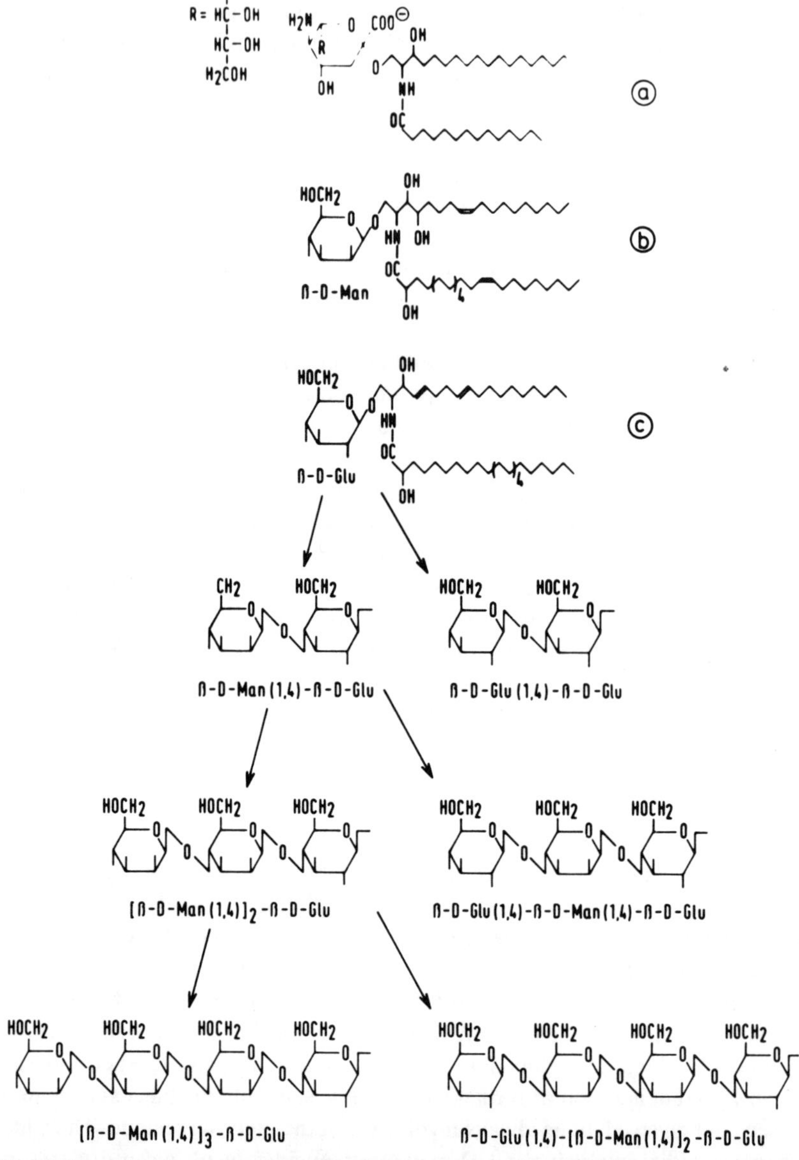

Fig. 6.6. Structures of ceramide glycosides. In contrast to steryl glycosides only gluco- and mannosyl derivatives have been identified as monoglycosyl derivatives (b,c). From these only the glucosyl derivatives are elongated by sequential addition of up to three β-1,4-mannosyl residues (series on the left side). Addition of a β-1,4-glucosyl residue prevents further elongation and results in a different series of oligosaccharide side chains as indicated by the oblique arrows pointing to the series on the right side. The long-chain bases shown include 8-*cis*-phytosphingosine (b) and 4,8-*trans*-sphingadienine, whereas the two amide-linked hydroxy fatty acids are C24:1-hydroxy (b, 2-hydroxynervonic acid) and C24:0-hydroxy acid (c, 2-hydroxylignoceric acid). Compound a represents the α-sialyl ceramide isolated from the photosynthetically active procaryote *Chlorobium* [199]. Sialic (neuraminic) acid-containing ceramides (gangliosides) have not been found in higher plants and are rare constituents in procaryotes.

minally capped either by a glucosyl or a mannosyl residue. In this series, therefore, the cellobiosyl ceramide is a dead-end product [149,150].

The second group represents the so-called phytoglycolipids (Figure 6.7), which comprise more than twenty different compounds containing 16 or more sugar residues [154]. Only a very few of these phosphoglycolipids from corn and tobacco leaves have been analysed in detail. In tobacco [154-157] three different, but related structures have been elucidated. A linear trisaccharide chain composed of β-D-galactopyranosyl(1→4)-α-D-2-acetamido-2-deoxy-glucopyranosyl(1→4)-α-D-glucuronopyranoside was attached to C2 of *myo*-inositol-1-phosphate, which was bound to ceramide via a phosphodiester linkage. Both of the other two components lacked the terminal galactose. In one the glucosamine residue was only partly *N*-acetylated, and the other had an additional α-D-mannopyranosyl residue attached in unknown position to inositol. In corn a similar structure had been proposed [158], but the glucuronoside was attached to C6 and the mannose to C2 of the inositol moiety. In compounds with extended oligosaccharide chains additional galactose and up to four arabinose residues have been found to be attached to the above mentioned core structures [154]. The biosynthesis of these compounds or of their core structure has not been studied in detail in plants, but in yeast the enzyme phosphatidylinositol:ceramide phosphoinositol transferase initiates the synthesis of related compounds [159], and it may well be this enzyme, which requires a long-chain fatty acid either as constituent of ceramides or in the *sn*-2 position of diacylglycerol (see above). In this context it should be mentioned that another enzyme of lipid metabolism, diacylglycerol kinase, can use both *sn*-1,2-diacylglycerol and ceramide for phosphorylation of a primary hydroxyl group, whereas the enantiomeric *sn*-2,3-diacylglycerol is not accepted [160]. This may be due to the fact that the four substituents at C2 of both substrates (H, CH_2OH and two different long-chain residues) have the same (*S*)-configuration.

The amide-linked fatty acids are generally saturated 2-D(*R*)-hydroxy fatty acids ranging from C_{14} to C_{26} and including significant, though not predominating chains of odd carbon numbers. Non-hydroxylated fatty acids account for 1-32%, and it is particularly in free ceramides, where the highest percentages are found. On the other hand, in ceramides from several plants 2-3% of 2,3-dihydroxy fatty acids have been detected [161-164]. In leaf glycosyl ceramides [165-167] non-hydroxy fatty acids are minor constituents ranging from 1 to 3%. In several tissues of different plants the profiles of fatty acids (and of long-chain bases, see below) have been compared in free ceramides and its glycosylated derivatives, but only a few generalisations can be made [161-166]. In free ceramides 2-hydroxylignoceric acid (C24:0-hydroxy or cerebronic acid) predominates accounting for about 50% in all tissues examined, and similar proportions were found in phytoglycolipids of different sources [154,168-170]. On the other hand, acyl patterns of ceramides differ from those found in cerebrosides, which in turn show considerable variation with increasing number of sugars in addition to tissue-specific variation. Often glucocerebrosides from leaves contain high proportions of 2-

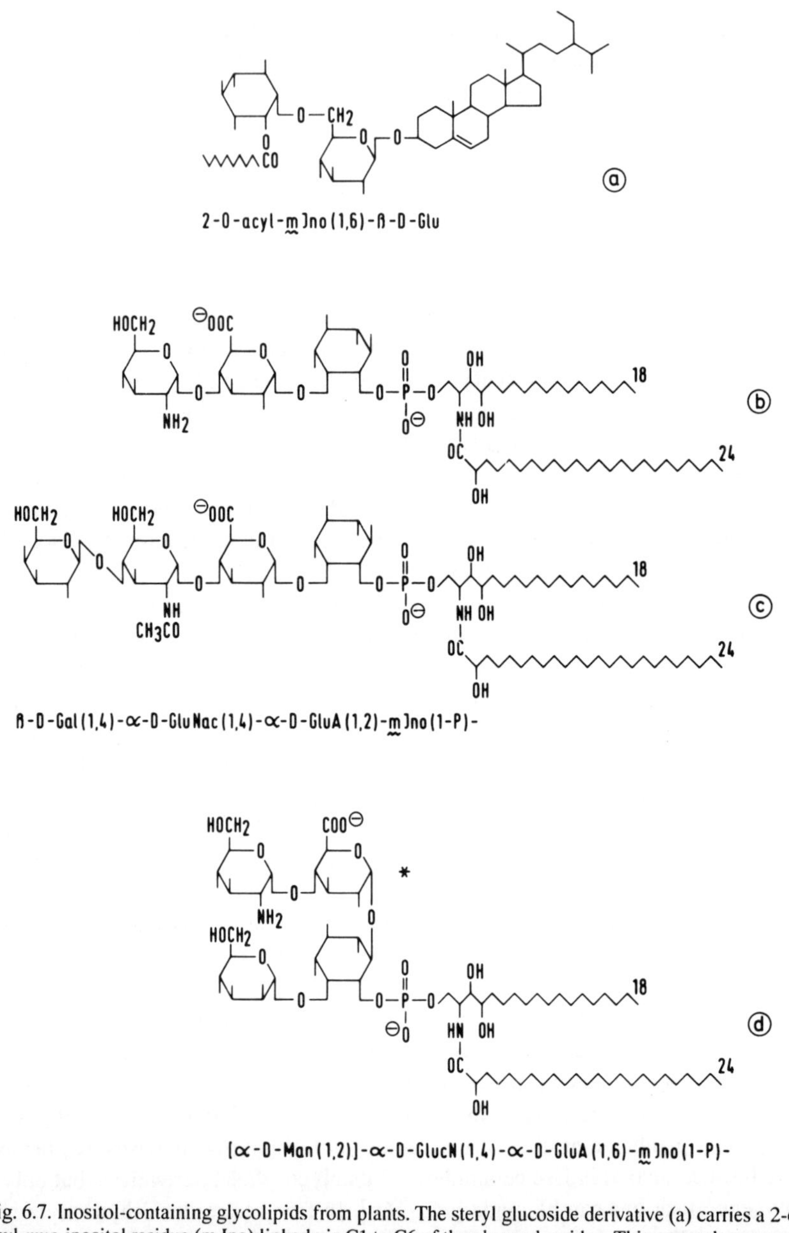

(a)

2-O-acyl-mIno(1,6)-β-D-Glu

(b)

(c)

β-D-Gal(1,4)-α-D-GluNac(1,4)-α-D-GluA(1,2)-mIno(1-P)-

(d)

[α-D-Man(1,2)]-α-D-GlucN(1,4)-α-D-GluA(1,6)-mIno(1-P)-

Fig. 6.7. Inositol-containing glycolipids from plants. The steryl glucoside derivative (a) carries a 2-*O*-acyl-*myo*-inositol residue (m Ino) linked via C1 to C6 of the glucosyl residue. This unusual compound has so far been isolated from banana only in free and palmitoylated form as shown [96]. The phytoglycolipid structures b,c were isolated from tobacco leaves [155,156]. The amino group of the glucosamine residue is partly acetylated (as shown). The phytoglycolipid d was isolated from corn kernels and differs from b,c by the substitution at the inositol residue and the α-mannopyranosyl branch [158]. The D-glucuronosyl residue (marked by an asterisk) may partly epimerize at C5 to an L-iduronosyl residue by refluxing in barium hydroxide solution. From c and d additional glycosyl derivatives with further galactosyl, mannosyl and fucosyl residues have been isolated, but their structures have not yet been assigned completely.

hydroxypalmitic acid [165,166], whereas in rye [167] exceptionally high proportions (52%) of hydroxynervonic acid (C24:1-hydroxy) are present, which was also most abundant in the glucosyl ceramide from plasma membranes of oat root cells [171]. It should be mentioned that saturated or monounsaturated fatty acids longer than C_{18} are rare in membrane lipids from higher plants, and the only other membrane constituent containing C_{20}-C_{26} fatty acids is phosphatidylserine [172].

The various long-chain bases from free ceramides and its glycosylated derivatives have been analysed in various plants [161-166] and allow some generalisations. Phytosphingosine derivatives, *i.e.* trihydroxy bases with or without an additional *cis/trans*-double bond at C8, predominate in free ceramides ranging from about 80-95%. Also in phytoglycolipids trihydroxy bases seem to predominate [168,170]. In glycosyl ceramides di- and trihydroxy bases occur in roughly equal proportions because of an increase of 8-sphingenines and 4,8-sphingadienines. In several tissues [161,166] there is a pronounced tendency to combine dihydroxy bases with a short-chain hydroxy fatty acid (C16:0-hydroxy), whereas trihydroxy bases carry very-long-chain hydroxy fatty acids (C24:0-hydroxy). In contrast to the *trans*-configuration of the C4-double bond, the double bond at C8 is always a mixture of *cis/trans*-isomers. In general one isomer predominates irrespective of the structure of the proximal part of the base, *i.e.* in 8-sphingenine, 4,8-sphingadienine and 4-hydroxy-8-sphingenine. A predominance of the *cis*-isomers is found in wheat [163] and rice [149], whereas in soybeans [164] the ratio is reversed in favour of the 8-*trans*-isomers. In spinach leaves the situation is exceptional, since in 4-hydroxy-8-sphingenine *cis*- and *trans*-isomers occur in about equal proportions, whereas in 4,8-sphingadienines the *trans*-isomer is most abundant [166]. It would be interesting to know, whether the proportions of the two isomers are controlled by two C8-desaturases of different stereoselectivities. From these general observations and the predominance of various long-chain bases and hydroxy fatty acids, the following combinations represent major sphingolipid molecular species: in free ceramides these are *N*-2′-hydroxylignoceroyl-4-hydroxy-sphinganine and -4-hydroxy-8-sphingenine. In monoglycosyl derivatives the ceramide part contains mainly *N*-2′-hydroxypalmitoyl-8-sphingenine, *N*-2′-hydroxypalmitoyl-4,8-sphingadienine and *N*-2′-hydroxylignoceroyl-4-hydroxy-sphinganine or -4-hydroxy-8-sphingenine. Upon glycosylation these species are transferred into higher glycosylated forms, although further variations may occur. The efforts to present the predominant species as practised in this survey necessarily result in a neglecting of the minor ones, which hopefully will not have vital "sparking" functions. In rye plasma membranes about 30 molecular species of monoglycosyl ceramides have been separated [167].

3. Isolation and Separation

In many investigations plant tissues were extracted with chloroform/methanol (2:1) and (1:2) [153], occasionally followed by water-saturated butanol [170]. These solvents are suitable for extraction of most membrane lipids including glycosphingolipids. On the other hand, glycophosphosphingolipids (phytoglycolipids) will not be extracted under these conditions [169] and, therefore, the analysis of these most complex lipids of plant origin is handicapped at the very beginning by a complicated extraction and purification procedure.

The extraction of phytoglycolipids was started with tissue material, from which the normal lipids had already been removed with chloroform/methanol or acetone [168,170]. The resultant residue was refluxed in 50-100 mM hydrochloric acid in aqueous (70%) ethanol for 10-45 min [168-170], apparently without hydrolysis of sugar residues. The yield of phytoglycolipids drops with hydrochloric acid concentrations below 50 mM [169]. The acidic extracts were stored at subzero temperatures (-10° to -25°C for up to 4 days) to precipitate a lipid fraction, which after washing (acidic ethanol, chloroform/methanol and acetone) was dried with diethyl ether to yield a non-hygroscopic powder. This material is insoluble in the usual solvents used for lipid handling, but is soluble in dimethyl sulphoxide, pyridine/water mixtures, the two phases of butanol/acetic acid/water (4:1:5), tetrahydrofuran/water (3:1) or even in pure water (for the most polar components) [168,173]. The lipid residue was dissolved in pyridine/water and passed over Chelex 100 (Na^+-form) for preparation of the sodium salt, which was subsequently used for fractionation on DEAE columns by elution with chloroform/methanol/1M ammonium acetate (30:60:8). Further details can be found in the original publications. Even more complicated extraction procedures have been applied to tobacco leaves [154,168-170,173]. 550 kg were homogenized with trichloroacetic acid in ethanol followed by many extraction and separation steps repeatedly carried out with 20 l batches. About 100 μmoles of phytoglycolipid phosphate/kg leaf fresh weight were isolated representing an estimated recovery of 50 ± 20%. The comparison of a correspondingly corrected value of roughly 200 μmoles with 2-3 mmoles of phospholipid phosphate/kg leaf fresh weight [11] shows that phytoglycolipids may contribute in the order of 10 mole% to leaf phospholipids. This is an appreciable proportion, comparable to the quantity of cardiolipin, but in contrast to this lipid, phytoglycolipids await an assignment of subcellular localisation and function. TLC of phytoglycolipid fractions on silica gel plates requires rather polar solvent mixtures such as chloroform/methanol/4N ammonia (9:7:2) and results in many spots covering the whole distance from start to front [174] or in a diffuse spot at R_f 0.3-0.4 in the case of rice-bran phytoglycolipid [170].

The isolation of larger quantities of cerebrosides usually starts with column chromatography to obtain enriched fractions. On the other hand, reproducible conditions for the isolation of pure compounds either on silica gel or ion-exchange columns [175,176] have not been worked out. When starting with plant lipid mixtures ceramide glycosides are always obtained together with sterol glycosides, and when the elution solvents contain methanol, then even phospho- and glycoglycerolipids are present in those fractions. The last two lipid groups are usually removed by saponification [163], either of the total lipid extract before chromatography or of the separated fraction after chromatography (for example in

0.4M potassium hydroxide in methanol for 2 h at 38°C), in the hope that plants do not contain *O*-acylated glycosyl ceramides as found in animals [177]. From DEAE-cellulose columns (in the acetate form) ceramide monohexosides are eluted with chloroform/methanol (9:1) without separation from steryl glycosides and some other glycerolipids [175]. Similarly, silica gel column chromatography with chloroform/methanol mixtures will not separate sterol and ceramide glycosides, which also applies for pairs of dihexosides and higher homologues. For example, increasing proportions of methanol in chloroform of 5, 10, 20, 30 and 60% have been used for the isolation of both sterol and ceramide glycosides with increasing numbers of sugar residues [150,178]. 5% methanol was used for the elution of free ceramides, whereas synthetic work [179,180] has shown that such compounds will elute at significantly lower methanol proportions (chloroform/methanol 150:1) or with n-hexane/ethyl acetate (1:1). Similarly, ceramide monohexoside from wheat grain lipids was eluted with 10% methanol. Without prior saponification, a solvent of this polarity will in addition elute mono- and digalactosyl diacylglycerol as well as phosphatidylglycerol [181]. Again, synthetic work has shown that ceramide monohexosides can be eluted with just 6% methanol [180,182]. A useful alternative is silica gel chromatography with chloroform/acetone mixtures. They have the advantage that phospholipids stay on the column [183], whereas monogalactosyl diacylglycerol (20% acetone), sterol and ceramide glycosides (50% acetone) and digalactosyl diacylglycerol are well separated [148,184]. Prolonged elution with 30% acetone will even separate sterol from ceramide glycosides, but the success of an actual separation depends on column dimensions, lipid load, speed of elution and the patience of the experimenter before increasing the acetone concentration.

All these parameters have not been standardized, since usually the overlapping fractions obtained by column chromatography are pooled and subjected to preparative TLC (see below) for separation of the faster moving sterol glycosides from the ceramide hexosides with corresponding numbers of sugar residues [150,178]. Sometimes, the mixture of sterol and ceramide glycosides was first acetylated [150,178] in acetic anhydride/pyridine before separation into individual components by TLC in chloroform/benzene/acetone (80:20:5 or 80:20:20, depending on the number of sugar residues). This combination of silica gel column chromatography, saponification and preparative TLC resulted in the following quantities of sphingolipids (mg in brackets), when starting with 50 g of total lipids from wheat grain [163] or rice leafy stem [162] (first and second figures, respectively): free ceramides (100, 34), ceramide monohexoside (580,568), ceramide dihexosides (17, 5), ceramide trihexosides (38, 2) and ceramide tetrahexoside (4). Figure 6.5c gives an impression of the relative abundance of the various ceramide glycosides with increasing number of sugar residues. From these and additional data [164-166,185] it is evident that on a quantitative basis free ceramides rank next to ceramide monohexosides, but only a very few investigations have dealt with plant ceramides. As mentioned above, in animal cells this compound receives increasing attention due to its function as signal for apoptosis [186], which is the final goal of cellular differentiation also in many plant cells (*e.g.* xylem vessels).

HPLC has only occasionally been used for work with plant cerebrosides. As already mentioned in the section on sterol glycosides, HPLC on silica gel columns can be used for resolution of plant lipid extracts, but only in a few studies have cerebrosides been identified or looked for [106,187]. Eluted compounds were detected by the evaporative light-scattering detector or by flame ionisation with a moving-wire belt (Figure 6.5a,b). The resolution of ceramide monohexosides into multiple peaks [188] is ascribed to differences in the number of hydroxyl groups, which also affects their TLC behaviour (see below). On the other hand, the preparative separation of underivatized animal glycosphingolipids up to nonaosyl oligosaccharide side-chains is well established [189]. The separation is carried out on porous silica gel beads (Iatrobeads) of 10 or 60 μm diameter with ternary solvent gradients of isopropanol/hexane/water. This system has been combined with direct and continuous FAB-MS [190,191]. The liquid matrix required for ionization such as glycerol or triethanolamine is added before separation in low percentage (0.7-0.9%) to the eluting solvent, and part of the eluate (split ratio 20:1) is fed continuously into the evacuated and heated interface of the mass spectrometer. Continuous ionisation and on-line recording of selected ions will give a mass chromatogram, whereas extended parts of mass spectra may be recorded on the top of eluting peaks. By this method, rather detailed information on glycosphingolipid mixtures can be obtained, provided each component is present at about 0.2 μg. The full potential of this HPLC/FAB-MS combination comes into use when detailed analyses of molecular species are required (see below) [192].

HPLC separation of glycosphingolipids has also been established for benzoylated compounds on silica columns using gradients of dioxane/hexane, isopropanol/hexane or water-saturated ethyl acetate/hexane for elution and UV detection at 230 or 280 nm [193]. Depending on the benzoylation reagent used, the degree of benzoylation and accordingly the elution behaviour of individual components may vary. With benzoic anhydride (10%) in pyridine and 4-dimethylaminopyridine (5%) as catalyst, per-*O*-benzoylated derivatives are obtained (4 h at 37°C). In contrast, the use of benzoyl chloride (10%) in pyridine (16 h at 37°C) will produce per-*O,N*-benzoylated derivatives, but only from those compounds, which carry a non-hydroxy fatty acid in amide linkage, whereas an α-hydroxy fatty acid in this position prevents the benzoylation of the amide group by steric hindrance. The difference in elution patterns of per-*O*- and per-*O,N*-benzoylated samples prepared in parallel may be useful for identification and isolation of hydroxy- from non-hydroxy fatty acid-containing cerebrosides. On the other hand, only the per-*O*-benzoylated derivatives can be converted by mild saponification to the parent compounds, since mono- and diacyl-*N*-derivatives will not be split under these conditions and require stronger alkaline conditions (see below). The benzoyl derivatives are generally separated in the order of increasing number of sugar residues, but particularly in the group of ceramide monohexosides, further subfractionation is observed governed by the following criteria: gluco- elute ahead of galactocerebrosides, and this effect is superimposed on the increasing retention time with increasing numbers of hydroxyl groups in the long-chain base and the amide-bound acyl group. Shallow gradients will improve this separation and may even result in overlap between fully hydroxylated galactosyl ceramides with non-hydroxylated ceramide dihexosides. Such complications may interfere with the use of this method for quantification (see below). The behaviour of the rare plant mannosyl- as compared to gluco- and galactosyl cerebrosides has not been studied.

Reversed-phase HPLC is used for separation of glycosphingolipid molecular species, but prefractionation according to the number of sugar residues by any of the methods outlined above is required. RP-HPLC is carried out with C18-RP columns and pure methanol [192,194] or methanol/acetonitrile mixtures [167] for isocratic elution. Since the compounds are separated in underivatized form, their elution is monitored by absorption measurement at 210 nm (or below), which will not give a quantitative record. When directly coupled to FAB/MS, this fractionation will give the most detailed information on the composition of glycosphingolipid mixtures. As an example for the application of this strategy the recent analysis of molecular species from ceramide mono- up to pentahexosides from animal origin may be consulted [192]. In the plant field, only one detailed analysis of this kind has been carried out to analyse the molecular species of glucocerebrosides from plasma membranes of rye leaves [167]. This mixture of ceramide monoglucosides was separated into more than 30 individual species (Figure 6.8a). Since a HPLC/MS-combination was not available for these studies, the peaks were collected, derivatized and structurally identified by direct-inlet MS. Seven structurally different groups were identified, and the plots of log retention times versus carbon numbers of amide-bound acyl groups resulted in seven parallel straight lines. From these results it is possible to deduce the influence of various structural parameters on the elution behaviour of ceramide glucosides in addition to those resulting from variation of carbon numbers in long-chain base and acyl group. The magnitude in reducing the elution time results in the following ranking of structural motifs (data were extracted from the original recording; apart from the first line, all other modifications affect the long-chain base; the peak pairs selected to demonstrate the various shifts are shown in Figure 6.8a and are numbered in the same way as in the original publication [167]. The decrease in retention times is exemplified with a cerebroside series, in which the compound with a saturated 24:0-hydroxy (hydroxylignoceric) acid and a 1,3-dihydroxy-4,8-sphingadienine base has the longest retention time):

structural modification	% reduction of elution time	peaks
insertion of a double bond into a saturated acyl group	43	28→20
insertion of the third C4-hydroxyl group	39	26→17
insertion of a C8-*cis*-double bond	28	24→17
insertion of a C8-*trans*-double bond	24	24→18
replacement of a C4-*trans*-double bond by a C4-hydroxyl group	15	28→27
replacement of a C8-*trans* by a C8-*cis*-double bond	6	22→20

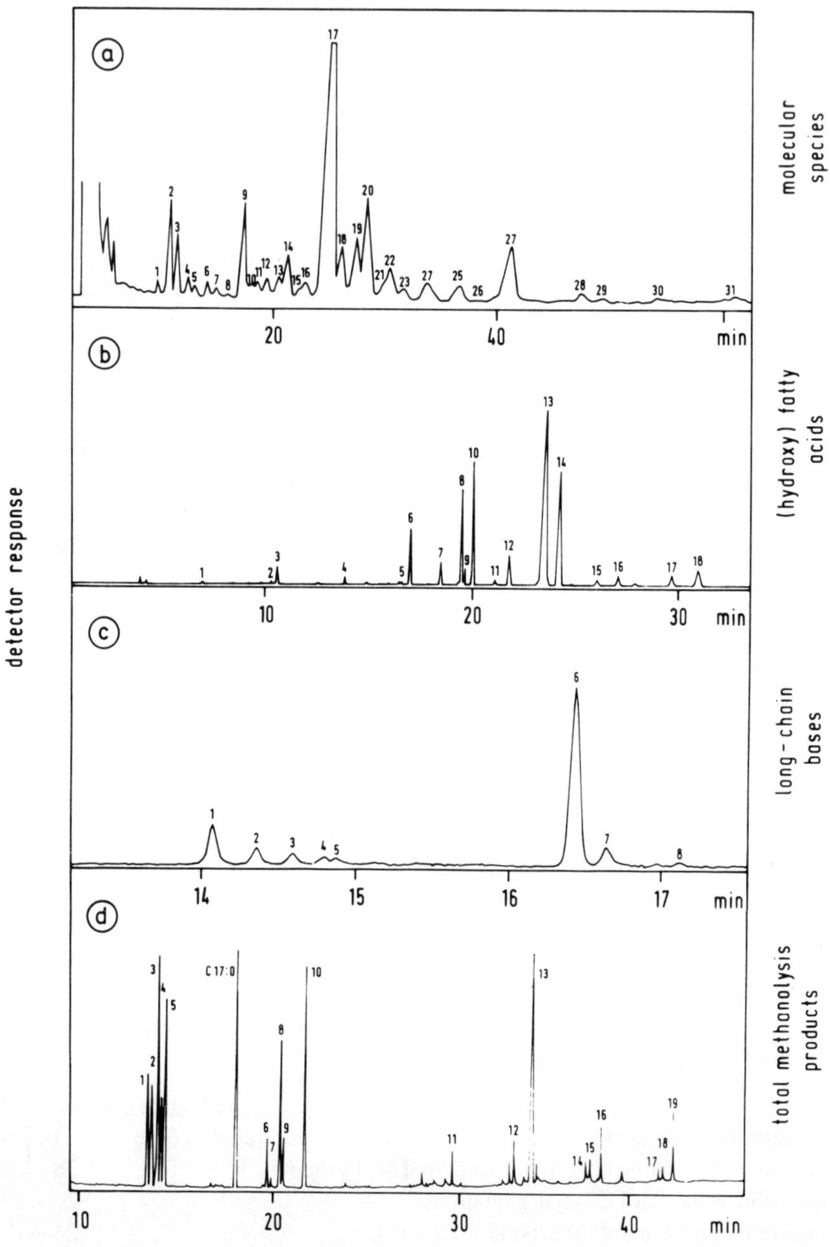

detector response
(a)
molecular species
20
40
min
(b)
(hydroxy) fatty acids
10
20
30
min
(c)
long-chain bases
14
15
16
17
min
(d)
total methanolysis products
C 17:0
10
20
30
40
min

Fig. 6.8. Analysis of cerebrosides by chromatographic identification of molecular species and constituent components released by hydrolysis. a: monoglucosyl ceramides from rye were separated by reversed-phase HPLC into molecular species [167] by isocratic elution (acetonitrile/methanol 6:4) and UV detection at 210 nm. The consequence of structural alterations on retention times is evident when comparing within a series of identical carbon numbers the elution time of the most polar ceramide (composed of C24:1-hydroxy acid amide-linked to 8-*cis*-phytosphingosine, peak 17) with the most apolar derivative present in this series (C24:0-hydroxy acid linked to 4,8-sphingadienine, peak 28). The other members of this series with intermediate polarity (18,20,22,24,26,27) have been described in the text.
b: identification of fatty acid methyl esters obtained by methanolysis and subsequent formation of *O*-TMS ethers by GLC/MS [167]. The only nonhydroxy fatty acids present are palmitic and stearic acid (peak 1 and 2). Peaks 8,9,10 demonstrate the resolution of *cis*-monoenoic from the corresponding *trans*-monoenoic and the saturated hydroxy acid, respectively, in the C_{22} series. The predominating component (peak 13) is C24:1-hydroxy acid (2-hydroxy nervonic acid).
c: analysis of the long-chain base fraction obtained by hydrolysis with 10% barium hydroxide in dioxane (24 h at 110°C) followed by *N*-acetylation and *O*-TMS ether formation. All components were identified by GLC/MS. The predominating component 6 is the derivative of 8-*cis*-phytosphingosine. A conversion of compounds with the labile C3-hydroxy group in allylic position to the C4 double bond (peaks 1,2,3) into artificial methoxy derivatives is not observed (see below); a-c are redrawn from [167].
d: GLC separation of the complete mixture of products obtained by acid methanolysis and subsequent acetylation of 8 nmoles of an animal ganglioside (GM_1). This compound has the structure of βGal (1→4)-βGalNac(1→4)-αNeuac (2→3)-βGal(1→4)-βGlu(1→1)ceramide and yields acetylated methyl glycosides of galactose (α/β-furanosides/pyranosides, peaks 1-3), glucose (α/β-pyranosides, 4 and 5), galactosamine (α/β-furanosides/pyranosides, 6-9) and sialic acid (13). The fatty acids are recovered as methyl stearate (10), arachidate (11) and behenate (12); C17:0 was added as internal standard, and hydroxy acids were absent. Peracetylated long-chain bases were recovered as 4-*trans*-sphingenine (14) in its genuine state and as its methanolysis artifacts (as 3-*O*- and 5-*O*-methyl derivatives, 15 and 16). The same series of genuine and artificial structures was obtained from 4-*trans*-eicosasphingenine (17-19); d is redrawn from [232] with omission of hydrocarbon references.

The largest effect results from introducing an additional double bond into the saturated acyl chain, exceeding the influence of an additional double bond in the long-chain base. Of similar magnitude is the effect of the third hydroxyl group at C4 of the long-chain base, and it may be anticipated that the C2-hydroxyl group in the amide-linked fatty acid may have similar consequences. But since the rye cerebrosides used for this investigation did not contain non-hydroxylated fatty acids, a direct comparison is not possible. A quantitative evaluation of the UV trace for the calculation of molar proportions required a tedious calibration with isolated individual species, which may prevent its general application.

Following prefractionation by column chromatography, TLC is usually the next step in cerebroside purification. Many solvent systems have been listed [2], particularly for compounds extracted from animal tissues [195,196], but only a very few will be mentioned that have been used repeatedly for plant cerebrosides. Acetone/benzene/water (91:30:8) [197] is particularly useful for ceramide (and sterol) monohexosides, since they are the only compounds running between the two galactolipids. Phospholipids are retained and, therefore, this solvent is also useful for cerebroside isolation directly from total lipid mixtures. For the separation of higher homologues, which usually are seen only after enrichment by column chromatography, chloroform/methanol/water (65:25:4) has been used [150,178]. With animal glycolipids, water is replaced by calcium chloride solution to prevent tailing of gangliosides, that is caused by the carboxyl group of the sialic acid residues. Sialic acid-containing glycosphingolipids (gangliosides) have not been detected in plants [198]. *Chlorobium limicola*, a photosynthetically active procaryote, is exceptional in two respects. It contains a membrane lipid [199], in which sialic acid is linked glycosidically directly to a ceramide (Figure 6.6a), which in addition is a rare component in procaryotic lipids [200].

With chloroform/methanol/water (65:25:4) spots of plant ceramide mono-, di-, tri- and tetrahexosides are diffuse and centred at R_f-values of about 0.65, 0.45, 0.25 and 0.15, respectively (Figure 6.5d), and have always lower R_f-values than the corresponding steryl glycosides [150,178]. From synthetic work R_f-values for plant *N*-hydroxyacyl ceramide mannosyl-glucoside (0.72) [201], mannosyl-mannosyl-glucoside (0.45) [202] and mannosyl-mannosyl-mannosyl-glucoside (0.28) [202] in chloroform/methanol/water (52:26:4) have been reported. As mentioned before, separation of ceramide from sterol glycosides may be achieved with acetylated compounds in solvents such as chloroform/benzene/acetone (80:20:5-20) [163] for higher homologues or in diethyl ether/hexane (6:4) for monohexosides [203] with R_f-values between 0.2-0.54. Sterol and ceramide monohexosides have also been separated in chloroform/methanol (10:2) in non-acetylated form due to a difference in R_f-values of 0.63 and 0.53, respectively [54].

The spreading of spots on TLC plates is ascribed to differences in sugar and ceramide parts. These differences in mobilities are significantly enhanced on borate-impregnated plates, which may be prepared by spraying commercially available plates with aqueous sodium tetraborate (1%, w/v) followed by air-drying and activation at 100° (a summary of different solvents for this application is given in [204]). In the context of separating gluco/galacto-epimeric cerebrosides by borate-TLC, it should be recalled that the same method was used for resolution of epimeric heterocyst glycolipids (see above, [42]). Also in this case, the glucosyl runs ahead of the galactosyl epimer, and the same separation is observed with gluco/galactosyl diacylglycerols (see below). But with glycosyl diacylglycerol derivatives, a clean separation is already observed on plain silica gel plates without borate impregnation, provided the solvents do not contain chloroform. Such solvents have not been tried for a resolution of glucosyl from galactosyl cerebrosides on non-impregnated plates, and a separation of the corresponding sterol glycosides has not been tried at all. With chloroform/methanol/water (100:30:2) ceramide monohexosides are separated with decreasing R_f-values according to the following structural details [205]:

dihydroxy base-normal fatty acid
dihydroxy base-hydroxy fatty acid
trihydroxy base-normal fatty acid
trihydroxy base-hydroxy fatty acid

The separated members of glucosyl ceramides form a group which runs ahead of the same group of galactoceramides. A similar resolution is achieved by silica gel HPLC (see above), but plant ceramide monohexosides do not contain galactose, which has only occasionally been detected as a minor sugar component (5%) in ceramide dihexosides [161]. Carbon and double-bond numbers in long-chain base and acyl group have hardly any influence on this resolution. On the other hand, TLC on borate-impregnated plates with chloroform/methanol/water (65:25:4) separated the faster moving diglucosyl from glucosylmannosyl ceramide of plant origin [150]. Borate-impregnated silica gel has also been used for column chromatography resulting in the same separations as described above [205]. These results suggest that upper and lower parts of diffuse TLC spots or the leading and trailing edges of column fractions must not be discarded, because they may contain molecular species of different structure and minor abundance.

4. Immuno-Thin-Layer Chromatography

The availability of an increasing number of mono- and polyclonal antibodies against various lipids and in particular against glycosphingolipids of animal origin [206,207] has added a new technology to lipid biochemistry, which combines handling of glycolipids in organic solvents with immunological procedures carried out in aqueous solutions. The specific interaction between the antibody and its glycolipid antigen is used for investigating various biochemical questions. The antibody binding is detected by different methods based on the use of primary or secondary antibodies or lectins and toxins carrying a fluorescent, radioactive, enzymatically active or a gold label. Some of the applications include: immunolocalisation of glycolipids in tissues and subcellular membranes (immunocytochemistry by gold or fluorescence labelling [208]; glycolipid identification or quantification in lipid extracts after TLC separation (immuno-TLC, see below); studies on glycolipids as receptors for bacterial and viral toxins and adhesins (microtiter plate enzyme-linked immunosorbent assay (ELISA) [209] or TLC overlay assays, sometimes using whole cells instead of isolated adhesins [210]; identification of glycolipid glycosyltransferase activities (microtiter plate ELISA based on the conversion of a bound glycolipid precursor into an immunologically active product [211,212].

Some details will be outlined for immuno-TLC [196]. TLC separation is carried out on small HPTLC plates including control and reference lipids in separate lanes. A separate plate should be used for colorimetric staining. After drying, the plate is dipped shortly (up to 30 sec) into a dilute solution of a polymer (0.01-0.3%, w/v, of polyisobutylmethacrylate in hexane), drained and thoroughly dried. The plate is then incubated with an aqueous blocking solution (1-2% of bovine serum albumin, ovalbumin or fish gelatine in buffered salt solution) for 2 h. Subsequently the plate is taken out and immediately (or after an intervening wash with salt solution) bathed or covered with antibody solution (60-100 $\mu l/cm^2$) of appropriate dilution. The plate is enclosed in a Petri dish and gently shaken for 2 h followed by several washings with buffer (five times for 5 min). Thereafter, the plate is incubated with the second antibody (as above) and washed again (as above). The final treatment is the incubation of the plate with the chromogenic substrates in the case that phosphatase- or peroxidase-coupled second antibodies were used. Appropriate controls may be carried out with a second plate (use of preimmune serum, inclusion of sugars specifically competing for binding, etc.). Instead of second antibodies, radio-iodinated ^{125}I-protein A can be used followed by drying and radioautography. When ^{125}I-labelled lectins and toxins are used instead of primary antibodies [213,214], the procedure is further abbreviated. A problem often encountered with this technique is a flaking of the sorbent layer during one of the repeated washings and swirlings in aqueous solution. This cannot be overcome by a heavier impregnation with polymer, since this will also mask the antigenic epitopes and reduce the wettability. A solution of this problem is offered by the recently developed lipid blotting technique, by which lipids after TLC separation are transferred from the HPTLC plate to a plastic membrane [215,216]. After chromatography, the plate is briefly (20 sec) dipped into the blotting solvent of isopropanol/methanol/aqueous calcium chloride (40:20:7) and covered with the blotting membrane of PVDF (polyvinylidene difluoride). This sandwich is topped by a glass microfiber sheet and the triple pack pressed with an household-iron (heated to 180°C) for 30 sec. If necessary, this hot ironing can be repeated for complete transfer of residual lipids. By this procedure both phospho- and glycolipids were transferred with high efficiency. The yields measured after elution from the PVDF membrane are in the range of 70-90% including mono-, di-, tri- and tetraglycosyl ceramides, and are actually higher than after direct extraction from the HPTLC plate. The lipid-spotted PVDF membrane can be handled for immunostaining without the risk of flaking. This procedure perfectly parallels the well-known western blotting technique developed for proteins.

The detection limit of immunostaining either on TLC plates or after blotting to PVDF membranes is in the range of 3-10 ng [215] and thus is one or two orders of magnitude more sensitive than colorimetric staining. The intensity of the immunostain depends on the size of the TLC spot at a given glycolipid quantity and, therefore, with simple dot-blots as little as 1 ng can be detected. Several monoclonal antibodies are available that react with galactosyl ceramides [217-219]. This sphingolipid is also a specific ligand and receptor for the human HIV envelope glycoprotein gp120 [220], which is available in recombinant form. On the other hand, all these antibodies do not recognize the glucosyl ceramide, which is predominant in plants and for which no antibodies are available. Because of this lack immunological techniques have not been used in the field of plant glycosphingolipid biochemistry. For the production of antibodies against cerebrosides, these compounds can be coupled as haptens to proteins. Ozonolytic opening of the C4-double bond creates an aldehyde or carboxylic acid group [221,222], which can be used for coupling of the sugar-containing sphingolipid fragment to proteins such as limpet hemocyanin, for example.

5. *Quantification*

The quantitative determination of cerebrosides in plant lipid extracts faces problems similar to those encountered with steryl glycosides and heterocyst glycolipids, which all are present in low proportions (at most a few percent) in total extracts [47,56,111]. Before quantification they have to be separated by TLC, which due to their low abundance often has to be carried out with the upper limit of lipid loading. This is turn results in quantities of more abundant lipids exceeding their linear range and, therefore, may necessitate two separate determinations and referring to a common reference to be measured with satisfactorily precision at both concentrations. The more sensitive methods usually require an extraction from the TLC scrapings, which are also required for direct colorimetry, followed by derivatisation and taking of an aliquot before the actual measurement. Such a procedure requires the inclusion of an internal standard as early as possible during the whole process.

The most direct method is colorimetric determination of sphingolipids (and other glycolipids) after TLC separation by adding the colour reagent directly onto the TLC scrapings without extraction. Glycosyl residues will react with reagents such as phenol, orcinol (3-hydroxy-phenol), resorcinol (3-hydroxy-5-methyl-phenol) or anthrone while heating under strongly acidic conditions (> 50% sulphuric acid) to form orange or green coloured products suitable for extinction measurement at 430, 480 or 625 nm after removing the silica gel by centrifugation. The colorimetric determination of sugars in aqueous solution is started by the rapid addition (1-2 sec) of concentrated sulphuric acid to produce the heat required for glycoside hydrolysis and reaction with the colour reagent. It has been repeatedly shown that these conditions are harsh enough for quantitative release of monosaccharides from methyl glycosides, oligo- and polysaccharides [223] or even acetylated glycogen [224]. On the other hand, aglyca such as proteins or lipids do not produce interfering colours. After TLC separation, the plate is sprayed lightly with water to improve the removal of spots by scraping. Subsequently the colour reagent is added in appropriately diluted sulphuric acid (for example 66%, v/v), followed by sonication and heating (20 min at 85°C) [225]. We found that the minimum volume of reagent required for this procedure is 1.5 ml, which is sufficient to disperse the silica gel scrapings of variously sized TLC spots, and to use a centrifuged aliquot for optical measurement in 1 ml cuvettes. A critical step in this procedure is the spraying and scraping of TLC plates, since colour yield may decrease with increasing water content in such small reaction volumes. Part of the excess water on scrapings may be removed by drying the tubes for some time in horizontal position before adding the reagent.

We used anthrone in sulphuric acid for calibration covering a range of 40-200 nmoles of sugar spotted to a TLC plate and carried through the spraying and scraping steps. The phenol or resorcinol reagents may be somewhat more sensitive, but it is important to point out that with all reagents the colour yield varies significantly with the sugar. With phenol [223] the relative extinctions for mannose, glucose and galactose are 1.0, 0.72 and 0.57, respectively, and with resorcinol they are 1.0, 0.58 and 0.72 [226], respectively.

The quantification of plant cerebrosides by a combination of HPLC and flame ionization [106] or evaporative light-scattering detection has only occasionally been carried out, but not evaluated in detail. As discussed above for steryl glycosides, the sensitivity of the light-scattering detector may be somewhat better than TLC/colorimetry as is evident from the quantities (10-200 μg) required for calibration [107]. On the other hand, in a study of animal lipids using a laser light-scattering detector [188], calibration required only 2-20 μg per component. But in this case, cerebroside subfractions, which were hardly separated by HPLC, showed a significantly different detector response, varying by a factor of about 2.

HPLC in combination with UV detection has been used many times for quantification of perbenzoylated glycosphingolipids of animal origin [193]. *N*/*O*-Benzoylation is effected with benzoyl chloride in dry pyridine (1 h at 60°C) and after several solvent partitioning steps for purification, the benzoyl derivatives are ready for gradient elution from silica columns. This method can be used with unfractionated lipid extracts, and the recovery of ^{14}C-labelled cerebrosides was 95% [227]. With water-saturated ethyl acetate in hexane, UV detection has to be carried out at 280 nm, which is not the absorption maximum of benzoate groups. At this wavelength, the detection limit is reached at about 70 pmol of cerebroside (in the order of 50 ng). Recording in the absorption maximum at 230 nm is possible with gradients of dioxane/hexane or isopropanol/hexane, which increases the sensitivity by a factor of about 7 to a detection limit of 10 pmol [227]. Calibration curves covered a range of 0.5-15 nmoles. The evaluation of UV-traces has to take into account the different numbers of benzoyl groups carried by the various compounds, especially in view of the fact that *N*-benzoylation is prevented by the presence of an amide-linked 2-hydroxy acyl group (see above). Whether total plant lipid extracts can be benzoylated and separated for quantification of cerebrosides or any of the other glycolipids present in this mixture has not been studied.

An alternative to TLC/direct colorimetry of sugars would be TLC/cerebroside hydrolysis/direct colorimetry or fluorimetry of liberated sphingosine. Sphingosines can be separated and quantified by GLC and HPLC (see below), but for a rapid and simple quantitative determination, direct hydrolysis of TLC scrapings and derivatisation without extraction or solvent partitioning would be the method of choice. Quantitative release of sphingosines from cerebrosides has been proven to occur by hydrolysis with 4N hydrochloric acid in methanol for 4 h at 105°C [228] or in 1N hydrochloric acid in methanol/water (82:18) at 70°C for 18 h [229]. After adjusting the hydrolysis mixture (0.5 ml, [229]) to pH 8.0 by addition of aqueous solutions of sodium hydroxide (2N, 0.25 ml) and borate buffer (0.2M, 0.75 ml, pH 8.0), fluorescamine is added (0.015% in acetone, 0.5 ml), the reaction mixture vortexed (assuming that silica gel scrapings will not interfere) and centrifuged. The clear supernatant is used for fluorescence measurement at 480 nm with excitation at 380 nm. Neither free fluorescamine nor cerebroside hydrolysis products show any fluorescence emission at this wavelength. The total volume of the assay (2 ml) is comparable to the above mentioned anthrone method (1.5 ml), but the sensitivity and range is far better as seen from the calibration curves, which depending on the fluorimeter set-up cover ranges of 5 to 60 or 1 to 6 nmoles of sphingosine or cerebroside [229].

In a similar procedure [228], methyl orange as acidic dye is complexed with sphingosine to form a complex which can be quantified by photometry at 415 nm (instead of fluorimetry). The method requires extraction of the released sphingosine (after alkalinization of the acidic hydrolysis mixture) into organic solvent such as ethyl acetate and mixing with a methyl orange solution, including several washing steps. With 5 ml of ethyl acetate as described in the original procedure calibration curves cover a range of 10-100 nmoles. This method is less sensitive than fluorimetry, but a fluorimeter may not be available in all laboratories.

The processing of glycolipids on TLC plates, as if they were SDS-PAGE separated proteins (see above), and the possibility to convert glycolipids similar to glycoproteins to fluorescent derivatives has created a new technology of glycolipid detection and quantification, which in addition is supported by the commercial supply of various staining kits. In combination with the above mentioned blotting technique, they may develop further potential. Although these methods have not yet been applied to plant glycolipids, they should be outlined briefly. After chromatography on a HPTLC plate (and anticipating their blotting), the separated glycolipids were oxidized *in situ* with sodium periodate in aqueous buffer solution. The resultant aldehyde groups (trihydroxy long-chain bases will produce an additional pair) are then linked covalently to 7-amino-4-methyl-coumarin by incubation in the presence of sodium cyanoborohydride [230]. After washing and drying, the fluorescence intensity was measured with a chromatoscanner. Several steps of the procedure have been optimized and the final calibration curves cover a range of 1 to 100 pmoles of glycolipid. The sensitivity is so high that only 1-2 mg of material is required for lipid extraction. The generation of aldehyde groups from glycolipids by periodate oxidation is also the basis for the general DIG-detection system, in which a digoxigenin derivative is coupled (mostly via a hydrazide group) to the aldehyde group. A selection of antibodies against the digoxigenin hapten is available commercially. They are either coupled to fluorescent chromophores or to enzymes such as alkaline phosphatase or peroxidase for subsequent detection and quantification of the glycolipid-digoxigenin derivatives on the plates.

6. Structural Analysis

The elucidation of cerebroside structures requires identification of building units (sugars, long-chain bases, fatty acids) as well as analysis of the linkages between these components regarding positions and stereochemistry. Hydrolysis followed by chromatographic analysis will identify the different components, whereas modern techniques of mass spectrometry and NMR spectroscopy will in addition yield information on structural details. Accordingly, the following part of this chapter will be divided into a first section on hydrolysis/chromatography followed by an overview on MS fragmentation patterns and ^{1}H- and ^{3}C-NMR spectra.

i) Hydrolysis and Degradation. The three different building blocks of cerebrosides cannot be released by a single hydrolysis procedure in satisfactory yield and structural integrity of the original components. This is mostly due to the sensitivity of the C3-hydroxyl group in allylic position to the C4-*trans*-double bond in sphingosines. This system is particularly labile and acidic methanolysis results in the introduction of 3-*O*- and 5-*O*-methyl groups as well as in a shift of the original double bond to the C3- and C5-positions, respectively. Such alterations are not observed with dihydrosphingosine and phytosphingosine, which points to the C3-ol-C4-ene structure as the critical part of the molecule.

A recent investigation on the liberation of sugars, sphingosine bases and fatty acids from animal gangliosides [232] showed that 2 h hydrolysis at 110°C with 0.75N hydrogen chloride in dry methanol (5 vol) and methyl acetate (1 vol) in a sealed tube was sufficient for maximal release of all compounds (Figure 6.8d). The kinetics of release demonstrate, that the acyl-amide linkage is the most resistant of all, whereas monosaccharides such as glucose, galactose and galactosamine were completely released after 30 min. Despite the short hydrolysis time of 2 h, only about 20% of the original sphingosine was recovered in unaltered form, whereas the major part was converted to 3- and 5-*O*-methyl derivatives (see Figure 6.8d). The stability of the amide linkage is also evident from studies on the degradation of bacterial lipid A. Methanolysis in 1N hydrochloric acid in methanol for 1 h was sufficient for complete splitting of all glycosidic and *O*-ester linkages [233], whereas the methyl glycoside of *N*-acyl glucosamine was isolated in quantitative yield. This allowed a separate analysis of *N*- and *O*-linked fatty acids, which of course is also possible after mild alkali treatment.

As already described above in the section on quantification, methanolysis followed by derivatisation and GLC is a useful and sensitive quantitative method. In the study just mentioned [232] methanolysis products were mixed with methyl heptadecanoate (internal standard), peracetylated (acetic anhydride in pyridine) and the whole mixture subjected to capillary GLC. In a single run with temperature programming from 150° to 300°C (4°C/min) all components were separated with only occasional overlap between groups (see Figure 6.8d). The elution sequence was acetylated methyl glycosides (multiple peaks for each sugar due to α/β- and furanose/pyranose isomers), fatty acid methyl esters (C18:0-C22:0, hydroxy fatty acids were absent), and *N,O*-acetylated long-chain bases (and their artifacts, see above). Despite the multiplicity of peaks, quantification with reference to the internal standard (C17:0) is possible after calibration of FID response factors, which vary for sugar derivatives, fatty acid methyl esters and long-chain base acetates. A linear response was found for the range of 10 pmol to 10 nmol with glycosphingolipid recovery of better than 95%.

In view of the above difficulties, it is consequent to use different hydrolysis procedures depending on the components to be analysed [163,167]. Aqueous hydrochloric acid will release free sugars suitable for direct reduction with sodium borohydride, acetylation and GLC/MS of alditol acetates. Methanolic hydrogen chloride will produce fatty acid methyl esters, which are extracted from the acidic hydrolysis mixture into petroleum ether followed by hydroxy group derivatisation and GLC/MS (see below). For release of long-chain bases with minimal structural alteration hydrolysis with barium hydroxide in aqueous or pure dioxane is recommended. Cerebrosides are heated in aqueous 10% barium hydroxide/dioxane (1:1) [234] or 10% barium hydroxide in dioxane [167] for up to 24 h at 110°C. It should be pointed out that these strongly alkaline conditions also result in a splitting of the glycosidic bond and most likely in an alteration of the sugar involved in this linkage. The remainder of the oligosaccharide chain (if present) may survive this treatment as evident from the recovery of oligosaccharides from phytoglycolipids after splitting the phosphate diester by a similar treatment [235] (but see below). Extraction of the alkaline hydrolysis mixture with chloroform or diethyl ether will yield the free bases in yields of 95% and in apparently unaltered form as evident from Figure 6.8c, where no indications of artifacts can be seen [236]. Nevertheless, several authors recommend various mixtures of methanol/water/hydrochloric acid for long-chain base release (summarized in [195,237]).

Apart from complete hydrolysis, several methods have been worked out for partial hydrolysis or degradation of cerebrosides that are useful for structural identification. Partial acid hydrolysis (reflux for 10 min in 0.1N hydrochloric acid in methanol [163]) or acetolysis (heating for 1 h at 80°C in 0.5N sulphuric acid in 90% aqueous acetic acid [153]) will yield a series of partially deglycosylated derivatives. These are useful for sequence localization of individual hexoses (if not carried out by 2D NMR).

Completely deglycosylated ceramides can be released by several methods. Ceramide glycanase from leechs [238] will remove disaccharides and longer oligosaccharides, but not monosaccharides. Glucosyl ceramides are hydrolysed by human glucosylceramide β-glucosidase [239]. The common carbohydrates can be eliminated by Smith degradation, which will yield ceramides from cerebrosides in high yield. Cerebrosides are first oxidized with periodate in solvents such as chloroform/ethanol/water [236], methanol or pyridine [240]. After a few hours, excess reagent is destroyed by addition of ethylene glycol. The oxidized compound is recovered by extraction with chloroform/methanol for subsequent reduction by sodium borohydride in methanol. After a few hours and decomposition of excess reagent by acetone, the reduced compound is extracted into chloroform/methanol and hydrolysed at room temperature in 0.1N hydrochloric acid in 90% aqueous tetrahydrofuran overnight. The free ceramide is recovered in yields above 90% [236]. They can be used for determination of the stereochemistry of the amide-linked 2-hydroxyacyl groups by TLC (see below). It is noteworthy that the periodate-oxidized compound before reduction is resistant to mild acid hydrolysis, and this is ascribed to the formation of cyclic structures from the original dialdehydes [236]. On the other hand, the periodate-generated aldehydes yield ceramides by β-elimination on incubation with cyclohexylamine [236] or dimethylhydrazine. Unfortunately, all these methods cannot be applied to phytosphingosine derivatives, since its 3,4-diol structure will be oxidized by periodate as well. On the other hand, the resultant C_{15}-aldehydes can be used for identification of phytosphingosine components.

The amide-linked acyl group can be released selectively under alkaline conditions (for example by heating for 4 h at 125°C in 1M potassium hydroxide in 90% butanol [241]) or with 2% hydrazine sulphate in anhydrous hydrazine (15 h at 150°C [242]). In contrast to the parent compounds, the lyso-derivatives are ninhydrin-positive. They can be recovered by preparative TLC in solvents, the polarity of which depends on the number of sugar residues in the oligosaccharide chain. The reaction has been carried out with mono- up to trisaccharide side chains [242] and was used to prepare a selection of reference molecular species of lactosylceramides with defined amide-linked fatty acids, which were introduced by reacylation with fatty acid succinimidyl esters [241].

At this point a few references should be made to additional reactions used for hydrolysis and degradation of phytoglycolipids, which differ from cerebrosides by the presence of phosphodiester bonds, glucosaminyl and glucuronosyl residues. As mentioned above, the phosphodiester was split by reflux with saturated aqueous barium hydroxide to produce oligosaccharides [235]. Subsequent studies have shown that the α-D-glucuronopyranosyl residues were partially converted to an L-iduronosyl residues by alkali-induced epimerization at C5 [243]. This reaction was prevented by first converting the C6-carboxyl group to its methyl ester followed by sodium borohydride reduction [82,84] to a stable glucopyranosyl residue (see also [155,170]). More recently, aqueous hydrofluoric acid (48%) is used for selective hydrolysis of phosphodiester bonds [244]. During incubation at 0° for 60 h (in polypropylene vials) all *O-P* esters are cleaved. Glycosidic bonds are stable under these conditions with the exception of furanosides, which are quantitatively split [245]. Hydrofluoric acid can be removed by neutralization with lithium hydroxide and sedimentation of the insoluble lithium fluoride. 2-Deoxy-2-amino sugars such as glucosamine react specifically with nitrous acid (incubation with sodium nitrite in aqueous solution at slightly acid pH 3-4, 3 h at 0°C [243,245]). The unstable carbonium ion formed at C2 via a diazonium salt is attacked by the ring oxygen resulting in ring contraction, concomitant opening of the glycosidic linkage at C1 and formation of a new 2,5-anhydro-mannose. Further reduction with sodium borohydride results in a new terminal 2,5-anhydro-mannitol at a position which was occupied by glucosamine in the original oligosaccharide chain [245]. The analysis of the substitution pattern and the stereochemistry of inositols will not be covered.

For the preparative isolation of fatty acid methyl esters obtained by methanolysis, TLC in hexane/diethyl ether (85:15) can be used [184], which separates normal (R_f 0.90) from hydroxy esters (R_f 0.33), although in most cases, the unfractionated mixture will be used for GLC/MS. Capillary GLC on a non-polar stationary phase will separate the *O*-TMS ether derivatives according to the same criteria as known from the resolution of normal fatty acids: at a given chain-length, *cis*-monoenoic acids elute slightly ahead of the *trans*-isomers, and both emerge before the saturated compound. Introduction of the *O*-TMS group results in an increase of retention time roughly equivalent to a C_2H_4 group (see Figure 6.8b). The TMS ethers of 2-hydroxy fatty acid methyl esters are easily recognized by their typical MS fragmentation, which results in spectra different from those of the other isomers regarding the position of the *O*-TMS group along the hydrocarbon chain [234]. The predominant ion is $[M-59]^+$ resulting from loss of the $COOCH_3$-fragment (for hydroxylignoceric acid at *m/z* 411) followed in intensity by $[M-15]^+$ due to loss of the ester methyl group. These two ions dominate the spectrum recorded at an ionization voltage of 70 eV [167].

Because only 2-D(*R*)-enantiomers of 2-hydroxy fatty acids have been found in cerebrosides, this stereochemical detail is now rarely confirmed. A very simple method relies on Smith degradation of intact cerebrosides (see above) followed by TLC of the resultant ceramides on normal silica gel TLC plates in chloroform/methanol (95:5) [246]. In this solvent ceramides with amide-linked 2-hydroxy acyl groups are separated according to the configuration of this 2-hydroxy group (at identical configuration of the long-chain base): the L-epimer always moves significantly faster than the D-isomer. This separation is seen for di- and trihydroxy long-chain bases irrespective of their degree of unsaturation, and the method has been successfully applied to the analysis of ceramide di- and trihexosides from wheat flour [153]. Unfortunately, phytosphingosine derivatives will not yield ceramides on Smith degradation, but even their fragments, which are in fact amide-analogues of 2-monoacylglycerols, may be separated into the epimeric diastereomers. For dihexosides and higher homologues, the use of the ceramide glycanase from leech (see above, [238]) may be useful for this purpose.

The enantiomers of hydrolytically released 2-hydroxy fatty acids (and of other positional isomers as well) can also be separated after conversion to diastereomers by reaction of the hydroxyl or carboxyl group with a chiral reagent. A whole collection of such reagents is available and has been used with long-chain hydroxy fatty acids including L-acetoxymandelic acid chloride, 1(-)*R*-phenyl (or naphthyl) ethyl isocyanate (formation of urethanes with the hydroxy group or amides with the carboxyl group), L-menthyl chloroformate, 2-D-phenyl propionyl chloride and MTPA chloride [247]. In most cases the derivatives can be separated into the two diastereomers by both TLC and GLC. For example, the L-acetoxymandelates of the 2-L-hydroxy isomers of fatty acid methyl esters run ahead of the 2-D-isomers and result in a clear separation during TLC in benzene/diethyl ether (97:3) as shown for C16:0 and C24:0 hydroxy acids [248,249]. Similarly, the 1(*R*)-phenyl ethylamides of 2-hydroxy fatty acids are resolved by TLC in hexane/ethyl acetate/acetic acid (70:30:2) into the diastereomers, and also in this case the 2-L-hydroxy derivatives run ahead of the 2-D-isomers [250]. GLC of these carboxyl group-derivatives, in which the hydroxyl group was protected by the trifluoroacetate, methyl or TMS ether group, results in a similar separation, but now the 2-D-hydroxy-derivatives elute first as shown for a variety of 2-hydroxy acids from C18:0 to C26:0 including C24:1 and C26:1. The use of the enantiomeric reagent results in an inversion of TLC and GLC elution order [250]. HPLC on normal-phase silica columns will also separate such diastereomeric pairs as shown for the (*S*)-MTPA esters [251] and the 1(*S*)-naphthyl ethyl amides [252] of long-chain hydroxy fatty acids. On the other hand, enantiomeric hydroxy fatty acids, after protecting the carboxyl and hydroxyl group with non-chiral reagents, have been resolved by reversed-phase HPLC on chiral-phase columns (for example silica bonded chemically to dinitrobenzoyl phenylglycine) [253]. Enantioselective resolution of hydroxy fatty acid methyl esters is also possible by GLC on capillary columns coated with a chiral stationary phase (for example hydrophobic derivatives of β-cyclodextrin [254]). A summary of methods applied for separation of enantiomeric lipids in general is given in [255].

The analysis of long-chain bases can be carried out by several methods after their release from cerebrosides. As mentioned above, the least artifacts are produced by hydrolysis with barium hydroxide in dioxane followed by direct extraction from the alkaline medium into an organic solvent. TLC purification of the extracted fraction is sometimes carried out despite the fact, that long-chain bases are separated into several subgroups. Chloroform/methanol/2N ammonia (40:40:1) produces five ninhydrin-positive spots [166,184] between R_f 0.63 (sphingosine), R_f 0.54 (dihydrosphingosine) and R_f 0.34 (phytosphingosine). The remarkable separation of dihydroxy bases with an allylic from those having a non-allylic hydroxyl group should be pointed out [203].

Periodate oxidation of long-chain bases will release the terminal segments as aldehydes with 16 (from dihydroxy bases) or 15 carbon atoms (from trihydroxy bases), which can be separated by GLC as free aldehydes, dimethyl acetals or as TMS ethers of alcohols (after sodium borohydride reduction) [200]. With free aldehydes, it was not possible to separate the 5- or 6-*cis/trans*-isomers originating from the original C8-*cis/trans*-double bonds in phytosphing-8-enine or sphing-8-enine, respectively, whereas a partial separation was obtained for *trans*-2-*trans*-6- and *trans*-2-*cis*-6-hexadecadienal originating from a mixture of *trans*-4-*cis/trans*-8-sphingadienine [166]. On the other hand, silver nitrate-TLC will separate these derivatives according to decreasing R_f-values into saturated, *trans*-unsaturated and the slowest-moving *cis*-unsaturated compounds [203]. The *cis/trans*-configuration can be determined by IR spectroscopy due to the characteristic absorption bands at about 10.3 μm (*cis*) and 14.0 μm (*trans*) [184]. With modern Fourier-transform IR instruments, suitable spectra can be recorded with as little as 10-20 μg of substance.

HPLC separation of long-chain bases has been carried out after derivatisation of the amino group by reaction with *o*-phthaldialdehyde. For this purpose the sphingosine mixture in methanol (for example 50 μl) is mixed with the *o*-phthaldialdehyde reagent (50 μl of an aqueous mixture containing in addition boric acid, ethanol and 2-mercaptoethanol). After 10 min at room temperature 400 μl of methanol/5 mM phosphate buffer pH 7.0 (9:1) is added, the sample cleared by short centrifugation and aliquots used for HPLC [256,257]. Reversed-phase separation was accomplished on a RP C18 column by isocratic elution with methanol/5 mM phosphate buffer pH 7.0 (9:1) and fluorimetric detection (excitation 340 nm, emission 455 nm). Long-chain base derivatives are separated according to number of carbon atoms, double bonds and hydroxyl groups, but according to the chromatograms shown, the separation of 8-*cis/trans* or of 4/8-mono-unsaturated isomers has not been tried [142,256,257]. On the other hand, the high resolution of molecular species of cerebrosides discussed above [167] may suggest, that similar separations can be obtained with the phthalaldehyde derivatives. This method has been used for quantitative determination of free or released sphingosine bases [142,256,257]. Due to the sensitivity of the fluorimetric detection, the lower limit measured was in the pmol range. As internal reference a C20-sphingosine was carried through the whole procedure.

For GLC analysis long-chain bases are in most cases used as *N*-acetyl-*O*-TMS derivatives. These are formed by first acetylating the amino group with methanol/acetic anhydride (4:1, overnight at room temperature). Under these conditions, the reagent is specific for amino groups and prevents the acetylation of hydroxyl groups, which subsequently are converted to TMS ethers. Extensive studies on the GLC separation of these derivatives have been documented [258-260] and a recent application of capillary GLC/MS is shown in Figure 6.8c [167]. This method separates all the configurational and positional isomers mentioned above and allows structural confirmation by MS. The EI-induced fragmentation patterns show characteristic ions, which in the case of phytosphingosine derivatives arise by fragmentation of the C3-C4 bond and charge retention on both fragments, $[M-76]^+$ comprising C1-C3, and $[M-297]^+$ representing the longer half (in this case from phytosphing-8-enine). In sphingosine derivatives, the fragmentation of the C2-C3 bond results in similar fragments, $[M-174]^+$ comprising C1-C2 and $[M\text{-}309]^+$ the longer part (in this case from 4,8-sphingadienine).

The location of double bonds in long-chain bases can be carried out by methods established for unsaturated fatty acids, and therefore will not be detailed here [237,262]. Protection of amino and hydroxyl groups by peracetylation, hydroxylation of double bonds with osmium tetroxide or potassium permanganate, conversion of the resultant diols to TMS ether derivatives and final GLC/MS analysis will yield fragments, which allow location of the original double bond. Similarly, von Rudloff oxidation (permanganate/periodate) will result in mono- or additional dicarboxylic acids, which can be identified by GLC/MS in the form of methyl esters. The original C8-double bond of the C18-bases will yield decanoic acid methyl ester. Examples for the application of these methods to plant ceramides and cerebrosides are described in [164,166,185] together with the reproduction of several representative mass spectra. The recently developed one-pot derivatisation of double bonds with dimethyl disulphide for subsequent GLC/MS analysis may also be useful in the field of long-chain bases [262].

The final details required for a complete description of the structure of long-chain bases concern the stereochemistry of the chiral centres at C2, C3 and C4 (only in phytosphingosines). The original assignment of the D-*erythro* and D-*ribo* configuration took several years (summarized in [195]) and was a major issue, but since then hardly any further efforts have been made, at least in the field of plant sphingolipids, to confirm this structure. GLC of *N*-acetyl-*O*-TMS derivatives of sphingosines will separate the faster eluting *threo*- from the slower *erythro*-derivatives, but an assignment of D- or L-configuration is not possible with these derivatives [195]. From synthetic work [263] the optical rotations for the four different bases forming the D/L-pairs of *erythro*- and *threo*-sphinganines are known ($[a]_{546} = \pm 6°$ and $\pm 13°$, respectively, in chloroform/methanol 10:1), which will allow the required assignment provided enough material is available. However, optical rotations are reported only very rarely in the course of studies on plant long-chain bases or ceramides [153].

The constituent sugars, their mode of linkage and sequence have been analysed by methods referred to above [117-120], *i.e.* acid hydrolysis or methanolysis followed by sodium borohydride reduction/acetylation and final GLC/MS. Methylation analysis will reveal the inter-subunit linkage points, and partial hydrolysis will identify the sequence of different monosaccharides, if present. Anomeric linkages have been identified by chromic oxide-oxidation of acetylated compounds. For this purpose peracetylated glycolipids (100-300 μg) together with peracetylated *myo*-inositol as internal standard are oxidized with chromic oxide (50 mg) in glacial acetic acid (0.5 ml) for 15 min at 40°C in an ultrasonic bath. After dilution with water (1 ml), the oxidized lipid is extracted with chloroform/methanol (2:1). The monosaccharides in the original and the oxidized sample are analysed after acetolysis/hydrolysis in 0.5N sulphuric acid (in 90% acetic acid, 0.3 ml, 16 h at 80°C) followed by water addition (0.3 ml) and another 5 h at 80°C [264]. Only α-linked pyranosyl residues will survive, since equatorially (β) exposed glycosidic linkages will be oxidized resulting in 5-hexulosonic acid

Table 6.5.
^{13}C- and ^{1}H-NMR data of glucocerebrosides with different long-chain bases (LCB) and amide-bound acyl (norm) or 2-hydroxy acyl groups (2-OH), whereas the sugar is always a β-D-glucopyranosyl residue. The spectra were recorded with underivatized (free) or acetylated compounds (acet, 2-OAc). The long-chain base from [276] was D-*erythro*-4-*trans*-sphingenine, from [147] 4-*trans*-8-*cis*-sphingadienine, from [284] saturated phytosphingosine and from [286] 4-*trans*-8-*trans*-sphingadienine. For non-equivalent protons of methylene groups, the different shifts are given (at C1 of long-chain bases, C3 of α-hydroxy acyl groups and C6 of glucose). For C4, C5 and C8, C9 slightly different ^{13}C-shifts may be found in the literature varying from 123-136 ppm, which may be ascribed to the different solvents used. A recent reinvestigation [287] led to a reassignment of the ^{13}C-signals for the olefinic carbons of D-*erythro*-sphingosine, i.e. of C4 to 129.1 and of C5 to 134.6 ppm in $CDCl_3$. This rearranges the attribution given in the first column of the table [276]. The various coupling constants can be found in the original references; shifts are given in δ ppm downfield from tetramethylsilane as internal standard. P = pyridine, C = chloroform, M = methanol, all deuterated.

	^{13}C-shifts								^{1}H-shifts						
carbon atom	**LCB**			**fatty acyl**			**glucose**		**LCB**		**fatty acyl**			**glucose**	
	free	free	acet	norm	2-OH	2-OAc	free	acet	free	acet	norm	2-OH	2-OAc	free	acet
1	63.7	70.0	66.7	175.0	175.6	169-171	105.5	100.6	3.52;4.18	3.37;3.65	-	-	-	4.22	4.47
2	57.0	54.5	47.9	36.9	72.4	74.3	75.0	71.2	3.96	4.31	2.17	3.97	5.10	3.20	4.95
3	74.8	72.2	70.6	26.2	35.6	31.7	78.3	72.7	4.08	5.27	1.58	1.67;1.57	1.81	3.36	5.18
4	134.9	132.0	73.3			25.0	71.4	68.2	5.47	4.87		1.27	1.25	3.30	5.08
5	129.9	132.1	27.4				78.4	71.9	5.71	1.60				3.24	3.68
6	32.9	32.9	25.7				62.6	61.9	2.08	1.25				3.67;3.82	4.13;4.23
7	29.4	27.3							2.08						
8		129.4							5.43						
9		130.6							5.43						
10		27.5							1.98						
n-1	23.1	22.9	22.6	22.9	22.9	22.6						1.27			
n	14.2	14.2	14.1	14.2	14.2	14.1			0.90	0,86	0.90	0.88			
solvent	C/M	P	C	C/M	P	C	P	C	M	C	M	M	C	C/M	C
ref	[276]	[147]	[284]	[285]	[147]	[284]	[147]	[282]	[286]	[284]	[286]	[283]	[284]	[282]	[282]

derivatives, which will escape the subsequent sugar analysis. This method works with all of the common monosaccharides (glucose, galactose, mannose) and their 2-deoxy-2-amino-drivatives, whereas α- and β-furanosides are oxidized at comparable rates. The applicability of this method has been demonstrated with a wide selection of glycolipids [264] and has been routinely used for the analysis of plant cerebrosides as well (for example [150,153]).

ii) Mass Spectrometry and NMR Spectroscopy. No attempt will be made to review the extended literature on MS analysis of glycosphingolipids and only a few references will be made to some of the excellent reviews of this topic. For many years, EI-MS was the main one in this field [195,265-267], and accordingly, methylated, acetylated or TMS ether derivatives are required, which have sufficient volatility during heating in the vacuum. From the various derivatives, permethylated compounds are preferred since the increase in molecular weight due to derivatisation is lowest. It should be mentioned, that not only hydroxyl, but also acyl amide groups are methylated with the dimethylsulphinyl carbanion/methyl iodide reagents. The resultant *N*-methyl acyl amides may be reduced further with lithium aluminium hydride, which removes the original carbonyl oxygen completely and results in tertiary amines. The same reaction sequence converts *N*-acetyl-glycosylamines into *N*-methyl-*N*-ethyl amine derivatives. The general fragmentation patterns of these derivatives have been reviewed in great detail and yield information on the number of sugars, their sequence and the structure and components of the ceramide part. On the other hand, linkage positions connecting the sugars and stereochemical details cannot be analysed. Molecular species can be identified, but quantification suffers from the fact that evaporation during heating in the vacuum represents a distillation, and vapour composition may change with temperature and heating time [268]. The same difficulties have been observed with galactolipids (see below).

The MS analysis of glycolipids is further extended and at the same time simplified by application of so-called "soft ionisation" methods such as fast-atom bombardment (FAB) and liquid secondary-ion mass spectrometry (SIMS), for which even underivatized compounds can be used [268,269]. Direct ionization from a matrix of low volatility (such as glycerol, thioglycerol, triethanolamine etc.) by accelerated atoms or ions of xenon, argon or caesium produces positive and negative ions. Both series can be recorded in the positive or negative mode and will yield complementary information due to differences in ion abundance and fragmentation patterns. Both ionisation methods can also be applied to TLC spots or lipids blotted to a PVDF membrane [270]. Relevant spots or areas from HPTLC plates are cut, wetted with a few μl of the ionisation matrix and introduced into the vacuum chamber of the instrument for measurement [271,272]. Upon recording serial scans in the direction of migration (46 spectra per 10 mm), the heterogeneity of spots even on normal TLC plates becomes evident [273]. The spot of a ceramide monohexoside, for example, was dissected into 27 spectra which showed a similar gradient of molecular species as known from borate-TLC (see above). The difference between these spectra demonstrates at the same time that the polar solvents used as ionisation matrix do not result in extensive elution of adsorbed compounds from the silica gel, since diffusion would randomize such profiles.

NMR spectroscopy has become one of the most powerful techniques in glycolipid research, and the chemical methods outlined above often seem to play a more confirming and supporting role. By working through homo- and heteronuclear short- and long-range connectivity pathways the elucidation of nearly all structural and stereochemical details is possible [117,274,275]. An exception is the absolute assignment of D- and L-stereoisomers, whereas the analysis of relative orientations (*erythro*/*threo*) has been carried out by ^{13}C-NMR of long-chain bases [276]. A limitation is the quantity of a compound required for ^{13}C-NMR spectroscopy, which is in the order of 10 mg [277] with modern instruments. Spectra of underivatized glycosphingolipids have been recorded in all the common deuterated solvents, but pyridine is of particular use. It can be used at 65-90°C [201,202,277] and spectra recorded at this temperature do not suffer from low resolution due to intermolecular associations, which may form with compounds of extended sugar chains. Acetylated derivatives are routinely measured in deuterochloroform. This field has been covered by several reviews [267,278-280] and only a few characteristics of ^{1}H- and ^{13}C-spectra will be pointed out (Table 6.5).

When plant cerebrosides were analysed for the first time, sophisticated NMR techniques were not in common use and accordingly, only partial and one-dimensional spectra have been published. In the meantime, some of these compounds have been synthesised, and in these papers more (though not all) details of ^{1}H- and ^{13}C-spectra can be found [201,202]. A characteristic feature of plant cerebrosides is the presence of 4-*O*-β-D-mannopyranosyl residues, and mono-, di- and trimannosyl derivatives of the ubiquitous glucocerebroside have been synthesised (but apparently not the simple β-D-mannopyranosyl ceramide). In the ^{1}H-spectra the anomeric protons show up as a separate group of well-resolved signals between 4.0 and 5.2 ppm [278]. H1 of the β-D-glucopyranosyl residue is seen at about 4.7 ppm (J = 7.8 Hz) accompanied by up to three additional signals for H1 of β-D-mannopyranosyl residues at 5.08, 5.07 and 5.04 ppm [201,202]. Because of the equatorial position of H2 and the small and similar dihedral angles between H1 and H2 of α- and β-mannosyl residues, the vicinal coupling constants are so small that the characteristic splitting usually used for anomeric differentiation cannot be used. The β-H1 signals show up as broad singlets. On the other hand, in the spectrum of a synthetic 4-*O*-α-D-mannopyranosyl derivative of glucocerebroside the α-H1 signal shows up as a doublet (J = 1.8 Hz) at 6.14 ppm [201]. This feature and additional details in shifts and coupling constants of ring protons allow a clear differentiation between the anomeric mannosyl residues.

Partially overlapping with anomeric protons (particularly from mannose), but usually further downfield olefinic protons from long-chain bases (H4-H5 at 5.4-5.8 ppm and H8-H9 at 5.1-5.4 ppm) and fatty acyl residues (if unsaturated) may show up. Their coupling constants are useful for assigning *trans* (15 Hz)- or *cis*

(9-11 Hz)-configurations in the long-chain bases [281], that may be supported by nuclear Overhauser enhancement and other NMR experiments [283]. The most downfield signal is ascribed to NH at 7.5-8.4 ppm (if not exchanged by deuterium).

Acetylation results in the downfield shift of the geminal CH-protons in the sugar ring, the long-chain base (to 5.30 ppm and 4.9 ppm for H3 and H4 of acetylated sphingosine and phytosphingosine derivatives, respectively) and the 2-hydroxy fatty acyl group (to 5.1 ppm), but also neighbouring positions are affected (see Table 6.5). Acetate groups exert a stronger deshielding effect than glycosyl residues which is evident from the shifts of H4 in glucocerebroside (3.3 ppm), its peracetylated form (5.1 ppm) and the 4-*O*-mannosyl derivative (4.2 ppm; but all in different solvents) [201,282]. In the spectrum of a peracetylated glucocerebroside containing a 2-hydroxy fatty acid all 6 acetate signals (4 from the sugar and one each from the long-chain base and the α-hydroxy acyl group) have been assigned (between 2.00 and 2.30 ppm), from which the α-acetoxy group was the most downfield (2.30 ppm) [281].

In the ^{13}C-NMR spectra of glycosphingolipids several signal groups are observed spread over a range of 10-180 ppm [277]. The most downfield signals are assigned to the carbonyl carbons of amide-bound fatty acids showing up between 173-176 ppm. Their exact position depends on the presence of a neighbouring hydroxy group at C2, which results in deshielding and a slightly lower position. Therefore, normal acyl group carbonyls show up at 173-174 ppm, whereas 2-hydroxy acyl carbonyls are separated at 175-176 ppm [277]. In acetylated derivatives the normal acyl carbonyls are not affected (174 ppm), whereas the acyl carbonyls of 2-acetoxy acyl ceramides are shifted upfield to 169-171 (β-shift), falling into the group of acetyl carbonyls [282,284]. Olefinic carbons from long-chain bases and unsaturated fatty acids are seen at 128-135 ppm.

The next group of signals further upfield comprise the sugar carbons spread between about 106 ppm (anomeric) and 62 ppm (C6) in patterns characteristic for the different hexoses in α/β- and pyranose/furanose-forms. The anomeric carbons form a clearly separated group at the downfield side of all sugar carbons [117,274,275,285]. Acetylation affects all of these signals, but does not displace them from this part of the spectrum as typical for ^{1}H-spectra. In this region of the spectrum, also C2 of the 2-hydroxy acyl amide (72.4 ppm free, 74.3 ppm acetylated) as well as C1, C3 and C4 of sphingosine and phytosphingosine show up (66-73 ppm, free and acetylated, see Table 6.5). The above mentioned differentiation between *erythro*- and *threo*-isomers of sphingosines [276] makes use of differences in ^{13}C-shifts of C3 and C4 (1.9 and 1.2 ppm upfield for the corresponding signals of the *threo*-isomer) and the shift difference between the olefinic carbons C4 and C5 (5 ppm for *erythro*- and 4 ppm for *threo*-orientation). These differences are only seen in chloroform/methanol (see Table 6.5), but persist in free sphingosines, ceramides and various glycosyl derivatives of ceramides [276].

The presence of an α-hydroxy group in the acyl amide residue does not only deshield the acyl carbonyl (3 ppm downfield), it also affects C2 of sphingosine

(0.6 ppm upfield) and even the anomeric sugar carbon (0.1-0.2 ppm upfield) [277]. The largest downfield shift (more than 30 ppm) due to hydroxyl group substitution is seen of course at the substituted carbon itself, *i.e.* at C2 of the 2-hydroxy acyl group (to 72 ppm) and C4 of phytosphingosine (to 73 ppm). Substitution of a sugar hydroxyl group by another sugar results in characteristic glycosylation shifts as exemplified by comparing the spectra of glucosyl- and lactosylceramide, *i.e.* its 4-*O*-β-D-galactopyranosyl derivative [277]. In glucosyl ceramide C4 is seen at 71.5 ppm and in lactosyl ceramide at 81.8 ppm. This downfield α-shift is accompanied by smaller upfield β-shifts (1-2 ppm) of the adjacent C3 and C5. These correlations and the analysis of connectivities in different spin systems can be used for a complete linkage and sequence elucidation. The 24 sugar carbons of a tetrahexoside ceramide of animal origin have been completely assigned [277], and this spectrum illustrates the clear separation of the group of four anomeric carbons. For the plant mannosyl derivatives of glucosylceramides only the shifts of the anomeric carbons have been given, which for the additional β-mannosyl residues are found at about 97 ppm and thus at slightly higher field than C1 of glucose at about 100 ppm of the peracetylated compounds [201,202].

F. GLYCOSYL DIACYLGLYCEROLS

1. Biosynthesis and Possible Functions

Galactosyl and sulphoquinovosyl diacylglycerols are the predominant glycolipid components of the different membrane systems found in chloroplasts and other forms of plastids [288]. Due to the cellular predominance of plastidial membranes, these glycolipids represent the prevalent lipids of all photosynthetic tissues and organs. In non-photosynthetic tissues, where plastid membrane systems are often confined to envelope membranes, the proportion of glycosyl diacylglycerols is reduced accordingly. The biosynthesis of the various glycosyl diacylglycerols is confined to plastids and follows a series of common steps up to diacylglycerol [289]. Many or most of the individual reactions and enzymes including fatty acid biosynthesis, acyl and glycosyl transferase reactions as well as the final desaturations have been investigated in great detail. Peculiarities regarding the positional distribution and pairing of fatty acids in glycolipids will be described in the context of molecular species analysis (see below). The enzymes catalysing the final steps of glycolipid assembly and desaturation are concentrated in envelope membranes [288,289]. Continuing and most recent studies in this field are directed on further characterization, purification and cloning of individual enzymes such as phosphatidate phosphohydrolase [290], UDP-galactose:diacylglycerol galactosyltransferase EC 2.4.1.46 [291], UDP-sulphoquinovose:diacylglycerol sulphoquinovosyltransferase [292] and the various desaturases [293]. Other areas of active research try to ascribe specific functions to the various glycolipids in thylakoid and envelope membranes and to understand the process of galactolipid export from envelope into thylakoid membranes [294,295]. The recent cloning of a plastidial protein involved in vesicle fusion or translocation of membrane proteins between envelope vesicles [296] has opened new perspectives in this field.

The galactosyltransferase catalysing MGD synthesis is a small protein (19-22 kD) as isolated from envelope membranes of spinach chloroplasts [297,298], whereas the corresponding enzyme from cucumber [299] is significantly larger (47 kD). Cloning and functional expression will finally prove the identity of these enzymes. Cyanobacteria contain the same set of thylakoid lipids as found in chloroplasts of higher plants, but the synthesis of MGD involves two different steps [300]. A UDP-glucose:diacylglycerol glucosyltransferase forms monoglucosyl diacylglycerol [301], which is subsequently epimerized at C4 to MGD. The intermediate monoglucosyl diacylglycerol usually represents only a minor component of cyanobacterial membrane lipids (0.6%), but upon addition of glucose [302] its level may increase to 12%. The advantage of this epimerization in terms of changes at the membrane surface or its consequences for lipid/protein interactions are not clear. Studies on the biosynthesis of DGD discovered a surprising mechanism [303]. In contrast to original expectations [304], the UDP-galactose-independent galactosyltransfer between two MGD molecules is the only reaction which results experimentally in the synthesis of DGD and of its higher homologues [305]. This enzyme has not been purified, and it is not clear whether the anomerically isomeric DGDs (and the correspondingly changed higher homologues, see below) are all formed by the same enzyme. Similar glycosyltransfer reactions without participation of sugar nucleotides are catalysed by glycosidases, which usually catalyse the hydrolysis of glycosidic bonds by glycosyltransfer to water. A well studied example is in fact β-galactosidase. It has been used to synthesise pure 3-*O*-β-D-galactopyranosyl-*sn*-glycerol from free glycerol or racemic isopropylidene glycerol and lactose or nitrophenyl-β-D-galactoside [306,307]. The resulting galactosylglycerol is an excellent and natural inductor of the lac-operon in *E. coli* and thus initiates its own degradation in the digestive tract after release from galactolipids (see below) [308]. Also in water-limited systems, which may reflect the situation in envelope membranes, this enzyme accepts a wide variety of acceptors (monosaccharides, alcohols, steroids, amino acids) to form β-galactosides and even oligogalactosides with lactose as donor [309-311]. Similar dismutative transfer reactions, but this time involving acyl groups, are responsible for the formation of acylated derivatives of MGD and DGD (see below). They are formed as enzymatic artefacts by an enzyme, which has been demonstrated in chloroplast envelope membranes [312,313].

Experiments on cloning the UDP-sulphoquinovose:diacylglycerol sulphoquinovosyltransferase have made significant progress [314]. Inactivation of the gene putatively coding for this enzyme in *Rhodobacter sphaeroides* has resulted in the accumulation of UDP-sulphoquinovose, which was isolated in labelled form and characterized [315]. In the meantime, this sulphosugar nucleotide has also been isolated from eucaryotic organisms (Tietje and Heinz, in preparation).

The orientation of these glycosyltransferases in envelope membranes may have relevance with respect to the subcellular origin of the sugar nucleotide and for the subsequent export of glycolipids into thylakoid membranes. Similar to other biomembranes the thylakoid membrane is characterized by an asymmetric distribution of glycolipids between the two leaflets [316,317]. Most pronounced is the enrichment of DGD on the luminal leaflet. It is not known how these asymmetries are established and maintained or whether they are required for specific functions of these leaflets [318]. It has been suggested that the larger polar head of DGD may be particularly well suited for the shuffling of protons along the luminal membrane surface to the F_0-part of the ATPase [319].

Various other functions have been attributed to the different glycolipids, which may be associated specifically with isolated membrane proteins such as ATPase and various components of the photosystems [320,321]. This association is strong enough to survive detergent solubilization, electrophoresis and final blotting for immunological detection of glycolipids [322]. The attribution of functions tries to take into account the physicochemical behaviour of these lipids and in particular the ability of MGD to form inverted micelles. The functions listed include bilayer formation providing ion-impermeable barriers, proper sealing of protein/lipid boundaries, activation of ATPase and supporting efficient energy-transfer in the process of light energy-harvesting. The presence of the various glycolipids and of PG in all oxygen-evolving photosynthetically active organisms does not necessarily indicate a causal connection between these lipids and the capacity of oxygen evolution. This correlation could simply be due to the history of the cyanobacteria, which in phylogenetic terms are the first organisms having this set of lipids and which evolve oxygen. The cyanobacterial cell may have evolved from anaerobic photosynthetically active procaryotes by a precyanobacterial symbiosis or gene transfer including information for membrane lipid biosynthesis from both ancestors [323].

Recently, some additional functions of glycolipids were discovered. The crystallization of the light-harvesting protein required the presence of DGD, which may have relevance for the interaction of these proteins *in vivo* in the thylakoid membrane [324]. The lipid composition of the outer leaflets of envelope and thylakoid membranes with the enrichment of MGD (at least in the case of thylakoids) may play a vital role in protein passage through these membranes [316,317]. It is initiated by binding of the transit sequence of the corresponding preproteins to the negatively charged lipids such as SQD and PG resulting in a conformational change to an α-helical structure [325-327]. This induces the formation of destabilizing non-bilayer structures of MGD and may be required as a first step in the process of protein insertion and passage through the target membranes. In this context recent experiments on the variation of membrane lipids in *Acholeplasma* should be mentioned that produces glucosyl diacylglycerols as major components together with diglucosyl diacylglycerol. Fatty acid feeding experiments have shown that this organism tries to keep the proportion of the non-bilayer forming glucosyl diacylglycerol at an upper limit just compatible with the bilayer structure [328]. This was interpreted as indicating that a delicate balance between destabilizing and bilayer-forming lipids is maintained that enables the membrane or its monolayers to respond to triggering events by the formation of localized non-bilayer structures.

Further approaches to identify specific functions of membrane lipids make use of lipolytic enzymes to reduce the proportion of glyco- or phospholipids in isolated thylakoids [295,316,317], or are based on the isolation of mutants deficient in individual lipids. Such mutants have been isolated from *R. sphaeroides*, where SQD-deficiency has no functional consequence [329,330], whereas a similar mutant of *Chlamydomonas* was detected by its high chlorophyll fluorescence indicating a defect in energy transfer [331]. Recently, an *Arabidopsis* mutant with severely reduced levels of DGD has been isolated (C. Benning, personal communication), and a closer analysis of structural and functional consequences, regarding ATP yield and antenna size and structure especially, may yield interesting results. Finally, recent studies on the permeability and electrical properties of planar lipid membranes from thylakoid lipids including isolated glycolipids [332] and reviews covering the increasing knowledge of physical properties of glycolipids [320,321,333,334] should be mentioned, since this aspect of glycolipid analysis will not be covered in the present context.

2. Chemical Structures

The predominant glycosyl diacylglycerols comprise 1,2-di-*O*-acyl-3-*O*-β-D-galactopyranosyl-*sn*-glycerol (MGD) and a series of higher homologues formed by sequential addition of α-D-galactopyranosyl residues to C6 of the terminal galactose of the preceding homologue [335-337]. In this way digalactosyl diacylglycerol (DGD) or 1,2-di-*O*-acyl-3-*O*-(6′-*O*-α-D-galactopyranosyl-β-D-galactopyranosyl)-*sn*-glycerol, trigalactosyl diacylglycerol (TGD) and tetragalactosyl diacylglycerol (TeGD) are formed (Figure 6.9, left series) [338-344]. In addition, a second series of fully β-linked homologues has been identified. These linear 1→6-linked *all*-β-galactolipids include DGD, TGD and TeGD (Figure 6.9, right series) [345]. Even a third series exists that is derived from the normal a→β-linked DGD by sequential addition of 6-*O*-β-D-galactopyranosyl residues resulting in alternative types of TGD and TeGD (Figure 6.9, middle series), although spectroscopic evidence is not yet available for these compounds [345]. The different isomers have not been separated in the form of intact lipids, but the heterogeneity becomes evident after alkaline deacylation (see below). The *all*-β-DGD has been detected frequently in many plants [340,345-348] (angiosperms, gymnosperms, ferns, mosses, algae), in bacteria [349] and in the anaerobic prokaryote *Rhodopseudomonas viridis* (Linscheid *et al.*, in preparation). In higher plants its proportion in DGD accounts for only a few percent with the exception of mature seeds of several species and genera of beans, where it may represent 13-36% of the total DGD. *Vigna angularis* (Adzuki bean) contains the highest percentage [347], and the third series of galactolipids (see Figure 6.9) has so far only been detected in the seeds of this plant [345]. If plants have a general capacity to syn-

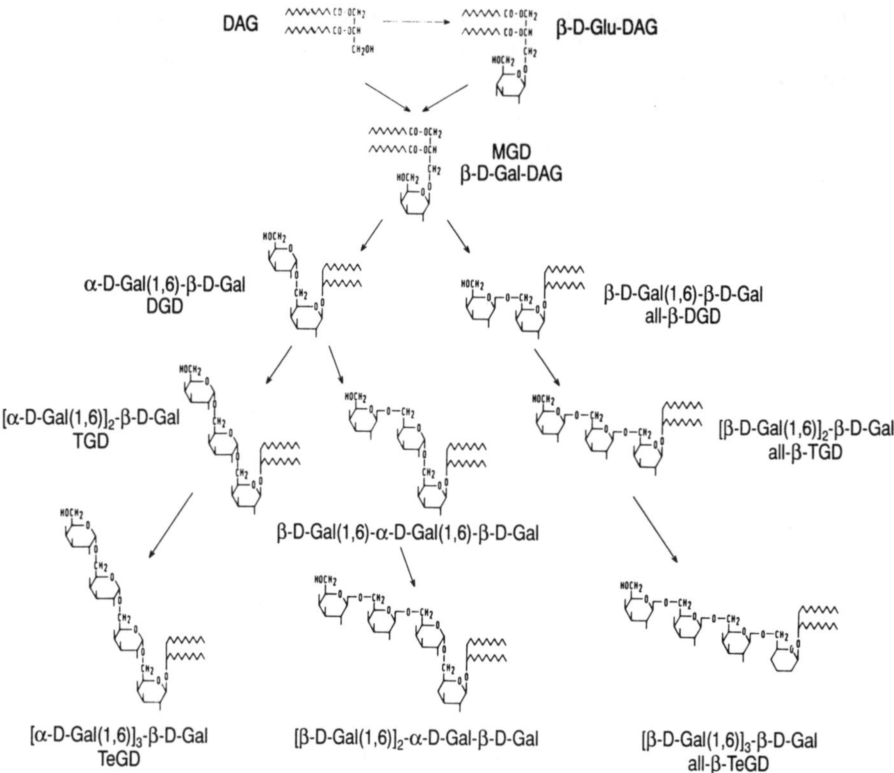

Fig. 6.9. Monogalactosyl diacylglycerol MGD and its derivatives. Sequential addition of α-D-galactopyranosyl residues to C6 of the preceeding galactose results in di-, tri- and tetragalactosyl diacylglycerol DGD, TGD and TeGD (left branch). Addition of β-D-galactopyranosyl residues to C6 results in *all*-β-DGD, *all*-β-TGD and *all*-β-TeGD (right branch). Chain extension by β-D-galactopyranosyl residues at C6 of DGD results in a third type of TGD and TeGD (branch in the middle). It is not clear whether all compounds are located in plastid membranes. In cyanobacteria diacylglycerol (DAG) is first converted to glucosyl diacylglycerol, which is occasionally also found in higher plants, followed by epimerization to MGD. References are given in the text.

thesise different DGDs involving different enzymes, it may not be surprising that a single mutation is not sufficient for a complete loss of DGD. On the other hand, the location of these *all*-β-galactolipids in plastids has not yet been ascertained.

In addition to the prevailing galactolipids glucose has been found in this group of glyceroglycolipids. 1,2-Di-*O*-acyl-3-*O*-β-D-glucopyranosyl-*sn*-glycerol (Figure 6.9) is present in rice bran, where it represents 36% of the total MGD fraction [343]. It has the same structure as the cyanobacterial lipid, but since galactosyl- and glucosyl diacylglycerols differ significantly in fatty acid profiles in rice tissue, a biosynthetic correlation as observed in cyanobacteria seems unlikely. Glucosyl diacylglycerols have been detected in other plants (angiosperms, gymnosperms, ferns), although their proportion was always low and at most 7% of the total MGD [350]. Glucose is also present in DGD, TDG and TeGD [343]. A particularly high proportion of glucose (82%) has been found in TGD from rice grains [351]. Therefore, it has been pointed out that in addition to the structures shown in Figure 6.9 further compounds do exist in which one or even more galactosyl residues are replaced by glucosyl groups [352]. None of these compounds has been identified so far. It should be noted that galactosyl diacylglycerols of identical or similar structures are found in a wide variety of organisms including photosynthetic and heterotrophic bacteria, yeasts and fungi [2,353,354] as well as mammals and other animals [355,356]. In most cases, the pyranose form of galactose is present, although some organisms accumulate galactofuranosyl diacylglycerols.

Two other plant galactolipids have so far been isolated and characterized once only, and it is not clear, whether they have a widespread distribution. From corn kernels 1-*O*-acyl-2-*O*-β-D-galactopyranosyl-ethanediol has been isolated (Figure 6.10), and the corresponding β-D-glucopyranosyl analogue has been identified tentatively [357]. In young kernels (15 days after pollination) these glycosyl ethanediol lipids account for 50-60% of the ordinary glycosyl diacylglycerols.

In germinating mung beans (*Phaseolus mungo*) a galactolipid was identified as 1-*O*-palmitoyl-2-*O*-hexadecyl-3-*O*-[α-L-rhamnopyranosyl(1→4)-*O*-β-D-galactopyranosyl(1→4)-*O*-β-D-galactopyranosyl(1→4)-*O*-β-D-galactopyranosyl]-glycerol (Figure 6.10) [358]. This lipid is unique in having a 2-*O*-alkyl ether group and a β-1→4-linked linear trigalactoside chain capped by a terminal rhamnose. The configuration of the glycerol has not been assigned. It should be mentioned that in animal glycerogalactolipids the widespread occurrence of alkyl ether groups is confined to the *sn*-1 position of glycerol [355].

The compounds described so far have been isolated from intact tissues and after appropriate precautions (for example steaming) to prevent enzymatic alterations so that they are considered as natural constituents of membranes. Most of the lipids listed below are not present in such extracts, since most of them are formed in enzymatic reactions triggered by the elimination of subcellular compartmentation. This mixes enzymes and substrates, which in intact cells are separated. Such a situation is realized after tissue homogenization in the laboratory, after industrial milling to produce flours and other fractions from seeds or even during the normal development of endosperm reserve tissues, which in their final state are no longer alive, so that cellular compartmentation cannot be maintained. In this context the formation and occurrence of high proportions of TGD and TeGD [288,312,342] and even pentagalactosyl diacylglycerols [359] in isolated envelope membranes should be mentioned, that are ascribed to an unphysiological release of control mechanisms as a consequence of membrane isolation. These higher homologues of the galactolipid series are particularly abundant in reserve tissues in which envelopes and not thylakoids are the predominant plastidial membranes [343]. The total lipids extracted from cassava flour (*Manihot esculenta*) contain 10.2% TGD and 7.3% TeGD [360].

From deliberately homogenized leaf tissue several compounds were isolated that are formed by the action of acyl transferases and acyl ester hydrolases. These

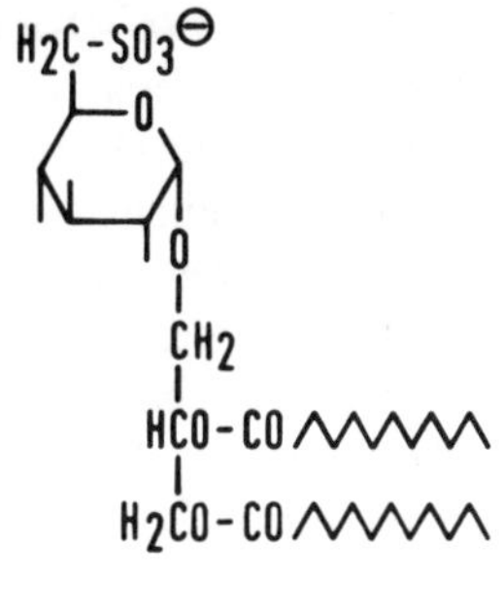

SQD

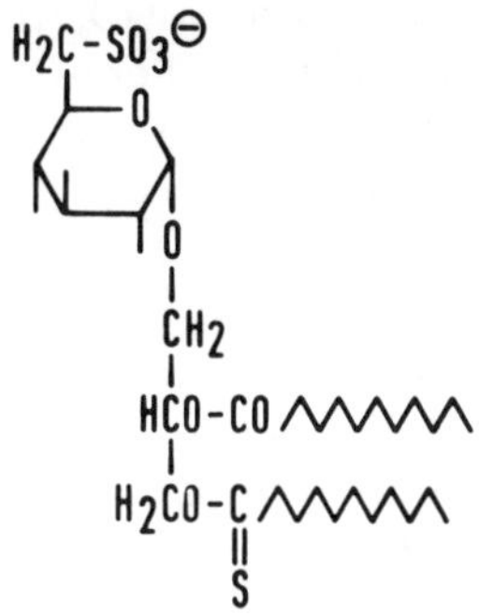

thion-SQD

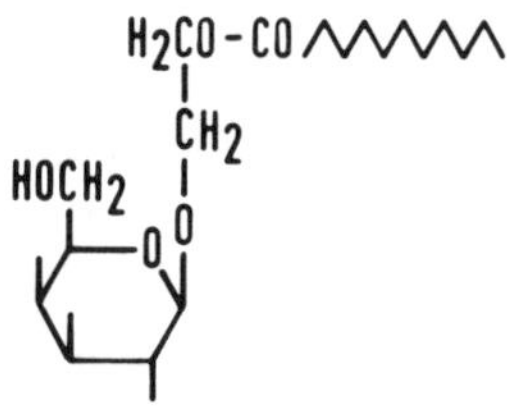

1-acyl-2-β-D-Gal-ethanediol

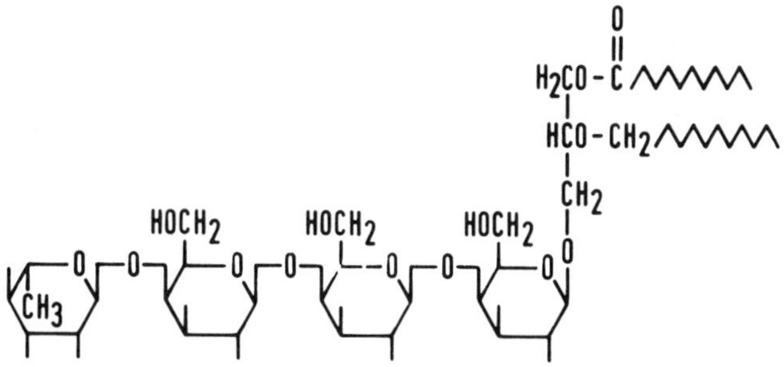

α-L-Rha(1,4)-[β-D-Gal(1,4)]$_2$-β-D-Gal(1,1)-

Fig. 6.10. Sulfoquinovosyl diacylglycerol (SQD) and a closely related compound thion-SQD carrying a S-carbonyl thioic acid ester in the *sn*-1 position. The ethanediol galactolipid (second line), from which the epimeric glucosyl derivative exists also, can be considered as a lower homologue with regard to the glycerol backbone. The rhamnosylated trigalactolipid at the bottom is unique in having a β-1,4-linked trigalactosyl chain attached to a very rare 1-acyl-2-alkyl glycerol backbone.

lipids are derived from the normal galactolipids by addition or removal of acyl groups resulting in acylated MGD, which is 1,2-di-*O*-acyl-3-*O*-(6′-*O*-acyl-β-D-galactopyranosyl)-*sn*-glycerol or AGD [361-363], the 2-lyso derivative of this compound, which is 1-*O*-acyl-3-*O*-(6′-*O*-acyl-β-D-galactopyranosyl)-*sn*-glycerol or AGM [364] and the 6″-*O*-acylated DGD or ADGD [365] (Figure 6.11). It is significant that some of these compounds (AGD and others, see below) are present in lipids of wheat flour [366,367], which may in fact represent a homogenized tissue. AGD has also been detected in fish brain where in addition to the 6′-*O*-acyl-MGD the 2′-*O*-acyl isomer is present [177]. Similar acylated glycosyl diacylglycerols occur in *Acholeplasma*, where the predominating glucosyl diacylglycerol may be converted to the 6′-*O*-acyl derivative. Because of the additional increase in the hydrophobic portion of this molecule, it has similar non-bilayer forming properties to its precursor [368,369].

Loss of a single ester group from the glycerol portion results in lyso-compounds, from which 1-*O*-acyl-3-*O*-β-D-galactopyranosyl-*sn*-glycerol or MGM and 1-*O*-acyl-3-*O*-[6″-*O*-α-D-galactopyranosyl-β-D-galactopyranosyl]-*sn*-glycerol or DGM have been identified in detail (Figure 6.11) after isolation from leaves and cyanobacteria (*Phormidium*) [370-372]. Also these lyso-compounds have been found in wheat flour [367] and rice bran [343,373], including the glucosyl form of MGM [343]. From a phytoflagellate (*Prymnesium*) the *all*-β-DGM or 1-*O*-acyl-3-*O*-[6″-*O*-β-D-galactopyranosyl-β-D-galactopyranosyl]-glycerol (Figure 6.11) has been isolated and identified as a haemolytic agent contributing to fish mortality in brackish water cultures [374]. It is not clear if the isolated 1-*O*-acyl isomers are the primary hydrolysis products resulting from removal of the 2-*O*-acyl ester group. Experiments with reference compounds have shown isomerisation by acyl migration resulting in the more stable 1-*O*-acyl isomer [375,376]. It is an unsettled question whether these lyso-compounds play a physiological role in galactolipid turnover and fatty acid remodelling as is frequently suggested. Their reacylation has been demonstrated with chloroplast preparations and with membranes from cyanobacteria [377-380]. This capacity was used as an easy way for *in situ*-labelling of the *sn*-1-bound acyl group of MGD for desaturation studies in cyanobacteria [381].

Several enzyme preparations have been obtained from various sources in different stages of enrichment and purification that attack the ester groups of galactolipids and in many cases of other lipids (acyltranferases, galactolipases and acyl ester hydrolases) [382-385]. Most preparations hydrolyse both ester groups and only occasionally lyso-forms or acylated intermediates were detected [383,386]. In most cases it is not clear whether a single enzyme splits both *sn*-1- and *sn*-2-bound ester groups, since only a few of these preparations have been purified to homogeneity [386,387]. In ruminants, which take up large quantities of galactolipids with green forage, both fatty acids are split off from MGD and DGD in the rumen of the animals in the course of a few hours [388]. Ruminal bacteria convert the free polyenoic acids by sequential biohydrogenation into stearic acid. These reactions have been followed by feeding sheep with ^{14}C-labelled grass and

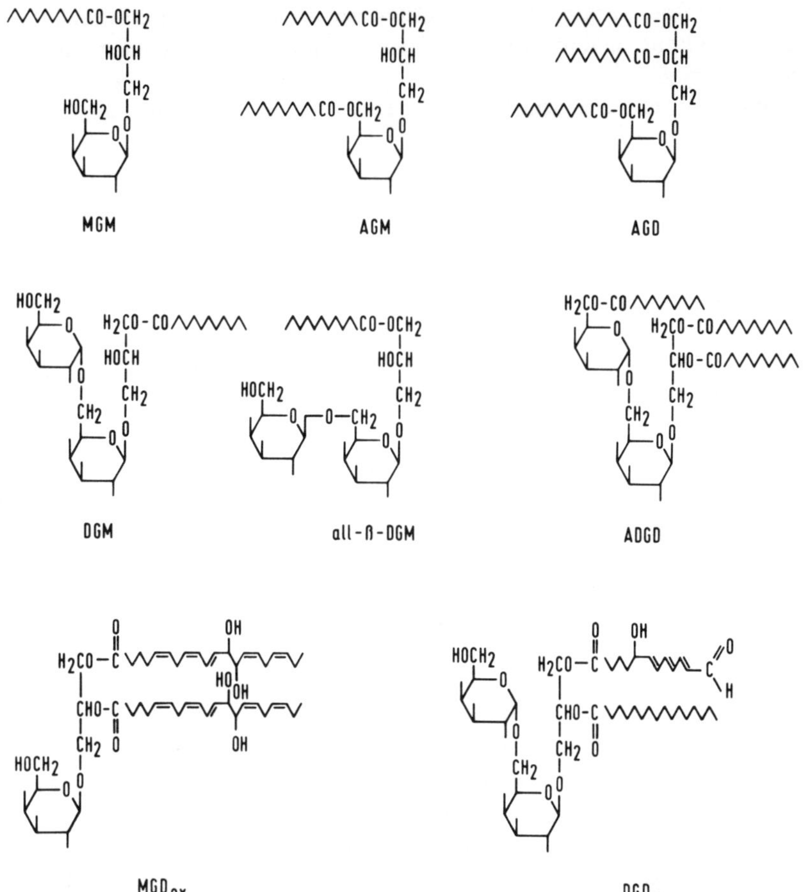

Fig. 6.11. Derivatives of mono- and digalactosyl diacylglycerol (MGD and DGD), which in most cases are considered to result from enzymatic alterations following cellular injury. Loss of an acyl group results in MGM, DGM and *all*-β-DGM, although it is not necessarily the *sn*-2 ester, which is lost first since the structures shown could result from acyl migration. Addition of an acyl group to C6 of the terminal galactose results in AGM, AGD and ADGD. The oxylipins MGD_{ox} and DGD_{ox} (bottom row) result from hydroxylation and even-chain cleavage of one or both polyunsaturated acyl groups. Whether these reactions involving the intact lipids are enzyme-catalysed or not is an open question.

sampling the ingested material from the rumen, or *in vitro* with partially purified pancreatic enzymes from sheep [389]. Similar experiments on the hydrolysis of MGD, DGD and SQD by human lipolytic enzymes have recently been published [390]. Chloroplasts of the alga *Dunaliella* contain an enzyme [391], which preferentially attacks the *sn*-1-ester of MGD and thus shows a specificity comparable to the cyanobacterial activity, although the MGM isolated from cyanobacteria is the 1-*O*-acyl isomer [372]. Also, galactolipase activities have been detected often in isolated thylakoids of higher plants and it is not understood what normally prevents them from carrying out their deleterious action [392,393]. Molecular biology via antisense inhibition may provide an approach to study the functions of these enzymes.

Completely deacylated backbones of DGD [394] and TGD [395] are present in algae and higher plants, but it is not clear whether they are formed via galactolipids or directly from water-soluble, non-lipid precursors. The deacylated product of MGD has not been found, but red algae contain 1-*O*-α-D-galactopyranosyl-glycerol or isofloridoside as a diastereomeric mixture with regard to the glycerol configuration [396].

Other galactolipid derivatives, so far only isolated from algae, are considered as first intermediates in a complex series of reactions resulting finally in the so-called oxylipins, which may have second messenger functions [397]. It has been suggested that intact MGD and DGD may be attacked by lipoxygenase activities and converted to compounds with hydroxylated and oxidized acyl residues such as those shown in Figure 6.11 [398,399]. Particularly remarkable is the aldehyde-containing DGD due to the presence of a 5-hydroxy-12-oxo-6,8,10-*all trans*-dodecatrienoyl residue as the *sn*-1-bound acyl group. The occurrence of these compounds suggests that in addition to free fatty acids intact lipids and in particular galactolipids may function as substrates for lipoxygenases [400]. In this context it should be mentioned that not only galactolipase but also lipoxygenase activities have been detected in chloroplasts [400,401].

The glycolipid most characteristic for photosynthetic organisms is the sulpholipid SQD (Figure 6.10), which has the structure of 1,2-di-*O*-acyl-3-*O*-(6′-deoxy-6′-sulpho-α-D-glucopyranosyl)-*sn*-glycerol [402]. In contrast to the biosynthesis of the above mentioned galactolipids, which involves common-place precursors formed by most organisms, the formation of the sulphoquinovose head-group (quinovose or 6-deoxyglucose) requires a set of specific enzymes, which serve only this purpose [403]. Until recently it was thought that SQD was confined to photosynthetically active organisms (anaerobic prokaryotes, cyanobacteria and eukaryotes), but the detection of SQD in several members of the genus *Rhizobium* [404] has expanded its occurrence to heterotrophic bacteria. The close interaction of the symbiotic *Rhizobium* with higher plants in root nodules has suggested that horizontal gene transfer from the plant may have created this remarkable capacity in the symbiotic bacterium.

A unique form of sulpholipid termed “thion-sulpholipid” was recently detected in cyanobacteria (*Synechococcus*) [405]. In this compound the normal *sn*-1-bound acyl group is replaced by an *O*-ester of a long-chain thiocarbonyl thioic acid (R-CS-O- or *O*-thioacyl) including hexadecane-, hexadecene- and tetradecanethioic acid (thion-SQD, Figure 6.10).

Similar to galactolipids, SQD is degraded by various enzymes from different organisms, including algae, higher plants and animals. Removal of one fatty acid yields a lyso-SQD [370,406,407], which is further deacylated to sulphoquinovosyl glycerol [407,408]. This compound, which represents a major sulphur-containing metabolite in algae [402], has been crystallized as the rubidium salt and analysed by X-ray crystallography [409]. On the other hand, in *Chlorella* an acyl-CoA:lyso-SQD acyltransferase activity has been described that reforms SQD and thus may be compared to the similar activity in MGD metabolism (see above) [407]. Further degradation of sulphoquinovosyl glycerol yields sulphoquinovose, sulpholactaldehyde, sulpholactic acid, sulphoacetate and free sulphate [410]. Some of these compounds have been detected in the gastrointestinal tract of guinea pigs after feeding with ^{35}S-labelled SQD [411]. Some years ago SQD from cyanobacteria was found to have AIDS-antiviral activity as shown by assays *in vitro* [412], but further publications on this subject have not appeared.

3. Isolation and Separation

Before extraction, homogenization or milling the plant tissue should be steamed or placed in boiling water to inactivate the different glycolipid-modifying enzymes. In contrast to enzymes glycolipids are thermally stable as evident from the fact that they can be subjected to GLC as TMS ethers of intact molecules [413-415] at temperatures of about 300°C. The inactivated tissue is extracted as described above for cerebrosides and steryl glycosides. A useful solvent mixture is chloroform/methanol (1:1), which dissolves all glycolipids and residual water of the tissue without forming a separate phase. Phase separation is achieved by addition of a second volume of chloroform and 0.25 volumes of sodium chloride solution (0.45%, w/v) yielding a washed lipid solution, from which most non-lipid contaminants are removed. With smaller quantities of lipids, complete removal of non-lipid contaminants may be achieved by chromatographic partitioning steps on Sephadex [416-418] or on prepacked solid-phase extraction cartridges [419].

It has been reported that with some algae the addition of two drops of 11 M hydrochloric acid per 100 ml of chloroform/methanol (2:1) can increase dramatically the yield of total lipids extracted [420]. On the other hand, attempts have been made to replace the cancerogenic chloroform by alternative solvents. Petroleum ether/methanol (1:1) has been reported to be as efficient as chloroform/methanol [421], although in many subsequent studies the same authors continued to use chloroform/methanol. Similarly, isopropanol/hexane (3:2) will extract all lipids and partitioning against sodium sulphate solution will remove non-lipid contaminants [422].

For isolation of larger quantities of individual components the lipid mixture is fractionated by preparative column chromatography. From the various adsorbents used, silica gel is the most common. The separation of glycolipids on such columns may be combined with preceding or subsequent chromatography on ion-exchange columns (CM- or DEAE-groups attached to different matrices), which may be useful for removal of phospholipids in cases, where they coelute from silica gel-columns (see below). The most useful solvents for silica gel-column chromatography are mixtures of chloroform with increasing proportions of methanol or acetone. Inclusion of methanol will also elute phospholipids, which stay on the column when using chloroform/acetone mixtures, and this is a great advantage for purifying glycolipids [183]. In addition, due to the reduced polarity of acetone as compared to methanol, about five times higher acetone proportions are required to result in equivalent polarities. This allows a slower increase in polarity when approaching the elution of the desired compound. The following listing gives the approximate proportions of acetone and methanol in chloroform to elute individual glycolipids, which are ordered according to increasing polarity as shown by TLC in chloroform/methanol (85:15) [365]. These proportions as taken from the different references cited above vary to some extent and may depend on the lipid load of the column and the patience of the experimenter. In most cases, more or less enriched fractions will be obtained to be used for further purification by rechromatography, preparative TLC or HPLC.

Elution of glycolipids from silica gel columns

Lipid	% (v/v) of polar solvent in chloroform	
	acetone	methanol
AGD	2	
ASG	10	2
AGM	25	
MGD	25	2-4
ADGD	45	
MGM	50	10
DGD	60	4-15
SQD	100	11-25
TGD, TeGD	100	

Most of the lipophilic pigments (chlorophyll, carotenoids) are eluted with low proportions of acetone, which do not elute AGD or MGD [363]. On the other hand, neoxanthin as the most polar carotenoid is difficult to separate from MGD. For this purpose chromatography at 4°C on silica gel adjusted to pH 8.0 has been recommended [423]. AGD can be eluted with alternative mixtures such as chloroform/hexane/diethyl ether (5:5:1) [362] and ASG with hexane/diethyl ether (3:2) [88,89]. It may be recalled that SG will elute with about 5% methanol or 40% acetone [87,101] and cerebrosides at slightly higher polarity with about 50% acetone in chloroform [148,184]. 1-Acyl-2-galactosyl-ethanediol elutes with proportions of methanol similar to those required for MGM in agreement with the similar polarity of both compounds [343,357]. For the hydroxylated and carbonyl-group carrying galactolipid derivatives (Figure 6.11) no solvent ratios can be given, since they have been isolated as acetylated derivatives [398,399]. Usually, only MGD and DGD can be obtained in pure form during the first chromatography, but also in this case, only part of the numerous fractions being enriched in the particular lipid are completely pure. In this context it should be pointed out that leading and trailing parts of eluting peaks from columns or spots travelling on TLC plates differ in the composition of molecular species, since components with shorter-chain fatty acids are retained slightly more [424]. SQD, TGD and TeGD have been eluted by extended elution with acetone [343]. To save solvent, increasing proportions of methanol may be added to acetone, but this results in contamination by phospholipids. SQD and TGD will be eluted with 5-10% methanol in acetone [47,351,414].

Phospholipids can be removed from these fractions by rechromatography on ion-exchange columns. From DEAE-cellulose (acetate form) mono- up to tetrahexosyl diacylglycerols can be eluted sequentially with increasing proportions of methanol in chloroform (for MGD 3%, DGD 10%, TGD 25%, TeGD 33%) [425], whereas the elution of SQD (as of other acidic lipids) requires the inclusion of ammonium ions for displacement, for example chloroform/methanol/aqueous ammonia (20%, w/v) (50:50:3), which can be removed subsequently by phase partitioning [414,426]. Rich sources of SQD are lipid extracts from algae, which contain far more SQD than higher plants [403]. The detailed lists of solvent mixtures required for elution of different phospholipids should be consulted to make full use of the complementary separations possible by combination of DEAE- [175,176,425] and CM-ion exchange [427,428] media following a first separation on silica gel-columns [425]. The methods just described can easily yield several grams [429] of pure MGD and DGD from a single preparative separation carried out on an appropriately sized column. For industrial purposes, batches of 100-200 kg (!) of pure DGD are being isolated by preparative HPLC (see below and personal communication, Scotia LipidTeknik).

TLC is a convenient method for final purification of compounds prefractionated by column chromatography or for separation of small quantities from total lipid mixtures as required for analysis of fatty acids or labelling patterns. In these cases, the problem is not so much the separation of the individual glycolipids, but the separation of all components present in plant lipid extracts. Many one- and two-dimensional systems are in use [2,425] and only a very few will be mentioned as examples for acidic, alkaline and neutral solvent mixtures, which often are combined for two-dimensional separation. One-dimensional separations are obtained with chloroform/methanol/acetic acid/water (170:30:20:7) [430], chloroform/methanol/aqueous ammonia (25%, w/v)/2-propylamine (130:70:10:1) [431] or acetone/benzene/water (91:30:8) [197]. This last solvent has several unique properties. It moves MGD, DGD and SQD further up than phospholipids resembling the effect seen during column chromatography (see above). If used with ammonium sulphate-impregnated plates [330,432], it results in a dramatic increase in the mobility of phosphatidylglycerol (due to protonation, whereas SQD is not affected), which now is found between MGD and DGD without risk of contamination by other phospholipids. Finally, this solvent resolves the faster moving monoglucosyl from the epimeric monogalactosyl diacylglycerol on normal silica gel plates [300,301]. The same separation is also effected by diethyl ether/isopropanol/methanol (100:4.5:3.5) [300].

Chloroform/methanol/water (70:30:4) may serve as an example for the separation of MGD (R_f 0.84), DGD (R_f 0.52), SQD (R_f 0.38), TGD (R_f 0.26) and TeGD

(R_f 0.10) [342]. Similar separations will be obtained with the same solvents in slightly different ratios (65:25:4) as shown by the R_f-values for MGD (R_f 0.76), DGD (R_f 0.53), TGD (R_f 0.31) and TeGD (R_f 0.15) [345]. This solvent has also been used for separation of sterol glycosides and cerebrosides (Figure 6.5c,d) [150,180]. All these solvents do not separate the normal DGD from the *all*-β-DGD. At this point, it should be noted that the thion-SQD is not separated from normal SQD by TLC [405]. Enzymatically altered lipids including AGD, ASG, AGM, MGD, ADGD, MGM and DGD are separated in this sequence in chloroform/methanol (85:15) [365], and such simple solvent mixtures are of general use for preparative purposes after prefractionation by column chromatography.

A critical point is the recovery of higher homologues of glycolipids from TLC plates. Extraction of the silica gel scrapings by shaking for 1 h with chloroform/methanol (2:1) releases different proportions into the supernatant after centrifugation as checked with labelled compounds (Sieberts and Heinz, unpublished): MGD (100%), DGD (90%), TGD (45%) and TeGD (8%). In contrast, extraction by phase partitioning of scrapings between chloroform/methanol (2:1, 4 volumes) and aqueous sodium chloride solution (0.45%, w/v, 1 volume) resulted in increased recovery of DGD (96%), TGD (94%) and TeGD (79%). SQD is also efficiently extracted by phase partitioning [426].

Another feature of TLC separation of glycolipids became evident during analysis of these compounds from algae. MGD, DGD and SQD may all separate into two more or less well resolved spots, which according to fatty acid analysis contain fatty acid pairings of different chain lengths [433-436]. Those with longer fatty acids move slightly higher than those with the shorter pairs, which may be explained by differences in the interaction with the polar matrix. A short description of possible acyl pairings is given in the next section (below). If two groups of species differing by at least four carbon atoms predominate in a glycolipid (20/18 + 18/16, 20/16 + 16/16 or 20/20 + 20/16), two spots may result. On the other hand, when a complex mixture of pairings is present, they are spread in a continuous gradient along the spot as shown for cerebrosides (see above [273]) and bacterial glycosyl diacylglycerols [424].

More recently, preparative HPLC has been used for the isolation of pure MGD and DGD (sometimes on the mg scale) by starting with prepurified fractions [437-440]. The purity of the resulting compounds can be expected to be superior to those obtained by column chromatography or preparative TLC (see below). They may be particularly useful for investigations going beyond an analysis of fatty acids, such as studies of physical parameters, permeability properties or monolayer characteristics [321,428,439]. The HPLC separations have to be divided into two groups (Table 6.6): normal-phase chromatography to obtain the individual lipid components and reversed-phase separations used for subsequent resolution of the molecular species (see below). Normal-phase separations are carried out on either underivatized or on chemically modified silica gel matrices. For recovery direct UV and RI detection or stream splitters with evaporative light-scattering detection or on-line thermospray mass spectrometry [441] have been used. Large quantities of DGD (up to 250 mg) have been subjected to isocratic elution from a preparative silica gel column (50 x 2.5 cm) with chloroform/methanol mixtures (85:15 or 93:7) at flow rates of 11.6 ml/min with retention times of 10 or 45 min [437]. MGD and DGD were eluted sequentially by isocratic chromatography with solvent mixtures such as n-hexane/isopropanol/water [439,440] or n-hexane/isopropanol/methanol/0.1M aqueous ammonium acetate [441], but also gradient systems have been used for the two galactolipids [438]. The lyso-galactolipids MGM and DGM have been eluted by the same systems as used for the intact lipids, with MGM eluting between MGD and DGD [438,440].

Instead of plain silica gel, chemically modified supports have been used, that may be advantageous in terms of yield and reproducibility. A diol phase representing 1,2-dihydroxypropyl-bonded silica gel was used for preparative isolation of highly purified DGD [442]. In this case, enriched DGD was obtained by solid-phase fractionation of a lipid extract (250 mg) on Sep-Pak^{TM} cartridges by sequential elution with hexane/acetone (1:1) (MGD fraction) and acetone (DGD fraction), from which 10 mg portions were subjected to isocratic elution on a semi-preparative DIOL-column (25 x 2.1 cm). The solvent, a mixture of n-hexane/isopropanol/n-butanol/water (60:30:7:3), was optimized for this separation and passed at a flow rate of 25 ml/min through a stream splitter (1:19) for evaporative light-scattering detection. DGD was eluted at about 11 min. This column can be used for other glycolipids, SQD and phospholipids as well, particularly in view of the powerful method for solvent selection and optimization.

The major advantage of normal-phase HPLC is the resolution of the different components present in the complex plant lipid mixtures. With increasing experience in this field, satisfying separations have been worked out, sometimes based on complicated solvent mixtures and gradients. Apart from preparative purposes, normal-phase HPLC is also used for qualitative and quantitative investigations of lipid mixtures. The only problem is the quantification of the different components in the effluent (see below). The upper part of Table 6.6 lists systems used for the separation of total lipids and in particular of glycolipids, some of which have already been mentioned before [443-451].

Not yet included is a cyclodextrin-bonded stationary phase, which shows a promising potential anticipating a similar resolution of glycolipids as obtained with soybean phospholipids [452]. It should be pointed out that there is rapid progress in this area and the major challenge is the separation of both glycolipid and phospholipid components in a single run (see Figure 6.5 a,b). The confinement of the present discussion to glycolipids is, therefore, admittedly somewhat artificial. The solvent gradients used in most of the above cited systems contain a basic set of a hydrocarbon (often n-hexane), isopropanol and water, to which various other components may be admixed (chloroform, methanol, acetonitrile, acetic acid, triethylamine). Irrespective of solvent gradient and stationary phase, the different glycolipids elute in the general sequence of ASG, MGD, SG, Cerebroside, DGD and SQD, although in most systems no reference is made to SQD. In addition, under these conditions all the common phospholipids including the most

Table 6.6.
HPLC separation of plant glycolipids. Normal-phase elution systems resolve glycolipid classes, whereas reversed-phase columns are used for separation of molecular species. The solvent mixtures for normal-phase chromatography are often so complex that they have not been included in the table, but a listing can be found in [442]. The C18-ODS-systems are in widespread use and many other references could be given.

normal-phase elution	reference
unmodified silicagel (Si)	106, 107, 443-446
1-cyanopropyl (CN)	447
polymerized polyvinylalcohol (PVA)	447
1,2-dihydroxypropyl (DIOL)	442, 448
1-aminopropyl (NH_2)	449-451
reversed-phase elution	
n-octadecyl (C18, ODS)	
methanol/water 90:10-95:5	345-348, 459, 462, 464, 472-474
methanol/phosphate buffer 91.5:8.5-95:5	467, 469
methanol/acetonitrile/water 90:5:10-90.5:2.5:7.5	469, 475
+ 20 mM choline chloride	464
acetonitrile/isopropanol 8:2 →100 % acetonitrile	440
n-octyl (C8)	
methanol/water 9:1 (plus $8\cdot 10^{-3}$M $AgClO_4$)	465
n-hexyl (C6)	
acetonitrile/water 1:1→100 % acetonitrile	468

mobile ones such as *N*-acyl-phosphatidylethanolamine, cardiolipin and phosphatidic acid are eluted subsequently to DGD. An exception is the aminopropyl-bonded phase, which retains acidic lipids and from which PC emerges in front of DGD, whereas in all other systems PC is eluted as the last intact lipid [449]. This is also the only system, for which the elution of TGD is indicated [449]. SQD always seems to elute in the phospholipid group.

4. Separation of Molecular Species

Reversed-phase HPLC is the method of choice for separation of molecular species, extending TLC on silver nitrate-impregnated plates (see below). In higher plants diacylglycerol backbones are assembled in plastidial and ER membranes by acyltransferase groups of characteristic selectivities [453]. The resultant diacylglycerol moieties in the various lipids have pro- or eucaryotic structures, which are recognized by the acyl substituent in the *sn*-2 position: In higher plants procaryotic combinations have C_{16}- and eucaryotic species have C_{18}-fatty acids in this position, whereas in the *sn*-1 position both C16:0 and C_{18}-fatty acids are present [454]. This simple pattern is extended by additional combinations found in various groups of lower plants, particularly in algae. However, in these organisms, an exact attribution to pro- or eucaryotic origin has not been made, that would require labelling studies combined with subcellular fractionation. On the other hand, pro- and eucaryotic acyl group localizations may be deduced from positional analyses of PG (always procaryotic) and PC or PE (always eucaryotic), if generalisations from higher plants [454,455] can be extrapolated to lower plants.

Acyl groups in diacylglycerol backbones

	higher plants		algae	
	procaryotic	eucaryotic	procaryotic	eucaryotic
sn-1	16:0, 18	16:0, 18	14:0, 16:0, 18, 20,	14:0, 16:0, 18, 20
sn-2	16	18	16	14:0, 16:0, 18, 20

Unfortunately, positional data are not given in all publications for PG, but the combinations shown above may be of general occurrence [456-462]. It is evident that even the classical absence of C16:0 at *sn*-2 of eucaryotic lipids such as PC or PE cannot be generalized [456,461]. However, in eucaryotic plants procaryotic lipids are invariably characterized by the exclusion of fatty acids with chain lengths exceeding C_{16} from the *sn*-2-position. This could indicate that of all the acyltransferases involved in lipid biosynthesis, only the 1-acyl-glycerol-3-phosphate acyltransferase from plastids has retained its acyl selectivity for C_{16}-acyl groups as already observed in cyanobacteria [47]. On the other hand, the anaerobic photosynthetically active procaryote *Rhodopseudomonas viridis* directs both C_{16} and C_{18} fatty acids into the *sn*-2 position of its glycerolipids (Linscheid *et al.*, in preparation). This suggests a cautious use of the term "procaryotic" with regard to acyl group distribution in glycerolipids.

In addition to differences in chain length, different numbers of double bonds exist in the different acyl groups at both positions, which will not be detailed here. It may be sufficient to mention that in MGD this range comprises highly desaturated C14:3 [463], C16:4 [451,462,464,465], C18:4 [461] and C20:5 [456,461] in addition to the corresponding desaturation intermediates and the more common C_{16} and C_{18} fatty acids. This may result in very complex mixtures of molecular species. Their separation and identification is of interest with respect to desaturation, since these intact molecules are the actual substrates for various desaturases. A selection of reversed-phase HPLC systems, which are most suitable for this separation, is given in the lower half of Table 6.6.

In general, isocratic systems are preferred over gradient elution. All systems are useful for separation of species from MGD and DGD, whereas a resolution of TGD and TeGD has only been mentioned once using a C18-column and isocratic elution with methanol/water (95:5) at a column temperature of 50°C [345]. SQD has been separated in underivatized form on C18-columns with methanol/water mixtures (9:1 or 92:8) [462,466], methanol/water/acetonitrile (90.5:7.5:2.5 + 20mM choline chloride) [464] and in methanol/aqueous disodium hydrogen phosphate (pH 9.0; no molarity given; 91.5:8.5) [467]. The high pH of this solution eliminated the peak tailing observed at neutral pH, that was ascribed to the mixture of dissociated and protonated sulphonic acid groups present at the lower pH. Protection of the sulphonic acid group as methyl ester was required for suitable resolution of SQD on the C6 column [468].

In a first approximation, the resolution of different molecular species is governed by the sum of acyl carbon atoms and (*cis*)-double bonds similar to the separation of fatty acids. Therefore, irrespective of the polar head group, shorter and more unsaturated acyl pairs will result in shorter elution times. A plot of retention times as a function of carbon and double bond numbers in a limited selection of pro- and eucaryotic species (18/16, 18/18) of glycolipids from higher plants resulted in two straight and parallel lines with no peak overlapping in the two series [468]. In contrast to this gradient elution, isocratic systems can be expected to yield similar figures when plotting the logarithm of retention times as compared to the direct values used for the gradient system. Such an analysis has only been mentioned briefly without being presented in detail [462]. The occurrence of additional species with 20/16, 20/18 and 20/20 combinations in algae may reward a more systematic analysis to help in identification of species by their retention times. At present the only reliable way is peak collection followed by analysis of constituent fatty acids or FAB/MS [465].

A nearly complete selection of procaryotic 18/16 MGD species occurs in *Chlamydomonas* without contamination by eucaryotic 18/18 pairings [464]. Their elution follows the expected sequence and demonstrates some important additional features. Procaryotic pairs of species with identical overall carbon and double bond numbers such as 18:3-16:3/18:2-16:4, 18:2-16:3/18:3-16:2, 18:2-16:1/18:3-16:0 and 18:2-16:2/18:1-16:3 were resolved into two peaks each, although no retention times are given [464]. A similar, but only partial resolution is shown in Figure 6.12 for the 18:2-16:1/18:3-16:0 pair in labelled form. In other systems such separations required the inclusion of silver ions in the eluant [465], or they could not be observed at all [462]. Similarly complete resolutions were obtained with eucaryotic 18:18-species [345-348], but because of the omission of experimental results, it is not always clear whether the species are listed according to elution sequences or not. In some cases, species pairs were separated that only differed by isomeric double bond systems [469]. This demonstrates that the resolution of species approaches the separation of fatty acids with regard to subtle structural differences.

In Table 6.7 pro- and eucaryotic MGD species are listed according to increasing elution time. The nearly balancing effects on retention times caused by one additional ethylene group (increase) and one additional *cis*-double bond (decrease) may result in critical pairs of pro- and eucaryotic structure in neighbouring columns, particularly in algae with their complex mixtures.

For monitoring eluted species, UV (200-205 nm), RI or FI detectors have been used, of which RI detection is not compatible with gradient elution. A comparison of UV (205 nm) and FI signals of galactolipid species demonstrates the expected differences of both systems regarding sensitivity towards more saturated species [459,470]. Similarly, UV detection (200 nm) of components from an equimolar

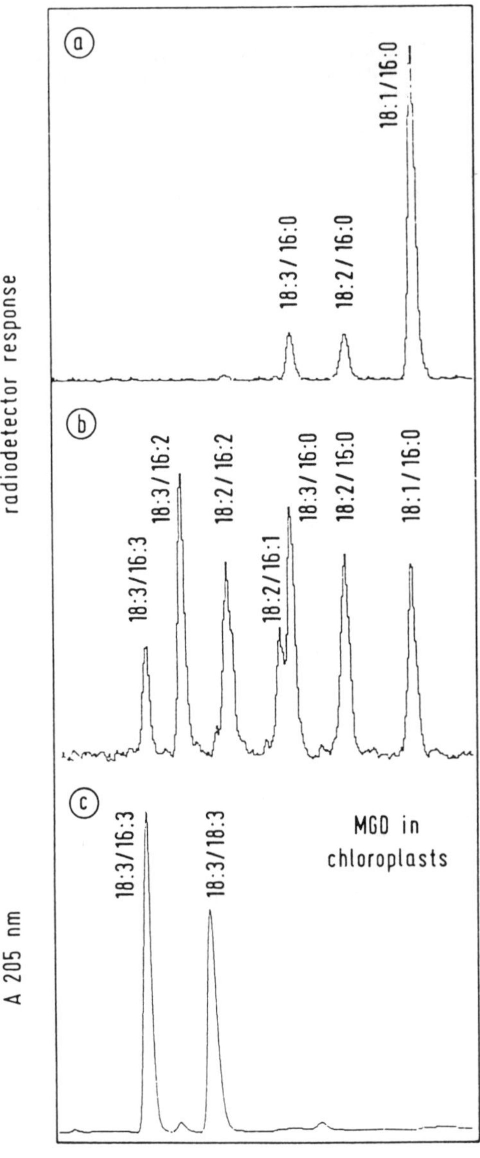

Fig. 6.12. Molecular species of MGD present and synthesized in isolated spinach chloroplasts. Separation was a achieved by RP-HPLC on a C18 column [475,476]. Eluted species were detected by UV monitoring (205 nm, c) or by flow-through scintillation (a, b). After a short incubation (3 min) with [^{14}C]acetate, the major species is 18:1-16:0. Subsequent desaturation during 60 min has produced many of the possible intermediates as well as the fully desaturated 18:3-16:3 species. The individual species, which all are of procaryotic structure, were identified by fatty acid analysis. The UV tracing (c) shows the steady state level of MGD desaturation in thylakoids, *i.e.* the two predominating and fully desaturated 18:3-16:3 (procaryotic) and 18:3-18:3 (eucaryotic) species. It should be pointed out that an almost identical sequence of species was resolved by silver nitrate-TLC [485].

Table 6.7.
Argentation-TLC (a) and reversed-phase HPLC (b-e) separation of molecular species of MGD. Examples have been selected to demonstrate the elution sequence of various pro- (18/16, 20/16) and eucaryotic (18/18, 20/18, 20/20) species, which are listed with increasing R_f-value (a) or increasing retention time (b-e). Species groups in b-e are placed into different columns, and the actual elution order is obtained by frequently changing between columns when going from top to bottom. In a, c-e all species listed were separated in the same run each, whereas in b the relative position of the 18/18 with respect to the 18/16 pairs was deduced from a separate analysis not including all 18/16 species. The listed sequences should not be considered as indicating similarly clear and equally spaced separations between all species. In different publications, slightly different sequences, inversions of pairs or overlapping positions may be found. In a-d underivatized MGD was separated, whereas in e benzoyl derivatives were used. The separations by $AgNO_3$-TLC and RP-HPLC follow remarkably similar criteria.

a	b		c			d			e			
18/16	18/16	18/18	18/16	18/18	20/18	20/16	20/18	20/20	18/16	18/18	20/16	20/18
18:3-16:3	18:3-18:3			18:4-18:4		20:5-16:4				18:3-18:2		
18:3-18:3	16:3-16:4			18:4-18:3			20:5-18:4			18:2-18:2		
18:3-16:2	18:2-16:4		18:3-16:3			20:5-16:3			18:3-16:0			
18:2-16:2	18:2-16:3			18:3-18:3				20:5-20:5			20:4-16:0	
18:3-16:1	18:3-16:2				20:3-18:3	20:5-16:2						20:1-18:3
18:3-16:0		18:3-18:3	18:3-16:2			20:5-16:1				18:2-18:1		
18:2-16:1	18:2-16:2			18:3-18:2			20:5-18:2		18:2-16:0			
18:2-16:0	18:2-16:1				20:2-18:3	20:4-16:1				18:0-18:3		
18:1-16:0	18:3-16:0			18:3-18:1						18:1-18:1		
16:0-16:0		18:2-18:2		18:2-18:1					18:1-16:0			
	18:2-16:0		18:3-16:0							18:0-18:2		
	18:1-16:0			18:3-18:0								20:0-18:3
		18:1-18:1	18:2-16:0							18:0-18:1		
												20:0-18:2
spinach chloroplasts [485]	Chlamydomonas [464] ([468])		gymnosperms [346]			diatom [462]			wheat flour [440]			

mixture of MGD species having 6, 4, 2 or no double bonds showed the dramatic decrease in UV detectability with decreasing unsaturation [468]. In addition, it was demonstrated that UV absorption is different for 18:1-18:1 and 18:2-16:0 species. From this investigation, relative UV response factors for the different acyl groups were deduced, which were 0.49 for 18:1, 2.05 for 18:2 and 3.45 for 18:3 and 16:3, whereas the corresponding factors for unsaturated acyl pairs were 1.00 for 18:1-18:1, 4.12 for 18:2/18:2 and 6.49 for 18:3-18:3 and 18:3/16:3 [468]. In a similar investigation [471], the resolution of underivatized phospholipid species by RP-HPLC was monitored by UV absorption at 205 nm followed by peak collection and quantification via fatty acid GLC. This allowed the calculation of relative conversion factors for different fatty acids, by which 205 nm absorptions have to be divided to give mole%. A recalculation of these relative factors for comparison with the above mentioned series results in 0.49 for 18:1, 1.99 for 18:2, 7.1 for 20:3, 6.0 for 20:4, 7.4 for 20:5 and 8.2 for 22:6. These data show that simple UV increments for increased unsaturation do not exist and fully saturated species as present in SQD can hardly be detected [464]. Therefore, quantification based on UV absorption has to be evaluated carefully. On the other hand, a reliable comparison enabled by UV detection is the ratio of the fully unsaturated and predominating 18:3/18:3 and 18:3/16:3 species of MGD from 16:3-plants (see Figure 6.12c). A generally more reliable quantification may be obtained by FI, RI or evaporative light-scattering detectors, but data on an evaluation have not been published. The quantities required for UV detection (at 200 nm about 0.5 nmol or 0.4 μg MGD [468]) are about 100 times smaller than those for RI detection [472]. It should be pointed out that UV absorption of double bonds decreases sharply when going from 200 nm to even slightly longer wavelengths (which decreases the price of solvents). Such problems are circumvented by peak collection and fatty acid analysis after addition of an internal standard.

A different approach for separation and quantification of glycolipid molecular species is based on the release of diacylglycerols from MGD, DGD and SQD by a simple chemical reaction sequence [477]. It involves oxidation with periodic acid in methanol (90 min) and, after decomposition of periodate with ethyleneglycol and extraction of the oxidized lipid, treatment with 1,1-dimethylhydrazine (4-24 h at room temperature) for release of diacylglycerols by β-elimination. This smooth reaction, which gives *sn*-1,2-diacylglycerols (see below) with unaltered fatty acid composition in high yields, can be carried out with as little as 100 μg of MGD. After release the diacylglycerols are esterified with an aromatic acyl chloride (benzoyl, anisoyl, nitrobenzoyl, dinitrobenzoyl chloride) or converted to aromatic urethanes by reaction with dinitrophenyl or naphthyl isocyanates. The separation of the resultant derivatives can be followed by UV absorbance at longer wavelength (250 nm) without interference from fatty acid double bonds. The resultant elution profiles are directly proportional to molar proportions as exemplified for plant glycolipids [468,478]. An extensive number of investigations has been devoted to the separation of these derivatives, particularly of dinitrobenzoyl derivatives by RP-HPLC [479,480] or of TMS ethers by GLC [481]. In most cases the diacylglycerols were obtained by enzymatic cleavage of phospholipids, and from the numerous investigations the two mentioned [479,480] provide extended lists of logarithm of retention times versus carbon/double bond numbers of many different species.

Instead of releasing diacylglycerols for attaching a chromophoric group to the free hydroxyl group, intact MGD and DGD have been converted to perbenzoylated derivatives by treating the glycolipids with benzoic anhydride in pyridine in the presence of dimethyl aminopyridine overnight [440]. The derivatives were resolved by gradient elution with 20% isopropanol in acetonitrile (0.5% increase/min), and were detected at 230 nm yielding molar proportions. The elution sequence of MGD species including 18/16, 18/18, 20/16 and 20/18 pairings is shown in Table 6.7e. MGM and DGM have also been separated as perbenzoylated derivatives into molecular species, although this separation does not provide more information than a corresponding analysis of fatty acids.

Argentation-TLC is another useful method for separation of molecular species, that was introduced into the glycolipid field by Nichols and Moorhouse [482]. This method was extensively used in studies on glycolipid desaturation in leaves [483,484], chloroplasts [485] and cyanobacteria [486]. For preparation of impregnated plates, commercial TLC plates are either immersed [487] in a solution of silver nitrate (5%, w/v, in acetonitrile) or sprayed [488] with such a solution (5-10% silver nitrate, w/v, in acetonitrile, 10 ml for a 20 x 20 cm plate) followed by air drying (poisonous) in the fume hood and activation for 3 h at 130°C. The solvents used for chromatography are the same as used on normal plates provided that the R_f value of the compound is at least 0.5. Therefore, with higher homologues of glycolipids the polarity of the solvents has to be increased. MGD has been separated with chloroform/methanol/water (60:21:4) [482,488], chloroform/methanol/acetone/acetic acid (80:15:5:1) [485] or acetone/benzene/water (91:30:8) [487]. A slightly more polar mixture of chloroform/methanol/water (60:30:4 [487] or 65:35:4 [488]) was used for DGD and SQD, whereas the polarity had to be increased significantly for TGD (chloroform/methanol/water/acetic acid (50:25:4:4)) and TeGD (25:15:4:2) [488]. Since increasing temperature worsens resolution, chromatography has been carried out in the refrigerator or even in the deep freeze. For quantification spots have to be visualized under UV light after spraying with any of the fluorescent dyes (primuline, fluoresceine, anilinonaphthalene), scraped off from the plate and transesterified to produce fatty acid methyl esters (for example in 0.5M sodium methoxide [484,489]) with inclusion of internal standard. Silver nitrate-TLC has also been used for preparative isolation of different molecular species of MGD, some in mg quantities, covering the degree of unsaturation from 18:1-16:0 to 18:4-16:4 [490]. Silver ions can be removed by phase partitioning against ammonia or aqueous chloride solution or by rechromatography on normal plates [490,491].

As an example of the potential of this technique, the separation of procaryotic MGD species obtained by incubating isolated spinach chloroplasts with [^{14}C]acetate is shown in Table 6.7a. A comparison with Figure 6.12b and column b of Table 6.7 demonstrates that resolution by argentation-TLC is very similar to HPLC separation and that both methods separate according to the same structural details. The separation of a large number of pro- and eucaryotic MGD species has been studied in labelling experiments with leaves from rapeseed. These and other analyses show that during silver nitrate-TLC also critical pairs do occur which are difficult to separate [484,488,490].

Before the general availability of HPLC facilities, GLC and MS had been used for analysis of fatty acid pairings or even of molecular species of different glycolipids [342,413-415]. GLC of TMS ethers had shown the existence of 18/16 to 20/18 pairings in MGD of ferns and even small proportions of "eucaryotic" 18/18 in cyanobacterial MGD. The occurrence of additional combinations such as 16/16 to 20/18 in DGD from mosses and 14/14 to 24/16 in SQD from brown algae had already been realized. Direct inlet-MS of acetylated glycolipids differentiated the various molecular species from MGD up to TeGD and demonstrated the existence of 16:0/16:0-SQD in spinach envelopes. On the other hand, these species patterns showed large and systematic variations with increasing probe temperature and heating time due to differences in volatility of the different species [342]. Similar effects have been described for cerebrosides (see above [268]).

5. *Quantification and Immunology*

The different glycosyl diacylglycerols can all be measured colorimetrically after TLC separation by direct addition of colour reagent to TLC scrapings. The method is the same as described above for cerebrosides. From the various reagents used, phenol is most popular and apart from its high sensitivity, a further advantage is the equal response of galactose and sulphoquinovose when measuring the absorbance at 485 nm [492]. In contrast, the colour developed with anthrone has different absorption maxima for galactose (620 nm) and sulphoquinovose (590 nm) [493]. The reagent volumes added to the TLC scrapings vary from 1.5-5.0 ml and accordingly the absorptions per sugar vary as well. A consequent reduction in volume is described in [226], in which a microplate assay in a total volume of 140 μl resulted in an 430 nm absorbance of about 1.0 for 3 μg of mannose. But as mentioned in the section on cerebrosides, a minimal volume of 1.5 ml of reagent solution is required for dispersion of TLC scrapings.

A practical range covered by all these methods is 20-200 nmoles of sugar, which parallels the sensitivity of the general methods for phospholipid phosphate quantification. An equivalent quantity of glycolipid is present in a lipid extract from 0.1-0.2 g of leaves containing in the order of 100-200 μg of chlorophyll [11,403]. This will often result in upper limits for DGD, whereas SQD, cerebrosides and steryl glycosides will fall into the lower end of the calibration curves with correspondingly high variation of individual measurements. Usually, TGD and TeGD are only minor components, but in some cyanobacteria [47], in chloroplast envelopes [342] and in some storage tissues [360] they occur in quantities high enough to permit direct quantification together with the other components.

Very often quantification is carried out by fatty acid analysis after TLC separation and transesterification in the presence of internal standard. The quantity of this internal standard is only rarely given in the published data, and in contrast to colorimetric studies, detailed analyses on range, linearity and variation of the whole procedure have hardly been published. Undoubtedly, GLC analysis of fatty acid methyl esters may be carried out at very high sensitivity, but the data given occasionally on quantities of internal standard used (C15:0 or C17:0, 20-200 nmoles, [483,494,495]) show that in many cases it is not the sensitivity of GLC but of the TLC separation and detection as well as the occasional presence of contaminant, which limit this method and increase its sensitivity to less than the expected level. Similar considerations apply to the analysis of sugar constituents after release and appropriate derivatisation for GLC or HPLC quantification, which also are very sensitive *per se* but require the same separation, detection and recovery on and from TLC systems. On the other hand, as outlined above in the case of cerebrosides, careful handling of this method can indeed increase the sensitivity by about two orders of magnitude to cover the range of 10 pmoles to 10 nmoles [252].

A different approach for quantification is based on TLC separation, visualization of glycolipids as well as of other lipids by charring or specific staining followed by densitometry. This method has been applied to total lipids [360,496,497] as well as to MGD, DGD and SQD, for which calibration curves covered the range from 1 to 10 μg, *i.e.* from about 1 to 10 nmoles [498]. In the meantime laser-operated densitometers are available, but this method does not seem to be in use for plant lipids, whereas encouraging results have been obtained with animal phospholipid mixtures even after two-dimensional TLC separation and charring with copper sulphate/phosphoric acid [499].

A way out of this unsatisfactory status of plant lipid quantification would be highly desirable, and it is expected that the high resolution and sensitivity of normal-phase HPLC may solve the problems outlined above, particularly by avoiding the various steps of TLC separation, detection and scraping of the spots followed by derivatisation and the actual quantification, which all increase the variation and decrease the sensitivity. But as frequently pointed out, the elegance of HPLC separation and detection suffers from the fact that the detector systems presently in use respond with different sensitivity to identical quantities of different lipids. This necessitates different and sometimes repeated calibration with different lipids, and the resulting calibration curves may be hyperbolic and not linear [447]. Compared to the FID system, the evaporative light-scattering detector may be the more promising one because of comparatively care-free operation [500]. With increasing refinement of instrumentation, reliable quantification is now possible with 1-5 μg per component [447], but calibration curves up to 200 μg have been published [107,444]. In many of these studies, MGD and DGD have been included, whereas an evaluation of the response of SQD is still missing.

The potential use of immuno-TLC methodology, both for qualitative and quantitative purposes, has been outlined above (section on cerebrosides). Antibodies against a whole spectrum of plant lipids and pigments, including MGD, DGD, TGD and SQD antisera, have been prepared by Radunz [501-503]. These antibodies were elicited in rabbits by repeatedly injecting (in some cases up to seven times) a mixture of glycolipid/methylated bovine serum albumin/Freund's adjuvant dispersed in saline solution. Sometimes other lipids such as lecithin or cholesterol were included to help in dispersion [504]. Various techniques including dot blots [322] were used to evaluate the monospecificity of the resulting sera. Only occasionally as in the case of MGD [504] and TGD [503], some of the sera showed cross-reactions with DGD, but the corresponding DGD antibodies could be removed yielding monospecific antisera [503]. Apart from polyclonal also monoclonal antibodies are available against MGD [505,506], and this is due to the fact that antibodies raised against animal sulphatide (galactocerebroside sulphate) and seminolipid (monogalactolipid sulphate) show cross-reactions with plant MGD despite the fact that they are monoclonals. The quantitative evalution of the different glycolipid antisera showed that it is possible to detect as little as 5 ng or 7 pmoles of the plant glycolipids [322] in agreement with the sensitivity of the cerebroside antibodies (see above, [215]). The glycolipid antibodies have been used in studies on the localization of galactolipids in the outer surface of the outer membrane of chloroplast envelopes [507] as well as on their sidedness in thylakoid membranes. Furthermore, the effects of their binding to thylakoid membranes on the partial reactions of photosynthetic electron transport have been investigated in detail [508]. Recently their use has been extended to western blotting, by which it was shown that the various protein complexes solubilized from thylakoid membranes in the presence of detergents retain specific glycolipids throughout the whole procedure involving electrophoresis and blotting [509]. On the other hand, the various new techniques for quantification involving TLC followed by direct immunostaining or after *in situ* conversion to an appropriate immuno-detectable or fluorescent derivative, as outlined above for cerebrosides, have not been applied to plant glycosyl diacylglycerols.

6. Structural Analysis

The elucidation of the structure of glycosyl diacylglycerols requires identification and quantification of the different constituents (usually sugar, glycerol, fatty acids) and the analysis of linkages between these building blocks regarding type, position and stereochemistry. Various hydrolytic and degradative reactions in combination with different chromatographic systems as well as spectroscopic studies of intact molecules by NMR spectroscopy and MS contribute to the final picture. Biochemical and metabolic studies rely heavily on the first set of methods, particularly with regard to localisation of radioisotopes in the various parts of the molecules, whereas a rigorous structural identification will include spectroscopic data. The methods resemble those already described for the other glycolipids and, therefore, discussion will be concentrated on items specific for the group of glycosyl diacylglycerols. As practised above, data resulting from the application of highly sophisticated instrumentation not available or accessible in every laboratory will be supplemented by approaches easily carried out in laboratories primarily engaged in the investigation of metabolically oriented problems.

i) Hydrolysis and deacylation. Depending on the constituents to be analysed different hydrolysis procedures have to be selected. When only fatty acids are of interest, transesterification in anhydrous methanol catalysed by hydrogen chloride, sulphuric acid, boron trifluoride or sodium methoxide yields methyl esters, which after extraction into petroleum ether and subsequent washing and drying (sodium or magnesium sulphate) are ready for GLC or HPLC analysis (not to be detailed here). These methods are carried out routinely in the presence of silica gel from TLC scrapings. A simple method for easy handling of many samples is treatment with 0.5 M sodium methoxide for 15 min at room temperature in the presence of dimethoxypropane (5% v/v) to eliminate any water present [484,489,510]. For confirmation of GLC assignment of fatty acids, re-analysis after hydrogenation may be useful, particularly in the case of glycolipids from lower plants. For this purpose, a simple hydrogenation procedure for intact lipids and fatty acids has been developed, which is carried out in sovirel tubes. The inverted tube containing the dried lipid sample in gassed for 1 min with hydrogen gas and capped followed by injection through the Teflon™ seal of the catalyst suspended in a suitable solvent (for example, 0.5 ml of methanol) [511]. This method is particularly convenient for radioactive samples.

Complete cleavage of glycosidic bonds for release of sugars and glycerol requires heating with mineral acids (usually 1-2N) either in anhydrous methanol or in aqueous media for a few hours at 90°C (see also the experiments on optimal hydrolysis conditions of cerebrosides discussed above [232]). Before chromatography or further derivatisation the mineral acids have to be removed. After fatty acid extraction, the aqueous phase is neutralized by addition of barium carbonate (for sulphuric acid), silver carbonate (for hydrochloric acid) or by incubation with anion-exchange resin (in hydroxide, formate or acetate form) [425]. After centrifugation, the supernatant fractions are taken to dryness after repeated addition of benzene/carbon tetrachloride/methanol (1:1:1). Hydrolysis mixtures with hydrogen chloride, particularly anhydrous solutions, are also simply blown to dryness and stored over potassium hydroxide pellets [370].

Whenever aqueous media were used for hydrolysis, the residue contains free sugars and glycerol and can be used directly for TLC separation or for further derivatisation (see below). TLC on silica gel in n-propanol/pyridine/water (7:4:2) will separate the following compounds, which are of particular interest for this group of glycolipids, in the sequence given [425,512]: glycerol (R_f 0.75), β-glucopyranosyl-3-glycerol, β-galactopyranosyl-3-glycerol, glucose, galactose, digalactosyl-glycerol, trigalactosyl-glycerol and tetragalactosyl-glycerol. Sulphoquinovose and sulphoquinovosyl-glycerol are also separated by this solvent, although the hydrolysis/deacylation products of SQD, *i.e.* glycerol, methyl sulphoquinovoside, free sulphoquinovose and sulphoquinovosylglycerol are more frequently separated by paper chromatography or cellulose TLC with *n*-butanol/pyridine/water (6:4:3) and methanol/formic acid/water (80:13:7) [351,436,513]. Glycerol has been separated from ethyleneglycol [514] with chloroform/methanol (85:15) or from glyceraldehyde [515] in acetone/chloroform/30% (w/v) ammonium hydroxide (80:10:10). Glycerol and ethyleneglycol will not react with the sugar-specific colour reagents, but may be detected by spraying with ammoniacal silver nitrate or alkaline permanganate solution [512,514]. Glycerol and glycerol-3-phosphate have also been separated by cellulose TLC in ethyl acetate/acetic acid/water (3:1:1), and identification may be based on enzymatic reduction of NAD on incubation of hydrolysis products with glycerol kinase, glycerol-3-phosphate dehydrogenase, ATP and NAD [516]. Commercial kits are available for enzymatic quantification of glycerol as well as of galactose (via galactose oxidase), and this at the same time is evidence for the D-configuration of the sugar [38].

Experiments on partial acid hydrolysis of the higher homologues of MGD have not been published, but our own studies (Wrage and Heinz, unpublished) showed that suitable conditions for release of MGD from DGD are 0.5% (w/v) lubrol and 0.5% (w/v) sulphuric acid in aqueous dispersion for 2 h at 100°C. This yields 20-30% of MGD as checked with labelled DGD (0.1-1.0 mg in 0.5 ml). The hydrolysis is unselective with regard to molecular species as demonstrated by RP-HPLC analysis [517]. Partial acid hydrolysis of deacylated glycolipid backbones has frequently been performed by reducing hydrolysis time under normal conditions to 10 min [90,518] or by extending hydrolysis time at both reduced temperature and reduced acid concentration (0.025N) [95,344]. It should be noted that partial hydrolysis results in complex mixtures of products due to the release of compounds from both ends of the backbone, including various reducing oligosaccharides (see below). Furanoside glycosides would also be split under these conditions, but in contrast to bacteria [2,353,354], this ring size has not been found in plant glycosyl diacylglycerols.

Apart from the identification by TLC as mentioned above, the free sugars released by aqueous acid hydrolysis can be converted to alditol acetates by sodium borohydride reduction/acetylation for subsequent GLC/MS analysis [125], whereas methyl glycosides obtained by methanolysis under anhydrous conditions are conveniently analysed by GLC as TMS ethers (see below). More recently, various HPLC methods have been developed for sugar separation and identification. Also in this case, a single derivative is formed from each free sugar by reacting the aldehyde group with a steadily increasing number of reagents such as 2-aminopyridine, phenylisothiocyanate or *p*-aminobenzoic ethyl ester [519-521]. The derivatives are detected by UV light absorption (236-254 nm) with high sensitivity and linear response between 0.1 and 1.25 nmols, but even 50 pmoles can be detected. Increased sensitivity is obtained by fluorescence detection. As well as derivatives, free sugars can be separated on various columns using RI or evaporative light-scattering detection [522]. All these methods have not been applied to the analysis of plant glycosyl diacylglycerols.

The different constituents of plant glycosyl diacylglycerols have been separated and quantified by a single GLC run [496,523]. After methanolysis in anhydrous methanol/hydrogen chloride and removal of hydrogen chloride by evaporation in a stream of nitrogen, the residue was mixed with reagents for TMS ether formation. GLC with temperature programming from 80 to 250°C on OV-17 resolved glycerol (4 min), multiple peaks for methyl galactosides and glucosides (10-12 min) and fatty acid methyl esters (12-14 min, C_{16} and C_{18}). The same procedure was used for methanolysis products from sterol glycosides, in which case sterols emerge after 22-25 min [496], and for deacylated glycolipids (see below). Depending on the speed of temperature rise and stationary phase, the various forms of methyl glycosides from galactose (usually 3 peaks) and glucose (2 peaks) are more or less well separated [351]. For quantitative and radioactive work it should be kept in mind that glycerol may be lost during evaporation of hydrolysis mixtures at higher temperatures due to its volatility. The detector response for the three groups of hydrolysis products may require calibration due to differences in signal yield [523], and when mannitol has been used as internal standard, the neutralisation of hydrogen chloride by silver carbonate should be avoided, since this results in specific and significant losses of mannitol, but not of methyl glycosides [524].

Particularly important is the identification of glycosylglycerol backbones resulting from removal of fatty ester groups. Deacylation is achieved by alkaline methanolysis catalysed by sodium methoxide (5-500mM in anhydrous methanol, [108,510]) or with potassium or sodium hydroxide (usually about 0.1M) in methanol or methanol/chloroform mixtures, sometimes in the presence of small proportions of water. Incubation at room temperature for 5-30 min is followed by water addition (in the presence of chloroform) for phase separation and removal of fatty acids (or methyl esters). The upper aqueous phase is used for removal of cations by addition of a cation exchange resin in the $H^{\oplus}$or $NH_4^{\oplus}$ form [425]. After centrifugation, washing and solvent removal, the residue is ready for analysis by TLC, GLC or enzymatic hydrolysis with various glycosyl hydrolases.

For TLC separation the solvent system *n*-propanol/pyridine/water (7:4:2) given above separates the series of galactosylglycerols with increasing number of galactosyl residues [425]. As already mentioned, it also separates β-glucosyl- (higher R_f) from β-galactosylglycerol as well as α- (higher R_f) from β-galactosylglycerol. A similar resolution is observed for the α/β-glucosylglycerols. Since α-glycosyl (1→2)glycerol (higher R_f) is separated from α-glycosyl(1→3)glycerol (shown for both glucose and galactose), an analogous separation can be expected for β(1→2)- and β(1→3)glycosylglycerols. On the other hand, we observed that βgal(1→6)βgal(1→3)glycerol (from *all*-β-DGD) was not separated from αgal(1→6)βgal(1→3)glycerol (from normal DGD), whereas *all*-α-digalactosylglycerol was separated from the normal α,β-digalactosyl glycerol (from normal DGD) [425]. A solvent of similar resolving power is n-propanol/ethyl acetate/water (7:2:2), which requires double development [425]. The detailed R_f-values for a multiplicity of glycosylglycerols have been tabulated for TLC on silica gel plates and for paper chromatography in [2,425]. The higher R_f-value of glucosyl- and galactosylglycerol as compared to free glucose and galactose is somewhat surprising. If this behaviour can be extended to higher homologues, as for example the products of partial or enzymatic hydrolysis, the following sequence may be observed for products released from trigalactosylglycerol, when subjected to silica gel TLC: glycerol, galactosylglycerol, galactose, digalactosylglycerol, galactosylgalactose, trigalactosylglycerol, digalactosylgalactose. These compounds will react differently with reagents specific for sugar (α-naphthol/sulphuric acid), hydroxylated compounds (alkaline silver nitrate or permanganate) or reducing sugars (anilinephthalate). The partial hydrolysis of an isomeric trigalactosylglycerol, *i.e.* 3-digalactosyl-2-galactosylglycerol of bacterial origin, did in fact give all the expected products when kept for 10 h at 100°C in 15 mM aqueous hydrochloric acid [525]. The somewhat simpler mixture resulting from partial hydrolysis of normal digalactosylglycerol (from plant DGD) was resolved by GLC as TMS ethers (see below, [518]).

It may be anticipated that the silica gel TLC systems given above will also separate α- from β-galactofuranosyl(1→3)glycerol and the α/β-galactofuranosyl from the corresponding α/β-pyranosyl forms as expected from the separation observed by paper chromatography in *n*-butanol/pyridine/water (6:4:3), in which the β-galactofuranosyl(1→3)glycerol (R_f 0.43) migrated well ahead of the β-galactopyranosyl (1→3)glycerol (R_f 0.27) [526]. Therefore, identical or different R_f-values of unknown and reference compounds in these solvents will already suggest many structural details, and this applies even more for GLC analyses.

GLC of TMS ethers is the method of choice for separation and identification of deacylated glycolipids, particularly regarding the resolution possible on capillary columns, although most data were obtained with packed glass columns using OV-17 or SE-30 as stationary phases and temperature programs covering the span between 180 and 320°C. The sequence of pyranosyl glycosides according to increasing retention times of TMS ethers is the following ([357,425]); the relative position of the *all*-α-digalactosylglycerol* has been taken from an analysis of trifluoroacetyl derivatives [525]):

β-gal-ethanediol < β-gluc-ethanediol < α-gal(1→2)-glycerol < α-gluc(1→2)-glycerol < α-gal(1→3)-glycerol < α-gluc(1→3)-glycerol < β-gal(1→3)-glycerol < β-gluc(1→3)-glycerol < α-gal(1→6)-α-gal(1→3)-glycerol* < α-gal(1→6)-β-gal(1→3)-glycerol < β-gal(1→6)-β-gal(1→3)-glycerol

In this collection furanosylglycerols are missing, although α/β-galactofuranosyl(1→3)glycerols have been subjected to GLC analysis as TMS ethers several times [526,527]. However, the experimental data given do not allow their integration into the list given above. This sequence indicates that the elution time increases, when changing from compact to more extended structures (galacto→gluco, α→β, 2→3-position at glycerol), and this also holds true for higher homologues. It is the basis for separating the deacylated backbone pairs from normal and *all*-β-DGD, TGD and TeGD [345]. In this context it should be mentioned that normal and *all*-β-backbones of DGD-TeGD were also separated as peracetylated derivatives by TLC on borax-impregnated TLC plates with chloroform/benzene/acetone (4:1:1) [345]. In this system the *all*-β-structures have lower R_f-values than the normal compounds, which covered the range from R_f 0.55 (normal DGD) to R_f 0.18 (*all*-β-TeGD). After deacylation and GLC separation of TMS ethers of MGD, DGD, TGD and TeGD from rice bran [343], broadening and partial splitting of the various peaks were observed, but in this case it may be attributed to the presence of glucosyl residues at different locations in the different lipids. GLC of TMS ethers was also used for the separation of products obtained by partial hydrolysis of digalactosylglycerol (5% hydrogen chloride in methanol at 75°C for 10 min), and all possible hydrolysis products from glycerol to galactosylgalactose were separated and identified [518]. In this case free galactose and galactosylgalactose always had shorter retention times than their glycerol glycosides as expected, in contrast to their TLC behaviour as discussed above.

Apart from TMS ethers, methyl, acetyl and trifluoroacetyl derivatives of glycolipid backbones have been used for GLC analysis [523] including capillary columns [528], but most data are available for TMS ethers as listed above.

Glycosylglycerol backbones have been used for enzymatic analysis of glycosidic linkages with regard to anomeric configuration. Because of the high purity and sugar selectivity of the commercially available enzymes, such an analysis proves the nature of the sugar forming this bond at the same time. Such an experiment led to the discovery of the glucosyldiacylglycerol in cyanobacteria [300]. Experimental details for incubations (100 μg of substrate in 300 μl assay volume) with α- and β-galactosidase and -glucosidase are given in [343] and [512]. The reactions are sometimes run for 72 h before being stopped by addition of methanol and followed by deionisation of the supernatant obtained by centrifugation. It should be pointed out that β-galactosidase from almond emulsin splits off both pyranosyl and furanosyl residues, whereas the β-galactosidase from *E. coli* is limited to pyranosides [529]. Therefore, the use of this enzyme from two different sources may support assignment of both anomeric structure and ring size of the galactosyl residues. Products can be analysed by TLC and GLC as described above. In contrast to steryl glycosides and cerebrosides (see above), no experiments on hydrolysis of intact MGD and DGD by galactosidases have been reported so far. Our own efforts in this direction always failed, even after presenting the galactolipids in mixed galactolipid/phospholipid vesicles to prevent formation of inverted micelles with MGD. But as outlined above, the most useful method for release of diacylglycerol portions from different glycolipids is a chemical reaction sequence involving periodate oxidation followed by β-elimination with dimethylhydrazine [477]. In digestive tracts, the sequence of degradation may start with lipolysis [388-390] followed by hydrolysis of deacylated backbones [308]. On the other hand, the α-glycosidic linkage in SQD is cleaved by β-galactosidase upon extended incubation [402,405], and recently SQD was shown to be a potent inhibitor of α-glucosidase from yeast with a K_i of about 3 μM [530]. The sequence of galactose oxidase/sodium borotritiide reduction for labelling of galactolipids at C6 of their terminal galactose is well established for cerebrosides [531], but with plant galactolipids labelling was not satisfactory [532].

The glycosylglycerol backbone of the glycolipid with an *O*-hexadecyl ether linked to C2 of glycerol (Figure 6.10) will not be released by alkali. The ether linkage was split by degradation of the lipid with boron trichloride in chloroform at -80°C [358]. The alkyl chloride extracted with hexane was analysed by GC/MS. Cleavage of ether linkages with hydriodic acid or boron trichloride is a common reaction in the analysis of lipids from archaebacteria [2].

ii) Positional analysis, glycerol configuration, methylation. The positional distribution of fatty acids, *i.e.* their attribution to the *sn*-1 and *sn*-2 position at the glycerol backbone, is analysed by enzymatic hydrolysis of *sn*-1-bound acyl groups followed by TLC separation of lyso-glycolipids (*sn*-2-bound acyl groups) and free fatty acids (*sn*-1-bound esters). A range of lipases EC 3.1.1.3 with regioselectivity for the *sn*-1-ester group can be used for neutral glycosyl diacylglycerols, but so far, the only lipase suitable for splitting the primary ester group of SQD is the extracellular lipase from *Rhizopus arrhizus (delemar)* [533]. Despite the increasing number of lipases, which are isolated, characterized and cloned from different organisms [534,535], an additional SQD-hydrolysing activity has not been reported (probably because this activity has not been looked for). The diacylglycerol portion of thion-SQD, released by incubation with *E. coli* β-galactosidase, was not "utterly" hydrolysed by pancreatic lipase, apparently due to the presence of the thioic acid ester in the *sn*-1 position [405]. Since the lipase from *Rhizopus* is useful for glycolipids as well as for phospholipids and works without acyl group selectivity, it is of general use for positional analysis. The lipid sample (quantity depending on the sensitivity of subsequent fatty acid analysis, 0.2-1 mg) is dispersed in buffer (0.5 ml) containing Triton X-100 (0.5%, w/v), mixed with *Rhizopus* lipase (for example from Sigma), briefly sonicated and incubated for 30-60 min at room temperature. The reaction mixture is taken to dryness after addition of isopropanol (4 ml), redissolved in chloroform/methanol (1:1) and separated by TLC in chloroform/methanol/ammonia (25% w/v) (65:35:5). In this solvent Triton travels with the front, whereas free fatty acids are retained (R_f 0.6) and separated from the lyso-compounds. Several solvent systems have been listed for various glyco- and phospholipids [342], sometimes including double development: first with diethyl ether/petroleum ether/acetic acid (50:50:4) to separate Triton from fatty acids, followed by a polar solvent mixture to move the lyso-compounds from the start [424]. The increasing water-solubility of lyso-TGD [342] and lyso-TeGD may pose problems during recovery from reaction mixtures (see above discussion on phase partitioning of intact lipids), but a positional analysis of the plant TeGD has apparently not been carried out so far. To prevent losses of lyso-compounds during phase partitioning, the use of water-saturated butanol has been recommended for extraction of reaction mixtures containing short-chain (C12:0, C14:0) lyso-glycolipids [424].

Recently, it has been shown that in the absence of borate in the hydrolysis buffer, the *Rhizopus* lipase was reported to catalyse a complete acyl migration of the *sn*-2-bound acyl group into the *sn*-1 position during an incubation at 37°C for 17 h followed by silica gel column chromatographic purification [376]. It should be pointed out that during a reinvestigation of the regioselectivity of the *Rhizopus* lipase significant acyl migration could not be observed, since characteristic *sn*-2-bound fatty acids such as C16:3 and C18:4 were not split off during a 1 h incubation of MGD [536]. On the other hand, we had shown previously that silica gel column chromatography may induce acyl migration in lyso-compounds [375]. Independent confirmation of the regioselectivity of the *Rhizopus* lipase (or its selectivity for primary ester groups) was provided by mass [537] and ^{13}C-NMR spectroscopy [536], which allowed localization of specific fatty acids at the *sn*-1/2-position (see below).

The configuration at C2 of the glycerol backbone has been assigned by several methods. In the following, the *sn*-nomenclature will be used for glycerol derivatives with C1 or *sn*-1 at the top of a Fischer projection and the hydroxyl group at C2 or *sn*-2 showing to the left. The *R*/*S*-nomenclature is unambiguous, but changes from *R* in the completely deacylated 3-galactosyl-(2*R*)-glycerol to *S* in the 1-acyl-3-galactosyl-(2*S*)-glycerol without change in the configuration at C2.

In the D/L-nomenclature the deacylated backbone of MGD is seen as a D-glycerol derivative substituted at C1 (top of a Fischer projection) by a galactosyl residue with the hydroxyl group at C2 showing to the right:

CH_2OH	CH_2OOCR	$CH_2O{-}gal$
HO—C—H	HO—C—H	H—C—OH
$CH_2O{-}gal$	$CH_2O{-}gal$	CH_2OH
3-O-galactosyl-*sn*-glycerol	1-O-acyl-3-O-galactosyl-*sn*-glycerol	1-O-galactosyl-D-glycerol
3-O-galactosyl-L-glycerol		
3-O-galactosyl-(2*R*)-glycerol	1-O-acyl-3-O-galactosyl-(2*S*)-glycerol	

Absolute assignment by X-ray crystallography is the most straightforward method not requiring references to previously assigned compounds. This approach has been exemplified with sulphoquinovosylglycerol, the deacylated backbone of SQD, which was crystallized as rubidium salt [409]. Unfortunately, in this publication only one figure (Figure 6.4) shows the right structure, whereas the others (Figures 6.3 and 6.7) represent the completely enantiomeric compound.

Benson and coworkers followed another approach, which as in all the others outlined in the following depends on reference compounds of known stereochemistry. They incubated the deacylated, radioactively labelled backbones of MGD, DGD and SQD with nitrogen dioxide in carbon tetrachloride for 5 days at room temperature to oxidize primary hydroxyl to carboxyl groups [538]. After acid hydrolysis (2N hydrochloric acid for 1-2 h at 100°C), the labelled glyceric acid originating from the glycerol moiety was isolated in 10-36% yield by preparative paper chromatography and subjected to co-crystallization with Ca^{2+} and quinine salts of authentic D- and L-glyceric acid. Since radioactivity was retained mainly in crystals of L-glyceric acid salts, the glycosyl-protected primary hydroxyl group in the glycerol portion of the oxidized compound is at *sn*-3. In some experiments, part of the radioactivity was retained in crystals of D-glyceric acid, and this was ascribed to acid-catalysed racemization.

Synthetic work had provided the diastereomeric β-D-galactopyranosyl-*sn*-3- and -*sn*-1-glycerols in crystalline form [539]. They differed significantly in melting point (~141°C for *sn*-3 versus ~106°C for *sn*-1) and IR-spectrum (950-650 cm^{-1}). In contrast their optical rotations were low and not too different ($[\alpha]_D$-7° for *sn*-3 and +2° for *sn*-1) due to the dominant contribution of identical galactosyl residues. Whenever the deacylated galactosylglycerol from MGD could be crystallized, melting point and IR-spectrum showed that it is the *sn*-3-derivative [335,337,361,371,429,540]. Physical parameters of all known isomeric/anomeric galactosylglycerols have been collected in reference [2].

Recently, a chromatographic method has been worked out for assignment of the glycerol configuration in glycosyldiacylglycerols [541]. The original glycolipids are deacylated, methylated and hydrolysed to yield 1,2-di-*O*-methyl-*sn*-glycerol (see below). This compound is reacted with a chiral fluorescent carboxylic acid chloride to yield a 3-*O*-ester derivative suitable for HPLC. The diastereomeric (*S*)-2-*tert*-butyl-2-methyl-1,3-benzodioxol-4-carboxyl esters of 1,2- and 2,3-di-*O*-methyl-*sn*-glycerol are resolved by HPLC on a normal silica column and detected by fluorescence at 370 nm. On elution with n-hexane/*tert*-butanol (100:1), the *sn*-2,3-diastereomer elutes before the *sn*-1,2-form at about 30 min. This method requires the availability of reference compounds.

We have devised an even simpler method (Linscheid *et al.*, in preparation). It involves release of diacylglycerols from glycolipids by the frequently quoted sequence of periodate oxidation/β-elimination [477] followed by enzymatic conversion to phosphatidic acid using diacylglycerol kinase from *E. coli* and ATP. One can start with ^{14}C-labelled glycolipids or use phosphate-labelled ATP. Following incubation the products are separated by TLC and detected by radioscanning. The diacylglycerol kinase from *E. coli* has high enantioselectivity and does not phosphorylate 2,3-diacyl-*sn*-glycerol [542]. Another possibility would be the chiral resolution of suitably derivatized diacylglycerols (see above) after their release from glycolipids [255]. The presence of different molecular species may result in complicated elution profiles not easily assigned to enantiomers or diastereomers (depending on the method and column used [255]) without the availability of reference compounds. The pattern may be simplified by hydrogenation of lipids before release of diacylglycerol.

It has been indicated frequently that ^{1}H-NMR spectroscopy can differentiate between 1-galactosylglycerol derivatives of 2*R*- and 2*S*-configuration, although shift data were not given [372,376]. Derivatisation of diacylglycerols with MTPA-Cl to give the corresponding Mosher ester (see above) represents another approach to determine the configuration at C2 of glycerol. ^{1}H-NMR spectroscopy can then differentiate between the diastereomeric derivatives due to clear shift differences of the signal for H-1a and H-3a in *sn*-1,2- or *sn*-2,3-diacylglycerol Mosher esters caused by the difference in proximity of the chiral substituent [543].

However, even non-covalent formation of a diastereomeric complex between the enantiomeric *sn*-1,2- or *sn*-2,3-dialkylglycerol and a chiral shift reagent (containing for example Europium III) has enabled clear ^{1}H-NMR assignment of the configuration of glycerol-C2. Whether this approach can be extended to diacylglycerol has not been investigated [544].

Methylation analysis is used to identify the hydroxyl groups blocked by ester or glycosidic bonds. For the location of glycosidic linkages, methylation under strongly alkaline conditions in strictly anhydrous dimethylsulphoxide is initiated by proton abstraction for alkoxide formation. The most effective reagent for this purpose is dimethylsulphinylcarbanion, which is prepared in very clean form with butyllithium in dimethylsulphoxide, replacing sodium hydride, potassium hydride or potassium *tert.*-butoxide used previously [123]. The actual methylation is affected with methyl iodide under sonication and finished in a short time (usually 1 h). Before further analysis the product should be purified by preparative TLC in chloroform/methanol (96:4 or 9:1) [512], in which permethylated cores of radioactive MGD-TeGD have been separated [488]. The purified product is subjected to acetolysis/hydrolysis as outlined above for subsequent formation of partially methylated alditol acetates, which are analysed by GLC/MS [123,125]. Alditol acetates resulting from terminal galactofuranosyl [2,3,5,6-tetra-*O*-methyl) or galactopyranosyl residues (2,3,4,6-tetra-*O*-methyl) are resolved as well as all four tri-*O*-methyl-isomers from substituted galactopyranosyl residues. Fragmentation during EI-MS occurs primarily between two methoxylated carbons with charge retention in either fragment. Less abundant is fission between methoxylated and acetoxylated carbons, which results in predominant charge retention in the fragment with the methoxylated carbon. These rules result in MS fragmentation patterns, which in addition to retention time identify individual components including furanosyl and pyranosyl residues. On the other hand, 4-*O*-substituted pyranosyl and 5-*O*-substituted furanosyl residues will both yield the same 1,4,5-tri-*O*-acetyl-2,3,6-tri-*O*-methyl-alditol derivative.

For biochemical studies with radioactive lipids, it may be sufficient to differentiate between labelling of terminal and internal galactose residues. This has been accomplished by TLC separation of tetramethyl from trimethyl galactose with chloroform/methanol (96:4) [488]. It is even possible to separate 2,3,6- from 2,4,6- and 2,3,4- + 3,4,6-trimethyl galactose by TLC with acetone/water/concentrated ammonia (250:3:1.5) [545]. This system may also separate the epimeric glucose/galactose derivatives of identical methylation pattern as observed with a similar solvent, which resolved tetramethyl glucose from tetramethyl galactose [546].

For additional or even specific localization of acyl substituents alternative methods have to be used. Methylation is performed under conditions, which avoid deacylation by using silver oxide as mild alkoxide-forming reagent in dry solvents such as methanol, dimethylformamide or dimethylsulphoxide in the presence of methyl iodide [337,361,366,512]. These conditions require reaction times up to several days. The methylated, acyl-group carrying lipid should be purified by preparative TLC in solvents such as benzene/methanol (95:5) [512]. Despite the mild conditions, this procedure may result in migration of acyl substituents located on sugar rings [361], whereas their shift from the glycerol moiety to sugar hydroxyl groups may not occur. To eliminate or minimize the risk of acyl migration, two-step procedures have been worked out. In a first reaction the free hydroxyl groups are converted to alkali-stable *O*-1′-methoxyethyl or *O*-tetrahydropyranyl ether derivatives by reaction with methyl vinyl ether [512] or dihydropyran [177] in dichloromethane in the presence of a trace of *p*-toluenesulphonic acid. These reactions are nearly quantitative [429], and the acetal ether derivatives are subsequently subjected to deacylation by sodium methoxide in methanol or lithium aluminium hydride in diethyl ether. The deacylated *O*-1′-methoxyethyl ether can be purified by TLC in chloroform/methanol (10:1) [429]. The newly exposed hydroxyl groups at positions occupied in the parent lipid by acyl groups are now methylated under strongly alkaline conditions with dimethylsulphinylcarbanion/methyl iodide as described above. With normal glycosyl diacylglycerols, subsequent acetolysis/hydrolysis/sodium borohydride-reduction/acetylation will yield sugar building blocks with no or significantly less methoxy and correspondingly more acetoxy groups. An elegant extension of this procedure carries out the hydrolysis of the 1′-methoxyethyl acetal ether blocking groups under very mild conditions (50% aqueous acetic acid for 45 min at 100°C) without affecting glycosidic bonds [512]. The resultant partially methylated backbone with the original hydroxyl groups free again is purified by TLC in chloroform/methanol/water (140:40:3) and subjected to a second round of methylation, this time with deuterated methyl iodide (CD_3I). GLC/MS of the acetylated methoxy derivatives obtained finally can identify exactly those positions, which carry a deuterated or a normal methoxy group, which in turn identify the originally unsubstituted or acyl-carrying hydroxyl groups, respectively [512].

Chromic oxide oxidation of acetylated glycolipid derivatives has often been used for analysis of anomeric structure of glycosidic linkages. Details of this reaction sequence, which selectively destroys pyranosides in β-form, are given in [122,264,343,512] and have been discussed above. Surviving sugars can be analysed by GLC as TMS ethers.

iii) Mass and NMR spectroscopy. Intact and deacylated glycosyl diacylglycerols have been subjected to MS using different forms of derivatisation and ionisation. The easily prepared and stable acetylated derivatives were the first to be investigated [537]. They are purified by preparative TLC in diethyl ether/petroleum ether (2:1, for MGD and DGD) and chloroform/methanol (99:1, for TGD and TeGD) [342]. With regard to structural information, the most relevant fragments observed in EI-mass spectra arise by random fragmentation of all ester and glycosidic bonds, in most cases with charge retention in either fragment. For all compounds, molecular ions can be detected with low intensity. Therefore, the structure of the original molecule can be reassembled from its components like a puzzle, sometimes requiring the addition of an oxygen atom (16 *m/z*), particularly when reconstructing the oligosaccharide chain. The terminal galactose forms the base peak at 331 *m/z* in all spectra from MGD-TeGD [47,342,537]. In the case that it carries an acyl group, as in AGD, AGM or ADGD, this peak is replaced by an correspondingly larger fragment, for example at 555 *m/z*, if stearic acid is carried at C6 [364,365,537]. Longer oligosaccharide chains such as occur in TGD and TeGD are fragmented from both the reducing and the non-reducing end giving two series of fragments, which differ by 288 *m/z* each. This is due to the sequential loss of a sugar residue with three acetyl groups, and this is in turn proof for a linear chain. Branching of TeGD at the second galactose (measured from glycerol) would require loss of a second tetra-*O*-acetyl-sugar unit and result in an correspondingly smaller fragment (by one acetyl group) of the portion representing the residual DGD, whereas the next smaller fragment in this series, representing the residual MGD moiety, would be unchanged. In contrast, branching at the first galactose would show up as a similarly smaller fragment representing the residual MGD. Based on such considerations, it is also possible to identify the sugar residue, which carries an extra acyl group, whereas the actual localization at a specific carbon atom is not possible (at least not without very detailed comparison of signal intensities and reference compounds).

The diacylglycerol fragment is another abundant and structurally relevant ion. Dependent on the fatty acids esterified, it shows up at 551 *m/z* (dipalmitoyl), 579 *m/z* (stearoyl-palmitoyl) or 607 *m/z* (distearoyl), from which the naturally occurring polyunsaturated pairs can be calculated. Lyso-glycolipids can easily be recognized by a correspondingly smaller fragment carrying an acetyl instead of the long-chain acyl group at the glycerol portion (383 *m/z* for stearoyl-acetyl- and 355 for palmitoyl-acetyl-glycerol).

A fragment of low abundance but of high structural relevance results from cleavage of the bond between the *sn*-1 and *sn*-2 carbons in the glycerol moiety with charge retention in the larger glycolipid fragment $[M\text{-}CH_2OCOR]^+$ [537]. The fragments at *m/z* 629 and 657 are characteristic for *sn*-2-bound palmitate and stearate, respectively. This fragmentation was used to confirm the structure of a synthetic 1-*O*-oleoyl-2-*O*-palmitoleoyl-3-*O*-galactosyl-*sn*-glycerol and in turn the selectivity of the *Rhizopus* lipase for the *sn*-1-position of glycolipids [533].

As already discussed above, the diacylglycerol fragments have been used to identify the predominant molecular species in MGD-TeGD from many plants [415], chloroplast envelopes [342], cyanobacteria [47] and from similar glucosyl analogues of *Lactobacillus* [512]. It should be pointed out that these data may be influenced by the sequential evaporation of species with increasing probe temperature. This effect was most pronounced with MGD [342].

Mass spectra of deacylated and reacetylated backbones of MGD-TeGD have not been published, in contrast to spectra of the closely related 1-*O*-α-D-galactopyranosyl-*sn*-glycerol (L-form of isofloridoside), 3-*O*-α-D-galactopyranosyl-*sn*-glycerol (D-form of isofloridoside) and 2-*O*-α-D-galactopyranosyl-glycerol (floridoside). The spectra are very simple and dominated by just two fragments at 331 *m/z* (tetramethyl galactose) and 159 *m/z* (diacetylglycerol), but they differ from each other in the relative intensity of both fragments [396].

The preparation of an acetylated SQD derivative requires the additional conversion of the non-volatile sulphonate group into a methyl ester. This is achieved by treating the protonated SQD-sulphonic acid (obtained by simple phase-partitioning of SQD against dilute hydrochloric acid) with diazomethane followed by acetylation with acetic anhydride in pyridine. During this step the methyl ester may be hydrolysed again and, therefore, has to be reformed by the same treatment. The final tri-*O*-acetyl SQD-sulphonic acid methyl ester is purified by preparative TLC in chloroform/methanol (9:1) [414]. The EI-induced fragmentation of this compound resembles that of MGD with major fragments representing the sulphosugar methyl ester part at *m/z* 367 (instead of 331) and the normal diacylglycerol portion [537]. By this technique it was shown that the rare dipalmitoyl species (*m/z* 551) is present in SQD [342]. The acetylated ethyleneglycol homologue of MGD has also been subjected to EI-MS. Apart from M^+ at *m/z* 630 for the palmitoyl-containing lipid, fragments for the hexose unit (*m/z* 331) and ethyleneglycol palmitate at *m/z* 283 are observed [357].

Instead of acetates, TMS ethers of intact lipids have been used for MS [370,374,537]. These derivatives are sensitive towards moisture and do not offer any advantage regarding structural information. In the spectrum of the MGD derivative the tetra-*O*-TMS sugar part shows up at *m/z* 451, whereas the diacylglycerol fragment is unchanged and represents a major ion in a spectrum which is dominated by sugar fragments at *m/z* 204 and 217. From the intensity of these two signals (204:217 > 1) it may be concluded that the sugar has the pyranose ring form [127,396,547]. GC/MS of TMS ethers of lipase-produced lyso-MGD from cyanobacteria have been studied in great detail to prove lipid-linked desaturation of *sn*-2-bound 16:0 to 16:1 [548]. In these experiments, cyanobacteria were ^{13}C-labelled for some time, and then lipid synthesis, but not desaturation, was blocked by addition of cerulenin. The ^{13}C-content and ratio in glycerol and acyl fragments as analysed by MS did not change during conversion of 16:0 to 16:1, *i.e.* the desaturase substrate is the glycolipid ester group.

In contrast to intact glycolipids, TMS ethers of deacylated backbones have frequently been used for structural identification and confirmation of the parent compounds, mostly in combination with GC/MS (see above) [343,345]. The spectra of monogalactosyl- up to tetragalactosyl-glycerol TMS ethers are all dominated by the two fragments at *m/z* 204 and 217 in a ratio characteristic for pyranosyl residues [127,396] and increasing in intensity with increasing chain length of the oligosaccharide. Similarly, the fragment of the terminal hexose unit at *m/z* 451 is found in every spectrum. Diagnostically relevant ions are M^+, $[M\text{-}15]^+$, $[M\text{-}15\text{-}TMSOH]^+$ and $[M\text{-}467]^+$ (which is caused by the loss of the terminal hexose unit), but the larger ions of this series are not stable in the higher homologues. The first three ions are only found in the spectra of MGD (M^+ at *m/z* 686) and DGD derivatives (M^+ at *m/z* 1064), whereas in the spectrum of TGD M^+ is missing and the largest fragment is $[M\text{-}15\text{-}TMSOH]^+$ at *m/z* 1337, and this is also found in DGD (at *m/z* 959) and in MGD (at *m/z* 581). In the TeGD derivative not all of these ions are detected and the largest fragment is $[M\text{-}467]^+$ at *m/z* 1353, which is also found in TGD (*m/z* 597).

The ion representing the di-*O*-TMS glycerol fragment (at *m/z* 219) is not found in MGD or in any of the other homologues. The loss of a tetra-*O*-TMS hexose (-467 *m/z*) followed by sequential elimination of tri-*O*-TMS hexose units (-378 *m/z*) is proof of an unbranched oligosaccharide chain in TeGD for the same reasons as discussed above for the acetyl derivatives. In the spectra of TMS derivatives of floridoside/isofloridosides (see structures above) no fragments larger than *m/z* 491 were observed [396]. This points to a particular contribution of the β-anomeric linkage to the stability of M^+ and of other larger ions containing the β-galactosylglycerol unit [343,345]. On the other hand, the spectra of TMS derivatives of the *all*-β-DGD, -TGD and -TeGD seem to be very similar to those of the "normal" homologues (complete spectra or data not given [345]).

The diagnostically relevant EI-MS fragmentation of TGD (as TMS ether after deacylation [344]) and TeGD (as acetylated intact 18:3/18:3-species [342]) is summarized in the following scheme:

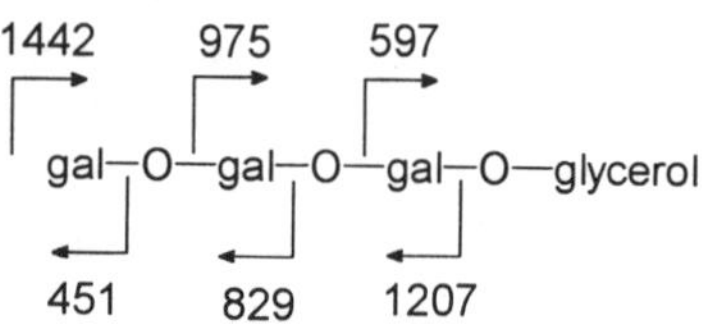

TGD TMS ether

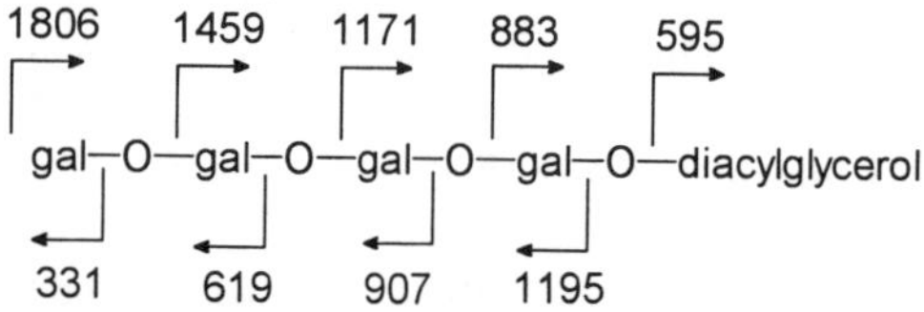

TeGD acetyl derivative

More recently, mass spectra of plant glycolipids have been obtained by other ionisation methods such as FAB, thermospray and electrospray [404,441,465]. These methods circumvent sample derivatisation, since the lipid (0.2-1 μg) is simply dissolved in an appropriate matrix (1-thioglycerol, triethanolamine). Spectra can be recorded by scanning positive or negative ions. In contrast to EI ionisation, the spectra obtained by these methods are characterized by comparatively less fragmentation of quasimolecular ions.

In FAB-MS spectra $[M-H]^-$ or $[M+Na]^+$ (additional forms are possible) are recorded with high intensity and can be used to study molecular species composition, although differences in volatility as discussed above have not been investigated. Extended series of $[M+Na]^+$ ions for different MGD and DGD species are listed in [465], for example 18:3/16:3-MGD at *m/z* 767 and 18:3/16:0-DGD at *m/z* 937, whereas the $[M-H]^-$ region of spinach SQD [549] is dominated by the 16:0/18:3-species at 815 *m/z*. Thermospray spectra of hydrogenated MGD and DGD are shown in [441]. Additional fragments arise by loss of acyl groups, which themselves may be recognized in the lower 250-300 *m/z* region together with sugar fragments such as mono- and diglycosyloxonium ions at *m/z* 163 and 325, respectively [550]. In SQD the sulphonyl anion (SO_3^-) shows up at *m/z* 80, whereas the sulphoquinovose sugar anion is seen at *m/z* 241 after additional loss of two hydrogens [549]. In general the spectra of underivatized glycolipids are of particular value for recognizing molecular weights and fatty acid pairings, and with regard to structural information they are comparable to and complement the EI-spectra of glycolipid derivatives.

NMR spectra of intact or derivatized glycolipids provide most and nearly all of the details required for a full description of a given structure. As the glycolipids discussed in the present section can be regarded as derivatives of glycosylglycerols, they can be identified by analysing the status of their various hydroxyl groups. These can be free or engaged in different types of linkages forming the hemiacetal sugar rings, glycosidic, ester or even ether bonds. ^{1}H- and ^{13}C-NMR spectroscopy of the corresponding methine and methylene carbinol protons and carbons is especially useful to recognize the position and type of bond involved in these linkages due to different deshielding effects caused by different substituents. Stereochemical correlations are reflected by the coupling constants of signal splitting, and with the aid of the newly developed two-dimensional methods complete assignment of MGD and DGD spectra was possible [117]. TGD and TeGD have not been subjected to such complete analyses, but the success with compounds of similar and even more complicated structure shows that this is possible. Even the positional distribution of fatty acids can be studied by NMR spectroscopy, although fatty acid analysis is a domain of GLC and HPLC. Tables 6.8 and 6.9 list ^{1}H- and ^{13}C-NMR data for MGD, DGD and SQD, from which the most relevant features will be outlined in the following.

In the ^{1}H-NMR spectra of lipids increasing deshielding of methine-CH- and methylene-CH_2-protons is observed by substituting the geminal hydroxyl group by alkyl ether [556], 1′-alkenyl ether [557], glycosyl (see Table 6.8), phosphodiester [557,558] or acyl groups (see Table 6.8). The acyl-induced shift is the largest, and it affects methine-signals more (~1.1 ppm downfield) than methylene protons (~0.5 ppm), whereas the glycosyl-induced deshielding is smaller (~0.2 ppm). These effects are seen by comparing the spectra of MGD, the two MGM isomers, AGM and the deacylated galactosylglycerol backbone. For convenience, only the protons at the glycerol moiety and at C1 and C6 of galactose of hydrogenated compounds are listed (in ppm) as recorded in pyridine [551,559]:

	galactosyl-glycerol	2 -acyl-MGM	1-acyl-MGM	diacyl-MGD	1,6-diacyl-AGM
sn-1	4.14; 4.13	4.20; 4.14	4.60; 4.59	4.72; 4.54	4.62
sn-2	4.45	5.61	4.52	5.71	4.52
sn-3	4.45; 4.27	4.49; 4.24	4.43; 4.11	4.40; 4.09	4.45; 4.14
H-1′	4.91	4.85	4.88	4.84	4.89
H-6′	4.45	4.42; 4.41	4.45	4.42	4.90; 4.80

Upon acylation the *sn*-2-proton experiences a larger shift than CH_2-signals from *sn*-1 and C6. In addition the deshielding by an acyl group is larger than that caused by glycosylation. This can be seen by comparing the *sn*-1 and *sn*-3 signals from various compounds in the listing. It is also obvious that substituents at neighbouring positions exert effects, but these will not be detailed. The effects just described are seen in other solvents (see Table 6.8) and are the basis for exact localization of acyl and glycosyl substituents at specific positions of glycerol and sugar moieties of intact or derivatized lipids.

Apart from the multiplet of *sn*-2-H (at 5.1-5.7 ppm, depending on the solvent)

Table 6.8.

^{1}H-NMR spectra of glycosyl diacylglycerols in free and acetylated (acet) form. The structures of the compounds are shown in Figures 6.9-11 with the exception of α-gluc-DAG, which is 1,2-diacyl-3-α-D-glucopyranosyl-glycerol of bacterial or synthetic origin. The spectra were recorded in deuterated pyridine (P), methanol (M), dimethylsulfoxide (D) or chloroform (C). Chemical shifts are given in δ ppm downfield from tetramethylsilane as internal standard. For non-equivalent protons in methylene groups the different shifts are given. Coupling constants can be found in the original references. The first 1-3 carbons refer to glycerol in *sn*-sequence, the next 1-6 to hexose, which is galactose, 6-deoxy-6-sulfo-glucose or glucose.The second galactose of DGD is represented as the right column in DGD. Protons from acyl groups have been omitted.

carbon	MGD	1-acyl-MGM	2-acyl-MGM	β-gluc-DAG	α-gluc-DAG	SQD	2-acyl-SQM	acet MGD	acet DGD	
1	4.72;4.54	4.17;4.13	3.66;3.79	4.34;4.14	4.32;4.15	4.69;4.18	3.74;3.71	4.31;4.15	4.32;4.14	
2	5.71	3.99	5.04	5.17	5.12	5.36	5.09	5.19	5.18	
3	4.40;4.09	3.91;3.65	3.96;3.66	3.86;3.66	3.70;3.49	4.10;3.57	4.07;3.53	3.95;3.68	3.99;3.64	
1	4.84	4.22	4.23	4.14	4.64	4.76	4.76	4.48	4.48	4.95
2	4.46	3.53	3.51	2.94	3.17	3.40	3.40	5.19	5.20	5.12
3	4.16	3.46	3.45	3.16	3.37	3.19	3.08	5.01	5.01	5.29
4	4.57	3.82	3.82	3.08	3.07	3.63	3.61	5.39	5 .42	5.42
5	4.08	3.51	3.50	3.11	3.30	4.06	4.07	3.90	3.85	4.16
6	4.42	3.76;3.69	3.79;3.66	3.65;3.43	3.58;3.43	3.34;2.91	3.35;2.92	4.15	3.77;3.43	4.18;4.12
solvent	P	M	M	D	D	M	M	C	C	
ref	[551]	[376]	[376]	[552]	[552]	[467]	[467]	[371]	[399]	

Table 6.9.
^{13}C-NMR spectra of glycosyl diacylglycerols in free and acetylated (acet) form. The spectra were recorded in deuterated methanol (M), benzene (B) or chloroform/methanol (C/M). For further details see legend of Table 6.8. Not included are signals for carbonyl carbons from acyl groups (172-174 ppm), acetyl carbonyl carbons (169-170 ppm), olefinic carbons (128-130 ppm) and the remaining aliphatic carbons of acyl groups.

carbon number	MGD	1-acyl-MGM	2-acyl-MGM	SQD	α-gluc-DAG	DGD		acet MGD	acet DGD	
1	68.7	66.6	61.7	64.3	62.4	64.4		62.7	62.9	
2	71.8	69.7	74.7	71.7	69.8	72.4		70.4	70.3	
3	64.0	71.9	68.8	67.1	66.2	68.8		67.8	67.8	
1	105.4	105.4	105.3	100.1	99.1	105.3	100.7	101.8	101.8	96.9
2	72.4	72.6	72.5	73.5	72.0	72.5	71.6	69.0	69.1	68.3
3	74.9	74.9	74.9	75.0	73.9	74.8	70.3	71.1	71.5	68.6
4	70.2	70.3	70.3	74.9	70.1	70.2	71.3	67.3	67.9	68.1
5	76.8	76.8	76.8	69.9	71.8	74.6	71.8	71.3	71.9	67.4
6	62.5	62.5	62.5	54.2	61.7	68.0	63.0	61.3	65.9	61.8
solvent	M	M	M	M	C/M	M		B	B	
ref	[551]	[376]	[376]	[467]	[554]	[553]		[398]	[398]	

the doublets of the anomeric protons are the most downfield signals in hydrogenated, non-acetylated lipids. Their position as well as their splitting is characteristic for α- (J = 3-4 Hz, further downfield than β-H1) and β-linkage (J = 7-8 Hz) [117]. In the α- and β-glucosyl diacylglycerols (Table 6.8) the anomeric signals are at 4.64 (α) and 4.14 ppm (β). From the glycerol part only the *sn*-3 protons are affected by the nature of the anomeric linkage in the intact lipid (see below for deacylated compounds) as shown with α- and β-glucosyl diacylglycerols (Table 6.8). In these compounds a β-linkage results in slightly more upfield *sn*-3 signals (by 0.16 ppm) than in α-anomers. In addition, the shift of the anomeric signal varies slightly with the *sn*-1/3 or *sn*-2 position of the sugar. For example, the α-anomeric signals in *sn*-1- and *sn*-2-*O*-α-D-galactopyranosylglycerol isomers were at 4.93 and 5.14 ppm, respectively [396]. Even the remote and small effects of 2*R*/*S*-configuration of the glycerol portion on the anomeric signal can be used to assign the glycerol configuration. This has not only been done with β-galactosylglycerol as referred to above [372,376] (but no data were given), but also for *sn*-1/3-α-D-galactopyranosylglycerol diastereomers [396]. From these two, the *sn*-1 form had an anomeric doublet at 4.93 ppm and the *sn*-3 at 4.94 ppm, whereas a mixture of both showed overlapping of the signals resulting in a triplet.

In the spectrum of SQD and lyso-SQD (Table 6.8) similar effects are observed to those just discussed. Most characteristic is the α-anomeric proton (at 4.77 ppm, J = 3.8 Hz) and the relatively highfield signal for the C6-H_2 protons. This is ascribed to reduced deshielding by S (in the sulphonic acid -CH_2-SO_3^-) as compared to O (in -CH_2-OH) [404]. Apart from the equatorial H1, the coupling constants of all the other ring protons in SQD are 8-10 Hz, which indicates *trans*-axial orientation and therefore gluco-configuration, whereas the equatorial H4 of MGD disrupts the equivalent series by its small coupling constant of 3 Hz [117]. It should be recalled, that in contrast to the β-isomer, the 1,2-di-*O*-acyl-3-*O*-α-D-glucopyranosylglycerol (Table 6.8) has not been found in plants and cyanobacteria, but in normal bacteria [424,512,554,555]. An illustrative example for the actual tracing of connectivities from H1 to H6 in a two-dimensional spectrum is given in [555] for an acylated diglucosyldiacylglycerol of bacterial origin.

From underivatized DGD no highly resolved and completely assigned spectrum has been published.

In addition to glycerol and sugar protons, the spectra contain signals from acyl groups, which will not be detailed. Olefinic signals show up between 5.3-6.0 ppm and may result in partial overlap with *sn*-2-H of glycerol. Hydrogenation will eliminate these as well as highfield signals from polyunsaturated acyl groups such as those from double allylic (=CH-CH_2-CH=) at 2.8 ppm and allylic (=CH-CH_2-) protons at 2.0 ppm. Other signals characteristic for fatty acyl esters are α-CH_2-groups (at about 2.3 ppm, triplets or overlapping AB-doublets) and β-CH_2-groups (at about 1.6 ppm). Terminal methyl groups show up as triplets at about 0.8 ppm.

The spectra of acetylated derivatives (Table 6.8) are interpreted as discussed above. There is no interference from free hydroxyl groups, and the main difference is the acyl-induced signal shift of methine and methylene protons from those carbons, which originally carried free hydroxyl groups. Therefore, shifted methine ring protons show up at ppm > 5.0 and shifted C6-H_2 at about 4.1 ppm. On the other hand, C6-H_2 in DGD stays at the same position as in MGD (3.7 to 3.8 ppm) as does H-5 (3.5-3.9 ppm) in all compounds due to its involvement in hemiacetal ring formation. The anomeric protons show up at 4.49 ppm (β) and 4.96 ppm (α) in normal DGD, whereas in the *all*-β series of DGD-TeGD additional β-doublets show up, that all are concentrated, but separated in a small region between 4.40 and 4.54 ppm [345].

Another approach for localization of substituents is based on the assignment of acetyl groups. This has been done with the MGD/AGD pair [362], but becomes progressively more difficult with more acetates to assign to specific positions in the higher homologues [365]. It should be recalled that all acetates in sucrose acetate have been assigned [34], whereas in many of the references in Table 6.8, acetates are listed but not assigned.

In the spectra of the oxylipin derivatives of DGD (DGD_{ox} in Figure 6.11) additional signals are seen, from which the aldehyde proton at C12 at 9.57 ppm is especially characteristic [399].

Signals in ^{13}C-NMR spectra of glycosyl diacylglycerols can be separated into different groups according to increasing shielding: carbonyl, olefinic, anomeric, hydroxylated and aliphatic carbons, from which the last group will not be discussed. The most downfield signals are due to acyl group carbonyls and, if acetylated derivatives have been used, to the additional acetyl carbonyl carbons. The long-chain acyl carbonyls in diacyl compounds show up at 172-174 ppm ([560,561] and most references in Table 6.9), although occasionally the *sn*-1- and *sn*-2-bound carbonyls may have been interchanged (see below). The acetyl carbonyls are located at slightly higher field (169-170 ppm). In the 1-*O*- and 2-*O*-acyl-MGM isomers the acyl carbonyls show up at slightly lower field [376], but both at the same position of 175 ppm. High-resolution analysis of the carbonyl region has shown that the *sn*-1-bound carbonyl in diacyl compounds is always shifted further downfield (0.2-0.4 ppm) than the *sn*-2-bound carbonyl [562]. This effect has been used for a positional analysis (see below). The unique aldehyde carbonyl carbon at C12 of the oxidized DGD (Figure 6.11) shows up at very low field at 192 ppm [399].

The next group of signals further upfield between 126-132 ppm is attributed to olefinic carbons which will not be detailed here [562]. Partial attribution to *sn*-1 and *sn*-2-bound acyl groups is possible.

The third group of signals results from anomeric sugar C1-carbons absorbing in the range of 96-112 ppm [117]. As with the other diagnostically relevant ring carbons, C1 of α-pyranosides shows up at higher field (by about 6 ppm) than the corresponding β-C1 as evident from the anomeric carbons in MGD, DGD and SQD (see Table 6.9). In *all*-β-DGD the widely separated α-C1 (at about 100 ppm) is replaced by a second β-C1 resulting in two closely approached signals between 103-104 ppm (Linscheid *et al.*, to be submitted). Higher homologues such as TGD and TeGD of plant origin have not been subjected to ^{13}C-NMR spectroscopy analysis.

The next group comprises signals from hydroxylated carbons of the sugar and glycerol moieties, which are intermixed. The primary carbons at *sn*-1, *sn*-3 and C6 show up at highest field between 62-68 ppm, but their position varies depending on substitution by acyl or glycosyl groups. As already mentioned for the ^{1}H-signal, the ^{13}C-signal for C6 of SQD is seen at slightly higher field at 54 ppm due to its reduced deshielding [404,467,553,561]. Occasionally, the assignment of *sn*-1/3-signals may be interchanged (see in Table 6.9 for MGD and DGD and [560,561]). From the sugar carbons, only C5 and C6 show up at nearly invariant positions in both α- and β-pyranosides, whereas C2, C3 and C4 are always displaced to higher fields in α-anomers as already pointed out for C1. For further details the excellent and comprehensive compilation of ^{1}H- and ^{13}C-NMR data for α/β-pyranosides/furanosides of many different sugars should be consulted [117].

High-resolution ^{13}C-NMR spectroscopy has also been used for the analysis of the positional distribution of fatty acids in triacylglycerols [563,564], phospholipids [565] and plant glycosyl diacylglycerols [536]. This approach makes use of two effects contributing to the shift of C1-carbonyl signals of glycerol-bound acyl groups. Firstly, *sn*-1-bound carbonyls are seen slightly further downfield (by about 0.2-0.4 ppm) than *sn*-2-carbonyls in the region of 172-174 ppm. Secondly, the carbonyls sense the presence and proximity of a double bond, and their position increases from most downfield in saturated acyl groups to increasingly higher field with increasing proximity of the nearest double bond. Therefore, the sequence from low to higher field is saturated, Δ9, Δ7, Δ6, Δ5, Δ4. This sequence is seen for both *sn*-1- and *sn*-2-bound carbonyls, and overlapping of both series does only occur for Δ4 of *sn*-1, which falls into the *sn*-2-bound series at higher field. These differences have been used to assign Δ7,10,13-16:3 and Δ6,9,12,15-18:4 to *sn*-2 of MGD from higher plants supporting previous enzymatic data [536].

Finally, it should be pointed out that ^{13}C- and ^{1}H-NMR spectroscopy may be a useful alternative for quantification of individual phospho- and glycolipids in the complex plant lipid extracts. The first experiments in this direction have been started [566] inspired by the equivalent approach for phospholipid mixtures based on ^{31}P-NMR spectroscopy analysis [567] (see also Chapter 2 in this volume). The full potential of this method, particularly the use of high-resolution analysis of selected and different parts of the spectra, has still to be explored. It requires the use of solvent mixtures, which result in molecular dispersion of lipids preventing any associations to avoid deterioration of spectral resolution. A solvent mixture optimized for this purpose is pyridine/aqueous hydrochloric acid/methanol/chloroform (1:1:2:10) [568].

Abbreviations

ASG, acylated steryl glycoside(s); CD, circular dichroism; DEAE, diethylaminoethyl; DGD, digalactosyl diacylglycerol; DIG, Digoxigenin; EI-MS, electron-impact MS; ELISA, enzyme-linked immunosorbent assay; ER, endoplasmic reticulum; FD-MS, field-desorption MS; FAB-MS, fast-atom bombardment MS; FID, flame-ionisation detector; GLC, gas-liquid chromatography; HPLC, high-performance liquid chromatography; HPTLC, high-performance thin-layer chromatography; MGD, monogalactosyl diacylglycerol; MS, mass spectrometry; MTPA-Cl, α-methoxy-α-trifluoromethylphenylacetyl chloride; NMR, nuclear magnetic resonance; PVDF, polyvinylidene difluoride; RI, refractive index; RP-HPLC, reverse-phase HPLC; SG, steryl glycoside(s); SQD, sulphoquinovosyl diacylglycerol; TGD, trigalactosyl diacylglycerol; TeGD, tetragalactosyl diacylglycerol; TLC, thin-layer chromatography; TMS, trimethylsilyl; UV, ultraviolet.

REFERENCES

1. Stults,C.L.M., Sweeley,.C. and Macher, B.A., *Methods Enzymol.*, **179**, 167-214 (1989).
2. Kates,M., in *Handbook of Lipid Research 6, Glycolipids, Phosphoglycolipids, and Sulfoglycolipids*, pp. 1-122 and 235-320 (1990) (edited by M. Kates, Plenum Press, New York).
3. Fobes,J.F., Mudd,J.B. and Marsden,P.F., *Plant Physiol.*, **77**, 567-570 (1985).
4. Wagner,G.J., *Plant Physiol.*, **96**, 675-679 (1991).
5. Goffreda,J.C., Szymkowiak,E.J., Sussex,J.M. and Mutschler,M.A., *Plant Cell*, **2**, 643-649 (1990).
6. Shapiro,J.A., Steffens,J.C. and Mutschler,M.A., *Biochem. Syst. Ecol.*, **22**, 545-561 (1994).
7. Nielsen,M.T., Akers,C.P., Järflors,U.E., Wagner,G.J. and Berger,S., *Bot. Gaz.*, **152**, 13-22 (1991).
8. Walters,D.S. and Steffens,J.C., *Plant Physiol.*, **93**, 1544-1551 (1990).
9. Kandra,L. and Wagner,G.J., *Plant Physiol.*, **94**, 906-912 (1990).
10. Kandra,L., Severson,R. and Wagner,G.J., *Eur. J. Biochem.*, **188**, 385-391 (1990).
11. Roughan,P.G. and Batt,R.D., *Phytochemistry*, **8**, 363-369 (1969).
12. King,R.R., Calhoun,L.A. and Singh,R.P., *Phytochemistry*, **27**, 3765-3768 (1988).
13. King,R.R. and Calhoun,L.A., *Phytochemistry*, **27**, 3761-3763 (1988).

14. Oscarson,S. and Ritzén,H., *Carbohydr. Res.,* **205**, 67-70 (1990).
15. Matsuzaki,T., Shinozaki,Y., Suhara,S., Shigematsu,H. and Koiwai,A., *Agric. Biol. Chem.*, **53**, 3343-3345 (1989).
16. Burke,B.A., Goldsby,G. and Mudd,J.B., *Phytochemistry,* **26**, 2567-2571 (1987).
17. Grellet,F., Cooke,R., Raynal,M., Laudié,M. and Delseny,M., *Plant Physiol. Biochem.*, **31**, 599-602 (1993).
18. Dörmann,P., Voelker,T.A. and Ohlrogge,J.B., *Arch. Biochem. Biophys.*, **316**, 612-618 (1995).
19. Kroumova,A.B., Xie,Z. and Wagner,G.J., *Proc. Natl. Acad. Sci. USA*, **91**, 11437-11441 (1994).
20. Ghangas,G.S. and Steffens,J.C., *Proc. Natl. Acad. Sci. USA*, **90**, 9911-9915 (1993).
21. Ghangas,G.S. and Steffens,J.C., *Arch. Biochem. Biophys.*, **316**, 370-377 (1995).
22. King,R.R., Calhoun,L.A., Singh,R.P. and Boucher,A., *Phytochemistry,* **29**, 2115-2118 (1990).
23. King,R.R., Singh,R.P. and Calhoun,L.A., *Carbohydr. Chem.*, **173**, 235-241 (1988).
24. Shinozaki,Y., Matsuzaki,T., Suhara,S., Tobita,T., Shigematsu,H. and Koiwai,A., *Agric. Biol. Chem.*, **55**, 751-756 (1991).
25. King,R.R., Calhoun,L.A., Singh,R.P. and Boucher,A., *J. Agric. Food Chem.*, **41**, 469-473 (1993).
26. Matsuzaki,T., Shinozaki,Y., Hagimori,M., Tobita,T., Shigematsu,H. and Koiwai,A., *Biosci. Biotech. Biochem.*, **56**, 1565-1569 (1992).
27. Matsuzaki,T., Shinozaki,Y., Suhara,S., Tobita,T., Shigematsu,H. and Koiwai,A., *Agric. Biol. Chem.*, **55**, 1417-1419 (1991).
28. King,R.R., Singh,R.P. and Calhoun,L.A., *Carbohydr. Res.,* **166**, 113-121 (1987).
29. Matsuzaki,T., Koseki,K. and Koiwai,A., *Agric. Biol. Chem.*, **52**, 1889-1897 (1988).
30. Shinozaki,Y., Matsuzaki,T., Suzuki,H., Shigematsu,H. and Koiwai,A., *Biosci. Biotech. Biochem.*, **56**, 1482-1483 (1992).
31. Severson,R.F., Arrendale,R.F., Chortyk,O.T., Green,C.R., Thome,F.A., Stewart,J.L. and Johnson,A.W., *J. Agric. Food Chem.*, **33**, 870-875 (1985).
32. Lin,Y. and Wagner,G.J., *J. Agric. Food Chem.*, **42**, 1709-1712 (1994).
33. Garegg,P.J., Oscarson,S. and Ritzén,H., *Carbohydr. Res.,* **181**, 89-96 (1988).
34. Buta,J.G., Lusby,W.R., Neal,J.W., Waters,R.M. and Pittarelli,G.W., *Phytochemistry,* **32**, 859-864 (1993).
35. Murray,M.A. and Wolk,C.P., *Arch. Microbiol.*, **151**, 469-479 (1989).
36. Nichols,B.W. and Wood,B.J.B., *Nature,* **217**, 767 (1968).
37. Bryce,T.A., Welti,D., Walsby,A.E. and Nichols,B.W., *Phytochemistry,* **11**, 295-302 (1972).
38. Lambein,F. and Wolk,C.P., *Biochemistry*, **12**, 791-798 (1973).
39. Soriente,A., Sodano,G., Gambacorta,A. and Trincone,A., *Tetrahedron*, **48**, 5375-5384 (1992).
40. Soriente,A., Gambacorta,A., Trincone, A., Sill,C., Vincenzini,M. and Sodano,G., *Phytochemistry,* **33**, 393-396 (1993).
41. Soriente,A., Bisogno,T., Gambacorta,A., Romano,I., Sill,C., Trincone,A. and Sodano,G., *Phytochemistry,* **38**, 641-645 (1995).
42. Davey,M.W. and Lambein,F., *Anal. Biochem.*, **206**, 323-327 (1992).
43. Davey,M.W. and Lambein,F., *Anal. Biochem.,* **206**, 226-230 (1992).
44. Feige,G.B., *Z. Pflanzenphysiol.*, **80**, 377-385 (1976).
45. Pfander,H., *Pure Appl. Chem.*, **47**, 121-128 (1976).
46. Hertzberg,S. and Liaaen-Jensen,S., *Phytochemistry,* **8**, 1259-1280 (1969).
47. Zepke,H.D., Heinz,E., Radunz,A., Linscheid,M. and Pesch,R., *Arch. Microbiol.*, **119**, 157-162 (1978).
48. Mori,K. and Qian,Z.-H., *Liebigs Ann. Chem.*, 35-39 (1994).
49. Abreu-Grobois,F.A., Billyard,T.C. and Walton,T.J., *Phytochemistry,* **16**, 351-354 (1977).
50. Parker,D., *Chem. Rev.*, **91**, 1441-1457 (1991).
51. Wojciechowski,Z.A., in *Physiology and Biochemistry of Sterols*, pp. 361-396 (1991) (edited by G.W. Patterson and W.D. Nes, Am. Oil Chem. Soc., Champaign).
52. Demel,R.A. and de Kruyff, B., *Biochim. Biophys. Acta,* **457**, 109-132 (1976).
53. Bretscher,M.S. and Munro,S., Science, **261**, 1280-1281 (1993).
54. Brown,D.J. and DuPont,F.M., *Plant Physiol.*, **90**, 955-961 (1989).
55. Lynch,D.V. and Steponkus,P.L., *Plant Physiol.*, **83**, 761-767 (1987).
56. Haschke,H.-P., Kaiser,G., Martinoia,E., Hammer,U., Teucher,T., Dorne,A.J. and Heinz,E., *Bot. Acta*, **103**, 32-38 (1990).
57. Schroeder,F., Nemecz,G., Wood,W.G., Joiner,C., Morrot,G., Ayraut-Jarrier,M. and Devaux,P.F.,

Biochim. Biophys. Acta, **1066**, 183-192 (1991).
58. Rodriguez,R.J., Low,C., Bottema,C.D.K. and Parks,L.W., *Biochim. Biophys. Acta,* **837**, 336-343 (1985).
59. Keesler,G.A., Laster,S.M. and Parks,L.W., *Biochim. Biophys. Acta,* **1123**, 127-132 (1992).
60. Nes,W.D., Janssen,G.G., Crumley,F.G., Kalinowska,M. and Akihisa,T., *Arch. Biochem. Biophys.,* **300**, 724-733 (1993).
61. Whitaker,B.D., *Phytochemistry,* **27**, 3411-3416 (1988).
62. Dyas,L. and Goad,L.J., *Phytochemistry,* **34**, 17-29 (1993).
63. Wertz,P.W., Stover,P.M., Abraham,W. and Downing,D.T., *J. Lipid Res.,* **27**, 427-435 (1986).
64. Abraham,W., Wertz,P.W., Burken,R.R. and Downing,D.T., *J. Lipid Res.,* **28**, 446-449 (1987).
65. Weber,N., *Lipids,* **23**, 42-47 (1988).
66. McKersie,B.D. and Thompson,J.E., *Plant Physiol.,* **63**, 802-805 (1979).
67. Mudd,J.B. and McManus,T.T., *Plant Physiol.,* **65**, 78-80 (1980).
68. Schuler,J., Milon,A., Nakatani,Y., Ourisson,G., Albrecht,A.-M., Benveniste,P. and Hartmann,M.-A., *Proc. Natl. Acad. Sci. USA,* **88**, 6926-6930 (1991).
69. Chye,M.-L., Tan,C.-T. and Chua,N.-H., *Plant Mol. Biol.,* **19**, 473-484 (1992).
70. Gondet,L., Bronner,R. and Benveniste,P., *Plant Pyhsiol.,* **105**, 509-518 (1994).
71. Abraham,P.R., Mulder,A., van't Riet,J., Planta,R.J. and Raué,H.A., *Yeast,* **8**, 227-238 (1992).
72. Kalinowska,M. and Wojciechowski,Z.A., *Phytochemistry,* **25**, 45-49 (1986).
73. Zimowski,J., *Phytochemistry,* **31**, 2977-2981 (1992).
74. Ullmann,P., Ury,A., Rimmele,D., Benveniste,P. and Bouvier-Navé,P., *Biochimie,* **75**, 713-723 (1993).
75. Warnecke,D.C. and Heinz,E., *Plant Physiol.,* **105**, 1067-1073 (1994).
76. Benveniste,P., Ann. Rev. *Plant Physiol.,* **37**, 275-308 (1986).
77. Ourisson,G., *J. Plant Physiol.,* **143**, 434-439 (1994).
78. Fujino,Y., Ohnishi,M. and Ito,S., *Proc. 7th World Cereal and Bread Congress, Prague,* pp. 145-150, 1982.
79. Ahmad,V.U., Aliya,R., Perveen,S. and Shameel,M., *Phytochemistry,* **31**, 1429-1431 (1992).
80. Srivastava,S.K., *Phytochemistry,* **19**, 2510-2511 (1980).
81. Iribarren,A.M. and Pomilio,A.B., *Phytochemistry,* **23**, 2087-2088 (1984).
82. Jares,E.A., Tettamanzi,M.C. and Pomilio,A.B., *Phytochemistry,* **29**, 340-341 (1990).
83. Iribarren,A.M. and Pomilio,A.B., *Phytochemistry,* **24**, 360-361 (1985).
84. Iribarren,A.M. and Pomilio,A.B., *Phytochemistry,* **26**, 857-858 (1987).
85. Eichenberger,W. and Menke,W., *Z. Naturforsch.,* **21 b**, 859-867 (1966).
86. Kamperdick,C., Sing,T.V., Thuy,T.T., Tri,M.V. and Adam,G., *Phytochemistry,* **38**, 699-701 (1995).
87. Lepage,M., *J. Lipid Res.,* **5**, 587-592 (1964).
88. Hashimoto,T., Tori,M. and Asakawa,Y., *Phytochemistry,* **30**, 2927-2931 (1991).
89. Greca,M.D., Molinaro,A., Monaco,P. and Previtera,L., *Phytochemistry,* **30**, 2422-2424 (1991).
90. Tandon,M., Shukla,Y.N. and Thakur,R.S., *Phytochemistry,* **29**, 2957-2959 (1990).
91. Dupéron,R., Doireau,P., Verger,A. and Dupéron,P., in *Biogenesis and Function of Plant Lipids,* pp. 445-447 (1980) (edited by P. Mazliak, P. Benveniste, C. Costes and R. Douce, Elsevier/North-Holland Biomedical Press, Amsterdam).
92. Khanna,I., Seshadri,R. and Seshadri,T.R., *Phytochemistry,* **13**, 199-202 (1974).
93. Kojima,M., Ohnishi,M., Ito,S. and Fujino,Y., *Lipids,* **24**, 849-853 (1989).
94. Ohnishi,M. and Fujino,Y., *Agric. Biol. Chem.,* **42**, 2423-2425 (1978).
95. Tiwari,K.P. and Choudhary,R.N. *Phytochemistry,* **18**, 2044-2045 (1979).
96. Ghosal,S., *Phytochemistry,* **24**, 1807-1810 (1985).
97. Ohnishi,M. and Fujino,Y., *Phytochemistry,* **20**, 1357-1358 (1981).
98. Mimaki,Y., Ishibashi,N., Ori,K. and Sashida,Y., *Phytochemistry,* **31**, 1753-1758 (1992).
99. Hanada,R., Abe,F. and Yamauchi,T., *Phytochemistry,* **31**, 3183-3187 (1992).
100. Kesselmeier,J., Eichenberger,W. and Urban,B., *Physiol. Plant.,* **70**, 610-616 (1987).
101. Heinz,E., Dieler,H.P and Rullkötter,J., *Zeitschr. Pflanzenphysiol.,* **75**, 78-87 (1975).
102. Dyas,L., Threlfall,D.R. and Goad,L.J., *Phytochemistry,* **35**, 655-660 (1994).
103. Guevara,A.P., Lim-Sylianco,C.Y., Dayrit,F.M. and Finch,P., *Phytochemistry,* **28**, 1721-1724 (1989).
104. Hartmann,M.-A. and Benveniste,P., *Methods Enzymol.,* **148**, 632-650 (1987).
105. Adler,G. and Kasprzyk,Z., *Phytochemistry,* **14**, 627-631 (1975).

106. Privett,O.S., Dougherty,K.A., Erdahl,W.L. and Stolyhwo,A., *J. Am. Oil Chem. Soc.*, **50**, 516-520 (1973).
107. Conforti,F.D., Harris,C.H. and Rinehart,J.T., *J. Chromatogr.*, **645**, 83-88 (1993).
108. Kesselmeier,J., Eichenberger,W. and Urban,B., *Plant Cell Physiol.*, **26**, 463-471 (1985).
109. Murui,T. and Wanaka,K., *Biosci. Biotech. Biochem.*, **57**, 614-617 (1993).
110. Laine,R.A. and Elbein,A.D., *Biochemistry*, **10**, 2547-2553 (1971).
111. Eichenberger, W. and Grob, E.C., *FEBS Lett.*, **11**, 177-180 (1970).
112. Richmond,W., *Ann. Clin. Biochem.*, **29**, 577-597 (1992).
113. Hashimoto,S.-I., Yanagiya,Y., Honda,T., Harada,H. and Ikegami,S., *Tetrahedron Lett.*, **33**, 3523-3526 (1992).
114. Aneja,R. and Harries,P.C., *Chem. Phys. Lipids,* **12**, 351-362 (1974).
115. Uvarova,N.I., Oshitok,G.I. and Elyakov,G.B., *Carbohydr. Res.,* **27**, 79-87 (1973).
116. Weber,N., *Chem. Phys. Lipids,* **18**, 145-148 (1977).
117. Agrawal,P.K., *Phytochemistry,* **31**, 3307-3330 (1992).
118. Kojima,H., Sato,N., Hatano,A. and Ogura,H., *Phytochemistry,* **29**, 2351-2355 (1990).
119. Leitao,S.G., Kaplan,M.A.C., Monache,F.D., Akihisa,T. and Tamura,T., *Phytochemistry,* **31**, 2813-2817 (1992).
120. Ahmad,V.U., Aliya,R., Perveen,S. and Shameel,M., *Phytochemistry,* **33**, 1189-1192 (1993).
121. Leitao,S.G., Kaplan,M.A.C. and Monache,F.D., *Phytochemistry,* **36**, 167-170 (1994).
122. Dell,A., Reason,A.J., Khoo,K.-H., Panico,M., McDowell,R.A. and Morris,H.R., *Methods Enzymol.*, **230**, 108-132 (1994).
123. Geyer,H. and Geyer,H., *Methods Enzymol.*, **230**, 86-108 (1994).
124. Gray,G.R., *Methods Enzymol.*, **193**, 573-587 (1990).
125. Hellerqvist,C.G., *Methods Enzymol.*, **193**, 554-573 (1990).
126. Schmitt,P. and Benveniste,P., *Phytochemistry,* **18**, 445-450 (1979).
127. Gerwig,G.J., Kamerling,J.P. and Vliegenhart,J.F.G., *Carbohydr. Res.,* **62**, 349-357 (1978).
128. Mimaki,Y., Kubo,S., Kinoshita,Y., Sashida,Y., Song,L.-G., Nikaido,T. and Ohmoto,T., *Phytochemistry,* **34**, 791-797 (1993).
129. Zweerink,M.M., Edison,A.M., Wells,G.B., Punto,W. and Lester,R.L., *J. Biol. Chem.*, **267**, 25032-25038 (1992).
130. Lynch,D.V., in *Lipid Metabolism in Plants*, pp. 285-308 (1993) (edited by T.S. Moore, Jr., CRC Press, Boca Raton).
131. Nelson,P.E., Desjardins,A.E. and Plattner,R.D., *Annu. Rev. Phytopathol.*, **31**, 233-252 (1993).
132. Mandon,E.C., Ehses,I., Rother,J., van Echten,G. and Sandhoff,K., *J. Biol. Chem.*, **267**, 11144-11148 (1992).
133. Hoekstra,D. and Kok,W.J., *Biochim. Biophys. Acta,* **1113**, 277-294 (1992).
134. Buede,R., Rinker-Schaffer,C., Pinto,W.J., Lester,R.L. and Dickson,R.C., *J. Bacteriol.*, **173**, 4325-4332 (1991).
135. Nagiec,M.M., Baltisberger,J.A., Wells,G.B., Lester,R.L. and Dickson,R.C., *Proc. Natl. Acad. Sci. USA*, **91**, 7899-7902 (1994).
136. Schulte,S. and Stoffel,W., *Proc. Natl. Acad. Sci. USA*, **90**, 10265-10269 (1993).
137. Carstea,E.D., Murray,G.J. and O'Neill,R.R., *Biochem. Biophys. Res. Commun.*, **184**, 1477-1483 (1992).
138. Sakai,N., Inui,K., Fujii,N., Fukushima,H., Nishimoto,J., Yamagihara,I., Isegawa,Y., Iwamatsu,A. and Okada,S., *Biochem. Biophys. Res. Commun.*, **198**, 485-491 (1994).
139. Lynch,D.V. and Fairfeld,S.R., *Plant Physiol.*, **103**, 1421-1429 (1993).
140. Rother,J., van Echten,G., Schwarzmann,G. and Sandhoff,K., *Biochem. Biophys. Res. Commun.*, **189**, 14-20 (1992).
141. Hannun,Y.A. and Bell,R.M., *Science*, **243**, 500-507 (1989).
142. Abbas,H.K., Tanaka,T., Duke,S.O., Porter,J.K., Wray,E.M., Hodges,L., Sessions,A.E., Wang,E., Merrill,A.H. and Riley,R.T., *Plant Physiol.*, **106**, 1085-1093 (1994).
143. Nagiec,M.M., Wells,G.B., Lester,R.L. and Dickson,R.C., *J. Biol. Chem.*, **268**, 22156-22163 (1993).
144. Lester,R.L., Wells,G.B., Oxford,G. and Dickson,R.C., *J. Biol. Chem.*, **268**, 845-856 (1993).
145. Kawai,G., Ikeda,Y. and Tubaki,K., *Agric. Biol. Chem.*, **49**, 2137-2146 (1985).
146. Mineki,S., Jida,M. and Tsutsumi,T., *J. Ferm. Bioeng.*, **78**, 327-330 (1994).
147. Shibuya,H., Kurosu,M., Minagawa,K., Katayama,S. and Kitagawa,I., *Chem. Pharm. Bull.*, **41**, 1534-1544 (1993).

148. Lynch,D.V., Caffrey,M., Hogan,J.L. and Steponkus,P.L., *Biophys. J.*, **61**, 1289-1300 (1992).
149. Fujino,Y. and Ohnishi,M., *Proc. Jap. Acad.*, **58B**, 36-39 (1982).
150. Fujino,Y., Ohnishi,M. and Ito,S., *Agric. Biol. Chem.*, **49**, 2753-2762 (1985).
151. Fujino,Y., Ohnishi,M. and Ito,S., *Lipids,* **20**, 337-342 (1985).
152. Sastry,P.S. and Kates,M., *Biochemistry*, **3**, 1271-1280 (1964).
153. Laine,R.A. and Renkonen,O., *Biochemistry*, **13**, 2837-2843 (1974).
154. Kaul,K. and Lester,R.L., *Biochemistry*, **17**, 3569-3575 (1978).
155. Hsieh,T.C.Y., Kaul,K., Laine,R.A. and Lester,R.L., *Biochemistry*, **17**, 3575-3581 (1978).
156. Hsieh,T.C.Y., Lester,R.L. and Laine,R.A., *J. Biol. Chem.*, **256**, 7747-7755 (1981)
157. Laine,R.A. and Hsieh,T.C.Y., *Methods Enzymol.*, **138**, 186-195 (1987).
158. Carter,H.E., Strobach,D.R. and Hawthorne,J.N., *Biochemistry*, **8**, 383-388 (1969).
159. Ko,J., Cheah,S. and Fischl,A.S., *J. Bacteriol.*, **176**, 5181-5183 (1994).
160. Schneider,E.G. and Kennedy,E.P., *Biochim. Biophys. Acta,* **441**, 201-212 (1976).
161. Ito,S., Ohnishi,M. and Fujino,Y., *Agric. Biol. Chem.*, **49**, 539-540 (1985).
162. Ohnishi,M., Ito,S. and Fujino,Y., *Agric. Biol. Chem.*, **49**, 3327-3329 (1985).
163. Fujino,Y. and Ohnishi,M., *J. Cereal Sci.*, **1**, 159-168 (1983).
164. Ohnishi,M. and Fujino,Y., *Lipids,* **17**, 803-810 (1982).
165. Ohnishi,M., Ito,S. and Fujino,Y., *Agric. Biol. Chem.*, **46**, 2855-2856 (1982).
166. Ohnishi,M. Ito,S. and Fujino,Y., *Biochim. Biophys. Acta,* **752**, 416-422 (1983).
167. Cahoon,E.B. and Lynch,D.V., *Plant Physiol.*, **95**, 58-68 (1991).
168. Carter,H.E. and Koob,J.L., *J. Lipid Res.,* **10**, 363-369 (1969).
169. Imre,Z., *Z. Naturforsch.*, **29c**, 195-200 (1974).
170. Ito,S., Kojima,M. and Fujino,Y., *Agric. Biol. Chem.*, **49**, 1873-1875 (1985).
171. Norberg,P., Månsson,J.-E. and Liljenberg,C., *Biochim. Biophys. Acta,* **1066**, 257-260 (1991).
172. Murata,N., Sato,N. and Takahashi,N., *Biochim. Biophys. Acta,* **795**, 147-150 (1984).
173. Carter,H.E., Celmer,W.D., Galanos,D.S., Gigg,R.H., Lands,W.E.M., Law,J.H., Mueller,K.L., Nakayama,T., Tomizawa,H.H. and Weber,E., *J. Am. Oil Chem. Soc.*, **35**, 335-343 (1958).
174. Kaul,K. and Lester,R.L., *Plant Physiol.*, **55**, 120-129 (1975).
175. Rouser,G., Kritchevsky,G., Yamamoto,A., Simon,G., Galli,C. and Bauman,A.J., *Methods Enzymol.*, **14**, 272-317 (1969).
176. Schnaar,R.L., *Methods Enzymol.*, **230**, 348-370 (1994).
177. Tamai,Y., Nakamura,K., Takayama-Abe,K., Uchida,K., Kasama,T. and Kobatake,H., *J. Lipid Res.,* **34**, 601-608 (1993).
178. Fujino,Y. and Ohnishi,M., *Biochim. Biophys. Acta,* **574**, 94-102 (1979).
179. Mori,K. and Funaki,Y., *Tetrahedron*, **41**, 2369-2377 (1985).
180. Mori, K. and Kinsho, T., *Ann. Chem.*, 1309-1315 (1991).
181. Haverkate,F. and van Deenen,L.L.M., *Biochim. Biophys. Acta,* **106**, 78-92 (1965).
182. Mori,K. and Funaki,Y., *Tetrahedron*, **41**, 2379-2386 (1985).
183. Vorbeck,M.L. and Marinetti,G.V., *J. Lipid Res.,* **6**, 3-6 (1965).
184. Ito,S. and Fujino,Y., *Canad. J. Biochem.*, **51**, 957-961 (1973).
185. Kraus,R. and Spiteller,G., *Ann. Chem.*, 125-128 (1991).
186. Hannun,Y.A. and Obeid,L.M., *TIBS*, **20**, 73-77 (1995).
187. Christie,W.W., *J. Chromatogr.*, **361**, 396-399 (1986).
188. Lutzke,B.S. and Braughler,J.M., *J. Lipid Res.,* **31**, 2127-2130 (1990).
189. Kannagi,R., Watanabe,K. and Hakomori,S.-I., *Methods Enzymol.*, **138**, 3-12 (1987).
190. Suzuki,M., Yamakawa,T. and Suzuki,A., *J. Biochem. (Tokyo)*, **109**, 503-506 (1991).
191. Kim,H.Y. and Salem,N., *Progr. Lipid Res.*, **32**, 221-245 (1993).
192. Suzuki,A.C., Nakamura,A. and Nishimura,K., *Glycoconj. J.*, **11**, 111-121 (1994).
193. McCluer,R.H., Ullman,M.D. and Jungalwala,F.B., *Methods Enzymol.*, **172**, 538-575 (1989).
194. Kawai,G., *Biochim. Biophys. Acta,* **1001**, 185-190 (1989).
195. Hakomori,S.-I., in *Handbook of Lipid Research 3, Sphingolipid Biochemistry, Chemistry of Glycosphingolipids,* pp. 1-165 (1983) (edited by Kanfer, J.N. and Hakomori, S.-I., Plenum Press, New York).
196. Schnaar,R.L. and Needham,L.K., *Methods Enzymol.*, **230**, 371-389 (1994).
197. Pohl,P., Glasl,H. and Wagner,H., *J. Chromatogr.*, **49**, 488-492 (1970).
198. Cherry,J.M., Buckhout,T.J. and Morré,D.J., *Experientia*, **34**, 1433 (1978).
199. Jensen,M.T., Knudsen,J. and Olson,J.M., *Arch. Microbiol.*, **156**, 248-254 (1991).
200. Tahora,Y. and Kawazu,M., *Biosci. Biotech. Biochem.*, **58**, 586-587 (1994).

201. Koike,K., Mori,M., Ito,Y., Nakahara,Y. and Ogawa,T., *Agric. Biol. Chem.*, **54**, 2931-2939 (1990).
202. Koike,K., Mori,M., Ito,Y., Nakahara,Y. and Ogawa,T., *Biosci. Biotech. Biochem.*, **57**, 698-702 (1993).
203. Morrison, W.R., *Chem. Phys. Lipids*, **11**, 99-102 (1973).
204. Saga,Y., Gasa,S. and Makita,A., *J. Chromatogr.*, **513**, 379-383 (1990).
205. Karlsson,K.-A., Samuelsson,B.E. and Steen,G.O., *Biochim. Biophys. Acta*, **306**, 317-328 (1973).
206. Hakomori,S.-I. and Young,W.W., in *Handbook of Lipid Research 3, Sphingolipid Biochemistry*, pp. 381-436 (1983) (edited by Kanfer, J.N. and Hakomori, S.-I., Plenum Press, New York).
207. Feizi,T., *Nature*, **314**, 53-57 (1985).
208. Kotani,M., Kawashima,I., Ozawa,H., Ogura,K., Ariga,T. and Tai,T., *Arch. Biochem. Biophys.*, **310**, 89-96 (1994).
209. Ångström,J., Teneberg,S. and Karlsson,K.-A., *Proc. Natl. Acad. Sci. USA*, **91**, 11859-11863 (1994).
210. Saitoh,T., Natomi,H., Zhao,W., Okuzumi,K., Sugano,K., Iwamori,M. and Nagai,Y., *FEBS Lett.*, **282**, 385-387 (1991).
211. Stults,C.L.M., Wilbur,B.J. and Macher,B.A., *Anal. Biochem.*, **174**, 151-156 (1988).
212. Stults,C.L.M., Sullivan,M.T., Macher,B.A., Johnston,R.F. and Stack,R.J., *Anal. Biochem.*, **219**, 61-70 (1994).
213. Magnani,J.L., Brockhaus,M., Smith,D.F. and Ginsberg,V., *Methods Enzymol.*, **83**, 235-241 (1982).
214. Lanne,B., Ciopraga,J., Bergström,J., Motas,C. and Karlsson,K.-A., *Glycoconj. J.*, **11**, 292-298 (1994).
215. Taki,T., Handa,S. and Ishikawa,D., *Anal. Biochem.*, **221**, 312-316 (1994).
216. Taki,T., Kasama,T., Handa,S. and Ishikawa,D., *Anal. Biochem.*, **223**, 232-238 (1994).
217. Bansal,R., Warrington,A.E., Gard,A.L., Rauscht,G.B. and Pfeiffer,S.E., *J. Neurosci. Res.*, **24**, 548-557 (1989).
218. Kalisiak,A., Oosterwijk,E., Minniti,J.G., Old,L.J. and Scheinberg,D.A., *Glycoconj. J.*, **8**, 55-62 (1991).
219. Gadella,B.M., Gadella,T.W.J., Colenbrander,B., van Golde,L.M.G. and Lopes-Cardozo,M., *J. Cell Sci.*, **107**, 2151-2163 (1994).
220. Bhat,S., Spitalnik,S.L., Gonzales-Scarano,F. and Silberberg,D.H., *Proc. Natl. Acad. Sci. USA*, **88**, 7131-7134 (1991).
221. Uchida,T. and Nagai,Y., *J. Biochem. (Tokyo)*, **87**, 1829-1841 (1980).
222. Jennemann,R., Gnewuch,C., Boßlet,S., Bauer,B.L. and Wiegandt,H., *J. Biochem. (Tokyo)*, **115**, 1047-1052 (1994).
223. Saha,S.K. and Brewer,C.F., *Carbohydr. Res.*, **254**, 157-167 (1994).
224. Morris,D.L., *Science*, **107**, 254-255 (1948).
225. Dorne,A.-J., Kappler,R., Kristen,U. and Heinz,E., *Phytochemistry*, **27**, 2027-2031 (1988).
226. Monsigny,M., Petit,C. and Roche,A.-C., *Anal. Biochem.*, **175**, 525-530 (1988).
227. Jungalwala,F.B., Haycs,L. and McCluer,R.H., *J. Lipid Res.*, **18**, 285-292 (1977).
228. Lauter,C.J. and Trams,E.G., *J. Lipid Res.*, **3**, 136-138 (1962).
229. Naoi,M., Lee,Y.C. and Roseman,S., *Anal. Biochem.*, **58**, 571-577 (1974).
230. Tomono,Y., Abe,K. and Watanabe,K., *Anal. Biochem.*, **184**, 360-368 (1990).
231. Kniep,B. and Mühlradt,P.F., *Anal. Biochem.*, **188**, 5-8 (1990).
232. Wiesner,D.A. and Sweeley,C.C., *Anal. Biochem.*, **217**, 316-322 (1994).
233. Bhat,U.R. and Carlson,R.W., *Glycobiology*, **2**, 535-539 (1992).
234. Morrison,W.R. and Hay,J.D., *Biochim. Biophys. Acta*, **202**, 460-467 (1970).
235. Carter,H.E., Betts,B.E. and Strobach,D.R., *Biochemistry*, **3**, 1103-1107 (1964).
236. Carter,H.E., Rothfus,J.A. and Gigg,R., *J. Lipid Res.*, **2**, 228-234 (1961).
237. Christie,W.W., *Lipid Analysis*, 2nd ed., Pergamon Press, Oxford 1982.
238. Li,Y.-T. and Li,S.-C., *Methods Enzymol.*, **179**, 479-487 (1989).
239. Sarmientos,F., Schwarzmann,G. and Sandhoff,K., *Eur. J. Biochem.*, **160**, 527-535 (1986).
240. Baumann,W.J., Schmid,H.H.O. and Mangold,H.K., *J. Lipid Res.*, 10, 132-133 (1969).
241. Kadowaki,H. and Grant,M.A., *Lipids*, **29**, 721-725 (1994).
242. Suzuki,Y., Hirabayashi,Y. and Matsumoto,M., *J. Biochem. (Tokyo)*, **95**, 1219-1222 (1984).
243. Carter,H.E., Kisic,A., Koob,J.L. and Martin,J.A., *Biochemistry*, **8**, 389-393 (1969).

244. Menon,A.K., *Methods Enzymol.*, **230**, 418-442 (1994).
245. deLederkremer,R.M., Lima,C.E., Ramirez,M.I., Goncalvez,M.F. and Colli,W., *Eur. J. Biochem.*, **218**, 929-936 (1993).
246. Karlsson,K.-A. and Pascher,I., *J. Lipid Res.*, **12**, 466-472 (1971).
247. Srinivas,N.R., Shyu,W.C. and Barbhaiya,R.H., *Biomed. Chromatogr.*, **9**, 1-9 (1995).
248. Bandi,P.C. and Schmid,H.H.O., *Chem. Phys. Lipids*, **17**, 267-274 (1976).
249. Tatsumi,K., Kishimoto,Y. and Hignite,C., *Arch. Biochem. Biophys.*, **165**, 656-664 (1974).
250. Karlsson,K.-A. and Pascher,I., *Chem. Phys. Lipids*, **12**, 65-74 (1974).
251. Abe,T. and Mori,K., *Biosci. Biotech. Biochem.*, **58**, 1671-1674 (1994).
252. Sonnet,P.E., Dudley,R.L., Osman,S., Pfeffer,P.E. and Schwartz,D., *J. Chromatogr.*, **586**, 255-258 (1991).
253. Laethem,R.M., Balazy,M., Falck,J.R., Laethem,C.L. and Koop,D.R., *J. Biol. Chem.*, **268**, 12912-12918 (1993).
254. Kaunzinger,A., Podebrad,F., Liske,R., Maas,B., Dietrich,A. and Mosandl,A., *J. High Resol. Chromatogr.*, **18**, 49-53 (1995).
255. Christie,W.W., in *Advances in Lipid Methodology-One*, pp. 121-148 (1992) (edited by W.W. Christie, The Oily Press, Ayr).
256. Kobayashi,T., Mitsuo,K. and Goto,I., *Eur. J. Biochem.*, **172**, 747-752 (1988).
257. Merrill,A.H., Wang,E., Mullins,R.E., Jamison,W.C.L., Nimkar,S. and Liotta,D.C., *Anal. Biochem.*, **171**, 373-381 (1988).
258. Carter,H.E. and Gaver,R.C., *J. Lipid Res.*, 8, 391-395 (1967).
259. Karlsson,K.-A., *Lipids*, **5**, 878-891 (1970).
260. Polito,A.J., Akita,T. and Sweeley,C.C., *Biochemistry*, **7**, 2609-2614 (1968).
261. Minnikin,D.E., *Chem. Phys. Lipids*, **21**, 313-348 (1978).
262. Yamamoto,K., Shibahara,A., Nakayama,T. and Kajimoto,G., *Chem. Phys. Lipids*, **60**, 39-50 (1991).
263. Stoffel,W., *Chem. Phys. Lipids*, **11**, 318-334 (1973).
264. Laine,R.A. and Renkonen,O., *J. Lipid Res.*, **16**, 102-106 (1975).
265. Karlsson,K.-A., Samuelsson,B.E. and Steen,G.O., *Biochim. Biophys. Acta*, **306**, 317-328 (1973).
266. Egge,H., *Chem. Phys. Lipids*, **21**, 349-360 (1978).
267. Sweeley,C.C. and Nunez,H.A., Ann. Rev. Biochem., **54**, 765-801 (1985).
268. Samuelsson,B.E., Pimlott,W. and Karlsson,K.-A., *Methods Enzymol.*, **193**, 623-646 (1990).
269. Peter-Katalinic,J. and Egge,H., *Methods Enzymol.*, **193**, 713-733 (1990).
270. Taki,T., Ishikawa,D., Handa,S. and Kasama,T., *Anal. Biochem.*, **225**, 24-27 (1995).
271. Pahlsson,P. and Nilsson,B., *Anal. Biochem.*, **168**, 115-120 (1988).
272. Kushi,Y., Rokukawa,C. and Handa,S., *Anal. Biochem.*, **175**, 167-176 (1988).
273. Karlsson,K.-A., Lanne,B., Pimlott,W. and Teneberg,S., *Carbohydr. Res.*, **221**, 49-61 (1991).
274. Koerner,T.A.W., Prestegard,J.H. and Yu,R.K., *Methods Enzymol.*, **138**, 38-59 (1987).
275. van Halbeck,H., *Methods Enzymol.*, **230**, 132-168 (1994).
276. Sarmientos,F., Schwarzmann,G. and Sandhoff,K., *Eur. J. Biochem.*, **146**, 59-64 (1985).
277. Tako,T., Kuroyanagi,M., Yoshioka,H. and Handa,S., *J. Biochem. (Tokyo)*, **111**, 614-619 (1992).
278. Falk,K.-E., Karlsson,K.-A. and Samuelsson,B.E., *Arch. Biochem. Biophys.*, **192**, 177-190 (1979).
279. Dabrowski,J., Haufland,P. and Egge,H., *Methods Enzymol.*, **83**, 65-86 (1982).
280. Yu,R.K., Koerner,T.A.W., Scardale,J. N. and Prestegard,J.H., *Chem. Phys. Lipids*, **42**, 27-48 (1986).
281. Mineki,S., Jida,M. and Tsutsumi,T., *J. Ferm. Bioeng.*, **78**, 327-330 (1994).
282. Boas,M.H.S.V., Egge,H., Pohlentz,G., Hartmann,R. and Bergter,E.B., *Chem. Phys. Lipids*, **70**, 11-19 (194).
283. Jin,W., Rinehart,K.L. and Jares-Erijman,E.A., *J. Org. Chem.*, **59**, 144-147 (1994).
284. Cafieri,F., Fattorusso,E., Mahajnah,Y. and Mangoni,A., *Liebigs Ann. Chem.*, 1187-1189 (1994).
285. Koerner,T.A.W., Cary,L.W., Li,S.-C. and Li,Y.-T., *J. Biol. Chem.*, **254**, 2326-2328 (1979).
286. Noda,N., Tanaka,R., Tsujino,K., Takasaki,Y., Nakano,M., Nishi,M. and Miyuahara,K., *J. Biochem. (Tokyo)*, **116**, 435-442 (1994).
287. Kisic,A., Tsuda,M., Kulmacz,R.J., Wilson,W.K. and Schroepfer,G.J., *J. Lipid Res.*, **36**, 787-803 (1995).
288. Joyard,J., Block,M.A. and Douce,R., *Eur. J. Biochem.*, **199**, 489-509 (1991).

289. Joyard,J., Block,M.A., Malherbe,A., Maréchal,E. and Douce,R., in *Lipid Metabolism in Plants, Origin and Synthesis of Galactolipid and Sulfolipid Headgroups*, pp. 231-258 (1993) (edited by T.S. Moore, CRC Press, Boca Raton).
290. Malherbe,A., Block,M.A., Douce,R. and Joyard,J., *Plant Physiol. Biochem.*, **33**, 149-161 (1995).
291. Maréchal,E., Block,M.A., Joyard,J. and Douce,R., *J. Biol. Chem.*, **269**, 5788-5798 (1994).
292. Seifert,U. and Heinz,E., *Bot. Acta*, **105**, 197-205 (1992).
293. Schmidt,H., Dresselhaus,T., Buck,F. and Heinz,E., *Plant Mol. Biol.*, **26**, 631-642 (1994) and literature cited therein.
294. Morré,D.J., Morré,T.J., Morré,S.R., Sundqvist,C. and Sandelius,S., *Biochim. Biophys. Acta*, **1070**, 437-445 (1991).
295. Rawyler,A., Meylan-Bettex,M. and Siegenthaler,P.A., *Biochim. Biophys. Acta*, **1233**, 123-133 (1995).
296. Hugueney,P., Bouvier,F., Badillo,A., d'Harlingue,A., Kuntz,M. and Camara,B., *Proc. Natl. Acad. Sci. USA*, **92**, 5630-5634 (1995).
297. Teucher,T. and Heinz,E., *Planta*, **184**, 319-326 (1991).
298. Maréchal,E., Block,M.A., Joyard,J. and Douce,R., *C.R. Acad. Sci. Paris*, **313**, 521-528 (1991).
299. Ohta,H., Shimojima,M., Arai,T., Masuda,T., Shioi,Y. and Takamiya,K.-I., in *Plant Lipid Metabolism*, pp. 152-155 (1995) (edited by J.-C. Kader and P. Mazliak, Kluwer Acad. Publ., Dordrecht).
300. Feige,G.B., Heinz,E., Wrage, K., Cochems,N. and Ponzelar,E., in *Biogenesis and Function of Plant Lipids*, pp. 135-140 (1980) (edited by P. Mazliak, P. Benveniste, C. Costes and R. Douce, Elsevier Biomedical Press, Amsterdam).
301. Sato,N. and Murata,N., *Plant Cell Physiol.*, **23**, 1115-1120 (1982).
302. Sato,N., *Plant Physiol. Biochem.*, **32**, 121-126 (1994).
303. van Besouw,A. and Wintermans,J.F.G.M., *Biochim. Biophys. Acta*, **529**, 44-53 (1978).
304. Ferrari,R.A. and Benson,A.A., *Arch. Biochem. Biophys.*, **93**, 185-192 (1961).
305. Heemskerk,J.W.M., Storz,T., Schmidt,R.R. and Heinz,E., *Plant Physiol.*, **93**, 1286-1294 (1990).
306. Boos,W., Lehmann,J. and Wallenfels,K., *Carbohydr. Res.*, **1**, 419-420 (1966).
307. Boos,W., *Methods Enzymol.*, **89**, 59-64 (1982).
308. Egel,R., *J. Theor. Biol.*, **79**, 117-119 (1979).
309. Binder,W.H., Kählig,H. and Schmid,W., *Tetrahedron*, **50**, 10407-10418 (1994).
310. Dumortier,V., Brassart,C. and Bouquelet,S., *Biotechnol. Appl. Biochem.*, **19**, 341-354 (1994).
311. Shin,H.-J. and Yang,J.-W., *Biotech. Lett.*, **16**, 1157-1162 (1994).
312. Heinz,E., Bertrams,M., Joyard,J. and Douce,R., *Z. Pflanzenphysiol.*, **87**, 325-331 (1978).
313. Heemskerk,J.W.M., Wintermans,J.F.G.M., Joyard,J., Block,M.A., Dorne,A.J. and Douce,R., *Biochim. Biophys. Acta*, **877**, 281-289 (1986).
314. Benning,C. and Somerville,C.R., *J. Bacteriol.*, **174**, 6479-6487 (1992).
315. Rossak,M., Tietje,C., Heinz,E. and Benning,C., *J. Biol. Chem.*, **270**, in press, (1995).
316. Rawyler,A., Unitt,M.D., Giroud,C., Davies,H., Mayor,J.-P., Harwood,J.L. and Siegenthaler,P.A., *Photosynth. Res.*, **11**, 3-13 (1987).
317. Siegenthaler,P.A., Rawyler,A. and Mayor,J.-P., in *Biological Role of Plant Lipids*, pp. 171-180 (1989) (edited by P.A. Biacs, K. Gruiz and T. Kremmer, Plenum Publ. Corp., New York).
318. Williamson,P. and Schlegel,R.A., *Mol. Membr. Biol.*, **11**, 199-216 (1994).
319. Prats,M., Tocanne,J.-F. and Teissie,J., *Eur. J. Biochem.*, **162**, 379-385 (1987).
320. Quinn,P.J. and Williams,W.P., *Biochim. Biophys. Acta*, **737**, 223-266 (1983).
321. Webb,M.S. and Green,B.R., *Biochim. Biophys. Acta*, **1060**, 133-158 (1991).
322. Voß,R., Radunz,A. and Schmid,G.H., *Z. Naturforsch.*, **47c**, 406-415 (1992).
323. Blankenship,R.E., *Antonie van Leeuwenhoek*, **65**, 311-329 (1994).
324. Nußberger,S., Dörr,K., Wang,D.N. and Kühlbrandt,W., *J. Mol. Biol.*, **234**, 347-356 (1993).
325. Van't Hof,R., van Klompenburg,W., Pilon,M., Kozubek,A., de Korte-Kool,G., Demel,R.A., Weisbeek,P.J. and de Kruijff,B., *J. Biol. Chem.*, **268**, 4037-4042 (1993).
326. Chupin,V., van't Hof,R. and de Kruijff,B., *FEBS Lett.*, **350**, 104-108 (1994).
327. Van't Hof,R. and de Kruijff,B., *FEBS Lett.*, **361**, 35-40 (1995).
328. Thurmond,R.L., Niemi,A.R., Lindblom,G., Wieslander,A. and Rilfors,L., *Biochemistry*, **33**, 13178-13188 (1994).
329. Benning,C., Beatty,J.T., Prince,R.C. and Somerville,C.R., *Proc. Natl. Acad. Sci. USA*, **90**, 1561-1565 (1993).

330. Benning,C. and Somerville,C.R., *J. Bacteriol.*, **174**, 2352-2360 (1992).
331. Sato,N., Tsuzuki,M., Matsuda,Y., Ehara,T., Osafune,T. and Kawaguchi,A., *Eur. J. Biochem.*, **230**, 987-993 (1995).
332. Fuks,B. and Homblé,F., *Biophys. J.*, **66**, 1004-1414 (1994).
333. Curatolo,W., *Biochim. Biophys. Acta,* **906**, 111-136 (1987).
334. Koynova,R. and Caffrey,M., *Chem. Phys. Lipids,* **69**, 181-207 (1994).
335. Carter,H.E., McCluer,R.H. and Slifer,E.D., *J. Am. Chem. Soc.*, **78**, 3735-3738 (1956).
336. Carter,H.E., Ohno,K., Nojima,S., Tipton,C.L. and Stanacev,N.Z., *J. Lipid Res.,* **2**, 215-222 (1961).
337. Carter,H.E., Hendry,R.A. and Stanacev,N.Z., *J. Lipid Res.,* **2**, 223-227 (1961).
338. Webster,D.E. and Chang,S.B., *Plant Physiol.,* **44**, 1523-1527 (1969).
339. Galliard,T., *Biochem. J.*, **115**, 335-339 (1969).
340. Brush,P. and Percival,E., *Phytochemistry,* **11**, 1847-1849 (1972).
341. Miyazawa,T. and Fujino,Y., *Agric. Biol. Chem.*, **42**, 1979-1980 (1978).
342. Siebertz,H.P., Heinz,E., Linscheid,M., Joyard,J. and Douce,R., *Eur. J. Biochem.,* **101**, 429-438 (1979).
343. Fujino,Y. and Miazawa,T., *Biochim. Biophys. Acta,* **572**, 442-451 (1979).
344. Ito,S. and Fujino,Y., *Agric. Biol. Chem.*, **44**, 1181-1182 (1980).
345. Kojima,M., Seki,K., Ohnishi,M., Ito,S. and Fujino,Y., *Biochem. Cell Biol.*, **68**, 59-64 (1990).
346. Kojima,M., Shiraki,H., Ohnishi,M. and Sito,S., *Phytochemistry,* **29**, 1161-1163 (1990).
347. Kojima,M., Sasaki,S., Ohnishi,M., Mano,Y. and Ito,S., *Phytochemistry,* **29**, 2091-2096 (1990).
348. Kojima,M., Kimura,H., Ohnishi,M., Fujino,Y. and Ito,S., *Phytochemistry,* **30**, 1165-1168 (1991).
349. Walker,R.W. and Bastl,C.P., *Carbohydr. Res.*, **4**, 49-54 (1967).
350. Jamieson,G.R. and Reid,E.H., *Phytochemistry,* **15**, 135-136 (1976).
351. Kondo,Y., Ito,S. and Fujino,Y., *Agric. Biol. Chem.*, **38**, 2549-2552 (1974).
352. Fujino,Y., *Cereal Chem.*, **55**, 559-571 (1978).
353. Sastry,P.S., *Adv. Lipid Res.*, **12**, 251-310 (1974).
354. Ratledge,C. and Wilkinson,S.G., eds., *Microbial Lipids,* Volume 1, Academic Press, London, 1988.
355. Slomiany,B.L., Murty,V.L.N., Liau,Y.H. and Slomiany,A., *Progr. Lipid Res.*, **26**, 29-51 (1987).
356. Vos,J.P., Lopes-Cardozo,M. and Gadella,B.M., *Biochim. Biophys. Acta,* **1211**, 125-149 (1994).
357. Vaver,V.A., Todria,K.G., Prokazova,N.V., Rozynov,B.V. and Bergelson,L.D., *Biochim. Biophys. Acta,* **486**, 60-69 (1977).
358. Kondo,Y., *Biochim. Biophys. Acta,* **665**, 471-476 (1981).
359. Heemskerk,J.W., Bögemann,G. and Wintermans,J.F.G.M., *Biochim. Biophys. Acta,* **754**, 181-189 (1983).
360. Hudson,B.J.F. and Ogunsua,A.O., *J. Sci. Food Agric.*, **25**, 1503-1508 (1974).
361. Heinz,E., *Biochim. Biophys. Acta,* **144**, 321-332 (1967).
362. Heinz,E. and Tulloch,A.P., *Hoppe-Seyler's Z. Physiol. Chem.*, **350**, 493-498 (1969).
363. Matsuzaki,T., Koiwai,A., Kawashima,N. and Matsuyama,S., *Agric. Biol. Chem.*, **46**, 723-729 (1982).
364. Critchley,C. and Heinz,E., *Biochim. Biophys. Acta,* **326**, 184-193 (1973).
365. Heinz,E., Rullkötter,J. and Budzikiewicz,H., *Hoppe Seyler's Z. Physiol. Chem.*, **355**, 612-616 (1974).
366. Myhre,D.V., *Canad. J. Chem.*, **46**, 3071-3077 (1968).
367. MacMurray,T.A. and Morrison,W.R., *J. Sci. Food Agric.*, **21**, 520-528 (1970).
368. Demel,R., Lindblom,G. and Rilfors,L., *Biochim. Biophys. Acta,* **1190**, 416-420 (1994).
369. Hauksson,J.B., Lindblom,G. and Rilfors,L., *Biochim. Biophys. Acta,* **1215**, 341-345 (1994).
370. Fusetani,N. and Hashimoto,Y., *Agric. Biol. Chem.*, **39**, 2021-2025 (1975).
371. Baruah,P., Baruah,N.C., Sharma,R.P., Baruah,J.N., Kulanthaivel,P. and Herz,W., *Phytochemistry,* **22**, 1741-1744 (1983).
372. Murakami,N., Morimoto,T., Ueda,T., Nagai,S.-I., Sakakibara,J. and Yamad,N., *Phytochemistry,* **31**, 2641-2644 (1992).
373. Choudhury,N.H. and Juliano,B.O., *Phytochemistry,* **19**, 1063-1069 (1980).
374. Kozakai,H., Oshima,Y. and Yasumoto,T., *Agric. Biol. Chem.*, **46**, 233-236 (1982).
375. Heisig,O.M.R.A. and Heinz,E., *Phytochemistry,* **11**, 815-818 (1972).
376. Murakami,N., Morimoto,T., Imamura,H., Nagatsu,A. and Sakakibara,J., *Tetrahedron,* **50**, 1993-

2002 (1994).
377. Safford,R., Appleby,R.S. and Nichols,B.W., *Biochim. Biophys. Acta,* **239**, 509-512 (1971).
378. Bajwa,S.S. and Sastry,P.S., *Ind. J. Biochem. Biophys.*, **10**, 65-66 (1973).
379. Sauer,A. and Heise,K.-P., *Z. Naturforsch.*, **37c**, 218-225 (1982).
380. Chen,H.-H., Wickrema,A. and Jaworski,J.G., *Biochim. Biophys. Acta,* **963**, 493-500 (1988).
381. Wada,H., Schmidt,H., Heinz,E. and Murata,N., *J. Bacteriol.*, **175**, 544-547 (1993).
382. Sastry,P.S. and Kates,M., *Biochemistry*, **3**, 1280-1287 (1964).
383. Galliard,T., *Phytochemistry,* **9**, 1725-1734 (1970).
384. Galliard,T., *Biochem. J.*, **121**, 379-390 (1971).
385. Burns,D.D., Galliard,T. and Harwood,J.L., *Phytochemistry,* **18**, 1793-1797 (1979).
386. Hirayama,O., Matsuda,H., Takeda,H., Maenaka,K. and Takatsuka,H., *Biochim. Biophys. Acta,* **384**, 127-137 (1975).
387. Helmsing,P.J., *Biochim. Biophys. Acta,* **178**, 519-533 (1969).
388. Dawson,R.M.C., Hemington,N., Grime,D., Lander,D. and Kemp,P., *Biochem. J.*, **144**, 169-171 (1974).
389. Bajwa,S.S. and Sastry,P.S., *Biochem. J.*, **144**, 177-187 (1974).
390. Andersson,L., Bratt,C., Arnoldsson,K.C., Herslöf,B., Olsson,N.U., Sternby,B. and Nilsson,Å., *J. Lipid Res.*, **36**, 1392-1400 (1995).
391. Cho,S.H. and Thompson,G.A., *Biochim. Biophys. Acta,* **878**, 353-359 (1986).
392. Kaniuga,Z. and Gemel,J., *FEBS Lett.*, **171**, 55-58 (1984).
393. O'Sullivan,J.N., Warwick,N.W.M. and Dalling,M.J., *J. Plant Physiol.,* **131**, 393-404 (1987).
394. Wickberg,B., *Acta Chem. Scand.*, **12**, 1183-1186 (1958).
395. Urbas,B., *Canad. J. Chem.*, **46**, 49-53 (1968).
396. Meng,J., Rosell,K.-G. and Srivastava,L.M., *Carbohydr. Res.*, **161**, 171-180 (1987).
397. Gerwick,W.H., *Biochim. Biophys. Acta,* **1211**, 243-255 (1994).
398. Jiang,Z.D. and Gerwick,W.H., *Phytochemistry,* **29**, 1433-1440 (1990).
399. Jiang,Z.D. and Gerwick,W.H., *Lipids,* **26**, 960-963 (1991).
400. Vick,B.A., in *Lipid Metabolism in Plants*, pp. 167-191 (1993) (edited by T.S. Moore, CRC Press, Boca Raton).
401. Feussner,I., Hause,B., Vörös,K., Parthier,B. and Wasternack,C., *Plant J.*, **7**, 949-957 (1995)
402. Benson,A.A., *Adv. Lipid Res.*, **1**, 387-394 (1963).
403. Heinz,E., in *Sulfur Nutrition and Assimilation in Higher Plants*, pp. 163-178 (1993) (edited by L.J. deKok, I. Stulen, H. Rennenberg, C. Brunhold and W.E. Rauser, SPB Acad. Publ., The Hague).
404. Cedergreen,R.A. and Hollingsworth,R.I., *J. Lipid Res.,* **35**, 1452-1461 (1994).
405. Kaya,K., Sano,T. Watanabe,M.M., Shiraishi,F. and Ito,H., *Biochim. Biophys. Acta,* **1169**, 39-45 (1993).
406. Yagi,T. and Benson,A.A., *Biochim. Biophys. Acta,* **57**, 601-603 (1962).
407. Wolfersberger,M.G. and Pieringer,R.A., *J. Lipid Res.,* **15**, 1-10 (1974).
408. Burns,D.D., Galliard,T. and Harwood,J.L., *Phytochemistry,* **16**, 651-654 (1977).
409. Okaya,Y., *Acta Crystallogr.*, **17**, 1276-1282 (1964).
410. Lee,R.F. and Benson,A.A., *Biochim. Biophys. Acta,* **261**, 35-37 (1972).
411. Gupta,S.D. and Sastry,P.S., *Arch. Biochem. Biophys.,* **259**, 510-519 (1987).
412. Gustafson,K.R., Cardellina,J.H., Fuller,R.W., Weislow,O.S., Kiser,R.F., Snader,K.M., Patterson,G.M.L. and Boyd,M.R., *J. Natl. Cancer Inst.*, **81**, 1254-1258 (1989).
413. Auling,G., Heinz,E. and Tulloch,A.P., *Hoppe-Seyler's Z. Physiol. Chem.*, **352**, 905-912 (1971).
414. Tulloch,A.P., Heinz,E. and Fischer,W., *Hoppe-Seyler's Z. Physiol. Chem.*, **354**, 879-889 (1973).
415. Rullkötter,J., Heinz,E. and Tulloch,A.P., *Z. Pflanzenphysiol.*, **76**, 163-175 (1975).
416. Wells,M.A. and Dittmer,J.C., *Biochemistry*, **2**, 1259-1263 (1963).
417. Williams,J.P. and Merrilees,P.A., *Lipids,* **5**, 367-370 (1970).
418. Wuthier,R.E., *J. Lipid Res.,* **7**, 558-561 (1966).
419. Christie,W.W., in *Advances in Lipid Methodology-One*, pp. 1-18 (1992) (edited by W.W. Christie, The Oily Press, Ayr).
420. Dubinsky,Z. and Aaronson,S., *Phytochemistry,* **18**, 51-52 (1979).
421. Somersalo,S., Karunen,P. and Aro,E.-M., *Physiol. Plant.*, **68**, 467-470 (1986).
422. Hara,A. and Radin,N.S., *Anal. Biochem.*, **90**, 420-426 (1978).
423. Bratt,C.E. and Akerlund,H.-E., *Biochim. Biophys. Acta,* **1165**, 288-290 (1993).
424. Fischer,W., *Biochim. Biophys. Acta,* **487**, 89-104 (1977).

425. Fischer,W., in *CRC Handbook of Chromatography Lipids,* Volume I, pp. 555-587 (1984) (edited by H.K. Mangold, G. Zweig and J. Sherma, CRC Press, Boca Raton).
426. O'Brien,J.S. and Benson,A.A., *J. Lipid Res.*, **5**, 432-436 (1964).
427. Comfurius,P. and Zwaal,R.F.A., *Biochim. Biophys. Acta,* **488**, 36-42 (1977).
428. Webb,M.S., Tilcock,C.P.S. and Green,B.R., *Biochim. Biophys. Acta,* **938**, 323-333 (1988).
429. Heinz,E., *Biochim. Biophys. Acta,* **231**, 537-544 (1971).
430. Nichols,B.W., Harris,R.V. and James,A.T., *Biochem. Biophys. Res. Commun.*, **20**, 256-262 (1965).
431. Sato,N. and Murata,N., *Biochim. Biophys. Acta,* **619**, 353-366 (1980).
432. Kahn,M.U. and Williams,J.P., *J. Chromatogr.*, **140**, 179-185 (1977).
433. Araki,S., Sakurai,T., Omata,T., Kawaguchi,A. and Murata,N., *Jap. J. Phycol.*, **34**, 94-100 (1986).
434. Pettit,T.R., Jones,A.L. and Harwood,J.L., *Phytochemistry,* **28**, 399-405 (1989).
435. Jones,A.L. and Harwood,J.L., *J. Exptl. Bot.*, **44**, 1203-1210 (1993).
436. Araki,S., Sakurai,T., Oohusa,T., Kayama,M. and Sato,N., *Plant Cell Physiol.*, **30**, 775-781 (1989).
437. van Kessel,W.S.M.G., Tieman,M. and Demel,R.A., *Lipids,* **16**, 58-63 (1981).
438. Christie,W.W. and Morrison,W.R., *J. Chromatogr.*, **436**, 510-513 (1988).
439. Gallant,J. and Leblanc,R.M., *J. Chromatogr.*, **542**, 307-316 (1991).
440. Prieto,J.A., Ebri,A. and Collar,C., *J. Am. Oil Chem. Soc.*, **69**, 1019-1022 (1992).
441. Kim,H.Y., Yergey,J.A. and Salem,N., *J. Chromatogr.*, **394**, 155-170 (1987).
442. Bergquist,M.H.J. and Herslöf,B.G., *Chromatographia*, **40**, 129-133 (1995).
443. Demandre,C., Tremolières,A., Justin,A.-M. and Mazliak,P., *Phytochemistry*, **24**, 481-485 (1985).
444. Moreau,R.A., Asmann,P.T. and Norman,H.A., *Phytochemistry,* **29**, 2461-2466 (1990).
445. Moreau,R.A., in *Plant Lipid Biochemistry,, Structure and Utilization*, pp. 20-22 (1990) (edited by P.J. Quinn and J.L. Harwood, Portland Press, London).
446. Murata,N., Higashi,S.I. and Fujimura,Y., *Biochim. Biophys. Acta,* **1019**, 261-268 (1990).
447. Christie,W.W. and Urwin,R.A., *J. High Resol. Chromatogr.*, **18**, 97-100 (1995).
448. Arnoldsson,K.C. and Kaufmann,P., *Chromatographia*, **38**, 317-324 (1994).
449. Heemskerk,J.W.M., Bögemann,G., Scheijen,M.A.M. and Wintermans,J.F.G.M., *Anal. Biochem.*, **154**, 85-91 (1986).
450. Gut,H. and Matile,P., *Botanica Acta*, **102**, 31-36 (1989).
451. Giroud,C. and Eichenberger,W., *Plant Cell Physiol.*, **30**, 121-128 (1989).
452. Abidi,S.L., Mounts,T.L. and Rennick,K.A., *J. Liq. Chromatogr.*, **17**, 3705-3725 (1994).
453. Frentzen,M., in *Lipid Metabolism in Plants*, pp. 195-230 (1993) (edited by T.S. Moore, CRC Press, Boca Raton).
454. Roughan,P.G. and Slack,C.R., *Ann. Rev. Plant Physiol.,* **33**, 97-132 (1982).
455. Somerville,C. and Browse,J., *Science*, **252**, 80-87 (1991).
456. Arao,T., Kawaguchi,A. and Yamada,M., *Phytochemistry,* **26**, 2573-2576 (1987).
457. Araki,S., Sakurai,T., Kawaguchi,A. and Murata,N., *Plant Cell Physiol.*, **28**, 761-766 (1987).
458. Arao,T. and Yamada,M., *Phytochemistry,* **28**, 805-810 (1989).
459. Sukenik,A., Yamaguchi,Y. and Livne,A., *J. Phycol.*, **29**, 620-626 (1993).
460. Sato,N., Nemoto,Y. and Furuya,M., *Plant Physiol. Biochem.,* **26**, 93-98 (1988).
461. Jones,A.L. and Harwood,J.L., *Phytochemistry,* **31**, 3397-3403 (1992).
462. Yongmanitchai,W. and Ward,O.P., *J. Gen. Microbiol.*, **139**, 465-472 (1993).
463. Robinson,P.M., Smith,D.L., Safford,R. and Nichols,B.W., *Phytochemistry,* **12**, 1377-1381 (1973).
464. Giroud,C., Gerber,A. and Eichenberger,W., *Plant Cell Physiol.*, **29**, 587-595 (1988).
465. della Greca,M., Monaco,P., Pinto,G., Pollio,A. and Previtera,L., *Biochim. Biophys. Acta,* **1004**, 271-273 (1989).
466. Gordon,D.M. and Danishefsky,S.J., *J. Am. Chem. Soc.*, **114**, 659-663 (1992).
467. Morimoto,T., Murakami,N., Nagatsu,A. and Sakakibara J., *Chem. Pharm. Bull.*, **41**, 1545-1548 (1993).
468. Kesselmeier,J. and Heinz,E., *Anal. Biochem.*, **144**, 319-328 (1985).
469. Lynch,D.V., Gundersen,R.E. and Thompson,G.A., *Plant Physiol.,* **72**, 903-905 (1983).
470. Smith,L.A., Norman,H.A., Cho,S.H. and Thompson,G.A., *J. Chromatogr.*, **346**, 291-299 (1985).

471. Wiley,M.G., Przetakiewicz,M., Takahashi,M. and Lowenstein,J.M., *Lipids,* **27**, 295-301 (1992).
472. Yamauchi,R., Kojima,M., Isogai,M., Kato,K. and Ueno,Y., *Agric. Biol. Chem.*, **46**, 2847-2849 (1982).
473. Norman,H.A. and St. John,J.B., *Plant Physiol.,* **85**, 684-688 (1987).
474. Cho,S.H. and Thompson,G.A., *J. Biol. Chem.,* **262**, 7586-7593 (1987).
475. Schmidt,H. and Heinz,E., *Biochem. J.*, **289**, 777-782 (1993).
476. Andrews,J., Schmidt,H. and Heinz,E., in *Biological Role of Plant Lipids,* pp. 181-191 (1989) (edited by P.A. Biacs, K. Gruiz and T. Kremmer, Plenum Publ. Corp., New York).
477. Heinze,F.J., Linscheid,M. and Heinz,E., *Anal. Biochem.*, **139**, 126-133 (1984).
478. Bishop,D.G., *J. Liq. Chromatogr.*, **10**, 1497-1505 (1987).
479. Takamura,H., Narita,H., Urade,R. and Kito,M., *Lipids,* **21**, 356-361 (1986).
480. Bell,M.V. and Dick,J.R., *Lipids,* **26**, 565-573 (1991).
481. Myher,J.J. and Kuksis,A., *Canad. J. Biochem.*, **60**, 638-650 (1982).
482. Nichols,B.W. and Moorhouse,R., *Lipids,* **4**, 311-316 (1969).
483. Siebertz,H.P. and Heinz,E., *Z. Naturforsch.*, **32c**, 193-205 (1977).
484. Johnson,G. and Williams,J.P., *Plant Physiol.*, **91**, 924-929 (1989).
485. Heinz,E. and Roughan,P.G., *Plant Physiol.,* **72**, 273-279 (1983).
486. Sato,N. and Murata,N., *Biochim. Biophys. Acta,* **710**, 279-289 (1982).
487. Sato,N. and Murata,N., *Methods Enzymol.,* **167**, 251-259 (1988).
488. Siebertz,H.P., Heinz,E., Joyard,J. and Douce,R., *Eur. J. Biochem.,* **108**, 177-185 (1980).
489. Slack,C.R., Roughan,P.G. and Balasingham,N., *Biochem. J.*, **162**, 289-296 (1977).
490. Siebertz,M. and Heinz,E., *Hoppe-Seyler's Z. Physiol. Chem.*, **358**, 27-34 (1977).
491. Williams,J.P., *Biochim. Biophys. Acta,* **618**, 461-472 (1980).
492. Roughan,P.G. and Batt,R.D., *Anal. Biochem.*, **22**, 74-88 (1968).
493. Russel,G.B., *Anal. Biochem.*, **14**, 205-214 (1966).
494. Williams,J.P., Khan,M.U. and Mitchell,K., *Plant Cell Physiol.*, **29**, 849-854 (1988).
495. Sato,N. and Furuya,M., *Physiol. Plant.*, **62**, 139-147 (1984).
496. Hirayama,O. and Matsuda,H., *Agric. Biol. Chem.*, **36**, 2593-2596 (1972).
497. Henderson,R.J. and MacKinlay,E.E., *Phytochemistry,* **28**, 2943-2948 (1989).
498. Lütz,C., *Z. Pflanzenphysiol.*, **104**, 43-52 (1981).
499. Mallinger,A.G., Yao,J.K., Brown,A.S. and Dippold,C.S., *J. Chromatogr.*, **614**, 67-75 (1993).
500. Christie,W.W., in *Advances in Lipid Methodology-One*, pp. 239-272 (1992) (edited by W.W. Christie, The Oily Press, Ayr).
501. Radunz,A. and Berzborn,R., *Z. Naturforsch.*, **25b**, 412-419 (1970).
502. Radunz,A., *Z. Naturforsch.*, **27b**, 822-826 (1972).
503. Radunz,A., *Z. Naturforsch.*, **31c**, 589-593 (1976).
504. Billecocq,A., Douce,R. and Faure,M., *Compt. Rend. Acad. Sci. Paris,* **275**, 1135-1137 (1972).
505. Bansal,R., Warrington,A.E., Gard,A.L., Ranscht,B. and Pfeiffer,S.E., *J. Neurosci. Res.*, **24**, 548-557 (1989).
506. Gadella,B.M., Gadella,T.W.J., Colenbrander,B., van Golde,L.M.G. and Lopez-Cardozo,M., *J. Cell Sci.*, **107**, 2151-2163 (1994).
507. Billecocq,A., *Biochim. Biophys. Acta,*, **352**, 245-251 (1974).
508. Schmid,G.H., Radunz,A. and Gröschel-Stewart U., Immunologie und ihre Anwendung in der Biologie, Thieme Verlag, Stuttgart 1993.
509. Haase,M., Unthan,M., Couturier,P., Radunz,A. and Schmid,G.H., *Z. Naturforsch.*, **48c**, 623-631 (1993).
510. Slack,C.R., Roughan,P.G. and Terpstra,J., *Biochem. J.*, **155**, 71-80 (1976).
511. Appelqvist,L.-Å., *J. Lipid Res.,* **13**, 146-148 (1972).
512. Nakano,M. and Fischer,W., *Hoppe-Seyler's Z. Physiol. Chem.*, **358**, 1439-1453 (1977).
513. Joyard,J., Blée,E. and Douce,R., *Biochim. Biophys. Acta,* **879**, 78-87 (1986).
514. Bergelson,L.D., Vaver,V.A., Prokazova,N.V., Ushakov,A.N. and Popkova,G.A., *Biochim. Biophys. Acta,* **116**, 511-520 (1966).
515. Friedberg,S.J. and Alkek,R.D., *Biochemistry,* **16**, 5291-5294 (1977).
516. Hippmann,H. and Heinz,E., *Z. Pflanzenphys.*, **79**, 408-418 (1976).
517. Heinz,E., in *Lecithin and Health Care*, pp. 35-52 (1985) (edited by F. Paltauf and D. Lekim, Semmelweis-Verlag, Hoya).
518. Sakata,K. and Ina,K., *Bull. Jap. Soc. Sci. Fish.*, **51**, 659-665 (1985).
519. Hase,S., *Methods Enzymol.,* **230**, 225-237 (1994).

520. Spiro,M.J. and Spiro,R.G., *Anal. Biochem.*, **204**, 152-157 (1992).
521. Kwon,H. and Kim,J., *Anal. Biochem.*, **215**, 243-252 (1993).
522. Clement,A., Yong,D. and Brechet,C., *J. Liq. Chromatogr.*, **15**, 805-817 (1992).
523. Williams,J.P., Watson,G.R., Khan,M., Leung,S., Kuksis,A., Stachnyk,O. and Myher, J.J., *Anal. Biochem.*, **66**, 110-112 (1975).
524. Rickert,S.J. and Sweeley,C.C., *J. Chromatogr.*, **147**, 317-326 (1978).
525. Koch,H.U. and Fischer,W., *Biochemistry*, **17**, 5275-5281 (1978).
526. Veerkamp,J.H. and van Schaik,F.W., *Biochim. Biophys. Acta*, **348**, 370-387 (1974).
527. Mayberry,W.R. and Smith,P.F., *Biochim. Biophys. Acta*, **752**, 434-443 (1983).
528. Leopold,K. and Fischer,W., *Eur. J. Biochem.*, **196**, 475-482 (1991).
529. Veerkamp,J.H., *Biochim. Biophys. Acta*, **273**, 359-367 (1972).
530. Kurihara,H., Ando,J. and Hatano,M., *Bioorg. Med. Chem. Lett.*, **5**, 1241-1244 (1995).
531. Myers,R.L., Ullman,M.D., Ventura,R.F. and Yates,A.J., *Anal. Biochem.*, **192**, 156-164 (1991).
532. Lingwood,C.A., *Canad. J. Biochem.*, **57**, 1138-1143 (1979).
533. Fischer,W., Heinz,E. and Zeus,M., *Hoppe-Seyler's Z. Physiol. Chem.*, **354**, 1115-1123 (1973).
534. Kazlauskas,R.J., *TIBTECH*, **12**, 463-472 (1994).
535. Jaeger,K.E., Ransac,S., Dijkstra,B.W., Colson,C., van Heuvel,M. and Misset,O., *FEMS Microbiol. Rev.*, **15**, 29-63 (1994).
536. Diehl,N.W.K., Herling,H., Riedl,I. and Heinz,E., *Chem. Phys. Lipids*, **77**, 147-153 (1995).
537. Budzikiewicz,H., Rullkötter,J. and Heinz,E., *Z. Naturforsch.*, **28c**, 499-504 (1973).
538. Miyano,M. and Benson,A.A., *J. Am. Chem. Soc.*, **84**, 57-59 (1962).
539. Wickberg,B., *Acta Chem. Scand.*, **12**, 1187-1201 (1958).
540. Smith,C.R. and Wolff,I.A., *Lipids*, **1**, 123-127 (1966).
541. Kim,J.-H., Nishida,Y., Ohrui,H. and Meguro,H., *J. Carbohydr. Chem.*, **14**, 889-893 (1995).
542. Lands,W.E.M., Pieringer,R.A., Slakey,S.P.M. and Zschocke,A., *Lipids*, **1**, 444-448 (1966).
543. Sonnet,P.E., *Chem. Phys. Lipids*, **58**, 35-39 (1991).
544. Pinchuk,A.N., Mitsner,B.I. and Shvets,V.I., *Chem. Phys. Lipids*, **65**, 65-75 (1993).
545. Stoffyn,P., Stoffyn,A. and Hauser,G., *J. Lipid Res.*, **12**, 318-323 (1971).
546. van Dessel,G., Lagrou,A., Hilderson,H.J. and Dierick,W., *Biochim. Biophys. Acta*, **528**, 399-408 (1978).
547. Kamerling,J.P., Vliegenthart,J.F.G., Vink,J. and de Ridder,J.J., *Tetrahedron*, **27**, 4275-4288 (1971).
548. Sato,N., Seyama,Y. and Murata,N., *Plant Cell Physiol.*, **27**, 819-835 (1986).
549. Gage,D.A., Huang,Z.-H. and Benning,C., *Lipids*, **27**, 632-636 (1992).
550. Orgambide,G.G., Hollingsworth,R.I. and Dazzo,F.B., *Carbohydr. Res.*, **233**, 151-159 (1992).
551. Murakami,N., Imamura,H., Sakakibara,J. and Yamada,N., *Chem. Pharm. Bull.*, **38**, 3497-3499 (1990).
552. Mannock,D.A., Lewis,R.N.A.H. and McElhaney,R.N., *Chem. Phys. Lipids*, **55**, 309-321 (1990).
553. Son,B.W., *Phytochemistry*, **29**, 307-309 (1990).
554. Huang,Y. and Anderson,R., *J. Bacteriol.*, **177**, 2567-2571 (1995).
555. Hauksson,J.B., Rilfors,L., Lindblom,G. and Arvidson,G., *Biochim. Biophys. Acta*, **1258**, 1-9 (1995).
556. Ferrante,G., Ekiel,I., Patel,G.B. and Sprott,G.D., *Biochim. Biophys. Acta*, **963**, 162-172 (1988).
557. Stein,J. and Budzikiewicz,H., *Z. Naturforsch.*, **42b**, 1017-1020 (1987)
558. Convert,O., Michel,E., Heymans,F. and Godfroid,J.J., *Biochim. Biophys. Acta*, **794**, 320-325 (1984).
559. Morimoto,T., Nagatsu,A., Murakami,N. and Sakakibara,J., *Tetrahedron*, **51**, 6443-6450 (1995).
560. Johns,S.R., Leslie,D.R., Willing,R.I. and Bishop,D.G., *Aust. J. Chem.*, **30**, 823-834 (1977).
561. Johns,S.R., Leslie,D.R., Willing,R.I. and Bishop,D.G., *Aust. J. Chem.*, **31**, 65-72 (1978).
562. Gunstone,F.D., in *Advances in Lipid Methodology-Two*, pp. 1-68 (1993) (edited by W.W. Christie, The Oily Press, Dundee).
563. Gunstone,F.D. and Seth,S., *Chem. Phys. Lipids*, **72**, 119-126 (1994).
564. Aursand,M., Jørgensen,L. and Grasdalen,H., *J. Am. Oil. Chem. Soc.*, **72**, 293-297 (1995).
565. Medina,I. and Sacchi,R., *Chem. Phys. Lipids*, **70**, 53-61 (1994).
566. Pollesello,P., Toffanin,R., Eriksson,O., Kilpeläinen,I., Hynninen,P.H., Paoletti,S. and Saris,N.-E.L., *Anal. Biochem.*, **214**, 238-244 (1993).
567. Meneses,P. and Glonek,T., *J. Lipid Res.*, **29**, 679-689 (1988).
568. Wang,Y. and Hollingsworth,R.I., *Anal. Biochem.*, **225**, 242-251 (1995).

APPENDIX

Some Important References in Lipid Methodology - 1993

William W. Christie

The Scottish Crop Research Institute, Invergowrie, Dundee (DD2 5DA), Scotland

A. Introduction
B. The Isolation of Lipids from Tissues
C. Chromatographic and Spectroscopic Analysis of Lipids - General Principles.
D. The Analysis of Fatty Acids
E. The Analysis of Simple Lipid Classes
F. The Analysis of Complex Lipids
G. The Analysis of Molecular Species of Lipids
H. Structural Analysis of Lipids by means of Enzymatic Hydrolysis
I. The Analysis and Radioassay of Isotopically Labelled Lipids
J. The Separation of Plasma Lipoproteins
K. Some Miscellaneous Separations

A. INTRODUCTION

When the *Journal of Lipid Research* ceased its current awareness service for lipid methodology, it left a gap which this series is intended to fill. As for our first two volumes, the search of the literature has been done mainly to keep my own research up to date, and may not be entirely subjective in the selection. I have tried to list references with something new to say about lipid methodology rather than those that use tested methods, however competent or important these may be. Some papers may have been included simply because the title seemed apposite, when I have not had personal access to the journal to check them. Others may have been omitted quite unjustly, because it is impracticable for one person to read every paper that deals with lipids in a comprehensive manner.

In "*Advances in Lipid Methodology - Two*", the years 1991 and 1992 were covered so papers for 1993 and 1994 are listed as two appendices to this book. They are grouped in sections that correspond broadly to chapter headings in an earlier book (*"Lipid Analysis (second edition)"*, Pergamon Press, 1982). Often this has caused difficulties, as methods in papers can be relevant to many chapters, especially with review articles. Occasionally, papers have been listed twice but usually I have selected the single section that appeared most appropriate. The literature on prostaglandins and clinical chemistry (especially lipoprotein analysis) is not represented comprehensively, but these areas are covered by current awareness services in the journals *Prostaglandins, Leukotrienes and Essential Fatty acids* and *Current Topics in Lipidology* respectively.

Note that the titles of papers listed below may not be literal transcriptions of the originals. In particular, a number of abbreviations have been introduced. References are listed alphabetically according to the surname of the first author in each section.

In addition to the research papers and reviews listed below, there have been five books published dealing with aspects of lipid analysis, including the second volume of this series - a vintage year, *i.e.*

Christie,W.W. (ed.) *Advances in Lipid Methodology - Two* (Oily Press, Dundee) (1993).

Hammond,E.W. *Chromatography for the Analysis of Lipids* (CRC Press, Boca Raton, FL) (1993).

Mukherjee,K.D. and Weber,N. (eds) *CRC Handbook of Chromatography. Analysis of Lipids* (CRC Press, Boca Raton, FL) (1993).

Murphy,R.C. *Mass Spectrometry of Lipids (Handbook of Lipid Research, Vol. 7)* (Plenum Press, N.Y.) (1993).

Shibamoto,T. (ed.) *Lipid Chromatographic Analysis* (Dekker, NY) (1993).

I especially enjoyed the single author volumes by Hammond and Murphy. Hammond's book is a distinctive account of methodology in regular use in a major industrial laboratory, and I do not recall having seen anything comparable. Murphy's book is indispensable to any one with an interest in mass spectrometry of lipids. Individual chapters from the three review volumes are cited below. In addition, although the *Phospholipids Handbook* (ed. G. Cevc, Dekker, N.Y., 1993) contains little on the analysis of lipids, it is an immense compendium of valuable information.

B. THE ISOLATION OF LIPIDS FROM TISSUES

Beattie,S.E., Stafford,A.E. and King,A.D. Reevaluation of the neutral lipids of *Tilletia controversa* and *T. tritici*. *Lipids*, **28**, 1041-1043 (1993).

Berg,B.E., Hansen,E.M., Gjorven,S. and Greibrokk,T. On-line enzymatic reaction, extraction, and chromatography of fatty acids and triglycerides with supercritical carbon dioxide. *J. High Resolut. Chromatogr.*, **16**, 358-363 (1993).

Bratt,C.E. and Akerlund,H.-E. Isolation of pigment-free bulk lipids from thylakoids. *Biochim. Biophys. Acta*, **1165**, 288-290 (1993).

Buhlmann,R., Carmona,J., Donzel,A., Donzel,N., Gil,J., Haab,J.M. and Hamilton,S.D. A robotic system for the extraction and esterification of bacterial fatty acids. *American Lab.*, **25** (No. 9), 18-19 (1993).

Christie,W.W. Preparation of lipid extracts from tissues. In: *Advances in Lipid Methodology - Two*, pp. 195-213 (edited by W.W. Christie, Oily Press, Dundee) (1993).

Coene,J., Van den Eeckhout,E. and Herdewijn,P. GC determination of alkyl lysophospholipids after solid-phase extraction from culture media. *J. Chromatogr.*, **612**, 21-26 (1993).

Crans,D.C., Mikus,M. and Marshman,R.W. ^{31}P NMR examination of phosphorus metabolites in the aqueous, acidic and organic extracts of Phaseolus vulgaris seeds. *Anal. Biochem.*, **209**, 85-94 (1993).

Dunstan,G.A., Volkman,J.K. and Barrett,S.M. The effect of lyophilization on the solvent extraction of lipid classes, fatty acids and sterols from the oyster *Crassostrea gigas*. *Lipids*, **28**, 937-944 (1993).

Eder,K., Reichlmayer-Lais,A.M. and Kirchgessner,M. Studies on the extraction of phospholipids from erythrocyte membranes in the rat. *Clin. Chim. Acta*, **219**, 93-104 (1993).

Erickson,M.C. Lipid extraction from channel catfish muscle - comparison of solvent systems. *J. Food Sci.*, **58**, 84-89 (1993).

Fried,B. Obtaining and handling biological materials and prefractionating extracts for lipid analysis. In: *CRC Handbook of Chromatography. Analysis of Lipids*, pp. 1-10 (ed. K.D. Mukherjee and N. Weber, CRC Press, Boca Raton) (1993).

Garces,R. and Mancha,M. One-step lipid extraction and fatty acid methyl esters preparation from fresh plant tissues. *Anal. Biochem.*, **211**, 139-143 (1993).

Garcia-Barcelo,M., Luquin,M., Belda,F. and Ausina,V. GC whole-cell fatty acid analysis as an aid for the identification of mixed Mycobacterial cultures. *J. Chromatogr.*, **617**, 299-303 (1993).

Gunnlaugsdottir,H. and Ackman,R.G. Three extraction methods for determination of lipids in fish meal: evaluation of hexane/isopropanol methods as an alternative to chloroform/methanol methods. *J. Sci. Food Agric.*, **61**, 2235-240 (1993).

Meadows,J., Tillit,D., Huckins,J. and Schroeder,D. Large-scale dialysis of sample lipids using a semipermeable-membrane device. *Chemosphere*, **26**, 1993-2006 (1993).

Meneses,P., Navarro,J.N. and Glonek,T. Algal phospholipids by P-31 NMR - comparing isopropanol pretreatment with simple chloroform-methanol extraction. *Int. J. Biochem.*, **25**, 903-910 (1993).

Mirbod,F., Mori,S. and Nozawa,Y. Methods for phospholipid extraction in *Candida albicans* with high efficacy. *J. Med. Vet. Mycol.*, **31**, 405-409 (1993).

Mishra,V.K., Temelli,F. and Ooraikul,B. Extraction and purification of omega-3-fatty acids with an emphasis on supercritical-fluid extraction. *Food Res. Int.*, **26**, 217-226 (1993).

Osterroht,C. Extraction of dissolved fatty acids from sea water. *Fres. J. Anal. Chem.*, **345**, 773-779 (1993).

Staby,A. and Mollerup,J. Separation of constituents of fish-oil using supercritical fluids - a review of experimental solubility, extraction and chromatographic data. *Fluid Phase Equilibria*, **91**, 349-386 (1993).

Wolff,R.L. A simple and rapid method for the purification of fat from low-calorie spreads allowing the quantitative determination of their fatty acid composition. *Sci. Aliments*, **13**, 559-565 (1993).

C. CHROMATOGRAPHIC AND SPECTROSCOPIC ANALYSIS OF LIPIDS. GENERAL PRINCIPLES.

Demirbuker,M., Anderson,P.E. and Blomberg,L.G. Miniaturized light scattering detector for packed capillary supercritical fluid chromatography. *J. Microcol. Sep.*, **5**, 141-147 (1993).

Ettre,L.S. Nomenclature for chromatography. *Pure Appl. Chem.*, **65**, 819-872 (1993).

Ettre,L.S. The new IUPAC nomenclature for chromatography. *LC-GC Int.*, **6**, 544-548 (1993).

Rich,M.R. Conformational analysis of arachidonic and related fatty acids using molecular dynamics simulations. *Biochim. Biophys. Acta*, **1178**, 87-96 (1993).

Takashi,K. and Hirano,T. Theoretical aspects of the chromatographic behaviour of cis-trans isomers and omega isomers of lipid in polar stationary phase GLC and reversed phase HPLC. *Hokkaido Daigaku Suisangakuku Kenkyu Iho*, **44**, 209-219 (1993).

D. THE ANALYSIS OF FATTY ACIDS

This section contains references relevant to both Chapters 4 and 5 in *Lipid Analysis*.

Adams,J. and Songer,M.J. Charge-remote fragmentation for structural determination of lipids. *Trends Anal. Chem.*, **12**, 28-36 (1993).

Adelhardt,R. and Spiteller,G. Products of the dimerisation of unsaturated fatty acids. IX. Kinetic studies on the dimerisation of linoleic acid. *Fat Sci. Technol.*, **95**, 85-90 (1993).

Aitzetmuller,K. Capillary GLC fingerprints of seed lipids - a tool in plant taxonomy. *J. High Resolut. Chromatogr.*, **16**, 488-490 (1993).

Aitzetmuller,K., Werner,G. and Tsevegsuren,N. Screening of seed lipids for γ-linolenic acid - capillary GLC separation of 18:3 fatty acids with Δ-5 and Δ-6 double bonds. *Phytochemical Anal.*, **4**, 249-255 (1993).

Akasaka,A., Akama,T., Ohrui,H. and Meguro,H. Measurement of hydroxy and hydroperoxy fatty acids by HPLC with a column switching system. *Biosci. Biotech. Biochem.*, **57**, 2016-2019 (1993).

Akasaka,K., Ohrui,H. and Meguro,H. Determination of carboxylic acids by HPLC with 2-(2,3-anthracendicarboximido)ethyl trifluoromethanesulphonate as a highly sensitive fluorescent labelling reagent. *Analyst*, **118**, 765-768 (1993).

August,A., Dao,C.J., Jensen,D., Zhang,Q. and Dea,P. A facile catalytic deuteration of unsaturated fatty acids and phospholipids. *Microchem. J.*, **47**, 224-229 (1993).

Aursand,M., Rainuzzo,J.R. and Grasdalen,H. Quantitative high-resolution ^{13}C and ^{1}H NMR of ω3 fatty acids from white muscle of Atlantic salmon (*Salar salar*). *J. Am. Oil Chem. Soc.*, **70**, 971-981 (1993).

Baillet,A., Corbeau,L., Rafidison,P. and Ferrier,D. Separation of isomeric compounds by reversed-phase HPLC using Ag^+ complexation - application to *cis-trans* fatty acid methyl esters and retinoic acid photoisomers. *J. Chromatogr.*, **634**, 251-256 (1993).

Ballesteros,E., Gallego,M. and Valcarel,M. Automatic method for on-line preparation of fatty acid methyl esters from olive oil and other types of oil prior to their GC determination. *Anal. Chim. Acta*, **282**, 581-588 (1993).

Bortolomeazzi,R., Pizzale,L. and Lercker,G. Chromatographic determination of position and configurational isomers of methyl oleate hydroxides from corresponding hydroperoxides. *Chromatographia*, **36**, 61-64 (1993).

Brakstad,F. Accurate determination of double bond position in mono-unsaturated straight-chain fatty acid ethyl esters from conventional electron impact mass spectra by quantitative spectrum structure modelling. *Chemometric. Intelligent Lab. Systems*, **19**, 87-100 (1993).

Brodowsky,I.D. and Oliw,E.H. Biosynthesis of 8-*R*-hydroperoxylinoleic acid by the fungus *Laetisaria arvalis*. *Biochim. Biophys. Acta*, **1168**, 68-72 (1993).

Brumback,T.B., Hazebroek,J., Lamb,D., Danielson,L. and Orman,B. Automated fatty acid analysis from seeds - from field samples to data-bases. *Chemom. Intell. Lab. Systems*, **21**, 215-222 (1993).

Burhenne,J. and Parlar,H. A contribution to the identification of fatty acids from fish oils by mass- and FTIR-spectroscopy. *Fresenius Env. Bull.*, **2**, 119-124 (1993).

Carballeira,N.M. and Emiliano,A. Novel brominated phospholipid fatty acids from the Caribbean sponge *Agelas sp. Lipids*, **28**, 763-766 (1993).

Carballeira,N.M. and Shalabi,F. Novel brominated phospholipid fatty acids from the Caribbean sponge *Petrosia sp. J. Nat. Products*, **56**, 739-746 (1993).

Ching,C.B., Hidajat,K. and Rao,M.S. Reversed-phase HPLC studies for homologous series of polyunsaturated fatty acids on a commercial μ Bondapak free fatty acid column. *J. Liqu. Chromatogr.*, **16**, 527-540 (1993).

Ching,C.B., Hidajat,K. and Rao,M.S. Liquid chromatographic studies for essential fatty acids on a commercial alkyl phenyl bonded silica column. *Chromatographia*, **35**, 399-402 (1993).

Christie,W.W. Preparation of ester derivatives of fatty acids for chromatographic analysis. In: *Advances in Lipid Methodology - Two*, pp. 69-111 (edited by W.W. Christie, Oily Press, Dundee) (1993).

Christie,W.W., Brechany,E.Y., Sebedio,J.L. and Le Quere,J.L. Silver ion chromatography and gas chromatography-mass spectrometry in the structural analysis of cyclic monoenoic acids formed in frying oils. *Chem. Phys. Lipids*, **66**, 143-153 (1993).

Cominacini,L., Pastorino,A.M., McCarthy,A., Campagnola,M., Garbin,U., Davoli,A., De Santis,A. and Cascio,V.L. Determination of lipid hydroperoxides in native low-density lipoprotein by a chemiluminescent flow injection assay. *Biochim. Biophys. Acta*, **1165**, 279-287 (1993).

Craske,J.D. Separation of instrumental and chemical errors in the analysis of oils by GC - a collaborative evaluation. *J. Am. Oil Chem. Soc.*, **70**, 325-334 (1993).

de Waard,P., van der Wal,H., Huijberts,G.N.M. and Eggink,G. Heteronuclear NMR analysis of unsaturated fatty acids in poly(3-hydroxyalkanoates). *J. Biol. Chem.*, **268**, 315-319 (1993).

Dembitsky,V.M., Rezanka,T. and Kashin,A.G. Comparative study of the endemic freshwater fauna of Lake Baikal - II. Unusual lipid composition of two sponge species *Baicalospongia bacillifera* and *B. intermedia* (Family Lubomirskiidae, Class Demospongiae). *Comp. Biochem. Physiol.*, **106B**, 825-831 (1993).

Ebeler,S.E., Shibamoto,T. and Osawa,T. Gas and high-performance liquid chromatographic analysis of lipid peroxidation products. In: *Lipid Chromatographic Analysis*, pp. 223-249 (edited by T. Shibamoto, Dekker, NY) (1993).

Foglia,T.A., Silbert,L.S. and Vail,P.D. Peroxides. 12. GLC and HPLC analysis of aliphatic hydroperoxides and dialkyl peroxides. *J. Chromatogr.*, **637**, 157-166 (1993).

Fournier,F., Remaud,B., Blasco,T. and Tabet,J.C. Ion-dipole complex formation from deprotonated phenol fatty acid esters evidenced by using gas-phase labelling combined with tandem MS. *J. Am. Soc. Mass Spectrom.*, **4**, 343-351 (1993).

Garces,R. and Mancha,M. One-step lipid extraction and fatty acid methyl esters preparation from fresh plant tissues. *Anal. Biochem.*, **211**, 139-143 (1993).

Garson,M.J., Zimmermann,M.P., Hoberg,M., Larsen,R.M., Barttershill,C.N. and Murphy,P.T. Isolation of brominated fatty acids from the phospholipids of the tropical marine sponge *Amphimedon terpenensis*. *Lipids*, **28**, 1011-1014 (1993).

Grondin,I., Smadja,J., Farines,M. and Soulier,J. Lipids of Litchi and Longan seeds - study of cyclopropanoic acids by NMR. *Oleagineux*, **48**, 425-428 (1993).

Guido,D.M., McKenna,R. and Mathews,W.R. Quantitation of hydroperoxy-eicosatetraenoic acids and hydroxyeicosatetraenoic acids as indicators of lipid peroxidation using GC-MS. *Anal. Biochem.*, **209**, 123-129 (1993).

Gunstone,F.D. High resolution ^{13}C NMR spectroscopy of lipids. In: *Advances in Lipid Methodology - Two*, pp. 1-68 (edited by W.W. Christie, Oily Press, Dundee) (1993).

Gunstone,F.D. Information on the composition of fats from their high-resolution ^{13}C NMR spectra. *J. Am. Oil Chem. Soc.*, **70**, 361-366 (1993).

Gunstone,F.D. High resolution ^{13}C NMR study of synthetic branched-chain acids and of wool wax acids and isostearic acid. *Chem. Phys. Lipids*, **65**, 155-163 (1993).

Gunstone,F.D. The composition of hydrogenated fats by high-resolution ^{13}C NMR spectroscopy. *J. Am. Oil Chem. Soc.*, 70, 965-970 (1993).

Gunstone,F.D. The study of natural epoxy oils and epoxidized vegetable oils by ^{13}C NMR spectroscopy. *J. Am. Oil Chem. Soc.*, **70**, 1139-1144 (1993).

Gunstone,F.D. High-resolution ^{13}C NMR spectra of long-chain acids, methyl esters, glycerol esters, nitriles, amides, alcohols and acetates. *Chem. Phys. Lipids*, **66**, 189-193 (1993).

Hayakawa,M., Sugiyama,S. and Ozawa,T. HPLC analysis of lipids: analysis of fatty acids and their derivatives by a microcolumn HPLC system. In: *Lipid Chromatographic Analysis*, pp. 273-290 (edited by T. Shibamoto, Dekker, NY) (1993).

Heikes,D.L. Procedure for supercritical fluid extraction and GC determination of chlorinated fatty acid bleaching adducts in flour and flour-containing food items utilizing acid hydrolysis-methylation and Florisil column cleanup techniques. *J. Agric. Food Chem.*, **41**, 2034-2037 (1993).

Husain,S. and Devi,K.S. Separation and identification of isomeric conjugated fatty acids by HPLC with photodiode array detection. *Lipids*, **28**, 1037-1040 (1993).

Husek,P. Long-chain fatty acids esterified by action of alkyl chloroformates and analysed by capillary GC. *J. Chromatogr.*, **615**, 334-338 (1993).

Inamoto,Y., Hamanaka,S., Hamanaka,Y., Ariyama,S., Takemoto,T. and Okita,K. Unique fatty acids of *Helicobacter pylori* are methoxy fatty acids. *Proc. Jap. Acad. Ser. B. - Phys. Biol. Sci.*, **69**, 65-69 (1993).

Ioneda,T. Chromatographic and mass spectrometric characterization of 3-*O*-benzoyl methyl ester derivatives of mycolic acid fractions from *Corynobacter pseudotuberculosis, C. diphtheriae* and *Rhodococcus rhodochrous*. *Chem. Phys. Lipids*, **65**, 93-101 (1993).

Joseph,M., Kader,J.C., Dubacq,J.P. and Galle,A.M. Improved separation of fatty acid methyl esters by TLC - laboratory note. *Rev. Franc. Corps Gras*, **40**, 383-386 (1993).

Jung,S., Lowe,S.E., Hollingsworth,R.I. and Zeikus,J.G. *Sarcina ventriculi* synthesises very long chain dicarboxylic acids in response to different forms of environmental stress. *J. Biol. Chem.*, **268**, 2828-2835 (1993).

Khan,M.U. and Williams,J.P. Microwave-mediated methanolysis of lipids and activation of TLC plates. *Lipids*, **28**, 953-955 (1993).

Kim,H.Y. and Sawazaki,S. Structural analysis of hydroxy fatty acids by thermospray liquid chromatography/tandem MS. *Biol. Mass Spectrom.*, **22**, 302-310 (1993).

Kondo,Y., Kawai,Y., Miyazawa,T. and Mizutani,J. Chemiluminescence HPLC analysis of fatty acid hydroperoxide isomers. *Biosci. Biotech. Biochem.*, **57**, 1575-1576 (1993).

Konishi,H., Neff,W.E. and Mounts,T.L. Correlation of reversed-phase HPLC and GLC for fatty acid compositions of some vegetable oils. *J. Chromatogr.*, **629**, 237-242 (1993).

Konishi,H., Neff,W.E. and Mounts,T.L. Chemical interesterification with regioselectivity for edible oils. *J. Am. Oil Chem. Soc.*, **70**, 411-415 (1993).

Kusaka,T. and Ikeda,M. LC-MS of fatty acids including hydroxy and hydroperoxy acids as their 3-methyl-7-methoxy-1,4-benzoxazin-2-one derivatives. *J. Chromatogr.*, **639**, 165-173 (1993).

Lee,Y.M., Nakamura,H. and Nakajima,T. Retention behaviour of fatty acids labelled with monodansylcadaverine in reversed-phase HPLC. *Anal. Sci.*, **9**, 541-544 (1993).

Lehmann,W.D., Stephan,M., Fuerstenberger,G. and Marks,F. Profiling of monohydroxylated fatty acids in normal, hyperplastic and neoplastic mouse epidermis by GC-MS. *Dev. Oncol.,* **71**, 405-408 (1993).

Le Quere,J.L. Tandem mass spectrometry in the structural analysis of lipids. In: *Advances in Lipid Methodology - Two*, pp. 215-245 (edited by W.W. Christie, Oily Press, Dundee) (1993).

Luthria,D.L. and Sprecher,H. 2-Alkenyl-4,4-dimethyloxazolines as derivatives for the structural elucidation of isomeric unsaturated fatty acids. *Lipids*, **28**, 561-564 (1993).

Mayberry,W.R. and Lane,J.R. Sequential alkaline saponification/acid hydrolysis/esterification: a one-tube method with enhanced recovery of both cyclopropane and hydroxylated fatty acids. *J. Microbial Meth.*, **18**, 21-32 (1993).

McIlhinney,R.A. Acylated proteins: identification of the attached fatty acids and their linkages. In: *CRC Handbook of Chromatography. Analysis of Lipids*, pp. 241-248 (ed. K.D. Mukherjee and N. Weber, CRC Press, Boca Raton) (1993).

Metori,A., Ogamo,A. and Nakagawa,Y. Quantitation of monohydroxy fatty acids by HPLC with fluorescence detection. *J. Chromatogr.*, **622**, 147-151 (1993).

Mielniczuk,Z., Mielniczuk,E. and Larsson,L. GC-MS methods for analysis of 2- and 3-hydroxylated fatty acids: application for endotoxin measurement. *J. Microbiol. Methods*, **17**, 91-102 (1993).

Minnikin,D.E. Mycolic acids. In: *CRC Handbook of Chromatography. Analysis of Lipids*, pp. 339-347 (ed. K.D. Mukherjee and N. Weber, CRC Press, Boca Raton) (1993).

Minnikin,D.E., Bolton,R.C., Hartmann,S., Besra,G.S., Jenkins,P.A., Mallet,A.I. and Wilkins,E. An integrated procedure for the direct detection of characteristic lipids in tuberculosis patients. *Ann. Soc. Belge Med. Trop.*, **73**, 13-24 (1993).

Mittelbach,M. Ozonides and ozonolysis of lipids. In: *CRC Handbook of Chromatography. Analysis of Lipids*, pp. 163-172 (ed. K.D. Mukherjee and N. Weber, CRC Press, Boca Raton) (1993).

Mossoba,M.M., McDonald,R.E. and Prosser,E.R. GC matrix-isolation Fourier-transform infrared spectroscopic determination of *trans*-monounsaturated and saturated fatty acid methyl esters in partially hydrogenated menhaden oil. *J. Agric. Food Chem.*, **41**, 1998-2002 (1993).

Muller,K.D., Nalik,H.P., Schmid,E.N., Husmann,H. and Schomburg,G. Fast identification of Mycobacterium species by GC analysis with trimethylsulfonium hydroxide (TMSH) for transesterification. *J. High Resolut. Chromatogr.*, **16**, 161-165 (1993).

Nakagawa,Y. and Matsuyama,T. Chromatographic determination of optical configuration of 3-hydroxy fatty acids composing microbial surfactants. *FEMS Microbiol. Letts.*, **108**, 99-102 (1993).

Nakashima,K., Taguchi,Y., Kuroda,N., Akiyama,S. and Duan,G.L. 2-(4-Hydrazinocarbonylphenyl)-4,5-diphenylimidazole as a versatile fluorescent reagent for the HPLC of free fatty acids. *J. Chromatogr.*, **619**, 1-8 (1993).

Nikolova-Damyanova,B., Christie,W.W. and Herslof,B.G. High-performance liquid chromatography of fatty acid derivatives in the combined silver ion and reversed-phase modes. *J. Chromatogr.*, **653**, 15-23 (1993).

Nimz,E.L. and Morgan,S.L. On-line derivatization for complex fatty acid mixtures by capillary GC-MS. *J. Chromatogr. Sci.*, **31**, 145-149 (1993).

Olawoye,T.L., Minnikin,D. and Christie,W.W. HPLC separation of tertbutyldimethylsilyl (TBDMS) ether homologs of mycolic acids using a silver ion column. *Biokemistri*, **3**, 103-108 (1993).

Orgambide,G,G., Reusch,R.N. and Dazzo,F.B. Methoxylated fatty acids reported in *Rhizobium* isolates arise from chemical alterations of common fatty acids upon acid-catalyzed transesterification procedures. *J. Bact.*, **175**, 4922-4926 (1993).

Passi,S., Picardo,M., De Luca,C., Nazzarro-Porro,M., Rossi,L. and Rotilio,G. Saturated dicarboxylic acids as products of unsaturated fatty acid oxidation. *Biochim. Biophys. Acta*, **1168**, 190-198 (1993).

Polzer,J. and Bachmann,K. Sensitive determination of alkyl hydroperoxides by high-resolution GC-MS and high-resolution GC with flame ionization detection. *J. Chromatogr.*, **653**, 283-291 (1993).

Reynaud,D., Thickitt,C.P. and Pace-Asciak,C.R. Facile preparation and structure determination of monohydroxy derivatives of docosahexaenoic acid by α-tocopherol-directed autoxidation. *Anal.*

Biochem., **214**, 165-170 (1993).

Rezanka,T. and Dembitsky,V.M. Isoprenoid fatty acids from fresh-water sponges. *J. Nat. Prod.*, **56**, 1898-1904 (1993).

Rezanka,T. and Sokolov,M.Y. Rapid method for the enrichment of very long-chain fatty acids from microorganisms. *J. Chromatogr.*, **636**, 249-254 (1993).

Rozes,N., Garbay,S., Denayrolles,M. and Lonvaud-Funel,A. A rapid method for determination of bacterial fatty acid composition. *Letts. Appl. Microbiol.*, **17**, 126-131 (1993).

Sakaki,K. Supercritical fluid chromatographic separation of fatty acid methyl esters on aminopropyl-bonded silica stationary phases. *J. Chromatogr.*, **648**, 451-457 (1993).

Santana-Marques,M.G.O., Ferrer-Correia,A.J.V., Caldwell,K.A. and Gross,M.L. Charge-remote fragmentation and the 2-step elimination of alkanols from fast-atom bombardment-desorbed $(M+H)^+$, $(M+Cat)^+$ and $(M-H)^-$ ions of aromatic β-hydroxyoximes. *J. Am. Soc. Mass Spectrom.*, **4**, 819-827 (1993).

Schulte,E. GC of acylglycerols and fatty acids with capillary columns. In: *CRC Handbook of Chromatography. Analysis of Lipids*, pp. 139-148 (ed. K.D. Mukherjee and N. Weber, CRC Press, Boca Raton) (1993).

Sebedio,J.L. HPLC of fatty acids. In: *CRC Handbook of Chromatography. Analysis of Lipids*, pp. 57-70 (ed. K.D. Mukherjee and N. Weber, CRC Press, Boca Raton) (1993).

Sebedio,J.L. Mercury adduct formation in the analysis of lipids. In: *Advances in Lipid Methodology - Two*, pp. 139-155 (edited by W.W. Christie, Oily Press, Dundee) (1993).

Shantha,K.L., Rao Bhaskar,V.S. and Pratap,G. Mass spectra of 6-oxo and 7-oxo aliphatic acids. *Fat. Sci. Technol.*, **95**, 344-346 (1993).

Shantha,N.C., Decker,E.A. and Hennig,B. Comparison of methylation methods for the quantitation of conjugated linoleic acid isomers. *J. Assoc. Off. Anal. Chem.*, **76**, 644-649 (1993).

Shibahara,A., Yamamoto,K., Shinkai,K., Nakayama,T. and Kajimoto,G. *cis*-9,*cis*-15-Octadecadienoic acid: a novel fatty acid found in higher plants. *Biochim. Biophys. Acta*, **1170**, 245-252 (1993).

Simpson,T.D. and Gardner,H.W. Conversion of 13(*S*)-hydroperoxy-9(*Z*),11(*E*)-octadecadienoic acid to the corresponding hydroxy fatty acid by KOH: a kinetic study. *Lipids*, **28**, 325-330 (1993).

Sippola,E., David,F. and Sandra,P. Temperature program optimization by computer simulation for the capillary GC analysis of fatty acid methyl esters on biscyanopropyl siloxane phases. *J. High Resolut. Chromatogr.*, **16**, 95-100 (1993).

Staby,A., Borch-Jensen,C., Mollerup,J. and Jensen,B. Flame ionization detector responses to ethyl esters of sand eel (*Ammodytes lances*) fish oil compared for different gas and supercritical fluid chromatography systems. *J. Chromatogr.*, **648**, 221-232 (1993).

Suutari,M. and Laakso,S. Signature GLC-MS ions in identification of Δ-5- and Δ-9-unsaturated *iso*- and *anteiso*-branched fatty acids. *J. Microbiol. Methods*, **17**, 39-48 (1993).

Takadate,A., Masuda,T., Murata,C., Tanaka,T., Miyahara,H. and Goya,S. Crowned-coumarin as a new fluorescent derivatization reagent for carboxilic acids. *Chem. Letts.*, **5**, 811-814 (1993).

Takeda,S., Sim,P.G., Horrobin,D.F., Sanford,T., Chisholm,K.A. and Simmons,V. Mechanism of lipid peroxidation in cancer cells in response to γ-linolenic acid (GLA) analysed by GC-MS. 1. Conjugated dienes with peroxyl (or hydroperoxyl) groups and cell-killing effects. *Anticancer Res.*, **13**, 193-200 (1993).

Teng,J.I. and Gowda,N.M. Analysis of *n*-3 fatty acids in fish oils by HPLC. *Chromatographia*, 35, 627-630 (1993).

Thies,W. Determination of the petroselinic acid content in seeds of *Coriandrum sativum* by GLC. *Fat Sci. Technol.*, **95**, 20-23 (1993).

Toschi,T.G., Capella,P., Holt,C. and Christie,W.W. A comparison of silver ion HPLC plus GC with Fourier-transform IR spectroscopy for the determination of *trans* double bonds in unsaturated fatty acids. *J. Sci. Food Agric.*, **61**, 261-266 (1993).

Traitler,H. and Wille,H.J. Large-scale chromatography of lipids. In: *CRC Handbook of Chromatography. Analysis of Lipids*, pp. 487-495 (ed. K.D. Mukherjee and N. Weber, CRC Press, Boca Raton) (1993).

Turnipseed,S.B., Allentoff,A.J. and Thompson,J.A. Analysis of trimethylsilylperoxy derivatives of thermally labile hydroperoxides by GC-MS. *Anal. Biochem.*, **213**, 218-225 (1993).

van der Heijdt,L.M., van der Lecq,F., Lachmansingh,A., Versluis,K., van de Kerk-van Hoof,A., Veldink,G.A. and Vliegenthart,J.F.G. Formation of octadecadienoate dimers by soybean lipoxygenase. *Lipids*, **28**, 779-782 (1993).

Wheelan,P., Zirrolli,J.A. and Murphy,R.C. Low-energy fast atom bombardment tandem mass spectrometry of monohydroxy substituted unsaturated fatty acids. *Biol. Mass Spectrom.*, **22**, 465-473 (1993).

White,J.D. and Jensen,M.S. Biomimetic synthesis of a cyclopropane containing eicosanoid from the coral *Plexaura homomalla*. Assignment of relative configuration. *J. Am. Chem. Soc.*, **115**, 2970-2971 (1993).

Wiesner,R. and Kuhn,H. HPLC of oxygenated fatty acids including enantiomer separation. In: *CRC Handbook of Chromatography. Analysis of Lipids*, pp. 89-100 (ed. K.D. Mukherjee and N. Weber, CRC Press, Boca Raton) (1993).

Wilson,R., Henderson,R.J., Burkow,I.C. and Sargent,J.R. The enrichment of *n*-3 polyunsaturated fatty acids using aminopropyl solid phase extraction columns. *Lipids*, **28**, 51-54 (1993).

Wolff,R.L. Heat-induced geometrical isomerization of α-linolenic acid: effect of temperature and heating time on the appearance of individual isomers. *J. Am. Oil Chem. Soc.*, **70**, 425-430 (1993).

Yurawecz,M.P., Molina,A.M., Mossoba,M. and Ku,Y. Estimation of conjugated octadecatrienoates in edible oils and fats. *J. Am. Oil Chem. Soc.*, **70**, 1093-1099 (1993).

Zelles,L. and Bai,Q.Y. Fractionation of fatty acids derived from soil lipids by solid phase extraction and their quantitative analysis by GC-MS. *Soil Biol. Biochem.*, **25**, 495-507 (1993).

Zeman,I., Schwarz,W., Brat,J. and Zajic,J. Chromatographic analysis of the products from fatty acid dimerization. *Potravin. Vedy*, **11**, 1-12 (1993).

Zirrolli,J.A. and Murphy,R.C. Low energy tandem-MS of the molecular ion derived from fatty acid methyl esters - a novel method for the analysis of branched-chain fatty acids. *J. Am. Soc. Mass Spectrom.*, **4**, 223-229 (1993).

E. THE ANALYSIS OF SIMPLE LIPID CLASSES

This section corresponds to Chapter 6 in *Lipid Analysis* and deals mainly with chromatographic methods, especially TLC and HPLC, for the isolation and analysis of simple lipid classes. Separations of molecular species of simple lipids are listed in Section G below.

Akasaka,K., Ohrui,H. and Meguro,H. Simultaneous determination of hydroperoxides of phosphatidylcholine, cholesterol esters and triacylglycerols by column-switching HPLC with a postcolumn detection system. *J. Chromatogr.*, **622**, 153-159 (1993).

Akasaka,K., Ohrui,H., Meguro,H. and Tamura,M. Determination of triacylglycerol and cholesterol ester hydroperoxides in human plasma by HPLC with fluorometric postcolumn detection. *J. Chromatogr.*, **617**, 205-211 (1993).

Akasaka,K., Ohrui,H. and Meguro,H. Normal-phase HPLC with a fluorometric post-column detection system for lipid hydroperoxides. *J. Chromatogr.*, **628**, 31-35 (1993).

Al-Hasani,S.M., Hlavac,J. and Carpenter,M.W. Rapid determination of cholesterol in single and multicomponent prepared foods. *J. Assoc. Off. Anal. Chem., Int.*, **76**, 902-906 (1993).

Amelio,M., Rizzo,R. and Varazini,F. Separation of wax esters from olive oils by HPLC. *J. Am. Oil Chem. Soc.*, **70**, 793-796 (1993).

Annan,M., Lequesne,P.W. and Vouros,P. Trimethylsilyl group migration in the mass spectra of trimethylsilyl ethers of cholesterol oxidation products. Product ion characterization by linked-scan tandem MS. *J. Am. Soc. Mass Spectrom.*, **4**, 327-335 (1993).

Bhatt,B.D., Ali,S. and Prasad,J.V. GC/GC-MS studies on the determination of the position and geometry of the double bond in straight-chain olefins by derivatization techniques. *J. Chromatogr. Sci.*, **31**, 113-119 (1993).

Biedermann,M., Grob,K. and Mariani,C. Transesterification and on-line LC-GC for determining the sum of free and esterified sterols in edible oils and fats. *Fat Sci. Technol.*, **95**, 127-133 (1993).

Carr,T.P., Andresen,C.J. and Rudel,L.L. Enzymatic determination of triglyceride, free cholesterol, and total cholesterol in tissue lipid extracts. *Clin. Biochem.*, **26**, 39-42 (1993).

Chen,W.X., Li,P.Y., Wang,S., Dong,J. and Li,J.Z. Serum cholesterol determined by liquid chromatography with 6-chlorostigmasterol as internal standard. *Clin. Chem.*, **39**, 1602-1607 (1993).

Choi,G.T.Y., Casu,M. and Gibbons,W.A. NMR lipid profiles of cells, tissues and body fluids - neutral, non-acidic and acidic phospholipid analysis of Bond Elut chromatographic fractions. *Biochem. J.*, **290**, 717-721 (1993).

Conforti,F.D., Harris,C.H. and Rinehart,J.T. HPLC analysis of wheat-flour lipids using an evaporative light-scattering detector. *J. Chromatogr.*, **645**, 83-88 (1993).

Das,D.K., Gangopadyhyay,H. and Cordis,G.A. Capillary gas chromatography of myocardial cholesterol oxides. In: *Lipid Chromatographic Analysis*, pp. 75-101 (edited by T. Shibamoto, Dekker, NY) (1993).

Davila,A.M., Marchal,R., Monin,N. and Van de Casteele,J.P. Identification and determination of individual sophorolipids in fermentation products by gradient elution HPLC with evaporative light-scattering detection. *J. Chromatogr.*, **648**, 139-149 (1993).

Draper,H.H., Squires,E.J., Mahmoodi,H., Wu,J., Agarwal,S. and Hadley,M. A comparative evaluation of thiobarbituric acid methods for the determination of malondialdehyde in biological materials. *Free Radical Biol. Med.*, **15**, 353-363 (1993).

Ebeler,S.E. and Shibamoto,T. Overview and recent developments in solid-phase extraction for separation of lipid classes. In: *Lipid Chromatographic Analysis*, pp. 1-48 (edited by T. Shibamoto, Dekker, NY) (1993).

El-Hamdy,A.H. and Christie,W.W. Separation of non-polar lipids by HPLC on a cyanopropyl column. *J. High Resolut. Chromatogr.*, **16**, 55-57 (1993).

Filip,V. and Kleinova,M. The determination of mono-, di- and triglycerides by HPLC. *Zeitschr. Lebensmitt.-Unters. Forsch.*, **196**, 532-535 (1993).

Grob,K., Mariani,C., Lanfranchi,M., Artho,A. and Biedermann,M. Analysis of lipids by on-line coupled liquid chromatography-gas chromatography. In: *CRC Handbook of Chromatography. Analysis of Lipids*, pp. 71-82 (ed. K.D. Mukherjee and N. Weber, CRC Press, Boca Raton) (1993).

Hopia,A.I. and Ollilainen,V.M. Comparison of evaporative light-scattering detector (ELSD) and refractive index detector (RID) in lipid analysis. *J. Liqu. Chromatogr.*, **16**, 2469-2482 (1993).

Ingalls,S.T., Kriaris,M.S., Xu,Y., Dewulf,D.W., Tserng,K.Y. and Hoppel,C.L. Method for isolation of nonesterified fatty acids and several other classes of plasma lipids by column chromatography on silica gel. *J. Chromatogr.*, **619**, 9-19 (1993).

Ismail,A.A., van de Voort,F.R., Emo,G. and Sedman,J. Rapid quantitative determination of free fatty acids in fats and oils by Fourier transform infrared spectroscopy. *J. Am. Oil Chem. Soc.*, **70**, 335-341 (1993).

Karlsson,L., Jaremo,M., Emilsson,M., Mathiasson,L. and Jonsson,J.A. Retention in coupled capillary column supercritical fluid chromatography with lipids as model compounds. *Chromatographia*, **37**, 402-410 (1993).

Korytowski,W., Bachowski,G.J. and Girotti,A.W. Analysis of cholesterol and phospholipid hydroperoxides by HPLC with mercury drop electrochemical detection. *Anal. Biochem.*, **213**, 111-119 (1993).

Kritharides,L., Jessup,W., Gifford,J. and Dean,R.T. A method for defining the stages of low-density lipoprotein oxidation by the separation of cholesterol- and cholesteryl ester-oxidation products using HPLC. *Anal. Biochem.*, **213**, 79-89 (1993).

Kuksis,A., Myher,J.J. and Geher,K. Quantitation of plasma lipids by GLC on high-temperature polarizable capillary columns. *J. Lipid Res.*, **34**, 1029-1038 (1993).

Kwon,S.J., Lee,S.Y., Cho,S.W. and Rhee,J.S. A rapid gas chromatographic method for quantitation of free fatty acids, monoacylglycerols, diacylglycerols and triacylglycerols without derivatization. *Biotechnol. Techniques*, **7**, 727-732 (1993).

Larner,J.M., Pahuja,S.L., Shackleton,C.H., McMurray,W.J., Giordano,G. and Hochberg,R.B. The isolation and characterization of estradiol-fatty acid esters in human ovarian follicular fluid. Identification of an endogenous long-lived and potent family of estrogens. *J. Biol. Chem.*, **268**, 13893-13899 (1993).

Lercker,G., Cocchi,M., Turchetto,E., Capella,P., Zullo,C. and Caboni,M.F. Determination of total lipid components by a high resolution GC method. Application to the short-term fasting of rats. *Riv. Sci. Aliment.*, **22**, 147-152 (1993).

Lin,D.S., Connor,W.E., Wolf,D.P., Neuringer,M. and Hachey,D.L. Unique lipids of primate spermatozoa: demosterol and docosahexaenoic acid. *J. Lipid Res.*, **34**, 491-499 (1993).

Liu,J., Lee,T., Guzman-Harty,M. and Hastilow,C. Quantitative determination of monoglycerides and diglycerides by HPLC and evaporative light-scattering detection. *J. Am. Oil Chem. Soc.*, **70**, 343-347 (1993).

Matsumoto,K. and Taguchi,M. Supercritical fluid chromatographic analysis of lipids. In: *Lipid Chromatographic Analysis*, pp. 365-396 (edited by T. Shibamoto, Dekker, NY) (1993).

Medina,I., Aubourg,S.P. and Martin,R.I.P. Analysis of 1-*O*-alk-1-enylglycerophospholipids of albacore tuna (*Thunnus alalunga*) and their alterations during thermal processing. *J. Agric. Food*

Chem., **41**, 2395-2399 (1993).

Mingrone,G., Degaetano,A., Greco,A.V., Caprristo,E., Raguso,C., Tataranni,P.A. and Castagnetto,M. A rapid GLC method for direct analysis of plasma medium chain fatty acids. *Clin. Chim. Acta*, **214**, 21-30 (1993).

Moreau,R.A. Quantitative analysis of lipids by HPLC with a flame-ionization detector or an evaporative light-scattering detector. In: *Lipid Chromatographic Analysis*, pp. 251-272 (edited by T. Shibamoto, Dekker, NY) (1993).

Nakamura,T., Takazawa,T., Maruyama-Ohki,Y., Nagaki,H. and Kinoshita,T. Location of double bonds in unsaturated fatty alcohols by microderivatization and liquid secondary ion tandem mass spectrometry. *Anal. Chem.*, **65**, 837-840 (1993).

Nomura,A., Yamada,J., Takatsu,A., Horimoto,Y. and Yarita,T. Supercritical-fluid chromatographic determination of cholesterol and cholesteryl esters in serum on ODS silica gel column. *Anal. Chem.*, **65**, 1994-1997 (1993).

Ohshima,T., Miyamoto,K. and Sakai,R. Simultaneous separation and sensitive measurement of free fatty acids in ancient pottery by HPLC. *J. Liqu. Chromatogr.*, **16**, 3217-3227 (1993).

Perez-Ruiz,T., Martinez-Lozano,C., Tomas,V. and Val,O. A simple and sensitive photochemical method for the determination of organic peroxides and lipohydroperoxides. *Microchem. J.*, **48**, 151-157 (1993).

Pizzoferrato,L., Nicoli,S. and Lintas,C. GC-MS characterization and quantification of sterols and cholesterol oxidation products. *Chromatographia*, 35, 269-274 (1993).

Plank,C. and Lorbeer,E. Analysis of free and esterified sterols in vegetable oil methyl esters by capillary GC. *J. High Resolut. Chromatogr.*, **16**, 483-487 (1993).

Pollesello,P., Toffanin,R., Eriksson,O., Kilpelainen,I. Hynninen,P.H., Paoletti,S. and Saris,N.-E.L. Analysis of lipids in crude extracts by ^{13}C nuclear magnetic resonance. *Anal. Biochem.*, **214**, 238-244 (1993).

Puttmann,M., Krug,H., Von Ochsenstein,E. and Katterman,R. Fast HPLC determination of serum free fatty acids in the picomole range. *Clin. Chem.*, **39**, 825-832 (1993).

Raharjo,S., Sofos,J.N. and Schmidt,G.R. Solid-phase acid-extraction improves thiobarbituric acid method to determine lipid oxidation. *J. Food Sci.*, **58**, 921 (1993).

Rossi,M., Spedicato,E. and Albania,V. A normal-phase HPLC method for analysis of total cholesterol in eggs. *Ital. J. Food Sci.*, **5**, 151-155 (1993).

Shireman,R.B. and Durieux,J. Microplate methods for determination of serum cholesterol, high density lipoprotein cholesterol, triglyceride and apolipoproteins. *Lipids*, **28**, 151-155 (1993).

Smith,L.L. Analysis of oxysterols by liquid chromatography. *J. Liqu. Chromatogr.*, **16**, 1731-1747 (1993).

Stein,J., Milovic,V., Zeuzem,S. and Caspary,W.F. Fluorometric HPLC of free fatty acids using panacyl bromide. *J. Liqu. Chromatogr.*, **16**, 2915-2922 (1993).

Takagi,T., Aoyanagi,N., Nishimura,K., Ando,Y. and Ota,T. Enantiomer separations of secondary alkanols with little asymmetry by HPLC on chiral columns. *J. Chromatogr.*, **629**, 385-388 (1993).

Takatsu,A. and Nishi,S. Determination of serum cholesterol by stable isotope dilution method using discharge assisted thermospray liquid chromatography mass spectrometry. *Biol. Mass Spectrom.*, **22**, 247-250 (1993).

Thompson,R.H. and Merola,G.V. A simplified alternative to the AOAC official method for cholesterol in multicomponent foods. *J. Ass. Off. Anal. Chem. Int.*, **76**, 1057-1068 (1993).

Traitler,H. and Janchen,D.E. Analysis of lipids by planar chromatography. In: *CRC Handbook of Chromatography. Analysis of Lipids*, pp. 11-32 (ed. K.D. Mukherjee and N. Weber, CRC Press, Boca Raton) (1993).

Tvrzicka,E. and Votruba,M. Thin layer chromatography with flame-ionization detection. In: *Lipid Chromatographic Analysis*, pp. 51-73 (edited by T. Shibamoto, Dekker, NY) (1993).

Valeur,A., Michelsen,P. and Odham,G. On-line straight-phase liquid chromatography/plasmaspray tandem MS of glycerolipids. *Lipids*, **28**, 255-259 (1993).

van den Berg,J.J.M., Winterbourn,C.C. and Kuypers,F.A. Hypochlorous acid-mediated modification of cholesterol and phospholipid: analysis of reaction products by GC-MS. *J. Lipid Res.*, **34**, 2005-2012 (1993).

Wardas,W. and Pyka,A. New visualising agents for selected fatty derivatives in TLC. *J. Planar Chromatography - Modern TLC*, **6**, 320-322 (1993).

Yuan,G., He,M.Y. and He,X.R. Fragmentation pathways of isomeric dodecenols studied by electron impact MS. *Org. Mass Spectrom.*, **28**, 873-877 (1993).

Yuan,G., He,M.Y., He,X.R., Horike,M., Kim,C.S. and Hirano,C. Mass spectrometric location of double-bond position in isomeric dodecenols, without chemical derivatization. *Rapid Commun. Mass Spectrom.*, **7**, 591-593 (1993).

Zambon,A., Hashimoto,S.I. and Brunzell,J.D. Analysis of techniques to obtain plasma for measurement of levels of free fatty acids. *J. Lipid Res.*, **34**, 1021-1028 (1993).

F. THE ANALYSIS OF COMPLEX LIPIDS

This section corresponds to Chapter 7 in *Lipid Analysis* and deals mainly with chromatographic methods, especially TLC and HPLC, for the isolation and analysis of complex lipid classes including both phospholipids and glycolipids. Degradative procedures for the identification of polar moieties and spectrometric methods for intact lipids are also listed here. Separations of molecular species of complex lipids are listed in the next section.

Adams,J. and Ann,Q.H. Structure determination of sphingolipids by mass spectrometry. *Mass Spectrom. Rev.*, **12**, 51-85 (1993).

Adosraku,R.K., Anderson,M.M., Anderson,G.J., Choi,G., Croft,S.L., Yardley,V., Phillipson,J.D. and Gibbons,W.A. Proton NMR lipid profile of *Leishmania donovani*. *Mol. Biochem. Parasitol.*, **62**, 251-262 (1993).

Akasaka,K., Ohrui,H. and Meguro,H. Simultaneous determination of hydroperoxides of phosphatidylcholine, cholesterol esters and triacylglycerols by column-switching HPLC with a postcolumn detection system. *J. Chromatogr.*, **622**, 153-159 (1993).

Alvarez,J.G. and Ludmir,J. Semiautomated multisample analysis of amniotic fluid lipids by high-performace TLC-reflectance photodensitometry. *J. Chromatogr.*, **615**, 142-147 (1993).

Barker,C.J. HPLC separation of inositol phosphates. *HPLC Neurosci. Res.*, **15**, 195-216 (1993).

Berry,A.M., Harriott,O.T., Moreau,R.A., Osman,S.F., Benson,D.R. and Jones,A.D. Hopanoid lipids compose the *Frankia* vesicle envelope, presumptive barrier of oxygen diffusion to nitrogenase. *Proc. Natl. Acad Sci. USA*, **90**, 6091-6094 (1993).

Blank,M.L. and Snyder,F.L. Chromatographic analysis of ether-linked glycerolipids, including platelet-activating factor and related cell mediators. In: *Lipid Chromatographic Analysis*, pp. 291-316 (edited by T. Shibamoto, Dekker, NY) (1993).

Bonekamp,A. Glycerophospholipids. In: *CRC Handbook of Chromatography. Analysis of Lipids*, pp. 197-201 (ed. K.D. Mukherjee and N. Weber, CRC Press, Boca Raton) (1993).

Chapman,K.D. and Moore,T.S. *N*-Acylphosphatidylethanolamine synthesis in plants: occurrence, molecular composition and phospholipid origin. *Arch. Biochem. Biophys.*, **301**, 21-33 (1993).

Choi,G.T.Y., Casu,M. and Gibbons,W.A. NMR lipid profiles of cells, tissues and body fluids - neutral, non-acidic and acidic phospholipid analysis of Bond Elut chromatographic fractions. *Biochem. J.*, **290**, 717-721 (1993).

Compagnon,D., Lagos,N. and Vergara,J. Phosphoinositides in giant barnacle muscle fibres: a quantitative analysis at rest and following electrical stimulation. *Biochim. Biophys. Acta*, **1167**, 94-100 (1993).

Conforti,F.D., Harris,C.H. and Rinehart,J.T. HPLC analysis of wheat-flour lipids using an evaporative light-scattering detector. *J. Chromatogr.*, **645**, 83-88 (1993).

Curtis,J.M., Holgersson,J., Derrick,P.J. and Samuelsson,B.E. Electron-impact ionisation tandem MS of glycosphingolipids. 3. Study of the fragmentation modes of selected glycosphingolipids. *Org. Mass Spectrom.*, **28**, 883-891 (1993).

Das,D.K., Maulik,N., Jones,R.M. and Bagchi,D. Gas chromatography-mass spectrometric detection of plasmalogen phospholipids in mammalian heart. In: *Lipid Chromatographic Analysis*, pp. 317-345 (edited by T. Shibamoto, Dekker, NY) (1993).

Datta,A.K. and Takayama,K. Biosynthesis of a novel 3-oxo-2-tetradecyl-octadecanoate-containing phospholipid by a cell-free extract of *Corynebacterium diphtheriae*. *Biochim. Biophys. Acta*, **1169**, 135-145 (1993).

De Kock,J. The European analytical subgroup of ILPS - a joint effort to clarify lecithin and phospholipid analysis. *Fat Sci. Technol.*, **95**, 352-355 (1993).

Dethloff,L.A. and Hook,G.E.R. Phospholipids from pulmonary surfactants. In: *CRC Handbook of*

Chromatography. Analysis of Lipids, pp. 471-485 (ed. K.D. Mukherjee and N. Weber, CRC Press, Boca Raton) (1993).

Drucker,D.B., Aluyi,H.S., Boote,V., Wilson,J.M. and Ling,Y. Polar lipids of strains of Prevotella, Bacteriodes and Capnocytophaga analysed by fast-atom-bombardment MS *Microbios*, **75**, 45-56 (1993).

Ebeler,S.E. and Shibamoto,T. Overview and recent developments in solid-phase extraction for separation of lipid classes. In: *Lipid Chromatographic Analysis*, pp. 1-48 (edited by T. Shibamoto, Dekker, NY) (1993).

Elsner,A. and Lange,R. Gram-scale preparation of plant phospholipids by flash chromatography. *Fat Sci. Technol.*, **95**, 31-34 (1993).

Fraey,W.H., Schmalz,J.W., Perfetti,P.A., Norris,T.L., Emory,C.R. and Ala,T.A. Silica-ELISA method improves detection and quantitation of minor glycolipid components in lipid mixtures and of other antigens. *J. Immunol. Methods*, **164**, 275-283 (1993).

Gaudette,D.C., Aukema,H.M., Jolly,C.A., Chapkin,R.S. and Holub,B.J. Mass and fatty acid composition of the 3-phosphorylated phosphatidylinositol bisphosphate isomer in stimulated human platelets. *J. Biol. Chem.*, **268**, 13773-13776 (1993).

Gerin,C. and Goutx,M. Separation and quantification of phospholipids from marine bacteria with the Iatroscan Mark IV TLC-FID. *J. Planar Chromatography - Modern TLC.*, **6**, 307-312 (1993).

Haase,R., Unthan,M., Couturier,P., Radunz,A. and Schmid,G.H. Determination of glycolipids, sulpholipid and phospholipids in the thylakoid membrane. *Zeits. Naturforsch.*, **48**, 623-631 (1993).

Hamid,M.E., Minnikin,D.E., Goodfellow,M. and Ridell,M. TLC analysis of glycolipids and mycolic acids from *Mycobacterium farcinogenes, M. senegalense* and related taxa. *Zentralblatt Bakt.*, **279**, 354-367 (1993).

Ii,T., Ohashi,Y. and Nagai,Y. Observation of a novel adduct ion in negative-ion electrospray mass spectrometry of neutral glycolipids. *Org. Mass Spectrom.*, **28**, 927-928 (1993).

Ikarashi,Y. and Maruyama,Y. Liquid chromatography with electrochemical detection for quantitation of bound choline liberated by phospholipase D hydrolysis from phospholipids containing choline in rat plasma. *J. Chromatogr.*, **616**, 323-326 (1993).

Kaheki,K.,Ikeda,S., Hirose,A. and Honda,S. An approach to ganglioside analysis by capillary electrophoresis. *Kuromatogurafi*, **14**, 86-87 (1993).

Kates,M., Moldoveanu,N. and Stewart,L.C. On the novel structure of the major phospholipid of *Halobacterium salinarum. Biochim. Biophys. Acta*, **1169**, 46-53 (1993).

Kaya,K., Sano,T. and Watanabe,M.K. Thioic *O*-acid ester in sulfolipid isolated from fresh-water picoplankton cyanobacterium *Synechoccus sp. Biochim. Biophys. Acta*, **1169**, 39-45 (1993).

Korytowski,W., Bachowski,G.J. and Girotti,A.W. Analysis of cholesterol and phospholipid hydroperoxides by HPLC with mercury drop electrochemical detection. *Anal. Biochem.*, **213**, 111-119 (1993).

Kriat,M., Viondury,J., Confort-Gouny,S., Favre,R., Viout,P., Sari,H. and Cozzone,P.J. Analysis of plasma lipids by NMR spectroscopy - application to modifications induced by malignant tumors. *J. Lipid Res.*, **34**, 1009-1019 (1993).

Kuksis,A., Myher,J.J. and Geher,K. Quantitation of plasma lipids by GLC on high-temperature polarizable capillary columns. *J. Lipid Res.*, **34**, 1029-1038 (1993).

Kunze,H. and Bohn,E. Separation and analysis of amino alcohol-containing diacylglycerophospholipids and their hydrolytic metabolites. *J. Chromatogr.*, **636**, 221-229 (1993).

Lercker,G., Cocchi,M., Turchetto,E., Capella,P., Zullo,C. and Caboni,M.F. Determination of total lipid components by a high resolution GC method. Application to the short-term fasting of rats. *Riv. Sci. Aliment.*, **22**, 147-152 (1993).

Lester,R.L., Wells,G.B., Oxford,G. and Dickson,R.C. Mutant strains of *Saccharomyces cerevisiae* lacking sphingolipids synthesise novel inositol glycerophospholipids that mimic sphingolipid structures. *J. Biol. Chem.*, **268**, 645-856 (1993).

Linard,A., Guesnet,P. and Durand,G. Separation of the classes of phospholipid by overpressured layer chromatography (OPLC). *J. Planar Chromatogr. - Modern TLC.*, **6**, 322-323 (1993).

Mallinger,A.G., Yao,J.K., Brown,A.S. and Dippold,C.S. Analysis of complex mixtures of phospholipid classes from cell membranes using 2-dimensional TLC and scanning laser densitometry. *J. Chromatogr.*, **614**, 67-75 (1993).

Meneses,P., Navarro,J.N. and Glonek,T. Algal phospholipids by P-31 NMR - comparing isopropanol pretreatment with simple chloroform-methanol extraction. *Int. J. Biochem.*, **25**, 903-910 (1993).

Miller-Podraza,H., Andersson,C. and Karlsson,K.A. New method for the isolation of polyglycosylceramides from human erythrocyte membranes. *Biochim. Biophys. Acta*, **1168**, 330-339 (1993).

Moreau,R. and Gerard,H.C. HPLC as a tool for the lipid chemist and biochemist. In: *CRC Handbook of Chromatography. Analysis of Lipids*, pp. 41-55 (ed. K.D. Mukherjee and N. Weber, CRC Press, Boca Raton) (1993).

Moreau,R.A. Quantitative analysis of lipids by HPLC with a flame-ionization detector or an evaporative light-scattering detector. In: *Lipid Chromatographic Analysis*, pp. 251-272 (edited by T. Shibamoto, Dekker, NY) (1993).

Murphy,E.J., Stephens,R., Jurkowitz-Alexander,M. and Horrocks,L.A. Acidic hydrolysis of plasmalogens followed by HPLC. *Lipids*, **28**, 565-568 (1993).

Murphy,R.C. *Mass Spectrometry of Lipids (Handbook of Lipid Research, Vol. 7)* (Plenum Press, N.Y.) (1993).

Murui,T. and Wanaka,K. Measurement of sterylglycosides by HPLC with *O*-anthroylnitrile derivatives. *Biosci. Biotechnol. Biochem.*, **57**, 614-617 (1993).

Muthing,J. and Heitmann,D. Non-destructive detection of gangliosides with lipophilic fluorochromes and their employment for preparative HP TLC. *Anal. Biochem.*, **208**, 121-124 (1993).

Natomi,H., Saitoh,T., Sugano,K., Iwamori,M., Fukayama,M. and Nagai,Y. Systematic analysis of glycosphingolipids in the human gastrointestinal tract: enrichment of sulfatides with hydroxylated longer-chain fatty acids in the gastric and duodenal mucosa. *Lipids*, **28**, 737-742 (1993).

Nie,Y., He,L.J. and Hsia,S.L. A micro enzymic method for determination of choline-containing phospholipids in serum and high density lipoproteins. *Lipids*, **28**, 949-951 (1993).

Noda,N., Tanaka,R., Miyahara,K. and Kawasaki,T. Isolation and characterization of a novel type of glycosphingolipid from *Neanthes diversicolor*. *Biochim. Biophys. Acta*, **1169**, 30-38 (1993).

Ogiso,M., Irie,A., Kubo,H., Komoto,M., Matsuno,T., Koide,Y. and Hoshi,M. Characterization of neutral glycosphingolipids in human cataractous lens. *J. Biol. Chem.*, **268**, 13242-13247 (1993).

Ohashi,Y. Structural analyses of glycoconjugates by advanced MS. *J. Synth. Org. Chem. Japan*, **51**, 529-540 (1993).

Pajarron,A.M., de Koster,C.G., Heerma,W., Schmidt,M. and Haverkamp,J. Structure identification of natural rhamnolipid mixtures by fast atom bombardment tandem mass spectrometry. *Glycoconjugate J.*, **10**, 219-226 (1993).

Pearce,J.M. and Komorowski,R. Resolution of phospholipid molecular species by P-31 NMR. *Magnetic Resonance in Medicine*, **29**, 724-731 (1993).

Pietsch,A. and Lorenz,R.L. Rapid separation of the major phospholipid classes on a single aminopropyl cartridge. *Lipids*, **28**, 945-947 (1993).

Pollesello,P., Toffanin,R., Eriksson,O., Kilpelainen,I., Hynninen,P.H., Paoletti,S. and Saris,N.-E.L. Analysis of lipids in crude extracts by ^{13}C nuclear magnetic resonance. *Anal. Biochem.*, **214**, 238-244 (1993).

Prior,S.L., Cunliffe,B.W., Robson,G.D. and Trinci,A.P.J. Multiple isomers of phosphatidylinositol monophosphate and inositol bis- and trisphosphates from filamentous fungi. *FEMS Microbiol. Letts*, **110**, 147-152 (1993).

Ramanadham,S., Bohrer,A., Gross,R.W. and Turk,J. Mass spectrometric characterization of arachidonate-containing plasmalogens in human pancreatic-islets and in rat islet beta-cells and subcellular membranes. *Biochemistry*, **32**, 13499-13509 (1993).

Rana,F.R., Harwood,J.S., Mautone,A.J. and Dluhy,R.A. Identification of phosphocholine plasmalogen as a lipid component in mamalian pulmonary surfactant using high-resolution ^{31}P NMR spectroscopy. *Biochemistry*, **32**, 27-31 (1993).

Rother,J., van Echten,G., Schwarzmann,G. and Sandhoff,K. Biosynthesis of sphingolipids: dihydroceramide and not sphinganine is desaturated by cultured cells. *Biochem. Biophys. Res. Commun.*, **189**, 14-20 (1993).

Silvestro,L., Dacol,R., Scappaticci,E., Libertucci,D., Biancone,L. and Camussi,G. Development of a HPLC-MS technique, with an ionspray interface, for the determination of platelet-activating factor (PAF) and lyso-PAF in biological samples. *J. Chromatogr.*, **647**, 261-269 (1993).

Skrivanek,J.A., King,D., Dasilva,D., Jhang,Y.J., Phelps,R. and Schwartz,E. Isolation and identification of gangliosides of human epidermis. *Microchem. J.*, **47**, 240-244 (1993).

Sperling,P., Linscheid,M., Stocker,S., Muhlbach,P.-M. and Heinz,E. *In vivo* desaturation of *cis*-9-monounsaturated to *cis*-9,12-diunsaturated alkenylether glycerolipids. *J. Biol. Chem.*, **268**, 26935-26940 (1993).

Sperling,P.and Heinz,,E. Isomeric *sn*-1-octadecenyl and *sn*-2-octadecenyl analogues of lysophosphatidylcholine as substrates for acylation and desaturation by plant microsomal membranes. *Eur.*

J. Biochem., **213**, 965-971 (1993).
St Angelo,A.J. and James,C. Analysis of lipids from cooked beef by TLC with flame-ionization detection. *J. Am. Oil Chem. Soc.*, **70**, 1245-1254 (1993).
Suzuki,E., Sano,A., Kuriki,T. and Miki,T. Separation and determination of phospholipids in plasma employing TLC plate with concentration zone or solid phase extraction. *Biol. Pharm. Bull.*, **16**, 77-80 (1993).
Tamai,Y., Nakamura,K., Takayama-Abe,K., Uchida,K., Kasama,T. and Kobatake,H. Less polar glycolipids in Alaskan pollack brain: isolation and characterization of acyl galactosyl diacylglycerol, acyl galactosyl ceramide, and acyl glucosyl ceramide. *J. Lipid Res.*, **34**, 601-608 (1993).
Traitler,H. and Janchen,D.E. Analysis of lipids by planar chromatography. In: CRC Handbook of Chromatography. Analysis of *Lipids*, pp. 11-32 (ed. K.D. Mukherjee and N. Weber, CRC Press, Boca Raton) (1993).
Tvrzicka,E. and Votruba,M. Thin layer chromatography with flame-ionization detection. In: *Lipid Chromatographic Analysis*, pp. 51-73 (edited by T. Shibamoto, Dekker, NY) (1993).
Valeur,A., Michelsen,P. and Odham,G. On-line straight-phase liquid chromatography/plasmaspray tandem MS of glycerolipids. *Lipids*, **28**, 255-259 (1993).
Valsecchi,M., Palestini,P., Chigorno,V., Sonnino,S. and Tettamanti,G. Changes in the ganglioside long-chain base composition of rat cerebrullar granule cells during differentiation and aging in culture. *J. Neurochem.*, **60**, 193-196 (1993).
Valsecchi,M., Palestini,P., Chigorno,V., Sonnino,S. and Tettamanti,G. Changes in the ganglioside long-chain base composition of rat cerebellar granule cells during differentiation and aging in culture. *J. Neurochem.*, **60**, 193-196 (1993).
Weintraub,S.T., Satsangi,R.K., Simmons,A.M., Williams,R.F. and Pinckard,R.N. Synthesis of pentafluorobenzoic anhydride - a superior derivatizing agents for lipids. *Anal. Chem.*, **65**, 2400-2402 (1993).
Welsh,C.J. and Schmeichel,K. Improved method for TLC resolution and photodensitometric assesment of the major classes of phospholipids. *J. Liqu. Chromatogr.*, **16**, 1819-1831 (1993).
Wissing,J.B. and Behrbohm,H. Diacylglycerol pyrophosphate, a novel phospholipid compound. *FEBS Letts*, **315**, 95-99 (1993).
Wolff,R.L., Combe,N.A., Entressangles,B., Sebedio,J.L. and Grandgirard,A. Preferential incorporation of dietary *cis*-9,*cis*-12,*trans*-15 18:3 acid into rat cardiolipins. *Biochim. Biophys. Acta*, **1168**, 285-291 (1993).
Yassin,A.F., Haggenei,B., Budzikiewicz,H. and Schaal,K.P. Fatty acid and polar lipid composition of the genus Amycolatopsis - application of fast-atom-bombardment MS to structure analysis of underivatized phospholipids. *Int. J. System. Bact.*, **43**, 414-420 (1993).
Yoo,Y.S., Kim,Y,S., Jhon,G.-J. and Park,J. Separation of gangliosides using cyclodextrin in capillary zone electrophoresis. *J. Chromatogr.*, **652**, 431-439 (1993).

G. THE ANALYSIS OF MOLECULAR SPECIES OF LIPIDS

This section corresponds to Chapter 8 in *Lipid Analysis* and deals mainly with chromatographic methods for the isolation and analysis of molecular species of lipid classes, including simple lipids, phospholipids and glycolipids. Many of the references in the next section are relevant here also and *vice versa*.

Abidi,S.L. and Mounts,T.L. Reversed-phase separations of subcomponents of minor soybean phospholipids, glycerol esters of phosphatidic acids. *J. Chromatogr. Sci.*, **31**, 231-236 (1993).
Abidi,S.L., Mounts,T.I. and Rennick,K.A. Reversed-phase HPLC of phospholipids with fluorescence detection. *J. Chromatogr.*, **639**, 175-184 (1993).
Aitzetmuller,K. and Gronheim,M. Gradient elution HPLC of fats and oils with laser light-scattering detection. *Fat Sci. Technol.*, **95**, 164-168 (1993).
Aitzetmuller,K. HPLC of triglycerides (Fingerprint-method). *Fat Sci. Technol.*, **95**, 361-366 (1993).
Anderson,M.A., Collier,L., Dilliplane,R. and Ayorinde,F.O. Mass spectrometric characterization of *Vernonia galamensis* oil. *J. Am. Oil Chem. Soc.*, **70**, 905-908 (1993).
Ann,Q. and Adams,J. Collision-induced decomposition of sphingomyelins for structural elucidation. *Biol. Mass Spectrom.*, **22**, 285-294 (1993).
Ann,Q.H. and Adams,J. Structure-specific colision-induced fragmentations of ceramides cationized

with alkali-metal ions. *Anal. Chem.*, **65**, 7-13 (1993).

Artz,W.E. Supercritical fluid chromatographic analysis of lipids. In: *CRC Handbook of Chromatography. Analysis of Lipids*, pp. 83-87 (ed. K.D. Mukherjee and N. Weber, CRC Press, Boca Raton) (1993).

Baiocchi,C., Saini,G., Cocito,C., Giacosa,D., Roggero,M.A., Marengo,E. and Favale,M. Analysis of vegetable and fish oils by capillary supercritical fluid chromatography with flame ionization detection. *Chromatographia*, **37**, 525-533 (1993).

Bell,M.V. and Dick,J.R. 1-*O*-Alkyl-1'-enyl-2-acyl-glycerophosphoethanolamine content and molecular species composition of fish. *Lipids*, **28**, 19-22 (1993).

Berg,B.E., Hansen,E.M., Gjorven,S. and Greibrokk,T. On-line enzymatic-reaction, extraction, and chromatography of fatty acids and triglycerides with supercritical carbon dioxide. *J. High Resolut. Chromatogr.*, **16**, 358-363 (1993).

Blomberg,L.G., Demirbuker,M. and Andersson,P.E. Argentation supercritical fluid chromatography for quantitative analysis of triacylglycerols. *J. Am. Oil Chem. Soc.*, **70**, 939-946 (1993).

Borch-Jensen,C., Staby,A. and Mollerup,J. Supercritical-fluid chromatographic analysis of a fish oil of the sand eel (*Ammodytes sp.*). *J. High Resolut. Chromatogr.*, **16**, 621-623 (1993).

Bruhl,L., Schulte,E. and Thier,H.-P. Fractionation of triglycerides in human milk using argentation and reversed-phase HPLC with a light-scattering detector. *Fat Sci. Technol.*, **95**, 370-376 (1993).

Burdge,G.C., Kelly,F.J. and Postle,A.D. Mechanisms of hepatic phosphatidylcholine synthesis in the developing guinea pig: contributions of acyl remodelling and of *N*-methylation of phosphatidylethanolamine. *Biochem. J.*, **290**, 67-73 (1993).

Carelli,A,A. and Cert,A. Comparative study of the determination of triacylglycerols in vegetable oils using chromatographic techniques. *J. Chromatogr.*, **630**, 213-222 (1993).

Carlson,K.D. and Kleiman,R. Chemical survey and erucic acid content of commercial varieties of nasturtium, *Tropaeolum majus* L. *J. Am. Oil Chem. Soc.*, **70**, 1145-1148 (1993).

Coene,J., Van den Eeckhout,E. and Herdewijn,P. GC determination of alkyl lysophospholipids after solid-phase extraction from culture media. *J. Chromatogr.*, **612**, 21-26 (1993).

Currie,G.J. and Kallio,H. Triacylglycerols of human milk: rapid analysis by ammonia negative ion tandem MS. *Lipids*, **28**, 217-222 (1993).

Dasgupta,A. and Banerjee,P. Microwave induced rapid preparation of acetyl, trifluoroacetyl and *tert*-butyl dimethylsilyl derivatives of fatty alcohols and diacylglycerols for GC-MS analysis. *Chem. Phys. Lipids*, **65**, 217-224 (1993).

Deffense,E. A new analytical method for the separation *via* RP-HPLC of monounsaturated triglycerides according to their isomeric position 1-2 and 1-3. *Rev. Franc. Corps Gras*, **40**, 33-39 (1993).

Demirbuker,M., Anderson,P.E. and Blomberg,L.G. Miniaturized light scattering detector for packed capillary supercritical fluid chromatography. *J. Microcol. Sep.*, **5**, 141-147 (1993).

Dizabo,P. and Pepe,C. Study of wax esters from natural origin by fragmentometry. *J. Chim. Phys. Physico-Chim. Biol.*, **90**, 1787-1795 (1993).

Dobarganes,M.C. and Marquez-Ruiz,G. Size exclusion chromatography in the analysis of lipids. In: *Advances in Lipid Methodology - Two*, pp. 113-137 (edited by W.W. Christie, Oily Press, Dundee) (1993).

Fabien,R., Craske,J.D. and Wootton,M. Quantitative analysis of synthetic mixtures of triacylglycerols with fatty acids from capric to stearic. *J. Am. Oil Chem. Soc.*, **70**, 551-554 (1993).

Falardeau,P., Robillard,M. and Hui,R. Quantification of diacylglycerols by capillary GC-negative ion chemical ionization mass spectrometry. *Anal. Biochem.*, **208**, 311-316 (1993).

Frega,N., Bocci,F. and Lercker,G. High resolution GC determination of diacylglycerols in common vegetable oils. *J. Am. Oil Chem. Soc.*, **70**, 175-177 (1993).

Hay,D.W., Cahalane,M.J., Timofeyeva,N. and Carey,M.C. Molecular species of lecithins in human gallbladder bile. *J. Lipid Res.*, **34**, 759-768 (1993).

Hazen,S.L., Hall,C.R., Ford,D.A. and Gross,R.W. Isolation of a human myocardial phospholipase-A(2) isoform - fast-atom-bombardment MS and reversed-phase HPLC identification of choline and ethanolamine glycerophospholipid substrates. *J. Clin. Invest.*, **91**, 2513-2522 (1993).

Heron,S. and Tchapla,A. Role of the solvent in RPLC - influence of the nature of the modifier of the mobile phase in nonaqueous reverse-phase liquid chromatography. *Analusis*, **21**, 269-276 (1993).

Hopia,A.I. and Ollilainen,V.M. Comparison of evaporative light-scattering detector (ELSD) and refractive index detector (RID) in lipid analysis. *J. Liqu. Chromatogr.*, **16**, 2469-2482 (1993).

Hopia,A.I., Lampi,A.-M., Piironen,V.I., Hyvonen,L.E.T. and Koivistoinen,P.E. Application of high-performance size-exclusion chromatography to study the autoxidation of unsaturated triacylglycerols. *J. Am. Oil Chem. Soc.*, 70, 779-784 (1993).

Horutz,K., Layman,L.R., Fried,B. and Sherma,J. HPLC determination of triacylglycerols in the digestive gland gonad complex of *Biomphalaria glabrata* snails fed hens egg-yolk versus leaf lettuce. *J. Liqu. Chromatogr.*, **16**, 4009-4017 (1993).

Hudiyono,S., Adenier,H. and Chaveron,H. Determination of the triglyceride composition of a hydrogenated palm oil fraction. *Rev. Franc. Corps Gras*, **40**, 131-141 (1993).

Kadowaki,H., Grant,M.A. and Williams,L.A. Effect of membrane lipids on the lactosylceramide molecular species specificity of CMP-*N*-acetylneuraminate/lactosylceramine sialyltransferase. *J. Lipid Res.*, **34**, 905-914 (1993).

Kallio,H. and Currie,G. Analysis of low erucic acid turnip rapeseed oil (B. Campestris) by negative ion chemical ionization tandem MS. A method giving information on the fatty acid composition in positions *sn*-2 and *sn*-1/3 of triacylgl *Lipids*, **28**, 207-215 (1993).

Kalo,P. and Kemppinen,A. Mass spectrometric identification of triacylglycerols of enzymically modified butter fat separated on a polarizable phenylmethylsilicone column. *J. Am. Oil Chem. Soc.*, **70**, 1209-1217 (1993).

Kamido,H., Kuksis,A., Marai,L. and Myher,J.J. Identification of core aldehydes among the *in vitro* peroxidation products of cholesteryl esters. *Lipids*, **28**, 331-336 (1993).

Kaufmann,P. and Olsson,N.U. Determination of intact molecular species of bovine milk. 1,2-Diacyl-*sn*-glycero-3-phosphocholine and 1,2-diacyl-*sn*-glycero-3-phosphoethanolanine by reversed-phase HPLC, a multivariate optimization. *Chromatographia*, **35**, 517-523 (1993).

Kaufmann,P. Prediction of mixture composition by chromatographic characterization,multivariate classification and partial least-squares regression. A comparison of methods. *Anal. Chim. Acta*, **277**, 467-471 (1993).

Kemppinen,A. and Kaol,P. Fractionation of the triacylglycerols of lipase modified butter oil. *J. Am. Oil Chem. Soc.*, **70**, 1203-1207 (1993).

Khaled,M.Y., McNair,H.M. and Hanson,D.J. High-temperature chromatographic analysis of monoacylglycerols and diacylglycerols. *J. Chromatogr. Sci.*, **31**, 375-379 (1993).

Kim,H.Y. and Salem,N. Liquid chromatography mass-spectrometry of lipids. Prog. Lipid Res., **32**, 221-245 (1993).

Konishi,H., Neff,W.E. and Mounts,T.L. Correlation of reversed-phase HPLC and GLC for fatty acid compositions of some vegetable oils. *J. Chromatogr.*, **629**, 237-241 (1993).

Kuksis,A. GLC and HPLC of neutral lipids. In: *Lipid Chromatographic Analysis*, pp. 177-222 (edited by T. Shibamoto, Dekker, NY) (1993).

Laakso,P. and Kallio,H. Triacylglycerols of winter butter fat containing configurational isomers of monoenoic fatty acyl residues. 1. Disaturated monoenoic triacylglycerols. *J. Am. Oil Chem. Soc.*, **70**, 1161-1171 (1993).

Laakso,P. and Kallio,H. Triacylglycerols of winter butter fat containing configurational isomers of monoenoic fatty acyl residues. 1. Saturated dimonoenoic triacylglycerols. *J. Am. Oil Chem. Soc.*, **70**, 1173-1176 (1993).

Lee,C. and Hajra,A.K. Quantitative analysis of molecular species of diacylglycerol in biological samples. *Methods Neurosci.*, **18**, 190-199 (1993).

Lehner,R., Kuksis,A. and Itabashi,Y. Stereospecificity of monoacylglycerol and diacylglycerol acyltransferases from rat intestine as determined by chiral phase HPLC. *Lipids*, **28**, 29-34 (1993).

Letter,W.S. A qualitative method for triglyceride analysis by HPLC using an evaporative light-scattering detector. *J. Liqu. Chromatogr.*, 16, 225-239 (1993).

Liepkalns,V.A., Myher,J.J., Kuksis,A., Leli,U., Freysz,N. and Hauser,G. Molecular species of glycerophospholipids and diacylglycerols of cultured SH-SY-5Y human neuroblastoma cells. *Biochem. Cell Biol.*, **71**, 141-149 (1993).

Matsumoto,K. and Taguchi,M. Supercritical fluid chromatographic analysis of lipids. In: *Lipid Chromatographic Analysis*, pp. 365-396 (edited by T. Shibamoto, Dekker, NY) (1993).

Morimoto,T., Murakami,N., Nagatsu,A. and Sakakibara,J. Studies on glycolipids. 7. Isolation of two new sulfoquinovosyl diacylglycerols from the green-alga *Chlorella vulgaris*. Chem. Pharm. Bull., **41**, 1545-1548 (1993).

Myher,J.J., Kuksis,A. and Marai,L. Identification of the less common isologous short-chain triacylglycerols in the most volatile 2.5% molecular distillate of butter oil. *J. Am. Oil Chem. Soc.*, **70**, 1183-1191 (1993).

Nakagawa,Y. Application of paired-ion HPLC to the separation of molecular species of phosphatidylinositol. *Lipids*, **28**, 1033-1035 (1993).

Neff,W.E., Adlof,R.O., Konishi,H. and Weisleder,D. HPLC of the triacylglycerols of *Vernonia galamensis* and *Crepis alpina* seed oils. *J. Am. Oil Chem. Soc.*, **70**, 449-455 (1993).

Nichols,P.D., Shaw,P.M., Mancuso,C.A. and Franzmann,P.D. Analysis of archaeal phospholipid-derived di- and tetraether lipids by high temperature capillary GC. *J. Microbial Meth.*, **18**, 1-9 (1993).

Nikolova-Damyanova,B., Chobanov,D. and Dimov,S. Separation of isomeric triacylglycerols by silver ion TLC. *J. Liqu. Chromatogr.*, **16**, 3997-4008 (1993).

Parreno,M., Castellote,A.I. and Codony,R. HPLC determination of plasma triglyceride type composition in a normal population of Barcelona - relationship with age, sex and other plasma lipid parameters. *J. Chromatogr.*, **655**, 89-94 (1993).

Pchelkin,V.P. and Vereschchagin,A.G. Analysis of a mixture of *rac*-1,2-diacylglycerols by argento TLC. *J. Anal. Chem.*, **48**, 257-262 (1993).

Pearce,J.M., Krone,J.T., Papas,A.A. and Komorowski,R.A. Analysis of saturated phosphatidylcholine in amniotic fluid by P-31 NMR. *Magnetic Resonance Med.*, **30**, 476-484 (1993).

Pepe,C., Dagaut,J., Scribe,P. and Saliot,A. Double bond location in monounsaturated wax esters by GC-MS of their dimethyl disulphide derivatives. *Org. Mass Spectrom.*, **28**, 1365-1367 (1993).

Petersson,B., Podlaha,O. and Jirskoghed,B. Triacylglycerol analysis of partially hydrogenated fats using HPLC. *J. Chromatogr., A*, **653**, 25-35 (1993).

Petit,T.R. and Wakelam,M.J.O. Bombesin stimulates distinct time-dependent changes in the *sn*-1,2-diradylglycerol molecular species profile from Swiss 3T3 fibroblasts as analysed by 3,5-dinitrobenzoyl derivatization and HPLC separation. *Biochem. J.*, **289**, 487-495 (1993).

Precht,D. GC of triacylglyerols and other lipids on packed columns. In: *CRC Handbook of Chromatography. Analysis of Lipids*, pp. 123-138 (ed. K.D. Mukherjee and N. Weber, CRC Press, Boca Raton) (1993).

Prieto,J.A., Ebri,A. and Collar,C. Composition and distribution of individual molecular species of wheat flour phospholipids. *J. Chromatogr. Sci.*, **31**, 55-60 (1993).

Quoc,K.P., Dubacq,J.-P., Justin,A.-M., Demandre,C. and Mazliak,P. Biosynthesis of eukaryotic lipid molecular species by the cyanobacterium *Spirulina platensis*. *Biochim. Biophys. Acta*, **1168**, 94-99 (1993).

Ramachandran,S., Bohrer,A., Mueller,M., Jett,P., Gross,R.W. and Turk,J. Mass spectrometric identification and quantitation of arachidonate-containing phospholipids in pancreatic islets - prominence of plasmenylethanolamine molecular species. *Biochemistry*, **32**, 5339-5351 (1993).

Sacchi,R., Medina,I., Aubourg,S.P., Giudicianni,I., Paolillo,L. and Addeo,F. Quantitative high resolution ^{13}C NMR analysis of lipids extracted from the white muscle of Atlantic tuna (*Thunnus alalunga*). *J. Agr. Food Chem.*, **41**, 1247-1253 (1993).

Sassano,G.J. and Jeffrey,B.S.J. Gas chromatography of triacylglycerols in palm oil fractions with medium-polarity wide-bore columns. *J. Am. Oil Chem. Soc.*, **70**, 1111-1114 (1993).

Schlamme,M., Brody,S. and Hostetter,K.Y. Mitochondrial cardiolipin in diverse eukaryotes. Comparison of biosynthetic reactions and molecular acyl species. *Eur. J. Biochem.*, **212**, 727-735 (1993).

Schulz,R., Strynadka,K.D., Panas,D.L., Olley,P.M. and Lopaschuk,G.D. Analysis of myocardial plasmalogen and diacyl phospholipids and their arachidonic acid content using HPLC. *Anal. Biochem.*, **213**, 140-146 (1993).

Song,J.H., Chang,C.O., Terao,J.J. and Park,D.K. Electrochemical detection of triacylglycerol hydroperoxides by reversed phase HPLC. *Biosci. Biotech. Biochem.*, **57**, 479-480 (1993).

Sukenik,A., Yamaguchi,Y. and Livne,A. Alterations in lipid molecular species of the marine Eustigmatophyte *Nannochloropsis sp. J. Phycol.*, **29**, 620-626 (1993).

Takagi,T. and Suzuki,T. Enantiomeric resolution of diacylglycerol derivatives by HPLC on a chiral stationary phase at low temperatures. *Lipids*, **28**, 251-253 (1993).

Therond,P., Couturier,M., Demelier,J.-F. and Lemonnier,F. Simultaneous determination of the main molecular species of soybean phosphatidylcholine or phosphatidylethanolamine and their corresponding hydroperoxides obtained by lipoxygenase treatment. *Lipids*, **28**, 245-249 (1993).

Toschi,T.G., Christie,W.W. and Conte,L.S. Capillary GC combined with HPLC for the analysis of olive oil triacylglycerols. *J. High Resolut. Chromatogr.*, 16, 725-730 (1993).

Tsevegsuren,N. and Aitzetmuller,K. γ-Linolenic acid in Anemone spp. seed lipids. *Lipids*, **28**, 841-846 (1993).

Tvrzicka,E. and Mares,P. Gas-liquid chromatography of neutral lipids. In: *Lipid Chromatographic Analysis*, pp. 103-176 (edited by T. Shibamoto, Dekker, NY) (1993).

Weintraub,S.T., Lear,C. and Pinckard,R.N. Differential electron capture mass spectral response of pentafluorobenzoyl derivatives of platelet activating factor alkyl chain homologues. *Biol. Mass Spectrom.*, **22**, 559-564 (1993).

Winter,C.H., Hoving,E.B. and Muskiet,F.A.J. Fatty acid composition of human milk triglyceride species. Possible consequences for optimal structures of infant formula triglycerides. *J. Chromatogr.*, **616**, 9-24 (1993).

Zhu,X. and Eichberg,J. Molecular species composition of glycerophospholipids in rat sciatic nerve and its alteration in streptozotocin-induced diabetes. *Biochim. Biophys. Acta*, **1168**, 1-12 (1993).

H. STRUCTURAL ANALYSIS OF LIPIDS BY MEANS OF ENZYMATIC HYDROLYSIS

This section corresponds to Chapter 9 in *Lipid Analysis* and relates to simple lipids, phospholipids and glycolipids. Many of the references in the last section are relevant here also and *vice versa*. Some methods for the resolution of chiral lipids or involving mass spectrometry are listed here when they deal with methods for determining positional distributions of fatty acids within lipid classes.

Ando,Y. and Takagi,T. Micro method for stereospecific analysis of triacyl-*sn*-glycerols by chiral-phase HPLC. *J. Am. Oil Chem. Soc.*, 70, 1047-1049 (1993).

Becker,C.C., Rosenquist,A. and Holmer,G. Regiospecific analysis of triacylglycerols using allyl magnesium bromide. *Lipids*, **28**, 147-149 (1993).

Henderson,R.J., Burkow,I.C. and Millar,R.M. Hydrolysis of fish oils containing polymers of triacylglycerols by pancreatic lipase in vitro. *Lipids*, **28**, 313-319 (1993).

Hills,M.J. Products of lipolysis. In: *CRC Handbook of Chromatography. Analysis of Lipids*, pp. 497-504 (ed. K.D. Mukherjee and N. Weber, CRC Press, Boca Raton) (1993).

Itabashi,Y., Myher,J.J. and Kuksis,A. Determination of positional distribution of short-chain fatty acids in bovine milk fat on chiral columns. *J. Am. Oil Chem. Soc.*, **70**, 1177-1181 (1993).

Kallio,H. and Currie,G. Analysis of natural fats and oils by ammonia negative ion tandem mass spectrometry - triacylglycerols and positional distribution of their acyl groups. In: *CRC Handbook of Chromatography. Analysis of Lipids*. pp. 435-458 (ed. K.D. Mukherjee and N. Weber, CRC Press, Boca Raton) (1993).

Lehner,R. and Kuksis,A. Triacylglycerol synthesis by an *sn*-1,2(2,3)-diacylglycerol transacylase from rat intestinal microsomes. *J. Biol. Chem.*, **268**, 8781-8786 (1993).

Leray,C., Raclot,T. and Groscolas,R. Positional distribution of *n*-3 fatty acids in triacylglycerols from rat adipose tissue during fish oil feeding. *Lipids*, **28**, 279-284 (1993).

MacKenzie,S.L., Giblin,E.M. and Mazza,G. Stereospecific analysis of *Onosmodium hispidissimum* Mack. seed oil triglycerides. *J. Am. Oil Chem. Soc.*, 70, 629-631 (1993).

Rogalska,E., Cudrey,C., Ferrato,F. and Verger,R. Stereoselective hydrolysis of triglycerides by animal and microbial lipases. *Chirality*, **5**, 24-30 (1993).

Stoll,U. The influence of fatty acids in triglycerides on the digestion of fats by pancreatic lipase. *Fat Sci. Technol.*, **95**, 231-236 (1993).

Suzuki,T., Ota,T. and Takagi,T. Enantiomer separation of diacylglycerol derivatives of marine fish triacylglycerols using a chiral stationary phase. *J. Chromatogr. Sci.*, **31**, 461-464 (1993).

Takagi,T. HPLC of enantiomeric acylglycerols and alkylglycerols. In: CRC Handbook of Chromatography. Analysis of *Lipids*, pp. 115-121 (ed. K.D. Mukherjee and N. Weber, CRC Press, Boca Raton) (1993).

Uzawa,H., Noguchi,T., Nishida,Y., Ohrui,H. and Meguro,H. Determination of the lipase stereoselectivities using circular dichroism (CD) - lipases produce chiral di-*O*-acylglycerols from achiral tri-*O*-acylglycerol. *Biochim. Biophys. Acta*, **1168**, 253-260 (1993).

Uzawa,H., Ohrui,H., Meguro,H., Mase,T. and Ichida,A. A convenient evaluation of the stereoselectivity of lipase-catalysed hydrolysis of tri-*O*-acylglycerols on a chiral-phase liquid chromatograph. *Biochim. Biophys. Acta*, **1169**, 165-168 (1993).

Wolff,R.L., Combe,N.A., Entressangles,B., Sebedio,J.L. and Grandgirard,A. Preferential incorporation of dietary *cis*-9,*cis*-12,*trans*-15 18:3 acid into rat cardiolipins. *Biochim. Biophys. Acta*, **1168**, 285-291 (1993).

I. THE ANALYSIS AND RADIOASSAY OF ISOTOPICALLY LABELLED LIPIDS

This section corresponds to Chapter 10 in *Lipid Analysis*. Only papers in which the radioactivity of the sample appeared to be central to the analysis are listed. To confuse matters, papers dealing with analysis of lipids enriched in stable isotopes are also listed here as they appear more relevant to this than any other section.

Baba,S. and Akira,K. Radio-gas-chromatography of lipids. In: *CRC Handbook of Chromatography. Analysis of Lipids*, pp. 149-162 (ed. K.D. Mukherjee and N. Weber, CRC Press, Boca Raton) (1993).

Cunnane,S.C., McDonagh,R.J., Narayan,S. and Kyle,D.J. Detection of [U-^{13}C]eicosapentaenoic acid in rat liver lipids using ^{13}C NMR spectroscopy. *Lipids*, **28**, 273-277 (1993).

Jones,A.D. Quantitative capillary GC-MS of lipids using stable isotope dilution methods. In: *Lipid Chromatographic Analysis*, pp. 347-364 (edited by T. Shibamoto, Dekker, NY) (1993).

Llado,I., Palou,A. and Pons,A. Combined enzymatic and chromatographic techniques to determine specific radioactivity in free and triglyceride fatty acid plasma fractions. *J. Chromatogr.*, **619**, 21-28 (1993).

Magni,F., Arnoldi,L., Monti,L., Piatti,P., Pozza,G. and Galli Kienle,M. Determination of plasma glycerol isotopic enrichment by GC-MS: an alternative glycerol derivative. *Anal. Biochem.*, 211, 327-328 (1993).

Nakajima,E. Image-plate system for radioluminographic detection of lipids on thin-layer plates. In: *CRC Handbook of Chromatography. Analysis of Lipids*. pp. 33-40 (ed. K.D. Mukherjee and N. Weber, CRC Press, Boca Raton) (1993).

Patterson,B.W. and Wolfe,R.R. Concentration dependence of methyl palmitate isotope ratios by electron inpact ionization GC/MS. *Biol. Mass Spectrom.*, **22**, 481-486 (1993).

J. THE SEPARATION OF PLASMA LIPOPROTEINS

This section corresponds to Chapter 11 in *Lipid Analysis*, and a only few key papers of particular interest are listed. *Current Topics in Lipidology* should be consulted for further listings

Brousseau,T., Clavey,V., Bard,J.M. and Fruchart,J.C. Sequential ultracentrifugation micromethod for separation of serum lipoproteins and assays of lipids, apolipoproteins, and lipoprotein particles. *Clin. Chem.*, **39**, 960-964 (1993).

Dobarganes,M.C. and Marquez-Ruiz,G. Size exclusion chromatography in the analysis of lipids. In: *Advances in Lipid Methodology - Two*, pp. 113-137 (edited by W.W. Christie, Oily Press, Dundee) (1993).

Okazaki,M., Muramatsu,T., Makino,K. and Hara,I. HPLC of serum lipoproteins. In: *CRC Handbook of Chromatography. Analysis of Lipids*, pp. 101-114 (ed. K.D. Mukherjee and N. Weber, CRC Press, Boca Raton) (1993).

Schmitz,G., Nowicka,G. and Mollers,C. Capillary isotachophoresis in the analysis of lipoproteins. In: *Advances in Lipid Methodology - Two*, pp. 157-193 (edited by W.W. Christie, Oily Press, Dundee) (1993).

Tschantz,J.C. and Sunahara,G.I. Microaffinity chromatographic separation and characterization of lipoprotein fractions in rat and mongolian gerbil serum. *Clin. Chem.*, **39**, 1861-1867 (1993).

K. SOME MISCELLANEOUS SEPARATIONS

Analyses of lipids such as prostaglandins, acylcarnitines, coenzyme A esters and so forth that do not fit conveniently into other sections are listed here. More complete listings for prostaglandins are available elsewhere (*Prostaglandins, Leukotrienes and Essential Fatty acids*). The decision on whether to list papers on eicosenoids here or in Section E was sometimes arbitrary.

Attygalle,A.B., Jham,G.N. and Meinwald,J. Determination of double bond position in some unsaturated terpenes and other branched compounds by alkylthiolation. *Anal. Chem.*, **65**, 2528-2533 (1993).

Claeys,M., van den Heuvel,H., Claereboudt,J., Corthout,J., Pieters,L. and Vlietinck,A.J. Determination of the double bond position in long-chain 6-alkenyl salicylic acids by collisional activation. *Biol. Mass Spectrom.*, **22**, 647-653 (1993).

Demoz,A., Netteland,B., Svardal,A., Mansoor,M.A. and Berge,R.K. Separation and detection of tissue CoASH and long-chain acyl-CoA by reversed-phase HPLC after precolumn derivatization with monobromobimane. *J. Chromatogr.*, **635**, 251-256 (1993).

Harrata,A.K., Domelsmith,L.N. and Cole,R.B. Electrospray mass spectrometry for characterisation of Lipid A from *Enterobacter agglomerans*. *Biol. Mass Spectrom.*, **22**, 59-67 (1993).

Henden,T., Strand,H., Borde,E., Aemb,A.G. and Larsen,T.S. Meaurement of leukotrienes in human plasma by solid-phase extraction and HPLC. *Prostaglandins Leukotrienes Essential Fatty Acids*, **49**, 851-854 (1993).

Hotter,G., Ramis,I., Bioque,G., Sarmiento,C., Fernandez,J.M., Rosello-Catafau,J. and Gelpi,E. Application of totally automated on-line sample clean-up for prostanoid extraction and HPLC separation. *Chromatographia*, **36**, 33-38 (1993).

Kelly,B.M., Rose,M.E. and Millington,D.S. The analysis of acylcarnitines. In: *Advances in Lipid Methodology - Two*, pp. 247-289 (edited by W.W. Christie, Oily Press, Dundee) (1993).

Lang,J.K. Characterisation of liposomes. In: *CRC Handbook of Chromatography. Analysis of Lipids*. pp. 459-470 (ed. K.D. Mukherjee and N. Weber, CRC Press, Boca Raton) (1993).

Matsumoto,K., Takahashi,M., Takiyama,N., Misaki,H., Matsuo,N., Murano,S. and Yuki,H. Enzyme reactor for urinary acylcarnitine assays by reversed-phase HPLC. *Clin. Chim. Acta*, **216**, 135-143 (1993).

Matsumoto,K., Takahashi,M., Takuyama,N., Misaki,H., Matsuo,N., Murano,S. and Yuki,H. Enzyme reactor for urinary acylcarnitine assay for reversed-phase HPLC. *Clin. Chim. Acta*, **216**, 135-143 (1993).

Minkler,P.E. and Hoppel,C.L. Quantification of carnitine and specific acylcarnitines by HPLC - application to normal human urine and urine from patients with methylmalonic aciduria, isovaleric acidemia or medium-chain acyl-CoA dehydrogenase deficiency. *J. Chromatogr.*, **613**, 203-222 (1993).

Minkler,P.E. and Hoppel,C.L. Quantification of free carnitine, individual short- and medium-chain acylcarnitines, and total carnitine in plasma by HPLC. *Anal. Biochem.*, **212**, 510-518 (1993).

Poorthuis,B.J.H.M., Jillevlckova,T. and Onkenhout,W. Determinations of acylcarnitines in urine of patients with inborn-errors of metabolism using HPLC with derivatization with 4'-bromophenacylbromide. *Clin. Chim. Acta*, **216**, 53-61 (1993).

Ramis,I., Hotter,G., Rosello-Catafau,J., Bulbena,O., Picado,C. and Gelpi,E. Application of totally automated on-line sample clean-up system for extraction and HPLC separation of peptide leukotrienes. *J. Pharm. Biomed. Anal.*, **11**, 1135-1139 (1993).

Schmidt-Sommerfeld,E., Penn,D., Duran,M., Bennett,M.J., Santer,R. and Stanley,C.A. Detection of inborn errors of fatty acid oxidation from acylcarnitine analysis of plasma and blood spots with the radioisotopic exchange-HPLC method. *J. Pediatr.*, **122**, 708-714 (1993).

Taylor,D.C., Weber,N. and MacKenzie,S.L. Acyl coenzyme A thioesters. In: *CRC Handbook of Chromatography. Analysis of Lipids*, pp. 285-320 (ed. K.D. Mukherjee and N. Weber, CRC Press, Boca Raton) (1993).

Tyman,J.H.P. Anacardic acids. In: *CRC Handbook of Chromatography. Analysis of Lipids*, pp. 173-195 (ed. K.D. Mukherjee and N. Weber, CRC Press, Boca Raton) (1993).

van Bocxlaer,J.F. and De Leenheer,A.P. Solid-phase extraction technique for GC profiling of acylcarnitines. *Clin. Chem.*, **39**, 1911-1917 (1993).

van Dessel,G.A.F., Lagrou,A.R., Hilderson,H.J.J. and Dierick,W.S.H. Dolichols and dolichyl derivatives. In: *CRC Handbook of Chromatography. Analysis of Lipids*, pp. 321-337 (ed. K.D. Mukherjee and N. Weber, CRC Press, Boca Raton) (1993).

Acknowledgement

This paper is published as part of a programme funded by the Scottish Office Agriculture and Fisheries Dept.

APPENDIX

Some Important References in Lipid Methodology - 1994

William W. Christie
The Scottish Crop Research Institute, Invergowrie, Dundee (DD2 5DA), Scotland

A. INTRODUCTION

The purpose of this chapter is the same as the previous, except that the year 1994 is covered. It has been compiled in the same way with sections corresponding to Chapters in *Lipid Analysis* (Second Edition, Pergamon Press, 1982) by the author, and the strengths and weaknesses are the same as in the previous listings. Again, note that the titles of papers listed below may not be literal transcriptions of the originals. In particular, a number of abbreviations have been introduced. References are listed alphabetically according to the surname of the first author in each section.

One new publication of special importance was the second edition of *The Lipid Handbook* (edited by F.D. Gustone, J.L. Harwood and F.B. Padley, Chapman & Hall, London, 1994), which covers a wide range of topics relating to all aspects of lipid science, including analysis. In addition, the proceedings of a conference - *Developments in the Analysis of Lipids* (edited by J.H.P. Tyman and M.H. Gordon,

Royal Soc. Chem., Cambridge (1994)) - have been published and details of some of the chapters are given below. One new volume of *Methods* in Enzymology (Vol. 230) deals with glycobiology and a second (Vol. 233) with lipid peroxidation, and both contain chapters relevant to lipid analysis. *Lipids - Molecular Organization, Physical applications and Technical Functions* by Kåre Larsson (The Oily Press, Dundee, 1994) will also be of interest.

B. THE ISOLATION OF LIPIDS FROM TISSUES

Deutsch,J., Grange,E., Rapoport,S.I. and Purdon,A.D. Isolation and quantitation of long-chain acyl-coenzyme A esters in brain tissue by solid-phase extraction. *Anal. Biochem.*, **220**, 321-323 (1994).

Grenacher,S. and Guerin,P.M. Inadvertent introduction of squalene, cholesterol, and other skin products into a sample. *J. Chem. Ecol.*, **20**, 3017-3025 (1994).

House,S.D., Larson,P.A., Johnson,R.R., Devries,J.W. and Martin,D.L. Gas chromatographic determination of total fat extracted from food samples using hydrolysis in the presence of antioxidant. *J. AOAC Int.*, **77**, 960-965 (1994).

Schnaar,R.L. Isolation of glycosphingolipids. *Methods Enzymol.*, **230**, 348-370 (1994).

Shaikh,N.A. Assessment of various techniques for the quantitative extraction of lysophospholipids from myocardial tissues. *Anal. Biochem.*, **216**, 313-321 (1994).

Wang,W.-Q. and Gustafson,A. Lipid determination from monophasic solvent mixtures: influence of uneven distribution of lipids after filtration and centrifugation. *J. Lipid Res.*, **35**, 2143-2150 (1994).

C. CHROMATOGRAPHIC AND SPECTROSCOPIC ANALYSIS OF LIPIDS. GENERAL PRINCIPLES.

Kaufmann,P., Kowalski,B.R. and Alander,J. Multivariate optimisation strategy for liquid chromatography. I. Targeting the search area in the multidimensional solvent space. *Chemom. Intell. Lab. Systems*, **23**, 331-339 (1994).

Snyder,L.R., Carr,P.W. and Rutan,S.C. Solvatochromically based solvent-selectivity triangle. *J. Chromatogr. A*, **656**, 537-547 (1994).

D. THE ANALYSIS OF FATTY ACIDS

This section contains references relevant to both Chapters 4 and 5 in *Lipid Analysis*.

Abushufa,R., Reed,P. and Weinkove,C. Fatty acids in erythrocytes measured by isocratic HPLC. *Clin. Chem.*, **40**, 1701-1712 (1994).

Ackman,R.G. and MacPherson,E.J. Coincidence of *cis*-monoethylenic and *trans*-monoethylenic fatty acids simplifies the open-tubular gas-liquid chromatography of butyl esters of butter fatty acids. *Food Chem.*, **50**, 45-52 (1994).

Adlof,R.O. Separation of *cis* and *trans* fatty acid methyl esters by silver ion HPLC. *J. Chromatogr.*, **659**, 95-99 (1994).

Adlof,R.O. The preparative separation of lipids by silver resin chromatography. *Process Technol. Proc.*, **11**, 777-781 (1994).

Adlof,R.O. and Emken,E.A. Silver ion high-performance liquid chromatographic separation of fatty acid methyl esters labeled with deuterium atoms on double bonds. *J. Chromatogr. A*, **685**, 178-181 (1994).

Ballasteros,E., Cardenas,S., Gallego,M. and Valcarcel,M. Determination of free fatty acids in dairy products by direct coupling of a continuous pre-concentration ion-exchange-derivatization module to a gas chromatograph. *Anal. Chem.*, **66**, 628-634 (1994).

Banni,S., Day,B.W., Evans,R.W., Corongui,F.P. and Lombardi,B. Liquid chromatographic-mass spectrometric analysis of conjugated diene fatty acids in a partially hydrogenated fat. *J. Am. Oil Chem. Soc.*, **71**, 1321-1325 (1994).

Barnathan,G., Doumenq,P., Njinkoue,J.-M., Miralles,J., Debitus,C., Levi,C. and Kornprobst,J.-M.

Sponge fatty acids 3. Occurrence of series of n-7 monoenoic and iso5,9 dienoic long-chain fatty acids in the phospholipids of the marine sponge *Cinachyrella aff. schulzei* Keller. *Lipids*, **29**, 297-303 (1994).

Bordier,C.G., Sellier,N., Foucault,A.P. and Le Goffic,F. Characterization and purification of fatty acid methyl esters from the liver oil of the deep-sea shark (*Centrophorus squamosus*) by gas chromatography-mass spectrometry and countercurrent chromatography. *Chromatographia*, **39**, 329-338 (1994).

Bousquet,O., Sellier,N. and Le Goffic,F. Characterization and purification of polyunsaturated fatty-acids from microalgae by GC-MS and countercurrent chromatography. *Chromatographia*, **39**, 40-44 (1994).

Brechany,E.Y. and Christie,W.W. Identification of the unsaturated oxo fatty acids in cheese. *J. Dairy Res.*, **61**, 111-115 (1994).

Brutting,R. and Spitteler,G. Products of the dimerization of unsaturated fatty acids. XI. The fraction of alicyclic dimer fatty acids. *Fat Sci. Technol.*, **96**, 361-370 (1994).

Cheung,M., Young,A.B. and Harrison,A.G. O- and OH- chemical ionization of some fatty acid methyl esters and triacylglycerols. *J. Am. Soc. Mass Spectrom.*, **5**, 553-557 (1994).

Christie,W.W., Brechany,E.Y., Marekov,I.N., Stefanov,K.L. and Andreev,S.N. The fatty acids of the sponge *Hymeniacidon sanguinea* from the Black Sea. *Comp. Biochem. Physiol.*, **109B**, 245-252 (1994).

Classen,E., Marx,F. and Fabricius,H. Mass spectra of 4,4-dimethyloxazoline derivatives of oxooctadecanoic acids. *Fat Sci. Technol.*, **96**, 331-332 (1994).

Cocito,C. and Delfini,C. Simultaneous determination by GC of free and combined fatty acids and sterols in grape musts and yeasts as silanized compounds. *Food Chem.*, **50**, 297-305 (1994).

Cordero,M.M. and Wesdemiotis,C. Characterization of the neutral products formed upon the charge-remote fragmentation of fatty acid ions. *Anal. Chem.*, **66**, 861-866 (1994).

Corongiu,F.P. and Banni,S. Detection of conjugated dienes by second derivative UV spectroscopy. *Methods Enzymol.*, **233**, 303-310 (1994).

Franko,M., Bicanic,D. and Vandebovenkamp,P. Dual-beam infrared thermal lens spectrometry at 965 cm($^{-1}$) absorption band as a measure of nonconjugated *trans*-fatty acids content in margarine samples. *J. Physique IV*, **4**, 479-482 (1994).

Garrido,J.L. and Medina,I. One-step conversion of fatty-acids into their 2-alkenyl-4,4-dimethyloxazoline derivatives directly from total lipids. *J. Chromatogr.*, **673**, 101-105 (1994).

Gavva,S.R., Wiethoff,A.J., Zhao,P.Y., Malloy,C.R. and Sherry,A.D. A C-13 isotopomer NMR method for monitoring incomplete β-oxidation of fatty acids in intact tissue. *Biochem. J.*, **303**, 847-853 (1994).

Gradowska,W. and Larsson,L. Determination of absolute configuration of 2- and 3-hydroxy fatty acids in organic dusts by GC-MS. *J. Microbiol. Meth.*, **20**, 55-67 (1994).

Grav,H.J., Asiedu,D.K. and Berge,R.K. Gas chromatographic measurement of 3-thia and 4-thia fatty acids incorporated into various classes of rat liver lipids during feeding experiments. *J. Chromatogr.*, **658**, 1-10 (1994).

Gross,M.L. Tandem mass spectrometric strategies for determining structure of biologically interesting molecules. *Acc. Chem. Res.*, **27**, 361-369 (1994).

Gunstone,F.D. High resolution ^{13}C NMR. A technique for the study of lipid structure and composition. *Prog. Lipid Res.*, **33**, 19-28 (1994).

Hagglund, I., Demirbuker,M. and Blomberg,L.G. Performance of silica-bonded quinolinol as a selective stationary-phase for packed capillary supercritical-fluid chromatography. *J. Microcolumn Sep.*, **6**, 223-228 (1994).

Hartmann,S., Besra,G.S., Fraser,J.L., Konig,W.A., Minnikin,D.E. and Ridell,M. Stereochemistry of 2,4-dimethyleicos-2-enoate from the pyruvylated glycolipid of *Mycobacterium smegmatis*. *Biochim. Biophys. Acta*, **1201**, 339-344 (1994).

Hartmann,S., Mallet,A.I., Minnikin,D.E. and Ridell,M. TLC and MS of pentafluorobenzylidene derivatives of members of the phthiocerol family from *Mycobacteria tuberculosis and M. marinum. Biochim. Biophys. Acta*, **1201**, 129-134 (1994).

Huang,S.-Y. and Jin,J.-D. Purification of methyl esters of polyunsaturated fatty acids by displacement chromatography. *Bioseparation*, **4**, 343-351 (1994).

Jacobsen,S.S., Becker,C.C. and Holmer,G. A more accurate gas chromatographic method for the analysis of butter oil fatty acids by estimation of relative response factors. *Chemom. Intell. Lab. Systems*, **23**, 231-234 (1994).

Juaneda,P., Sebedio,J.L. and Christie,W.W. Complete separation of the geometrical isomers of linolenic acid by HPLC with a silver ion column. *J. High Resolut. Chromatogr.*, **17**, 321-324 (1994).

Jung,S.H., Zeikus,J.G. and Hollingsworth,R.I., A new family of very-long-chain α,ω-dicarboxylic acids is a major structural fatty acid component of the membrane lipids of *Thermanaerobacter ethanolicus* 39E. *J. Lipid Res.*, **35**, 1057-1065 (1994).

Lamberto,M. and Ackman,R.G. Confirmation by GC/MS of two unusual *trans*-3-monoethylenic fatty acids from the Nova Scotian seaweeds *Palmaria palmata* and *Choridrus crispus*. *Lipids*, **29**, 441-444 (1994).

Liebich,H.M., Schmeider,N., Wahl,H.G. and Woll,J. Separation and identification of unsaturated fatty acid isomers in blood serum and therapeutic oil preparations in the form of their oxazoline derivatives by GC-MS. *J. High Resolut. Chromatogr.*, **17**, 519-521 (1994).

Liu,K.-S. Preparation of fatty acid methyl esters for gas chromatographic analysis of lipids. *J. Am. Oil Chem. Soc.*, **71**, 1179-1187 (1994).

Loupy,A. Microwaves: their potential applications in lipid chemistry. *Oleag. Corps gras Lipides*, **1**, 62-68 (1994).

Luche,J.-L. Sonochemistry: some basic principles and applications in lipid chemistry. *Oleag. Corps gras Lipides*, **1**, 69-74 (1994).

Major,C. and Wolf,B.A. Quantitation of the fatty acid composition of phosphatidic acid by capillary gas chromatography electron-capture detection with picomole sensitivity. *J. Chromatogr.*, **658**, 233-240 (1994).

Martinez-Lorenzo,J.L., Marzo,I., Naval,J. and Pineiro,A. Self-staining of polyunsaturated fatty acids in argentation chromatography. *Anal. Biochem.*, **220**, 210-212 (1994).

Marx,F. and Classen,E. Analysis of epoxy fatty acids by GC-MS of their dimethyloxazoline derivatives. *Fat Sci. Technol.*, **96**, 207-211 (1994).

Mason,S.R., Ward,L.C. and Reilly,P.E.B. Fluorimetric detection of microsomal lauric acid hydroxylations using HPLC after selective solvent partitioning and esterification with 1-pyrenyldiazomethane. *J. Liqu. Chromatogr.*, **17**, 619-632 (1994).

Mossoba,M.M., Yurawecz,M.P., Roach,J.A.G., Lin,H.S., McDonald,R.E., Flickinger,B.D. and Perkins,E.G. Rapid determination of double bond configuration and position along the hydrocarbon chain in cyclic fatty acid monomers. *Lipids*, **29**, 893-896 (1994).

Nikolova-Damyanova,B., Christie,W.W. and Herslof,B. Improved separation of some positional isomers of monounsaturated fatty acids, as their phenacyl derivatives, by silver-ion thin-layer chromatography. *J. Planar Chromatogr. - Modern TLC*, **7**, 382-385 (1994).

Park,P.W. and Goins,R.E. *In situ* preparation of fatty acid methyl esters for analysis of fatty acid composition in foods. *J. Food Sci.*, **59**, 1262-1266 (1994).

Procida,G., Gabrielli-Favretto,L., Pertoldi-Marletta,G. and Ceccon,L. Modified esterification procedure employing boron trifluoride/methanol complex for the determination of fats and oils. Riv. *Ital. Sostanze Grasse*, **71**, 547-553 (1994).

Ramos,L.S. Characterisation of mycobacteria species by HPLC and pattern recognition. *J. Chromatogr. Sci.*, **32**, 219-227 (1994).

Ratnayake,W.M.N., Chen,Z.Y., Pelletier,G. and Weber,D. Occurrence of 5*c*,8*c*,11*c*,15*t*-eicosatetraenoic acid and other unusual polyunsaturated fatty acids in rats fed partially hydrogenated canola oils. *Lipids*, **29**, 707-714 (1994).

Read,G., Richardson,N.R. and Wickens,D.G. Determination of octadecadienoic acids in human serum: a critical reappraisal. *Analyst*, **119**, 393-396 (1994).

Rezanka,T. and Dembitsky,V.M. Identification of unusual cyclopropane monounsaturated fatty acids from the deep-water lake invertebrate *Acanthogammarus grewingkii*. *Comp. Biochem. Physiol.*, **109B**, 407-413 (1994).

Sacchi,R., Medina,I., Paolillo,L. and Addeo,F., High-resolution ^{13}C-NMR olefinic spectra of DHA and EPA acids, methyl esters and triacylglycerols. *Chem. Phys. Lipids*, **69**, 65-73 (1994).

Sandmann,B.W. and Grayeski,M.L. Quinoxaline derivatization of biological carboxylic acids for detection by peroxyoxalate chemiluminescence with HPLC. *J. Chromatogr*, **653**, 123-130 (1994).

Schmid,P.C. and Schmid,H.H.O. Reaction of diazomethane with glycerolipids in the presence of serum or inorganic salts. *Lipids*, **29**, 883-887 (1994).

Sebedio,J.L., Prevost,J., Ribot,E. and Grandgirard,A. Utilization of HPLC as an enrichment step for the determination of cyclic fatty acid monomers in heated fats and biological samples. *J. Chromatogr.*, **659**, 101-109 (1994).

Sergent,O., Cillard,P. and Cillard,J. Ultraviolet and infrared methods for analysis of fatty acyl esters in cellular systems. *Methods Enzymol.*, **233**, 310-313 (1994).

Serres,V., Benjelloun-Mlayah,B. and Delmas,M. Separation of polyunsaturated fatty acids on silver resins. 2. Adsorption-desorption phenomena. *Rev. Franc. Corps Gras*, **41**, 3-8 (1994).

Shoemaker,M. and Spener,F. Enzymatic flow injection analysis for essential fatty acids. *Sensors and Activators B. Chem.*, **19**, 607-609 (1994).

Smith,R.M. and Cocks,S. Separation of saturated and unsaturated fatty acid methyl esters by supercritical-fluid chromatography on a silica column. *Analyst*, **119**, 921-924 (1994).

Spitzer,V., Bordignon,S.A. de L., Schenkel,E.P. and Marx,F. Identification of nine acetylenic fatty acids, 9-hydroxystearic acid and 9,10-epoxystearic acid in the seed oil of *Jodina rhombifolia* Hook et Arn. (Santalaceae). *J. Am. Oil Chem. Soc.*, **71**, 1343-1348 (1994).

Spitzer,V., Marx,F. and Pfeilsticker,K. Electron impact mass spectra of the oxazoline derivatives of some conjugated diene and triene C18 fatty acids. *J. Am. Oil Chem. Soc.*, **71**, 873-876 (1994).

Spitzer,V., Marx,F., Maia,J.G.S. and Pfeilsticker,K. The mass spectra of the picolinyl ester derivatives of malvalic and sterculic acid. *Fat Sci. Technol.*, **96**, 395-396 (1994).

Takagi,T. Resolution of vic-dihydroxy acid diastereomers to four enantiomers by HPLC. *J. Am. Oil Chem. Soc.*, **71**, 547-548 (1994).

Tassignon,P., de Waard,P., de Rijk,T., Tournois,H., de Wit,D. and de Buyck,L. An efficient countercurrent distribution method for the large-scale isolation of dimorphecolic acid methyl ester. *Chem. Phys. Lipids*, **71**, 187-196 (1994).

Ulberth,F. and Henninger,M. Quantitation of *trans*-fatty acids in milk fat using spectroscopic and chromatographic methods. *J. Dairy Res.*, **61**, 517-527 (1994).

Vaidyanathan,V.V. and Sastry,P.S. Preparation of hydroperoxy fatty acids. *Anal. Biochem.*, **219**, 381-383 (1994).

Vallance,H. and Applegarth,D. An improved method for quantification of very long-chain fatty acids in plasma. *Clin. Biochem.*, **27**, 183-186 (1994).

Vetter,W. and Walther,W. Pyrrolidides as derivatives for the determination of the fatty acids of triacylglycerols by gas-chromatography. *J. Chromatogr. A*, **686**, 149-154 (1994).

Vicanova,J., Tvrzicka,E. and Stulik,K. Capillary GC of underivatized fatty acids with a free fatty acid phase column and a programmed-temperature vaporizer detector *J. Chromatogr.*, **656**, 45-50 (1994).

Voinov,V.G., Boguslavskiy,V.M. and Elkin,Y.N. Resonance electron-capture for determining double bond and hydroxy group locations in fatty acids. *Org. Mass Spectrom.*, **29**, 641-646 (1994).

Wahl,H.G., Habel,S.-Y., Schmieder,N. and Liebich,H.M. Identification of *cis-trans*-isomers of methyl ester and oxazoline derivatives of unsaturated fatty acids using GC-FTIR-MS. *J. High Resolut. Chromatogr.*, **17**, 543-548 (1994).

Wahl,H.G., Liebich,H.M. and Hoffmann,A. Identification of fatty acid methyl esters as minor components of fish oil by multidimensional GC-MSD - new furan fatty acids. *J. High Resolut. Chromatogr.*, **17**, 308-311 (1994).

Wolff,R.L. and Sebedio,J.L. Characterization of gamma-linolenic acid geometrical isomers in borage oil subjected to heat treatments (deodorization). *J. Am. Oil Chem. Soc.*, **71**, 117-126 (1994).

Wolff,R.L. Contribution of *trans*-18:1 acids from dairy fat to European diets. *J. Am. Oil Chem. Soc.*, **71**, 277-283 (1994).

Wolff,R.L. *cis-trans* Isomerization of octadecatrienoic acids during heating. Study of pinolenic (*cis*-5,*cis*-9,*cis*-12 18:3) acid geometrical isomers in heated pine seed oil. *J. Am. Oil Chem. Soc.*, **71**, 1129-1134 (1994).

Wolff,R.L. Analysis of α-linolenic acid geometrical isomers in deodorized oils by capillary GC on cyanoalkyl polysiloxane stationary phases: a note of caution. *J. Am. Oil Chem. Soc.*, **71**, 907-909 (1994).

Yamane,M., Abe,A. and Nakajima,M. High-performance liquid chromatograpy thermospray mass-spectrometry of ω-hydroxy polyunsaturated fatty acids from rat brain homogenate. *J. Chromatogr. B*, **662**, 91-96 (1994).

Yamane,M., Abe,A. and Yamane,S. HPLC thermospray mass spectrometry of epoxy polyunsaturated fatty acids and epoxyhydroxy polyunsaturated fatty acids from an incubation mixture of rat tissue homogenate. *J. Chromatogr.*, **652**, 123-136 (1994).

Yasaka,Y. and Tanaka,M. Labelling of free carboxyl groups. *J. Chromatogr. B*, **659**, 139-155 (1994).

Zelles,L. and Bai,Q.Y. Fatty acid patterns of phospholipids and lipopolysaccharides in environmental samples. *Chemosphere*, **28**, 391-411 (1994).

Zhang,L.-Y. and Hamberg,M. A GLC method for steric analysis of 2-hydroxy, 3-hydroxy and 2,3-dihydroxy acids. *Chem. Phys. Lipids*, **74**, 151-161 (1994).

E. THE ANALYSIS OF SIMPLE LIPID CLASSES

This section corresponds to Chapter 6 in Lipid Analysis and deals mainly with chromatographic methods, especially TLC and HPLC, for the isolation and analysis of simple lipid classes. Separations of molecular species of simple lipids are listed in Section G below.

Andrisano,V., Gotti,R., Di Pietra,A.M. and Cavrini,V. Comparative evaluation of three chromatographic methods in the quality control of fatty alcohols for pharmaceutical and cosmetic use. *Farmaco*, **49**, 387-391 (1994).

Ansari,G.A.S. and Smith,L.L. Assay of cholesterol autoxidation. *Methods Enzymol.*, **233**, 332-338 (1994).

Antonopolou,S., Andrikopoulos,N.K. and Demopoulos,C.A. Separation of the main neutral lipids into classes and species by PR-HPLC and UV detection. *J. Liqu. Chromatogr.*, **17**, 633-648 (1994).

Arnoldsson,K.C. and Kaufmann,P. Lipid class analysis by normal-phase high-performance liquid-chromatography, development and optimization using multivariate methods. *Chromatographia*, **38**, 317-324 (1994).

Beijaars,P.R., van Dijk,R. and Houwen-Claasen,A.A.M. Determination of polymerized triglycerides in frying fats and oils by gel-permeation chromatography: interlaboratory study. *J. Ass. Off. Anal. Chem. Int.*, **77**, 667-671 (1994).

Bortolomeazzi,R., Pizzale,L., Conte,L.S. and Lercker,G. Identification of thermal oxidation products of cholesteryl acetate. *J. Chromatogr. A*, **683**, 75-85 (1994).

Bortolomeazzi,R., Pizzale,L., Vichi,S. and Lercker,G. Analysis of isomers of cholesteryl acetate hydroperoxides by HPLC and GC-ITDMS. *Chromatographia*, **39**, 577-580 (1994).

Cardenas,M.S., Ballesteros,E., Gallego,M. and Valcarel,M. Sequential determination of triglycerides and free fatty acids in biological fluids by use of a continuous pretreatment module coupled to a gas chromatograph. *Anal. Biochem.*, **222**, 332-341 (1994).

Chamoin,M.-C., Charbonnier,M., Lafont,H. and Ternaux,J.-P. Highly sensitive chemiluminescent assay for cholesterol. *Biochim. Biophys. Acta*, **1210**, 151-156 (1994).

Chen,B.H. and Chen,Y.C. Evaluation of the analysis of cholesterol oxides by liquid chromatography. *J. Chromatogr.*, **661**, 127-136 (1994).

Chen,X.J. and Simoneit,B.R.T. Epicuticular waxes from vascular plants and particles in the lower troposphere - analysis of lipid classes by iastroscan thin-layer chromatography with flame ionization detection. *J. Atmos. Chem.*, **18**, 17-31 (1994).

Cueto,R., Squadrito,G.L. and Pryor,W.A. Quantifying aldehydes and distinguishing aldehydic product profiles from autoxidation and ozonation of unsaturated acids. *Methods Enzymol.*, **233**, 174-182 (1994).

Dasgupta,A., Thompson,W.C. and Malik,S. Use of microwave irradiation for rapid synthesis of perfluorooctanoyl derivatives of fatty alcohols, a new derivative for gas chromatography-mass spectrometric and fast-atom-bombardment mass spectrometric study. *J. Chromatogr. A*, **685**, 279-285 (1994).

De Jong,C., Palma,K. and Neeter,R. Sample preparation before capillary gas chromatographic estimation of free fatty acids in fermented dairy products. *Neth. Milk Dairy J.*, **48**, 151-156 (1994).

Feldbrugge,R., Renneberg,R. and Spener,F. Development and practival evaluation of an amperometric triglyceride sensor. *Sensors and Activators B. Chem.*, **19**, 365-367 (1994).

Firestone,D. Gas chromatographic determination of monoglycerides and diglycerides in fats and oils - summary of a collaborative study. *J. Ass. Off. Anal. Chem. Int.*, **77**, 677-680 (1994).

Grob,K., Giuffre,A.M., Leuzzi,U. and Mincione,B. Recognition of adulterated oils by direct analysis of the minor components. *Fat Sci. Technol.*, **96**, 286-290 (1994).

Gu,Y.-F., Chen,Y. and Hammond,E.G. Use of cyclic anhydrides to remove cholesterol and other hydroxy compounds from fats and oils. *J. Am. Oil Chem. Soc.*, **71**, 1205-1209 (1994).

Guardiola,F., Codony,R., Rafecas,M. and Boatella,J. Analytical methods of oxysterol determination. *Grasas y Aceites*, **45**, 164-192 (1994).

Guardiola,F., Codony,R., Rafecas,M. and Boatella,J. Selective gas chromatographic determination of cholesterol in eggs. *J. Am. Oil Chem. Soc.*, **71**, 867-871 (1994).
Jaremo,M., Karlsson,L., Jonsson,J.A. and Mathiasson,L. The contribution of the pressure drop to analyte retention in coupled-column supercritical-fluid chromatography of lipids. *Chromatographia*, **38**, 17-21 (1994).
Jessup,W., Dean,R.T. and Gebicki,J.M. Iodometric determination of hydroperoxides in lipids and lipoproteins. *Methods Enzymol.*, **233**, 289-303 (1994).
Kermasha,S., Kubow,S. and Goetghebeur,M. Comparative HPLC analyses of cholesterol and its oxidation products using diode array ultraviolet and laser light-scattering detection. *J. Chromatogr. A.*, **685**, 229-235 (1994).
Li,X.-Y. and Chow,C.K. An improved method for the measurement of malondialdehyde in biological samples. *Lipids*, **29**, 73-75 (1994).
Linnet,K. An HPLC-GC/MS reference method for serum total cholesterol with control for ester hydrolysis. *Clin. Biochem.*, **27**, 177-182 (1994).
Matthaus,B., Wiezorek,C. and Eichner,K. Fast chemiluminescence method for detection of oxidized lipids. *Fat Sci. Technol.*, **96**, 95-99 (1994).
Norton,R.A. Isolation and identification of steryl cinnamic acid derivatives from corn bran. *Cereal Chem.*, **71**, 111-117 (1994).
Plank,C. and Lorbeer,E. On-line liquid-chromatography gas chromatography for the analysis of free and esterified sterols in vegetable oil methyl esters used as diesel fuel substitutes. *J. Chromatogr. A*, **683**, 95-104 (1994).
Przybylski,R. and Eskin,N.A.M. Two simplified approaches to the analysis of food lipids. *Food Chem.*, **51**, 231-235 (1994).
Regueiro,J.A.G., Gibert,J. and Diaz,I. Determination of neutral lipids from subcutaneous fat of cured ham by capillary GC and liquid chromatography. *J. Chromatogr.*, **667**, 225-233 (1994).
Sakai,K. and Yoshida,S. Quantitative and nondestructive analyses of fatty acid esters and cholesterol in brain tissues by Fourier-transform infrared-spectroscopy. *Vibrational Spectr.*, **7**, 163-167 (1994).
Sevanian,A., Seraglia,R., Traldi,P., Rossato,P., Ursini,F. and Hodis,H. Analysis of plasma cholesterol oxidation products using gas chromatography and high-performance liquid chromatography-mass spectrometry. *Free Radical Biol. Med.*, **17**, 397-409 (1994).
Shantha,N.C. and Decker,E.A. Rapid, sensitive, iron-based spectrophotometric methods for determination of peroxide values of food lipids. *J. Ass. Off. Anal. Chem. Int.*, **77**, 421-424 (1994).
Staby,A., Borch-Jensen,C., Balchen,S. and Mollerup,J. Quantitative-analysis of marine oils by capillary supercritical-fluid chromatography. *Chromatographia*, **39**, 697-705 (1994).
Suzuki,T. High-performance liquid chromatographic resolution of dinophysistoxin-1 and free fatty acids as 9-anthrylmethyl esters. *J. Chromatogr.*, **677**, 301-306 (1994).
Tazuma,S., Hatsushika,S., Yamashita,G., Aihara,N., Sasaki,M., Horikawa,K., Yamashita,Y., Teramen,K., Ochi,H., Hirano,N., Miura,H., Ohya,T., Hino,H. and Kajiyama,G. Simultaneous microanalysis of biliary cholesterol, bile acids and fatty acids in lecithin using capillary column GC - an advantage to assess bile lithogenicity. *J. Chromatogr.*, **653**, 1-7 (1994).
van de Voort,F.R., Ismail,A.A., Sedman,J., Dubois,J. and Nicodemo,T. The determination of peroxide value by Fourier transform infrared spectroscopy. *J. Am. Oil Chem. Soc.*, **71**, 921-926 (1994).
Wang,H., Hachey,D.L., Liu,D., Zhang,S. and Dudleu,M.A. Microanalysis of cholesterol, phospholipid and medium- and long-chain fatty acids in biological materials. *Anal. Biochem.*, **218**, 74-79 (1994).
Wang,W.Q. and Gustafson,A. Pigment interferes with cholesterol analysis in erythrocyte lipid extracts - a procedure for removal. *Acta Chem. Scand.*, **48**, 699-700 (1994).
Wardas,W. and Pyka,A. Visualizing agents for cholesterol in TLC. *J. Planar Chromatogr., Modern TLC*, **7**, 440-443 (1994).
Yamamoto,Y. Chemiluminescence-based HPLC assay of lipid hydroperoxides. *Methods Enzymol.*, **233**, 319-324 (1994).

F. THE ANALYSIS OF COMPLEX LIPIDS

This section corresponds to Chapter 7 in *Lipid Analysis* and deals mainly with chromatographic methods, especially TLC and HPLC, for the isolation and analysis of complex lipid classes including both phospholipids and glycolipids.

Degradative procedures for the identification of polar moieties and spectrometric methods for intact lipids are also listed here. Separations of molecular species of complex lipids are listed in the next section.

Abidi,S.L., Mounts,T.L. and Rennick,K.A. Separations of major soybean phospholipids on beta-cyclodextrin-bonded silica. *J. Liqu. Chromatogr.*, **17**, 3705-3725 (1994).

Adosraku,R.K., Choi,G.T.Y., Constantinoukokotos,V., Anderson,M.M. and Gibbons,W.A. NMR lipid profiles of cells, tissues, and body-fluids - proton NMR analysis of human erythrocyte lipids. *J. Lipid Res.*, **35**, 1925-1931 (1994).

Arnoldsson,K.C. and Kaufmann,P. Lipid class analysis by normal-phase high-performance liquid chromatography, development and optimization using multivariate methods. *Chromatographia*, **38**, 317-324 (1994).

Bernhard,W., Linck,M., Creutzberg,H., Postle,A.D., Arning,A., Martin-Carrera,I. and Sewing,K.-F. HPLC analysis of phospholipids from different sources with combined fluorescence and UV detection. *Anal. Biochem.*, **220**, 172-180 (1994).

Blomberg,L.G. and Andersson,P.E. Capillary electrophoresis for lipid analysis. INFORM, **5**, 1030-1037 (1994).

Borgeat,P., Picard,S., Braquet,P., Alien,M. and Shushan,B. LC-MS-MS with ion spray: a promising approach for analysis of underivatized platelet activating factor (PAF). *J. Lipid Mediators Cell Signal.*, **10**, 11-12 (1994).

Boyle,T., Lancaster,V., Hunt,R., Gemski,P. and Jett,M. Method for simultaneous isolation and quantitation of platelet activating factor and multiple arachidonate metabolites from small samples: analysis of effects of *Staphylococcus aureus* enterotoxin B in mice. *Anal. Biochem.*, **216**, 373-382 (1994).

Bruch,J., Gono,E., Malkusch,W. and Rehn,B. Improved method for quantitative analysis of lung surfactant phospholipids in bronchoalveolar lavage fluids by high-performance liquid chromatography. *Clin. Chim. Acta*, **231**, 193-204 (1994).

Caboni,M.F., Menotta,S. and Lercker,G. High-performance liquid chromatography separation acid light-scattering detection of phospholipids from cooked beef. *J. Chromatogr. A*, **683**, 59-65 (1994).

Cedergren,R.A. and Hollingsworth,R.I. Occurrence of sulfoquinovosyl diacylglycerol in some members of the family Rhizobiaceae. *J. Lipid Res.*, **35**, 1452-1461 (1994).

Cho,Y. and Ziboh,V.A. Incorporation of 13-hydroxyoctadecadienoic acid (13-HODE) into epidermal ceramides and phospholipids: phospholipase C-catalysed release of novel 13-HODE-containing diacylglycerol. *J. Lipid Res.*, **35**, 255-262 (1994).

Costello,C.E., Juhasz,P. and Perreault,H. New mass spectral approaches to ganglioside structure determinations. *Prog. Brain Res.*, **101**, 45-61 (1994).

Darrow,R.A. and Organisciak,D.T. An improved spectrophotometric triiodide assay for lipid hydroperoxides. *Lipids*, **29**, 591-594 (1994).

de Koster,C.G., Vos,B., Versluis,C., Heerma,W. and Haverkamp,J. High-performance TLC/fast atom bombardment (tandem) mass spectrometry of Pseudomonas rhamnolipids. *Biol. Mass Spectrom.*, **23**, 179-185 (1994).

Demopoulos,C.A., Andrikopoulos,N.K. and Antonopoulou,S. A simple and precise method for the routine determination of platelet-activating factor in blood and urine. *Lipids*, **29**, 305-309 (1994).

Evanochko,W.T. and Pohost,G.M. Structural studies of NMR detected lipids in myocardial ischemia. *NMR in Biomedicine*, **7**, 269-277 (1994).

Felde,R. and Spiteller,G. Search for plasmalogens in plants. *Chem. Phys. Lipids*, **71**, 109-113 (1994).

Fischer,W., Hartmann,R., Peter-Katalinic,J. and Egge,H. (S)-Amino-1,3-propanediol-3-phosphate-carrying diradylglyceroglycolipids - novel major membrane lipids of *Clostridium innocuum*. *Eur. J. Biochem.*, **223**, 879-892 (1994).

Flamand,N., Justine,P., Bernaud,F., Rougier,A. and Gaetani,Q. In vivo distribution of free long-chain sphingoid bases in the human stratum corneum by HPLC analysis of strippings. *J. Chromatogr.*, **656**, 65-71 (1994).

Gage,D.A., Huang,Z.-H. and Sweeley,C.C. Characterization of diacylglycerophospholipids by FAB-MS. *Mass Spectrom.*, **2**, 53-87 (1994).

Han,X. and Gross,R.W. Electrospray ionization mass spectrometric analysis of human erythrocyte plasma membrane phospholipids. *Proc. Natl. Acad. Sci.*, **91**, 10635-10639 (1994).

Hartmann,S., Minnikin,D.E., Mallet,A.I., Ridell,M., Rigouts,L. and Portaels,F. Fast atom bombardment mass spectrometry of mycobacterial phenolic glycolipids. *Biol. Mass Spectrom.*, **23**, 362-368 (1994).

Hathout,Y., Maume,G. and Maume,B.F., HPLC study of the regulation of phospholipid metabolism in cultured adrenocortical cells. *J. Chromatogr.*, 652, 1-8 (1994).

Hauksson,J.B., Lindblom,G. and Rilfors,L. Structures of glucolipids from the membrane of *Acholeplasma laidlawii* strain A-EF22. 1. Glycerophosphoryldiglucosyldiacylglycerol and monoacylbis-glycerophosphoryldiglucosyldiacylglycerol. *Biochim. Biophys. Acta*, **1214**, 124-130 (1994).

Hauksson,J.B., Lindblom,G. and Rilfors,L. Structures of glycolipids from the membranes of *Acholeplasma laidlawii* strain A-EF22. 2. Monoacylmonoglucosyldiacylglycerol. *Biochim. Biophys. Acta*, **1215**, 341-345 (1994).

Hechtberger,P., Zinser,E., Saf,R., Hummel,K., Paltauf,F. and Daum,G. Characterization, quantification and subcellular localization of inositol-containing sphingolipids of the yeast, *Saccharomyces cerevisiae*. *Eur. J. Biochem.*, **225**, 641-649 (1994).

Helmy,F.M. and Hack,M. Some TLC observations on the *in vitro* formation of *N*-acyl phosphatidylethanolamine by endogenous components of bovine and porcine retina. *J. Planar Chromatogr. - Modern TLC*, **7**, 14-17 (1994).

Ilinov,P.P., Deleva,D.D., Zaprianova,E.T. and Dimov,S.I. A method for determination of lipid-bound sialic-acid after chromatographic isolation of brain gangliosides. *J. Liquid Chromatogr.*, **17**, 2871-2879 (1994).

Ingvardsen,L., Michaelsen,S. and Soresen,H. Analysis of individual phospholipids by high-performance capillary electrophoresis. *J. Am. Oil Chem. Soc.*, **71**, 183-188 (1994).

Jaaskelainen,I. and Urtti,A. Liquid chromatography determination of liposome components using a light-scattering evaporative detector. *J. Pharm. Biomed. Anal.*, **12**, 977-982 (1994).

Jennemann,R. and Wiegandt,H. A rapid method for the preparation of ganglioside G_{lac2} (GD_3). *Lipids*, **29**, 365-368 (1994).

Jin,W.Z., Rinehart,K.L. and Jareserijman,E.A. Ophidiacerebrosides - cytotoxic glycosphingolipids containing a novel sphingosine from a sea star. *J. Org. Chem.*, **59**, 144-147 (1994).

Johnston,N.C. and Goldfine,H. Isolation and characterization of new phosphatidylglycerol acetals of plasmalogens - a family of ether lipids in Clostridia. *Eur. J. Biochem.*, **223**, 957-963 (1994).

Jones,G.W. and Ashwood,E.R. Enzymatic measurement of phosphatidylglycerol in amniotic fluid. *Clin. Chem.*, **40**, 518-525 (1994).

Kamata,T., Akasaka,K., Ohrui,H. and Meguro,H. Enzymatic assay of phosphatidylcholine hydroperoxide by phospholipase A(2) and glutathione peroxidase - based on fluorometry with *N*-(9-acridinyl)maleimide. *Biosci. Biotechnol. Biochem.*, **58**, 881-884 (1994).

Klein,B.H. and Dudenhausen,J.W. Simultaneous determination of phospholipid classes and the major molecular species of lecithin in human amniotic fluid by HPLC. *J. Liqu. Chromatogr.*, **17**, 981-998 (1994).

Klein,R.A. and Egge,H. Recent developments in the MS-NMR analysis of gangliosides. *Pergamon Studies Neurosci.*, **10**, 245-273 (1994).

Lingwood,C.A. and Nutikka,A. A novel chemical procedure for the selective removal of nonreducing terminal *N*-acetyl hexosamine residues from glycolipids. *Anal. Biochem.*, **217**, 119-123 (1994).

Major,C. and Wolf,B.A. Quantitation of the fatty acid composition of phosphatidic acid by capillary gas chromatography electron-capture detection with picomole sensitivity. *J. Chromatogr.*, **658**, 233-240 (1994).

Maliakal,M.A., Ravindranath,M.H., Irie,R.F. and Morton,D.L. An improved method for the measurement of total lipid-bound sialic acids after cleavage of α-2,8 sialic-acid linkage with *Vibrio cholerae* sialidase in the presence of cholic-acid, SDS and Ca^{2+}. *Glycoconjugate J.*, **11**, 97-104 (1994).

Matsubara,C., Ohyama,T. and Takamura,K. Densitometric quantitation of platelet-activating-factor and other phospholipids in human saliva using enzyme reaction on a silica plate. *J. Pharm. Soc. Japan*, **114**, 681-690 (1994).

Matsuda,K., Kasama,T., Ishizuka,I., Handa,S., Yamamoto,N, and Taki,T. Structure of a novel phosphocholine-containing glycerolipid from *Mycoplasma fermentans*. *J. Biol. Chem.*, **269**, 33123-33128 (1994).

Menon,A.K. Structural analysis of glycosylphosphatidylinositol anchors. *Methods Enzymol.*, **230**, 418-442 (1994).

Menzeleev,R.F., Krasnopolsky,Y.M., Zvonkova,E.N. and Shvets,V.I. Preparative separation of ganglioside GM_3 by HPLC. *J. Chromatogr. A*, **678**, 183-187 (1994).

Merchant,T.E., Van der Ven,L.T.M., Minsky,B.D., Diamantis,P.M., Delapaz,R., Galicich,J. and Glonek,T. P-31 NMR phospholipid characterization of intracranial tumors. *Brain Res.*, **649**, 1-6 (1994).

Miyazawa,T., Fujimoto,K., Suzuki,T. and Yasuda,K. Determination of phospholipid hydroperoxides using luminol chemiluminescence HPLC. *Methods Enzymol.*, **233**, 324-332 (1994).

Morrison,I.M. Glycolipids. In: *Carbohydrate Analysis: A Practical Approach*, pp. 295-317 (edited by M.F. Chapman & J.F. Kennedy, IRL Press, Oxford) (1994).

Murphy,R.C. GC/MS of PAF and related lipids. *J. Lipid Mediators Cell Signal.*, **10**, 183-184 (1994).

Muthing,J. Improved TLC separation of gangliosides by automated multiple development. *J. Chromatogr. B.*, **657**, 75-81 (1994).

Muthing,J. and Unland,F. Improved separation of isomeric gangliosides by anion-exchange HPLC. *J. Chromatogr. B.*, **658**, 39-45 (1994).

Nakanishi, K., Yasugi,E., Morita,H., Dohi,T. and Oshima,M. Plasmenylethanolamine in human intestinal mucosa detected by an improved method for analysis of phospholipid. *Biochem. Mol. Biol. Int.*, **33**, 457-462 (1994).

Noda,N., Tanaka,R., Tsujino,K., Takasaki,Y., Nakano,M., Nishi,M. and Miyahara,K. Phosphocholine-bonded galactosylceramides having a triunsaturated long-chain base from the clam worm, *Marphysa sanguinea*. *J. Biochem. (Tokyo)*, **116**, 435-442 (1994).

Ohta,H., Ruan,F., Hakomori,S.-i. and Igarashi,Y. Quantification of free sphingosine in cultured cells by acylation with radioactive acetic anhydride. *Anal. Biochem.*, **222**, 489-494 (1994).

Paige,D.G., Morse-Fisher,N. and Harper,J.I. Quantification of stratum corneum ceramides and lipid envelope ceramides in the hereditary ichthyoses. *Brit. J. Dermatol.*, **131**, 23-27 (1994).

Perez,M.K., Fried,B. and Sherma,J. Comparison of mobile phases and hptlc qualitative and quantitative analysis, on preadsorbent silica gel plates, of phospholipids in *Biomphalaria glabrata* (Gastropoda) infected with *Echinostoma caproni* (Trematoda). *J. Planar Chromatogr. - Modern TLC*, **7**, 340-343 (1994).

Perreault,H. and Costello,C.E. Liquid secondary ionization, tandem and matrix-assisted laser-desorption ionization time-of-flight mass-spectrometric characterization of glycosphingolipid derivatives. *Org. Mass Spectrom.*, **29**, 720-735 (1994).

Peter-Katalinic,J. Analysis of glycoconjugates by fast atom bombardment mass spectrometry and related MS techniques. *Mass Spectrom. Rev.*, **13**, 77-98 (1994).

Pivot,V., Bruneteau,M., Mas,P., Bompeix,G. and Michel,G. Isolation, characterization and biological activity of inositol sphingophospholipids from *Pythophora capsici*. *Lipids*, **29**, 21-25 (1994).

Pohlentz,G., Schlemm,S., Klima,B. and Egge,H. Fast atom bombardment mass spectrometry of *N*-acetylated neoglycolipids of the 1-deoxy-1-phosphatidylethanolamino-lactitol-type. *Chem. Phys. Lipids*, **70**, 83-94 (1994).

Redman,C.A., Green,B.N., Thomas-Oates,J.E., Reinhold,V.N. and Ferguson,M.A.J. Analysis of glycosylphosphatidylinositol membrane anchors by electrospray-ionization mass-spectrometry and collision-induced dissociation. *Glycoconjugate J.*, **11**, 187-193 (1994).

Reinhold,B.B., Chan,S.Y., Chan,S. and Reinhold,V.N. Profiling glycosphingolipid structural detail - periodate-oxidation, electrospray, collision-induced dissociation and tandem mass-spectrometry. *Org. Mass Spectrom.*, **29**, 736-746 (1994).

Robson,K.J., Stewart,M.E., Michelsen,S., Lazo,N.D. and Downing,D.T. 6-Hydroxy-4-sphingenine in human epidermal ceramides. *J. Lipid Res.*, **35**, 2060-2068 (1994).

Sajbidor,J., Certik,M. and Grego,J. Lipid analysis of baker's yeast. *J. Chromatogr.*, **665**, 191-195 (1994).

Sawabe,A., Morita,M., Okamoto,T. and Ouchi,S. The location of double bonds in a cerebroside from edible fungi (mushroom) estimated by B/E linked scan fast atom bombardment mass spectrometry. *Biol. Mass Spectrom.*, **23**, 660-664 (1994).

Schnaar,R.L. and Needham,L.K. Thin-layer chromatography of glycosphingolipids. *Methods Enzymol.*, **230**, 371-389 (1994).

Seijo,L., Merchant,T.E., van der Ven,L.T.M., Minsky,B.D. and Glonek,T. Meningioma phospholipid profiles measured by 31P NMR spectroscopy. *Lipids*, **29**, 359-364 (1994).

Sprott,G.D., Dicaire,C.J. and Patel,J.B. The ether lipids of *Methanosarcina mazei* and other *Methanosarcina* species, compared by fast-atom-bombardment mass-spectrometry. *Canad. J. Microbiol.*, **40**, 837-843 (1994).

Sprott,G.D., Ferrante,G. and Ekiel,I. Tetraether lipids of *Methanospirillum hungatei* with head groups consisting of phospho-*N,N*-dimethylaminopentanetetrol, phospho-*N,N,N*-trimethylaminopentanetetrol and carbohydrates. *Biochim. Biophys. Acta*, **1214**, 234-242 (1994).

Stanley,J.A., Williamson,P.C., Drost,D.J., Carr,T.J., Rylett,R.J., Morrison-Stewart,S. and Thompson,R.T. Membrane phospholipid metabolism and schizophrenia - an *in vivo* P-31-MR spectroscopy study. *Schizophrenia Res.*, **13**, 209-215 (1994).

Taki,T., Handa,S. and Ishikawa,D. Blotting of glycolipids and phospholipids from a HP-TLC chromatogram to a polyvinylidene difluoride membrane. *Anal. Biochem.*, **221**, 312-316 (1994).

Taki,T., Kasama,T., Handa,S. and Ishikawa,D. A simple and quantitative purification of glycosphingolipids and phospholipids by TLC blotting. *Anal. Biochem.*, **223**, 232-238 (1994).

Trumbach,B., Rogler,G., Lackner,K.J. and Schmitz,G. Improved separation of radioactively labelled cellular phospholipids by HPLC. *J. Chromatogr.*, **656**, 73-76 (1994).

Tsvetnitsky,V. and Gibbons,W.A. An assay for phosphatidylserine decarboxylase using BondElut columns. *Anal. Biochem.*, **217**, 157-158 (1994).

Uemura,M. and Steponkus,P.L. A contrast of the plasma membrane lipid composition of oats and rye leaves in relation to freezing tolerance. *Plant Physiol.*, **104**, 479-496 (1994).

Van Veldhoven,P.P., De Ceuster,P., Rozenberg,R., Mannaerts,G.P. and de Hoffmann,E. On the presence of phosphorylated sphingoid bases in rat tissues - a mass spectrometric approach. *FEBS Letters*, **350**, 91-95 (1994).

Villas Boas,M.H.S., Egge,H., Pohlentz,G., Hartmann,R. and Bergter,E.B. Structural determination of *N*-2'-hydroxyoctadecenoyl-1-*O*-beta-*D*-glucopyranosyl-9-methyl-4,8-sphingadiene from species of *Aspergillus*. *Chem. Phys. Lipids*, **70**, 11-19 (1994).

Wait,R., Jones,C., Routier,F.H., Previato,J.O. and Mendonca-Previato,L. Structure determination of phosphoinositol oligosaccharides from parasitic protozoa using fast-atom-bombardment mass-spectrometry. *Org. Mass Spectrom.*, **29**, 767-781 (1994).

Waki,H., Kon,K., Tanaka,Y. and Ando,S. Facile methods for isolation and determination of gangliosides in a small scale-age related changes of gangliosides in mouse brain synaptic plasma-membranes. *Anal. Biochem.*, **222**, 156-162 (1994).

Waki,H., Kon,K., Tanaka,Y. and Ando,S. Facile methods for isolation and determination of gangliosides in a small scale: age-related changes of gangliosides in mouse synaptic plasma membranes. *Anal. Biochem.*, **222**, 156-162 (1994).

Wieder,T., Fritsch,M., Haase,A. and Geilen,C.C. Determination of alkylphosphocholines by HPLC with light-scattering mass detection. *J. Chromatogr.*, **652**, 9-13 (1994).

Wiesner,D.A. and Sweeley,C.C. Microscale analysis of glycosphingolipids by methanolysis, peracetylation, and gas chromatography. *Anal. Biochem.*, **217**, 316-322 (1994).

Yamamoto,Y. Chemiluminescence-based HPLC assay of lipid hydroperoxides. *Methods Enzymol.*, **233**, 319-324 (1994).

Zellmer,S. and Lasch,J. Quantitative determination of phosphatidylethanolamine and phosphatidylserine in liposomes and on hptlc plates with *O*-phthalaldehyde. *Anal. Biochem.*, **218**, 229-231 (1994).

G. THE ANALYSIS OF MOLECULAR SPECIES OF LIPIDS

This section corresponds to Chapter 8 in *Lipid Analysis* and deals mainly with chromatographic methods for the isolation and analysis of molecular species of lipid classes, including simple lipids, phospholipids and glycolipids. Many of the references in the next section are relevant here also and *vice versa*.

Abidi,S.L. and Mounts,T.L. Reversed-phase retention behaviour of fluorescence-labelled phospholipids in ammonium acetate buffers. *J. Liqu. Chromatogr.*, **17**, 105-122 (1994).

Ariza,M.A. and Valero-Guillen,P.L. Delineation of molecular species of a family of diacyltrehaloses from *Mycobacterium fortuitum* by mass spectrometry. *FEMS Microbiol. Letters*, **119**, 279-282 (1994).

Blomberg,L.G. and Demibuker,M. Analysis of triacylglycerols by argentation supercritical fluid chromatography. In: *Developments in the Analysis of Lipids*, pp. 42-58 (edited by J.H.P. Tyman and M.H. Gordon, Royal Soc. Chem., Cambridge) (1994).

Boot,A.J. and Speek,A.J. Determination of the sum of dimer and polymer triglycerides and of acid value of used frying fats and oils by near-infrared reflectance spectroscopy. *J. Ass. Off. Anal. Chem. Int.*, **77**, 1184-1189 (1994).

Bruhl,L., Schulte,E. and Thier,H.-P. Composition and structures of triglycerides of human milk and some base components for infant milk formulas. *Fat Sci. Technol.*, **96**, 147-154 (1994).

Bruhl,L., Schulte,E. and Thier,H.-P. Triglycerides with polyunsaturated fatty acids in human milk and in ingredients for infant formulas. *Fat Sci. Technol.*, **96**, 223-227 (1994).

Burdge,G.C. and Postle,A.D. Hepatic phospolipid molecular species in the guniea pig. Adaptations to pregnancy. *Lipids*, **29**, 259-264 (1994).

Chang,M.-K., Conkerton,E.J., Chapital,D. and Wan,P.J. Behaviour of diglycerides and conjugated fatty acid triglycerides in reverse-phase chromatography. *J. Am. Oil Chem. Soc.*, **71**, 1173-1175 (1994).

Chen,S. and Li,K.W. Structural analysis of underivatized and derivatized aminophospholipids and phosphatidic acid by positive-ion liquid secondary-ion and collisionally induced dissociation tandem mass spectrometry. *J. Biochem. (Tokyo)*, **116**, 811-817 (1994).

Chen,S. Partial characterization of the molecular species of phosphatidylserine from human plasma by high-performance liquid chromatography and fast-atom-bombardment mass-spectrometry. *J. Chromatogr. B*, **661**, 1-5 (1994).

Cheng,B.L., Kowal,J. and Abraham,S. Analysis of adrenal cholesteryl esters by reversed-phase HPLC. *J. Lipid Res.*, **35**, 1115-1121 (1994).

Chicha,A., Demandre,C., Justin,A.M. and Mazliak,P. The molecular species of phosphatidylinositol and phosphatidylinositolphosphate present in the coleoptiles and the first leaves of maize (*Zea mays* L). *Comptes Rend. Acad. Sci. III.*, **317**, 419-423 (1994).

Christie,W.W. Silver ion and chiral chromatography in the analysis of triacylglycerols. *Prog. Lipid Res.*, **33**, 9-18 (1994).

Damiani, P., Santinelli,F. and Magnarini,C. Prediction of HGRC retention parameters and response factors of triacylglycerols. *J. Chromatogr. Sci.*, **32**, 21-24 (1994).

Dyas,L., Threlfall,D.R. and Goad,L.J. The sterol composition of five plant species grown as cell suspension cultures. *Phytochemistry*, **35**, 655-660 (1994).

Evershed,R.P., Application of modern mass spectrometric techniques to the analysis of lipids. In: *Developments in the Analysis of Lipids*, pp. 123-160 (edited by J.H.P. Tyman and M.H. Gordon, Royal Soc. Chem., Cambridge) (1994).

Felouati,B.-E., Pageaux,J.-F., Fayard,J.-M., Lagarde,M. and Laugier,C. Estradiol-induced changes in the composition of phospholipid classes of quail oviduct - specific replacement of arachidonic acid by docosahexaenoic acid in alkenylacyl-glycerophosphoethanolamine. *Biochem. J.*, **301**, 361-366 (1994).

Firestone,D. Gel-permeation liquid-chromatographic method for determination of polymerized triglycerides in oils and fats - summary of collaborative study. *J. Ass. Off. Anal. Chem. Int.*, **77**, 957-960 (1994).

Firestone,D. Liquid-chromatographic method for determination of triglycerides in vegetable oils in terms of their partition numbers - summary of collaborative study. *J. Ass. Off. Anal. Chem. Int.*, **77**, 954-957 (1994).

Ford,D.A. and Gross,D.W. The discordant rates of *sn*-1 aliphatic chain and polar head group incorporation into plasmalogen molecular species demonstrate the fundamental importance of polar head group remodeling in plasmalogen biosynthesis in rabbit myocardium. *Biochemistry*, **33**, 1216-1222 (1994).

Garcia Regueiro,J.A., Diaz,I., David,F. and Sandra,P. Possibilities of PTV injection for the analysis of triglycerides. *J. High Resolut. Chromatogr.*, **17**, 180-183 (1994).

Gelsema,W.J., Choma,I., den Ouden,T.A.F., Zander,R. and van den Bosch,H. Quantitation of the diacyl, alkylacyl, and alk-1-enylacyl subclasses of cholins glycerophospholipids by chemical dephosphorylation and benzoylation. *Anal. Biochem.*, **217**, 265-276 (1994).

Heron,S. and Tchlapa,A. Choice of stationary and mobile phases for separation of mixed triglycerides by liquid-phase chromatography. *Analusis*, **22**, 114-126 (1994).

Heron,S. and Tchapla,A. Using a molecular interaction model to optimize the separation of fatty triglycerides in reversed-phase liquid chromatography - fingerprinting the different types of fats. *Oleag. Corps Gras Lipides*, **1**, 219-228 (1994).

Hori,M., Sahashi,Y., Koike,S., Yamaoka,R. and Sato,M. Molecular species analysis of polyunsaturated fish triacylglycerol by high-performance liquid chromatography fast-atom-bombardment mass spectrometry. *Anal. Sci.*, **10**, 719-724 (1994).

Hradec,J. and Dufek,P. Determination of cholesteryl 14-methylhexadecanoate in blood serum by reversed-phase high-performance liquid chromatography. *J. Chromatogr. B*, **660**, 386-389 (1994).

Huang,A.S., Robinson,L.R., Gursky,L.G., Profita,R. and Sabidong,C.G. Identification and quantification of SALATRIM 23CA in foods by the combination of supercritical extraction, particle beam LC-MS and HPLC with light-scattering detector. J. Agric. *Food Chem.*, **42**, 468-473 (1994).

Ju,D.D., Wei,G.J. and Her,G.R. High-energy collision-induced dissociation of ceramide ions from permethylated glycosphingolipids. *J. Am. Soc. Mass Spectrom.*, **5**, 558-563 (1994).

Kadowaki,H. and Grant,M.A. Preparation of defined molecular species of lactosylceramide by chemical deacylation and reacylation with *N*-succinimidyl fatty acid esters. *Lipids*, **29**, 721-725 (1994).

Kadowaki,H., Grant,M.A. and Seyfried,T.N. Effect of golgi membrane phospholipid composition on the molecular species of GM_3 gangliosides synthesised by rat liver sialyltransferases. *J. Lipid Res.*, **35**, 1956-1964 (1994).

Kaplan,M., Davidson,G. and Poliakoff,M. Capillary supercritical-fluid chromatography FT-IR study of triglycerides and the qualitative analysis of normal and unsaturated cheeses. *J. Chromatogr.*, **673**, 231-237 (1994).

Kayganich-Harrison,K.A. and Murphy,R.C. Fast-atom-bombardment tandem-mass spectrometry of [C13]arachidonic acid-labelled phospholipid molecular species. *J. Am. Soc. Mass Spectrom.*, **5**, 144-150 (1994).

Kayganich-Harrison,K.A. and Murphy,R.C. Characterization of chain-shortened oxidized glycerophosphocholine lipids using fast-atom-bombardment and tandem mass spectrometry, *Anal. Biochem.*, **221**, 16-24 (1994).

Kayganich-Harrison,K.A. and Murphy,R.C. Incorporation of stable isotope-labelled arachidonic acid into cellular phospholipid molecular species and analysis by fast atom bombardment tandem MS. *Biol. Mass Spectrom.*, **23**, 562-571 (1994).

Kerwin,J.L., Tuininga,A.R. and Ericsson,L.H. Identification of molecular species of glycerophospholipids and sphingomyelin using electrospray MS. *J. Lipid Res.*, **35**, 1102-1114 (1994).

Kheifets,G.M., Alekseeva,E.M. and Vendik,O.O. Gas chromatographic microdetermination of blood plasma glycerophosphatides without their preseparation from a lipid extract. *J. Anal. Chem.*, **49**, 479-485 (1994).

Kim,D.H., Lee,K.J. and Heo,G.S. Analysis of cholesterol and cholesteryl esters in human serum using capillary supercritical fluid chromatography. *J. Chromatogr.*, **655**, 1-8 (1994).

Kim,H.Y., Wang,T.C.L. and Ma,Y.C. Liquid-chromatography mass spectrometry of phospholipids using electrospray ionization. *Anal. Chem.*, **66**, 3977-3982 (1994).

Kuksis,A., Myher,J.J. and Yang,L.Y. Glycerolipid metabolism with deuterated tracers. In: *Biol. Mass Spectrom. Present & Future. (Proceedings of Kyoto Conference)*, pp. 481-494 (1994).

Liepkalns,V.A., Myher,J.J., Kuksis,A., Leli,U., Freysz,N. and Hauser,G. Complementary chromatographic analysis of free diacylglycerols and potential glycerophospholipid precursors in human SH-SY5Y neuroblastoma cells following incubation with lithium chloride. *J. Chromatogr. B*, **658**, 223-232 (1994).

Lima,L.R. and Synovec,R.E. Isocratic mixed-mode liquid chromatographic separation of phospholipids with octadecylsilane-silica staionary phases. *Talanta*, **41**, 581-588 (1994).

Lin,S.W. and Lam,N.W. Analysis of lipids in palm oil by on-column capillary GLC. *J. Chromatogr. Sci.*, **32**, 185-189 (1994).

Marai,L., Kuksis,A. and Myher,J.J. Reversed-phase liquid-chromatography mass-spectrometry of the uncommon triacylglycerol structures generated by randomization of butteroil. *J. Chromatogr.*, **672**, 87-99 (1994).

McIntyre,T., Patel,K.D., Smiley,P.L., Stafforini,D., Prescott,S.M. and Zimmerman,G.A. Oxidized phospholipids with PAF-like bioactivity. *J. Lipid Mediators Cell Signal.*, **10**, 37-39 (1994).

Molkentin,J. and Precht,D. Comparison of packed and capillary columns for quantitative gas chromatography of triglycerides in milk-fat. *Chromatographia*, **39**, 265-270 (1994).

Murphy,R.C. and Harrison,K.A. Fast-atom-bombardment mass-spectrometry of phospholipids. *Mass Spectrom. Rev.*, **13**, 57-75 (1994).

Myher,J.J., Kuksis,A., Tilden,C. and Oftedal,O.T. A cross-species comparison of neutral lipid composition of milk fat of prosimian primates. *Lipids*, **29**, 411-419 (1994).

Neff,W.E., Adlof,R.O. and El-Agaimy,M. Silver ion HPLC of the triacylglycerols of *Crepis alpina* seed oil. *J. Am. Oil Chem. Soc.*, **71**, 853-855 (1994).

Neff,W.E., Adlof,R.O., List,G.R. and Elagaimy,M. Analyses of vegetable oil triacylglycerols by silver ion high-performance liquid chromatography with flame ionization detection. *J. Liquid Chromatogr.*, **17**, 3951-3968 (1994).

Patton,G.M., Fasulo,J.M. and Robins,S.J. Hepatic phosphatidylcholines - evidence for synthesis in the rat by extensive reutilization of endogenous acylglycerides. *J. Lipid Res.*, **35**, 1211-1221 (1994).

Pittenauer,E., Allmaier,G. and Schmid,E.R. Rapid molecular mass determination of monoglycerides, diglycerides and triglycerides by Cf plasma desorption mass specrometry using 3-(3-pyridyl)acrilic acid as matrix. *Org. Mass Spectrom.*, **29**, 108-111 (1994).

Rubino,F.M., Zecca,L. and Sonnino,S. Characterization of a complex mixture of ceramides by fast atom bombardment and precursor and fragment analysis mass spectrometry. *Biol. Mass Spectrom.*, **23**, 82-90 (1994).

Seenaiah,B. and Ellingson,J.S. High-performance liquid chromatographic method for determination of the metabolism of polyunsaturated molecular species of phosphatidylserine labeled in the polar group. *J. Chromatogr. B*, **660**, 380-385 (1994).

Smith,K.W., Perkins,J.M., Jeffrey,B.S.J. and Phillips,D.L. Separation of molecular species of *cis-* and *trans*-triacylglycerols in *trans*-hardened confectionery fats by silver-ion high-performance liquid chromatography. *J. Am. Oil Chem. Soc.*, **71**, 1219-1222 (1994).

Staby,A., Borch-Jensen,C., Balchen,S. and Mollerup,J. Supercritical fluid chromatographic analysis of fish oils. *J. Am. Oil Chem. Soc.*, **71**, 355-359 (1994).

Stoll,U. Techniques of phospholipid analyses applied to sera and egg yolk. *Fat Sci. Technol.*, **96**, 188-194 (1994).

Suzuki,A.C., Nakamura,A. and Nishimura,K. Molecular-species analysis of glycosphingolipids from small-intestine of japanese-quail, *Coturnix coturnix japonica*, by HPLC/FAB/MS. *Glycoconjugate J.*, **11**, 111-121 (1994).

Takagi,T. and Ando,Y. Separation of monoacylglycerols by HPLC on nitrile-bonded phase. *J. Am. Oil Chem. Soc.*, **71**, 459-460 (1994).

Tokumura,A., Tanaka,T. and Tsukatani,H. Characterization of PAF-like phospholipids formed by lipid-peroxidation. J. *Lipid Mediators Cell Signal.*, **10**, 179-181 (1994).

Valeur,A., Olsson,N.U., Kaufmann,P., Wada,S., Kroon,C.-G., Westerdahl,G. and Odham,G. Quantification and comparison of some natural sphingomyelins by on-line HPLC/discharge assisted thermospray mass spectrometry. *Biol. Mass Spectrom.*, **23**, 313-319 (1994).

Viinanen,E. and Hopia,A. Reversed-phase HPLC analysis of triacylglycerols autoxidation products with ultraviolet and evaporative light-scattering detection *J. Am. Oil Chem. Soc.*, **71**, 537-539 (1994).

Wagner,S. and Paltauf,F. Generation of glycerophospholipid molecular species in the yeast *Saccharomyces cerevisiae* - fatty acid pattern of phospholipid classes and selective acyl turnover at *sn*-1 and *sn*-2 positions. *Yeast*, **10**, 1429-1437 (1994).

Wasan,K.M., Hayman,A.C. and Lopez-Berestein,G. Determination of dimyristoylphosphatidylglycerol in human serum by liquid-liquid-extraction and reversed-phase liquid-chromatography. *J. Pharm. Biomed. Anal.*, **12**, 851-854 (1994).

Zhang,J.Y., Nobes,B.J., Wang,J. and Blair,I.A. Characterization of hydroxyeicosatetraenoic acids and hydroxyeicosatetraenoic acid phosphatidylcholines by liquid secondary ion tandem MS. *Biol. Mass Spectrom.*, **23**, 399-405 (1994).

Zollner,P., Lorbeer,E. and Remberg,G. Utility of nicotinyl derivatives in structural studies of monoacylglycerols and diacylglycerols by GC-MS. 1. *Org. Mass Spectrom.*, **29**, 253-259 (1994).

H. STRUCTURAL ANALYSIS OF LIPIDS BY MEANS OF ENZYMATIC HYDROLYSIS

This section corresponds to Chapter 9 in *Lipid Analysis* and relates to simple lipids, phospholipids and glycolipids. Many of the references in the last section are relevant here also and *vice versa*. Some methods for the resolution of chiral lipids or involving mass spectrometry are listed here when they deal with methods for determining positional distributions of fatty acids within lipid classes.

Basu,S.S., Dastgheib-Hosseini,S., Hoover, G., Li,Z.X. and Basu,S. Analysis of glycosphingolipids by fluorophore-assisted carbohydrate electrophoresis using ceramide glycanase from *Mercenaria mercenaria*. *Anal. Biochem.*, **222**, 270-274 (1994).

Carballeira,N.M., Emiliano,A. and Morales,R. Positional distribution of octadecadienoic acids in sponge phosphatidylethanolamines. *Lipids*, **29**, 523-525 (1994).

Damiani,P., Rosi,M., Castellini,M., Santinelli,F., Cossignani,L. and Simonetti.M.S. Stereospecific analysis of triacylglycerols by an enzymic procedure using a new *sn*-1,2-diacylglycerol kinase preparation - application to olive and sunflower oils. *Ital. J. Food Sci.*, **6**, 113-122 (1994).

Damiani,P., Santinelli,F., Simonetti,M.S., Castellini,M. and Rosi,M. Comparison between two procedures for stereospecific analysis of triacylglycerols from vegetable oils. 1. Olive oil. *J. Am. Oil Chem. Soc.*, **71**, 1157-1162 (1994).

Diez,E., Chilton,F.H., Stroup,G., Mayer,R.J., Winkler,J.D. and Fonteh,A.N. Fatty acid and phospholipid selectivity of different phospholipase A_2 enzymes studied by using a mammalian membrane as substrate. *Biochem. J.*, **301**, 721-726 (1994).

Fan,T.W.N., Clifford,A.J. and Higashi,R.M. *In vivo* C-13 NMR analysis of acyl chain composition and organization of perirenal triacylglycerides in rats fed vegetable oils. *J. Lipid Res.*, **35**, 678-689 (1994).

Gunstone,F.D. and Seth,S. A study of the distribution of eicosapentaenoic acid and docosahexaenoic acid between alpha and beta glycerol chains in fish oils by ^{13}C-NMR spectroscopy. *Chem. Phys. Lipids*, **72**, 119-126 (1994).

Henderson,J.M., Petersheim,M., Templeman,G.J. and Softly,B.J. Quantitation and structure elucidation of the positional isomers in a triacylglycerol mixture using proton and carbon one- and two-dimensional NMR. *J. Agric. Food Chem.*, **42**, 435-441 (1994).

Kallio,H. and Rua,P. Distribution of the major fatty acids of human milk between *sn*-2 and *sn*-1,3 positions of triacylglycerols. *J. Am. Oil Chem. Soc.*, **71**, 985-992 (1994).

Medina,I. and Sacchi,R. Acyl stereospecific analysis of tuna phospholipids via high resolution ^{13}C-NMR spectroscopy. *Chem. Phys. Lipids*, **70**, 53-61 (1994).

Murakami,N., Morimoto,T., Imamura,H., Nagatsu,A. and Sakakibara,J. Enzymatic transformation of glycerolipids into *sn*-1 and *sn*-2 lysoglyceroglycolipids by use of *Rhizopus arrhizus* lipase. *Tetrahedron*, **50**, 1993-2002 (1994).

Nachiappan,V. and Rajasekharan,R. Enzymatic synthesis of [P-32]acyl-*sn*-glycerol 3-phosphate using diacylglycerol kinase. *Anal. Biochem.*, **222**, 283-285 (1994).

Ota,T., Kawabata,Y. and Ando,Y. Positional distribution of 24:6(*n*-3) in triacyl-*sn*-glycerols from flathead flounder liver and flesh. *J. Am. Oil Chem. Soc.*, 71, 475-478 (1994).

Sempore,B.G. and Bezard,J.A., Separation of monoacylglycerol enantiomers as urethane derivatives by chiral-phase HPLC. *J. Liqu. Chromatogr.*, **17**, 1679-1694 (1994).

Taylor,D.C., MacKenzie,S.L., McCurdy,A.R., McVetty,P.B.E., Giblin,E.M., Pass,E.W., Stone,S.C., Scarth,R., Rimmer,S.T. and Pickard,M.D. Stereospecific analyses of seed triacylglycerols from high-erucic acid Brassicaceae: detection of erucic acid at the *sn*-2 position in *Brassica oleracea* L. genotypes. *J. Am. Oil Chem. Soc.*, **71**, 163-167 (1994).

Vesterqvist,O., Sargent,C.A., Grover,G.J., Warrack,B.M., DiDonato,G.C. and Ogletree,M.L. Characterization of rabbit myocardial phospholipase A_2 activity using endogenous phospholipid substrate. *Anal. Biochem.*, **217**, 210-219 (1994).

Wagner,S. and Paltauf,F. Generation of glycerophospholipid molecular species in the yeast Saccharomyces cerevisiae - fatty acid pattern of phospholipid classes and selective acyl turnover at sn-1 and sn-2 positions. *Yeast*, **10**, 1429-1437 (194).

I. THE ANALYSIS AND RADIOASSAY OF ISOTOPICALLY LABELLED LIPIDS

This section corresponds to Chapter 10 in *Lipid Analysis*. Only papers in which the radioactivity of the sample appeared to be central to the analysis are listed. To confuse matters, papers dealing with analysis of lipids enriched in stable isotopes are also listed here as they appear more relevant to this than any other section.

Brossard,N., Pachiaudi,C., Croset,M., Normand,S., Lecerf,J., Chirouze,V., Riou,J.P., Tayot,J.L. and Lagarde,M. Stable isotope tracer and GC combustion isotope ratio mass spectrometry to study the *in vivo* compartmental metabolism of docosahexaenoic acid. *Anal. Biochem.*, **220**, 192-199 (1994).

Chen,X.J. and Simoneit,B.R.T. Epicuticular waxes from vascular plants and particles in the lower troposphere - analysis of lipid classes by iastroscan thin-layer chromatography with flame ionization detection. *J. Atmos. Chem.*, **18**, 17-31 (1994).

Magni,F., Piatti,P.M., Monti,L.D., Lecchi,P., Pontiroli,A.E., Pozza,G. and Kienle,M.G. Fast GC-MS method for the evaluation of plasma fatty acid turnover using [1-^{13}C]-palmitate. *J. Chromatogr. B.*, **657**, 1-7 (1994).

Metges,C.C., Kempe,K. and Wolfram,G. Enrichment of selected serum fatty acids after a small oral dosage of (1-^{13}C)- and (8-^{13}C)-triolein in human volunteers analysed by gas chromatography/combustion isotope ratio mass spectrometry. *Biol. Mass Spectrom.*, **23**, 295-301 (1994).

Naraoka,H., Yamada,K. and Ishiwara,I. Stable carbon isotope measurement of individual fatty acids using GC/isotope ratio monitoring mass spectrometry (GC/IRMS). *J. Mass Spectrom. Soc. Japan*, **42**, 315-323 (1994).

Ohta,H., Ruan,F., Hakomori,S.-i. and Igarashi,Y. Quantification of free sphingosine in cultured cells by acylation with radioactive acetic anhydride. *Anal. Biochem.*, **222**, 489-494 (1994).

Trumbach,B., Rogler,G., Lackner,K.J. and Schmitz,G. Improved separation of radioactively labelled cellular phospholipids by HPLC. *J. Chromatogr.*, **656**, 73-76 (1994).

Yohe,H.C. Removal of borate from tritiated gangliosides via the mannoborate complex. *J. Lipid Res.*, **35**, 2100-2102 (1994).

Zamecnik,J., Fayolle,C. and Vallerand,A. Isotope enrichment of palmitic acid esters by EI GC/MS. *Biol. Mass Spectrom.*, **23**, 804-805 (1994).

J. THE SEPARATION OF PLASMA LIPOPROTEINS

This section corresponds to Chapter 11 in *Lipid Analysis*, and a only few key papers of particular interest are listed. *Current Topics in Lipidology* should be consulted for further listings

Ala-Korpela,M., Korhonen,A., Keisala,J., Horkko,S., Korpi,P., Ingman,L.P., Jokisaari,J., Savolainen,M.J. and Kesaniemi,Y.A. ^{1}H NMR-based absolute quantification of human lipoproteins and their lipid contents directly from plasma. *J. Lipid Res.*, **35**, 2292-2304 (1994).

Chirico,S., HPLC-based thiobarbituric acid tests. *Methods Enzymol.*, **233**, 314-318 (1994).

Jessup,W., Dean,R.T. and Gebicki,J.M. Iodometric determination of hydroperoxides in lipids and lipoproteins. *Methods Enzymol.*, **233**, 289-303 (1994).

Kulkarni,K.R., Garber,D.W., Marcovina,S.M. and Segrest,J.P. Quantification of cholesterol in all lipoprotein classes by the VAP-II method. *J. Lipid Res.*, **35**, 159-168 (1994).

Potts,J.L., Fisher,R.M., Humphreys,S.M., Gibbons,G.F. and Frayn,K.N. Separation of lipoprotein fractions by ultracentrifugation - investigation of analytical recovery with sequential flotation and density gradient procedures. *Clin. Chim. Acta*, **230**, 215-220 (1994).

Puppione,D.L. and Charugundla,S. A microprecipitation technique suitable for measuring α-lipoprotein cholesterol. *Lipids*, **29**, 595-597 (1994).

Sattler,W., Mohr,D. and Stocker,R. Rapid isolation of lipoproteins and assessment of their peroxidation by high-performance liquid chromatography postcolumn-chemiluminescence. *Methods Enzymol.*, **233**, 469-489 (1994).

Tallis,G.A., Shepherd,M.D.S. and Whiting,M.J. Lipoprotein profiling by high-performance gel chromatography. *Clin. Chim. Acta*, **228**, 171-179 (1994).

K. SOME MISCELLANEOUS SEPARATIONS

Analyses of lipids such as prostaglandins, acylcarnitines, coenzyme A esters and so forth that do not fit conveniently into other sections are listed here. More complete listings for prostaglandins are available elsewhere (*Prostaglandins, Leukotrienes and Essential Fatty acids*). The decision on whether to list papers on eicosenoids here or in Section E was sometimes arbitrary.

Boyle,T., Lancaster,V., Hunt,R., Gemski,P. and Jett,M. Method for simultaneous isolation and quantitation of platelet activating factor and multiple arachidonate metabolites from small samples: analysis of effects of *Staphylococcus aureus* enterotoxin B in mice. *Anal. Biochem.*, **216**, 373-382 (1994).

Farooqui,A.A., Yang,H.C. and Horrocks,L.A. Purification of lipases, phospholipases and kinases by heparin-Sepharose chromatography. *J. Chromatogr.*, **673**, 149-158 (1994).

Fritsch,H., Molnar,I. and Wurl,M. Separation of arachidonic acid metabolites by on-line extraction and reversed-phase high-performance liquid chromatography optimized by computer simulation. *J. Chromatogr. A*, **684**, 65-75 (1994).

Hirota,T., Minato,K., Ishii,K., Nishimura,N. and Sato,T. HPLC determination of the enantiomers of carnitine and acetylcarnitine on a chiral stationary-phase. *J. Chromatogr.*, **673**, 37-43 (1994).

Kamimori,H., Hamashima,Y. and Masaharu,K. Determination of carnitine and saturated-acyl group carnitines in human urine by HPLC with fluorescence detection. *Anal. Biochem.*, **218**, 417-424 (1994).

Kumps,A., Duez,P. and Mardens,Y. GC profiling and determination of urinary acylcarnitines. *J. Chromatogr. B.*, **658**, 241-248 (1994).

Matsumoto,K., Ichitani,Y., Ogasawara,N., Yuki,H. and Imai,K. Precolumn fluorescence derivatization of carnitine and acylcarnitines with 4-(2-aminoethylamino)-7-nitro-2,1,3-benzoxadiazole prior to high-performance liquid-chromatography. *J. Chromatogr.*, **678**, 241-247 (1994).

Mueller,M.J. and Brodschelm,W. Quantification of jasmonic acid by capillary GC-negative chemical ionization-mass spectrometry. *Anal. Biochem.*, **218**, 425-435 (1994).

Surette,M.E., Odeimat,A., Palmantier,R., Marleau,S., Poubelle,P.E. and Borgeat,P. Reversed-phase HPLC analysis of arachidonic acid metabolites in plasma after stimulation of whole blood *ex vivo*. *Anal. Biochem.*, **216**, 392-400 (1994).

Ueda,J., Miyamoto,K., and Kamisaka,S. Separation of a new type of plant growth regulator, jasmonates, by chromatographic procedures. *J. Chromatogr.*, **658**, 129-142 (1994).

Zhang,J.Y. and Blair,I.A. Direct resolution of epoxyeicosatrienoic acid enantiomers by chiral-phase HPLC. *J. Chromatogr.*, **657**, 23-29 (1994).

Zhang,J.Y., Nobes,B.J., Wang,J. and Blair,I.A. Characterization of hydroxyeicosatetraenoic acids and hydroxyeicosatetraenoic acid phosphatidylcholines by liquid secondary ion tandem MS. *Biol. Mass Spectrom.*, **23**, 399-405 (1994).

Zirolli,J.A., Wheelan,P. and Murphy,R.C. Analysis of long-chain fatty acyl-coenzyme A thioesters by negative-ion fast-atom-bombardment mass spectrometry and tandem mass spectrometry. *J. Am. Soc. Mass Spectrom.*, **5**, 416-424 (1994).

Acknowledgement

This paper is published as part of a programme funded by the Scottish Office Agriculture and Fisheries Dept.

INDEX